# TREATISE ON ANALYSIS

*Volume V*

This is Volume 10-V in
PURE AND APPLIED MATHEMATICS
A Series of Monographs and Textbooks

Editors: SAMUEL EILENBERG AND HYMAN BASS

A list of recent titles in this series appears at the end of this volume.

Volume 10

**TREATISE ON ANALYSIS**

10-I. Chapters I–XI, Foundations of Modern Analysis, enlarged and corrected printing, 1969
10-II. Chapters XII–XV, enlarged and corrected printing, 1976
10-III. Chapters XVI–XVII, 1972
10-IV. Chapters XVIII–XX, 1974
10-V. Chapter XXI, 1977

# TREATISE ON
# ANALYSIS

**J. DIEUDONNÉ**
Membre de l'Institut

*Volume V*

Translated by

**I. G. Macdonald**
Queen Mary College
University of London

**ACADEMIC PRESS** New York San Francisco London 1977
A Subsidiary of Harcourt Brace Jovanovich, Publishers

ACADEMIC PRESS, INC.
111 Fifth Avenue, New York, New York 10003

*United Kingdom Edition published by*
ACADEMIC PRESS, INC. (LONDON) LTD.
24/28 Oval Road, London NW1

**Library of Congress Cataloging in Publication Data** (Revised)

Dieudonne, Jean Alexandre, Date
Treatise on analysis.

(Pure and applied mathematics, a series of monographs and textbooks ; 10)
Except for v. 1, a translation of Elements d'analyse.
Vols. 2- translated by I. G. MacDonald.
Includes various editions of some volumes.
Includes bibliographies and indexes.
1. Mathematical analysis–Collected works.
I. Title. II. Series.
QA3.P8 vol. 10, 1969 510′.8s [515] 75-313532
ISBN 0–12–215505–X (v. 5)

PRINTED IN THE UNITED STATES OF AMERICA

"Treatise on Analysis," Volume V

First published in the French Language under the title "Éléments d'Analyse," tome 5 and copyrighted in 1975 by Gauthier-Villars, Éditeur, Paris, France.

# CONTENTS

# SCHEMATIC PLAN OF THE WORK

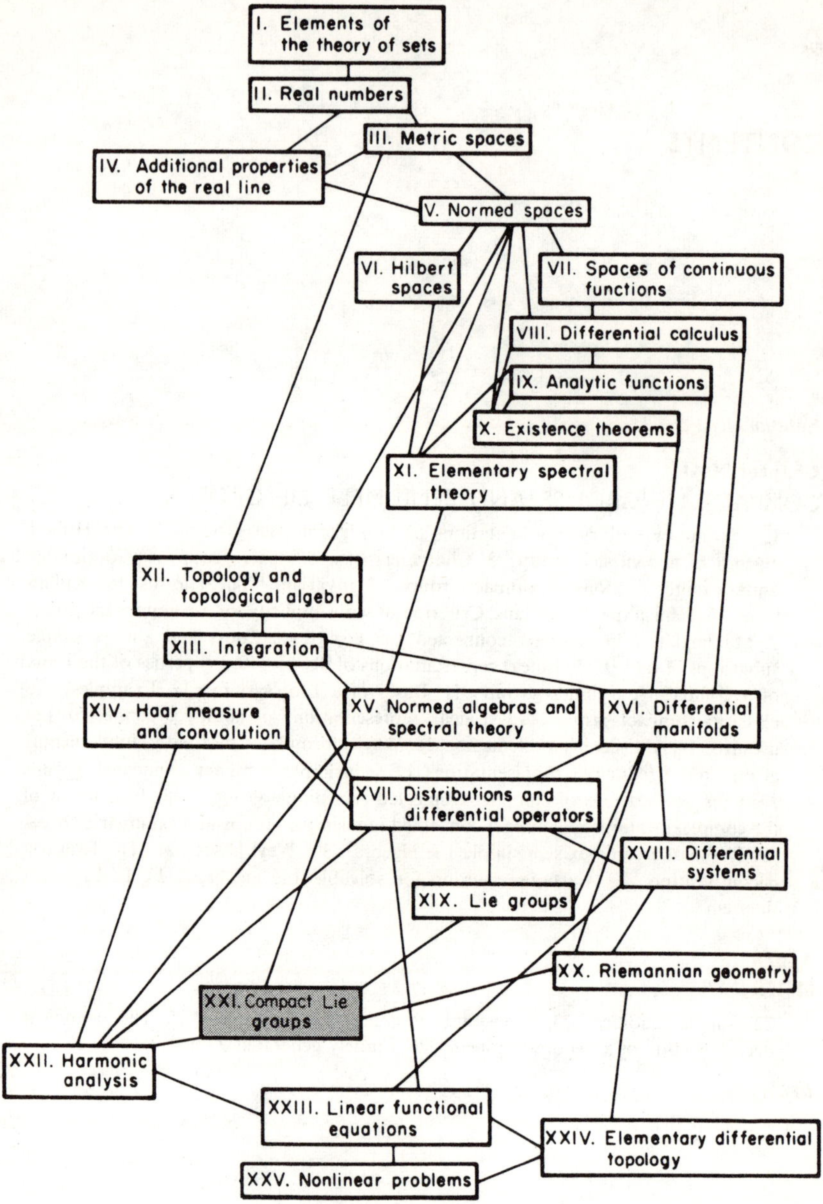

# NOTATION

In the following definitions, the first number indicates the chapter in which the notation is introduced, and the second number indicates the section within the chapter.

| | |
|---|---|
| $U(\mu)$ | $\int U(s)\,d\mu(s)$, where $U$ is a continuous unitary representation of a group G and $\mu$ is a bounded measure on G: 21.1 |
| $U(f)$, $U(\tilde{f})$ | $\int f(s)\,U(s)\,d\beta(s)$, where $\beta$ is a Haar measure on G, and $f \in \mathscr{L}^1_{\mathbf{C}}(\mathrm{G}, \beta)$: 21.1 |
| $U_{\mathrm{ext}}$ | mapping $\mu \mapsto U(\mu)$: 21.1 |
| $R(s)$ | left regular representation $\tilde{f} \mapsto (\varepsilon_s * f)^{\sim}$: 21.1 |
| $U_1 \oplus U_2$ | direct sum of two continuous linear representations: 21.1 |
| $U^{(\mathbf{R})}$, $U^{(\mathbf{H})}$ | real (resp. quaternionic) linear representation corresponding to a complex linear representation $U$: 21.1, Problem 9 |
| $\mathfrak{a}_\rho$ ($\rho \in \mathrm{R}(\mathrm{G})$) | minimal two-sided ideals of the complete Hilbert algebra $\mathrm{L}^2_{\mathbf{C}}(\mathrm{G})$, for G compact: 21.2 |
| $u_\rho$ | identity element of $\mathfrak{a}_\rho$: 21.2 |
| $n_\rho$ | the integer such that $\mathfrak{a}_\rho$ is isomorphic to $\mathbf{M}_{n_\rho}(\mathbf{C})$: 21.2 |
| $m^{(\rho)}_{jk}$ | elements of $\mathfrak{a}_\rho$: 21.2 |
| $M_\rho(s)$ | the matrix $(n_\rho^{-1} m^{(\rho)}_{ij}(s))$: 21.2 |
| $\rho_0$ | index of the trivial ideal $\mathfrak{a}_{\rho_0} = \mathbf{C}$: 21.2 |

| | |
|---|---|
| $\chi_{\boldsymbol{\rho}}$ | $n_{\boldsymbol{\rho}}^{-1} u_{\boldsymbol{\rho}}$: 21.3 |
| $\bar{\boldsymbol{\rho}}$ | the index such that $\chi_{\bar{\boldsymbol{\rho}}} = \overline{\chi_{\boldsymbol{\rho}}}$: 21.3 |
| $\mathrm{cl}(V)$ | class $\sum_{\boldsymbol{\rho} \in \mathbf{R}} d_{\boldsymbol{\rho}} \cdot \boldsymbol{\rho}$ of a finite-dimensional linear representation: 21.4 |
| $\mathbf{Z}^{(\mathrm{R})}$, $\mathbf{Z}^{(\mathrm{R}(\mathrm{G}))}$ | ring of classes of continuous linear repsentations of G: 21.4 |
| $\mathrm{B}_U$ | bilinear form $(\mathbf{u}, \mathbf{v}) \mapsto \mathrm{Tr}(U_*(\mathbf{u}) \circ U_*(\mathbf{v}))$ associated with a linear representation $U$ of a Lie group: 21.5 |
| $\mathrm{B}_\rho$ | bilinear form $(\mathbf{u}, \mathbf{v}) \mapsto \mathrm{Tr}(\rho(\mathbf{u}) \circ \rho(\mathbf{v}))$ associated with a homomorphism of Lie algebras $\rho : \mathfrak{g} \to \mathfrak{gl}(\mathrm{F})$: 21.5 |
| $\mathrm{B}_\mathfrak{a}$ | Killing form $(\mathbf{u}, \mathbf{v}) \mapsto \mathrm{Tr}(\mathrm{ad}(\mathbf{u}) \circ \mathrm{ad}(\mathbf{v}))$ of a Lie algebra $\mathfrak{a}$: 21.5 |
| $\Gamma_\mathrm{T}$ | kernel $\exp_\mathrm{T}^{-1}(e)$ of the exponential $\exp_\mathrm{T}$: $\mathfrak{t} \to \mathrm{T}$, where $\mathfrak{t}$ is the Lie algebra of the torus T: 21.7 |
| $\Gamma_\mathrm{T}^*$ | dual of the lattice $\Gamma_\mathrm{T}$, in $\mathfrak{t}^*$: 21.7 |
| W(G, T), W(G), W | Weyl group $\mathscr{N}(\mathrm{T})/\mathrm{T}$, where T is a maximal torus of G: 21.7 |
| $w.\lambda$ | ${}^t w^{-1}(\lambda)$, for $w \in \mathrm{W}$ and $\lambda \in \mathfrak{t}^*$: 21.8 |
| **S**(G, T), **S**(G), **S** | set of roots of G with respect to T: 21.8 |
| $\mathfrak{g}_\alpha$ | subspace of $\mathfrak{g}_{(\mathbf{C})}$ consisting of the vectors $\mathbf{x}$ such that $[\mathbf{u}, \mathbf{x}] = \alpha(\mathbf{u})\mathbf{x}$ for all $\mathbf{u} \in \mathfrak{t}$: 21.8 |
| $\mathrm{U}_\alpha$ | subgroup $\chi_\alpha^{-1}(1)$ of T, where $\chi_\alpha(\exp(\mathbf{u})) = e^{\alpha(\mathbf{u})}$ for $\mathbf{u} \in \mathfrak{t}$: 21.8 |
| $\mathfrak{u}_\alpha$ | hyperplane $\alpha^{-1}(0)$ in $\mathfrak{t}$: 21.8 |
| $s_\alpha$ | element of W acting on $\mathfrak{t}$ by reflection in the hyperplane $\mathfrak{u}_\alpha$: 21.8 |
| $\mathrm{L}_m$ | simple $\mathrm{U}(\mathfrak{sl}(2, \mathbf{C}))$-module of dimension $m + 1$: 21.9 |
| $\mathfrak{g} = \mathfrak{h} \oplus \bigoplus_{\alpha \in \mathbf{S}} \mathfrak{g}_\alpha$ | root decomposition of a complex semi-simple Lie algebra $\mathfrak{g}$: 21.10 and 21.20 |
| $\mathbf{h}_\alpha^0$ | element of $\mathfrak{h}$ such that $\alpha(\mathbf{h}) = \Phi(\mathbf{h}, \mathbf{h}_\alpha^0)$: 21.10 |
| $\mathbf{h}_\alpha$ | element of $\mathfrak{h}$ such that $\alpha(\mathbf{h}_\alpha) = 2$ and $\mathbf{h}_\alpha \in [\mathfrak{g}_\alpha, \mathfrak{g}_{-\alpha}]$: 21.10 |
| $\mathbf{x}_\alpha$, $\mathbf{x}_{-\alpha}$ | elements of $\mathfrak{g}_\alpha$, $\mathfrak{g}_{-\alpha}$, respectively, such that $[\mathbf{x}_\alpha, \mathbf{x}_{-\alpha}] = \mathbf{h}_\alpha$: 21.10 |

| | |
|---|---|
| $\sigma_\alpha$ | bijection $\lambda \mapsto \lambda - \lambda(\mathbf{h}_\alpha)\alpha$ of $\mathfrak{h}^*$ onto itself: 21.10 |
| $\mathfrak{s}_\alpha$ | Lie subalgebra $\mathbf{C}\mathbf{h}_\alpha \oplus \mathbf{C}\mathbf{x}_\alpha \oplus \mathbf{C}\mathbf{x}_{-\alpha}$: 21.10 |
| $N_{\alpha,\beta}$ | numbers such that $[\mathbf{x}_\alpha, \mathbf{x}_\beta] = N_{\alpha,\beta}\mathbf{x}_{\alpha+\beta}$ when $\alpha + \beta \in \mathbf{S}$: 21.10 |
| D(G) | union of the hyperplanes in $\mathfrak{t}$ with equations $\alpha(\mathbf{u}) = 2\pi i n$, $n \in \mathbf{Z}$: 21.10, Problem 2 |
| $\sigma_\alpha$ | bijection $\lambda \mapsto \lambda - v_\alpha(\lambda)\alpha$, for a reduced root system $\mathbf{S}$ in F: 21.11 |
| $W_{\mathbf{S}}$ | Weyl group of $\mathbf{S}$, generated by the $\sigma_\alpha$; 21.11 |
| $n(\alpha, \beta)$ | Cartan integers $v_\beta(\alpha) = 2(\beta\|\alpha)/(\beta\|\beta)$ for $\alpha, \beta \in \mathbf{S}$: 21.11 |
| $\mathbf{S}_{\mathbf{x}}^+$ | set of $\alpha \in \mathbf{S}$ such that $\alpha(\mathbf{x}) > 0$: 21.11 |
| $\mathbf{B}_{\mathbf{x}}$ | basis of $\mathbf{S}$, namely the set of indecomposable elements of $\mathbf{S}_{\mathbf{x}}^+$: 21.11 |
| $\mathbf{S}^+$ | set of positive roots, relative to a basis $\mathbf{B}$ of $\mathbf{S}$: 21.11 |
| $\mathbf{S}^\vee$ | root system formed by the $v_\alpha \in F^*$: 21.11 |
| $\mathbf{B}^\vee$ | basis of $\mathbf{S}^\vee$ consisting of the $v_\alpha$, $\alpha \in \mathbf{B}$: 21.11 |
| $\delta$ | $\frac{1}{2}\sum_{\lambda \in \mathbf{S}+} \lambda$: 21.11 |
| $\varepsilon_r$ | linear form on $\mathfrak{t} = \bigoplus_{s=1}^{n} \mathbf{R}iE_{ss} < \mathbf{M}_n(\mathbf{C})$ such that $\varepsilon_r(iE_{ss}) = i\delta_{rs}$: 21.12 |
| $\mathbf{Sp}(2n, \mathbf{C})$, $\mathfrak{sp}(2n, \mathbf{C})$ | complex symplectic group and its Lie algebra: 21.12 |
| $\mathbf{SO}(m, \mathbf{C})$, $\mathfrak{so}(m, \mathbf{C})$ | complex special orthogonal group and its Lie algebra: 21.12 |
| $A_n$, $B_n$, $C_n$, $D_n$ | Lie algebras of the classical groups: 21.12 |
| P(G, T), P(G), P | lattice $2\pi i\Gamma_T^*$ of weights of G with respect to T: 21.13 |
| $e^p$, $s \mapsto e^{p(s)}$ | character $\exp(\mathbf{u}) \mapsto e^{p(\mathbf{u})}$ of T, where $p \in P$: 21.13 |
| S(Π) | $\sum_{p \in \Pi} e^p$, where Π is an orbit of the Weyl group W in P: 21.13 |
| $\mathbf{Z}[P]^W$ | set of W-invariant elements of $\mathbf{Z}[P]$: 21.13 |

| | |
|---|---|
| $\mathbf{h}_j$ | $\mathbf{h}_{\beta_j}$, where $\{\beta_1, \ldots, \beta_l\}$ is a basis of S: 21.14 |
| P(g) | set of $\lambda \in \mathfrak{t}^*_{(\mathbf{C})}$ such that $\lambda(\mathbf{h}_\alpha) \in \mathbf{Z}$ for all $\alpha \in \mathbf{S}$, or equivalently such that $\lambda(\mathbf{h}_j) \in \mathbf{Z}$ for $1 \le j \le l$: 21.14 |
| C(g), C | Weyl chamber in $i\mathfrak{t}^*$, consisting of the $\lambda$ such that $\lambda(\mathbf{h}_j) > 0$ for $1 \le j \le l$: 21.14 |
| $\lambda \le \mu$ | order relation on $i\mathfrak{t}^*$, equivalent to $\lambda = \mu$ or $\mu - \lambda = \gamma + \sum_{j=1}^{l} c_j \beta_j$, with $\gamma \in i\mathfrak{c}^*$ and $c_j \ge 0$ and not all zero: 21.14 |
| $s_j$ | reflection $s_{\beta_j}$: $\lambda \mapsto \lambda - \lambda(\mathbf{h}_j)\beta_j$ for $1 \le j \le l$: 21.14 |
| $H_\alpha$ | hyperplane in $i\mathfrak{t}^*$ with equation $\lambda(\mathbf{h}_\alpha) = 0$: 21.14 |
| $\mathbf{Z}[P]^{aW}$ | set of W-anti-invariant elements of $\mathbf{Z}[P]$: 21.14 |
| $J(e^p)$ | $\sum_{w \in W} \det(w) e^{w \cdot p}$, where $p \in P$: 21.14 |
| $P_{reg}$ | set of weights $\lambda \in P$ which are regular linear forms: 21.14 |
| $S(p)$ | $S(\Pi)$, where $\Pi$ is the W-orbit of $p \in P \cap \overline{C}$: 21.14 |
| $\Delta$ | $J(e^\delta) = \prod_{\alpha \in \mathbf{S}^+} (e^{\alpha/2} - e^{-\alpha/2})$: 21.14 |
| $T_{reg}$ | set of regular points of the maximal torus $T \subset G$: 21.15 |
| $v_G, v_T, v_{G/T}$ | invariant volume-forms on G, T and G/T: 21.15 |
| $m_G, m_T, m_{G/T}$ | invariant measures corresponding to the volume-forms $v_G, v_T, v_{G/T}$: 21.15 |
| $\mu = n_1\beta_1 + \cdots + n_l\beta_l$ | highest root in $\mathbf{S}$, relative to the basis $\mathbf{B} = \{\beta_1, \ldots, \beta_l\}$: 21.15, Problem 10 |
| $W_a$ | affine Weyl group: 21.15, Problem 11 |
| $\mathfrak{u}_{\alpha, k}$ | hyperplane with equation $\alpha(\mathbf{u}) = 2\pi k$ in $i\mathfrak{t}$: 21.15, Problem 11 |
| $\{\mathbf{p}_1, \mathbf{p}_2, \ldots, \mathbf{p}_l\}$ | basis of $i\mathfrak{t}$ dual to $\{\beta_1, \beta_2, \ldots, \beta_l\}$: 21.15, Problem 11 |
| Q(g) | sublattice P(G/Z) of P(G) generated by the roots $\alpha \in \mathbf{S}$: 21.16 |
| $\varpi_j$ | fundamental weights $(1 \le j \le l)$ relative to the basis $\mathbf{B}$ of $\mathbf{S}$: 21.16 |
| **Spin** $(m)$ | simply connected covering group of $\mathbf{SO}(m)$ $(m \ge 3)$: 21.16 |

| | |
|---|---|
| $\mathfrak{a}(E)$ | set of self-adjoint automorphisms of E: 21.17 |
| $\mathfrak{a}_+(E)$ | set of positive self-adjoint automorphisms of E: 21.17 |
| $\tilde{G}_u$, $\mathfrak{g}_u$, $\mathfrak{g}$, $c_u$ | $\tilde{G}_u$ a simply connected compact semi-simple Lie group; $\mathfrak{g}_u = \text{Lie}(\tilde{G}_u)$; $\mathfrak{g} = (\mathfrak{g}_u)_{(\mathbf{C})}$; $c_u$ the conjugation of $\mathfrak{g}$ for which $\mathfrak{g}_u$ is the set of fixed vectors: 21.18 |
| $\tilde{G}$ | simply connected complex Lie group with Lie algebra $\mathfrak{g}$: 21.18 |
| $c_0$ | conjugation of $\mathfrak{g}$ which commutes with $c_u$: 21.18 |
| $\mathfrak{k}_0$, $i\mathfrak{p}_0$ | real vector subspaces of $\mathfrak{g}_u$ on which $c_0(\mathbf{x}) = \mathbf{x}$ and $c_0(\mathbf{x}) = -\mathbf{x}$, respectively: 21.18 |
| $\mathfrak{g}_0$ | subalgebra of invariants of $c_0$: 21.18 |
| $\tilde{P}$ | image of $i\mathfrak{g}_u$ under the mapping $i\mathbf{u} \mapsto \exp_{\tilde{G}}(i\mathbf{u})$: 21.18 |
| $G_0$, $K_0$, $P_0$ | $G_0$ the Lie subgroup of $\tilde{G}_{\|\mathbf{R}}$ consisting of the fixed points of $\sigma$ such that $\sigma_* = c_0$; $K_0 = G_0 \cap \tilde{G}_u$; $P_0 = G_0 \cap \tilde{P}$: 21.18 |
| $G_1$ | $\tilde{G}_0/D$, a group locally isomorphic to $\tilde{G}_0$: 21.18 |
| $K_1$, $P_1$ | $K_1 = \tilde{K}_0/D$; $P_1$ = image of $\mathfrak{p}_0$ under $\exp_{G_1}$: 21.18 |
| $G_2$ | $\tilde{G}_u/(C \cap G_0)$, C the centre of $\tilde{G}_u$: 21.18 |
| $K_2$ | $K_0/(C \cap G_0)$: 21.18 |
| $K'_2$ | subgroup of fixed points of $\sigma_2$, the automorphism of $G_2$ obtained from $\sigma$ on passing to the quotient: 21.18 |
| $P_2$ | image of $i\mathfrak{p}_0$ under $\exp_{G_2}$: 21.18 |
| $\alpha \prec \beta$ | lexicographic ordering: 21.20 |
| $\mathfrak{a}_0$ | maximal commutative subalgebra of $\mathfrak{p}_0$: 21.21 |
| $\mathfrak{t}$ | maximal commutative subalgebra of $\mathfrak{g}_u$ containing $\mathfrak{a}_0$: 21.21 |
| $\mathbf{S}'$ | subset of $\mathbf{S}$ consisting of the roots which vanish on $i\mathfrak{a}_0$: 21.21 |

$\mathbf{S}''_+$ — subset of $\mathbf{S}'' = \mathbf{S} - \mathbf{S}'$ consisting of the $\alpha$ such that $\alpha(\mathbf{z}_0) > 0$: 21.21

$\mathfrak{n}, \mathfrak{n}_0$ — $\mathfrak{n} = \bigoplus_{\alpha \in \mathbf{S}^+} \mathfrak{g}_\alpha$, $\mathfrak{n} = \mathfrak{n}_0 \cap \mathfrak{g}_0$: 21.21

$\mathfrak{T}_k$ — Lie algebra of matrices $(x_{hj})$ such that $x_{hj} = 0$ for $j + k > h$: 21.21

## CHAPTER XXI

# COMPACT LIE GROUPS AND SEMISIMPLE LIE GROUPS

It is rarely the case in mathematics that one can describe explicitly *all* the objects endowed with a structure that is characterized by a few simple axioms. A classical (and elementary) example is that of *finite commutative groups* (A.26.4). By contrast, in spite of more than a century of effort and an enormous accumulation of results, mathematics is still very far from being able to describe all *noncommutative finite groups*, even when supplementary restrictions (such as simplicity or nilpotency) are imposed.

It is therefore all the more remarkable that, in the theory of Lie groups, all the *compact simply connected* Lie groups are explicitly known, and that, starting from these groups, the structure of compact connected Lie groups is reduced to a simple problem in the theory of finitely generated commutative groups ((16.30.2) and (21.6.9)). The compact simply connected Lie groups are finite products of groups that are either the universal covering groups of the "classical groups" $\mathbf{SO}(n)$, $\mathbf{SU}(n)$, and $\mathbf{U}(n, \mathbf{H})$ (16.11) (and therefore depend on an integral parameter) or the five "exceptional" groups, of dimensions 14, 52, 78, 133, and 248. We shall not get as far as this final result, but we shall develop the methods leading to it, up to the point where what remains to be done is an enumeration (by successive exclusion) of certain algebraic objects related to Euclidean geometry, subjected to very restrictive conditions of an arithmetic nature, which allow only a small number of possibilities (21.10.3) (see [79] or [85] for a complete account).

These methods are based in part on the elementary theory of Lie groups in Chapter XIX, and in part on a fundamental new idea, which dominates this chapter and the next, and whose importance in present-day mathematics cannot be overemphasized; the notion of a *linear representation* of a group. The first essential fact is that where *compact* groups are concerned (whether they are Lie groups or not) we may restrict our attention to *finite-dimensional* linear representations (21.2.3). The second unexpected phenomenon is that where compact connected Lie groups are concerned, everything rests on the explicit knowledge of the representations of only *two* types of groups: the tori $\mathbf{T}^n$ and the group $\mathbf{SU}(2)$ **(21.9)**. Roughly speaking, these are the "building blocks" with which we can "construct" all the other compact connected Lie groups and obtain not only their explicit structure but also an enumeration of *all* their linear representations **(21.15.5)**.

The interest attached to the compact connected Lie groups arises not only from the esthetic attractions of the theory, which is one of the most beautiful and most satisfying in the whole of mathematics, but also from the central position they occupy in the welter of modern theories. In the first place, they are closely related to a capital notion in the theory of Lie groups, namely that of a *semisimple group* (compact or not), and in fact it turns out that a knowledge of the compact semisimple groups determines all the others **(21.18)**. Since the time of F. Klein it has been recognized that classical "geometry" is essentially the study of certain semisimple groups; and E. Cartan, in his development of the notions of fiber bundle and connection, showed that these groups play an equally important role in differential geometry (see Chapter XX). From then on, their influence has spread into differential topology and homological algebra. We shall see in Chapter XXII how—again following E. Cartan—it has been realized over the last twenty-five years that the study of representations of semisimple groups (but now on infinite-dimensional spaces) is fundamental in many questions of analysis, not to speak of applications to quantum mechanics. But the most unexpected turn has been the invasion of the theory of semisimple groups into regions that appear completely foreign: "abstract" algebraic geometry, number theory, and the theory of finite groups. It has been known since the work of S. Lie and E. Cartan that semisimple groups are *algebraic* (that is, they can be defined by polynomial equations); but it is only since 1950 that it has come to be realized that this is no accidental fact, but rather that the theory of semisimple groups has *two faces* of equal importance: the analytic aspect, which gave birth to the theory, and the purely algebraic aspect, which appears when one considers a ground field other than **R** or **C**. We have not, unfortunately, been able to take account of this second aspect; here we can only remark that its repercussions are increasingly numerous, and refer the reader to the works [80], [81], [74], [77], and [78] in the bibliography.

## 1. CONTINUOUS UNITARY REPRESENTATIONS OF LOCALLY COMPACT GROUPS

**(21.1.1)** Let G be a topological group, E a Hausdorff topological vector space over the field **C** of complex numbers. Generalizing the definition given in **(16.9.7)**, we define a *continuous linear representation of* G *on* E to be a mapping $s \mapsto U(s)$ of G into the group **GL**(E) of automorphisms of the *topological vector space* E, which satisfies the following conditions:

(a) $U(st) = U(s)U(t)$ for all $s, t \in G$;
(b) for each $x \in E$, the mapping $s \mapsto U(s) \cdot x$ of G into E is *continuous.*

It follows from (a) that $U(e) = 1_E$ (where $e$ is the identity element of G) and that, for all $s \in G$,

**(21.1.1.1)** $$U(s^{-1}) = U(s)^{-1}.$$

If E is of *finite* dimension $d$, the representation $U$ is said to be of *dimension* (or *degree*) $d$, and we sometimes write $d = \dim U$.

The mapping $U_0$ that sends each $s \in G$ to the identity automorphism $1_E$ is a continuous linear representation of G on E, called the *trivial representation.*

A vector subspace F of E is said to be *stable* under a continuous linear representation $U$ of G on E if $U(s)(F) \subset F$ for all $s \in G$; in that case, the mapping $s \mapsto U(s) | F$ is a continuous linear representation of G on F, called the *subrepresentation* of $U$ corresponding to F.

A continuous linear representation $U$ of G on E is said to be *irreducible* (or *topologically irreducible*) if the only *closed* vector subspaces F of E that are stable under $U$ are $\{0\}$ and E. For each $x \neq 0$ in E, the set $\{U(s) \cdot x : s \in G\}$ is then *total* in E **(12.13)**.

**(21.1.2)** In this chapter and the next, we shall be concerned especially with the case where E is a *separable Hilbert space.* A *continuous unitary representation* of G on E is then a continuous linear representation $U$ of G on E such that for each $s \in G$ the operator $U(s)$ is *unitary*, or in other words **(15.5)** is an automorphism of the Hilbert space structure of E. This means that the operators $U(s)$ satisfy conditions (a) and (b) of **(21.1.1)**, together with the following condition:

(c) $(U(s) \cdot x \,|\, U(s) \cdot y) = (x \,|\, y)$ for all $s \in G$ and all $x, y \in E$.

In particular, $U(s)$ is an *isometry* of E onto E, for all $s \in G$, and we have

**(21.1.2.1)** $$U(s)^{-1} = (U(s))^*$$

for all $s \in G$.

**(21.1.3)** (i) When E is *finite*-dimensional, condition (b) of **(21.1.1)** is equivalent to saying that $s \mapsto U(s)$ is a *continuous* mapping of G into the normed algebra $\mathscr{L}(E)$ (relative to any norm that defines the topology of E); for it is equivalent to saying that if $(u_{jk}(s))$ is the matrix of $U(s)$ relative to some basis of E, then the functions $u_{jk}$ are continuous on G. On the other hand, if E is a separable Hilbert space of infinite dimension and $U$ is a continuous unitary representation of G on E, then $U$ is not in general a continuous mapping of G into the normed algebra $\mathscr{L}(\mathrm{E})$ (Problem 3).

(ii) When E is finite-dimensional, a continuous linear representation $U$ of G on E is not necessarily a continuous *unitary* representation relative to any scalar product **(6.2)** on E. For example, if $\mathrm{G} = \mathbf{R}$, the continuous linear representation

$$U : x \mapsto \begin{pmatrix} 1 & 0 \\ x & 1 \end{pmatrix}$$

of G on $\mathbf{C}^2$ is not unitary, relative to any scalar product on $\mathbf{C}^2$, because any unitary matrix is similar to a diagonal matrix **(15.11.14)** (cf. Section **21.18**, Problem 1).

**(21.1.4)** *Throughout the rest of this chapter we shall consider only separable metrizable locally compact groups, and as in Chapter XIV the phrases "locally compact group" and "compact group" will mean "separable metrizable locally compact group" and "metrizable compact group," respectively.*

Let G be a locally compact group, $\mu$ a *bounded complex measure* **(13.20)** on G, and $U$ a continuous unitary representation of G on a separable Hilbert space E. For each pair of vectors $x, y$ in E, the function $s \mapsto (U(s) \cdot x \mid y)$ is continuous and bounded on G, because $\|U(s) \cdot x\| = \|x\|$; it is therefore $\mu$-integrable, and by **(13.20.5)** we have

**(21.1.4.1)**
$$\left| \int (U(s) \cdot x \mid y) \, d\mu(s) \right| \leqq \|\mu\| \cdot \|x\| \cdot \|y\|.$$

Since E may be identified with its dual, it follows that there exists a unique vector $U(\mu) \cdot x$ in E such that

$$\int (U(s) \cdot x \mid y) \, d\mu(s) = (U(\mu) \cdot x \mid y)$$

for all $y \in \mathrm{E}$, and this allows us to write **(13.10.6)**

**(21.1.4.2)**
$$U(\mu) \cdot x = \int (U(s) \cdot x) \, d\mu(s).$$

It is clear that this relation defines a continuous endomorphism $U(\mu)$ of E, since **(21.1.4.1)** implies that

**(21.1.4.3)**
$$\|U(\mu)\| \leqq \|\mu\|.$$

In particular, we have

(21.1.4.4) $$U(\varepsilon_s) = U(s)$$

for all $s \in G$.

The relation (21.1.4.2) is sometimes written in the abridged form

(21.1.4.5) $$U(\mu) = \int U(s)\, d\mu(s).$$

(21.1.5) We recall (15.4.9) that the set $M_C^1(G)$ of bounded complex measures on G is an *involutory Banach algebra* over **C**, the multiplication being convolution of measures, and the involution $\mu \mapsto \check{\mu}$. When a left Haar measure $\beta$ has been chosen on G, the normed space $L_C^1(G)$ may be canonically identified with a closed vector subspace of $M_C^1(G)$, by identifying the class $\tilde{f}$ of a $\beta$-integrable function $f$ with the bounded measure $f \cdot \beta$, since $\|f \cdot \beta\| = N_1(f)$ (13.20.3). By the definition of the convolution of two functions in $\mathscr{L}_C^1(G)$ (14.10.1), $L_C^1(G)$ is a subalgebra of $M_C^1(G)$ if we define the product of the classes of two functions $f, g \in \mathscr{L}_C^1(G)$ to be the class of $f * g$. If in addition G is *unimodular* (14.3), $L_C^1(G)$ is a two-sided ideal in $M_C^1(G)$, and the transform of the measure $f \cdot \beta$ under the involution $\mu \mapsto \check{\mu}$ is $\check{\tilde{f}} \cdot \beta$ (14.3.4.2). We may therefore consider $L_C^1(G)$ as an *involutory closed subalgebra* of $M_C^1(G)$, the involution being that which transforms the class of $f$ into the class of $\check{\tilde{f}}$.

We deduce from this that if G is *unimodular*, then for each *representation* (15.5) $V$ of the involutory Banach algebra $L_C^1(G)$ on a Hilbert space E, we have

(21.1.5.1) $$\|V(\tilde{f})\| \leqq N_1(f)$$

for all $f \in \mathscr{L}_C^1(G)$. For if G is discrete, this is just (15.5.7) because the identity element $\varepsilon_e$ of $M_C^1(G)$ then belongs to $L_C^1(G)$. If G is not discrete, it is immediately seen that $V$ may be extended to a representation on E of the involutory Banach subalgebra $A = L_C^1(G) \oplus C\varepsilon_e$ of $M_C^1(G)$ by putting $V(f \cdot \beta + \lambda\varepsilon_e) = V(\tilde{f}) + \lambda \cdot 1_E$, and (15.5.7) can then be applied to this algebra with identity element.

(21.1.6) *Under the assumptions of* (21.1.4), *the mapping* $\mu \mapsto U(\mu)$ *is a representation* (15.5) *of the involutory Banach algebra* $M_C^1(G)$ *on the Hilbert space* E. *If in addition* G *is unimodular, the restriction of* $\mu \mapsto U(\mu)$ *to* $L_C^1(G)$ *is nondegenerate.*

It follows immediately from (21.1.4.4) that $U(\varepsilon_e) = 1_E$. To prove the first assertion, it remains to show that $U(\mu * \nu) = U(\mu)U(\nu)$ and $U(\check{\mu}) = (U(\mu))^*$, where $\mu$, $\nu$ are any two bounded measures on G. If $x$, $y$ are any two vectors in E then by definition **(14.5)** we have

$$\begin{aligned}(U(\mu * \nu) \cdot x \mid y) &= \int (U(s) \cdot x \mid y)\, d(\mu * \nu)(s) \\ &= \iint (U(vw) \cdot x \mid y)\, d\mu(v)\, d\nu(w) \\ &= \iint (U(w) \cdot x \mid (U(v))^* \cdot y)\, d\mu(v)\, d\nu(w) \\ &= \int (U(\nu) \cdot x \mid (U(v))^* \cdot y)\, d\mu(v) \\ &= \int (U(v) \cdot (U(\nu) \cdot x) \mid y)\, d\mu(v) \\ &= (U(\mu)U(\nu) \cdot x \mid y)\end{aligned}$$

by virtue of the Lebesgue–Fubini theorem, and this proves the first relation. Next, using the fact that the operators $U(s)$ are unitary, we have

$$\begin{aligned}((U(\mu))^* \cdot x \mid y) &= \overline{(U(\mu) \cdot y \mid x)} \\ &= \int \overline{(U(s) \cdot y \mid x)\, d\mu(s)} \\ &= \int \overline{(U(s) \cdot y \mid x)}\, d\bar{\mu}(s) \\ &= \int (U(s^{-1}) \cdot x \mid y)\, d\bar{\mu}(s) \\ &= \int (U(t) \cdot x \mid y)\, d\check{\bar{\mu}}(t) \\ &= (U(\check{\bar{\mu}}) \cdot x \mid y)\end{aligned}$$

by the definition of the measure $\check{\bar{\mu}}$ **(15.4.9)**, and this proves the second relation.

*In particular, for each* $s \in$ G *and each bounded measure* $\mu$ *on* G, *we have*

**(21.1.6.1)** $$U(\varepsilon_s * \mu) = U(s)U(\mu), \qquad U(\mu * \varepsilon_s) = U(\mu)U(s).$$

Let $(V_n)$ be a decreasing sequence of neighborhoods of $e$ in G, forming a fundamental system of neighborhoods of $e$. For each $s \in$ G and each $n$, let $u_n$

be a positive-valued function belonging to $\mathscr{K}(G)$ with support contained in $sV_n$ and such that $\int u_n \, d\beta = 1$. For each $x \in E$ and each $\varepsilon > 0$, there exists an integer $n$ such that

**(21.1.6.2)** $$\|U(t) \cdot x - U(s) \cdot x\| \leqq \varepsilon$$

for all $t \in sV_n$. We have then, for all $y \in E$,

$$(U(u_n \cdot \beta)x - U(s) \cdot x \mid y) = \int (U(t) \cdot x - U(s) \cdot x \mid y) \, u_n(t) \, d\beta(t)$$

and the inequality **(21.1.6.2)** therefore implies that

$$\|U(u_n \cdot \beta) \cdot x - U(s) \cdot x\| \leqq \varepsilon.$$

If there existed a vector $x \neq 0$ such that $U(f \cdot \beta) \cdot x = 0$ for *all* functions $f \in \mathscr{L}^1_{\mathbf{C}}(G)$, we should therefore have $U(s) \cdot x = 0$ for *all* $s \in G$, which is absurd (take $s = e$). The restriction of the representation $\mu \mapsto U(\mu)$ to $L^1_{\mathbf{C}}(G)$ is therefore *nondegenerate.*

By abuse of language, we shall call the restriction of $\mu \mapsto U(\mu)$ to $L^1_{\mathbf{C}}(G)$ the *extension* of $U$ to $L^1_{\mathbf{C}}(G)$, and we shall denote it by $U_{\text{ext}}$. For $f \in \mathscr{L}^1_{\mathbf{C}}(G)$, we shall write $U(f)$ instead of $U(f \cdot \beta)$ or $U(\tilde{f})$.

**(21.1.7)** *Let* G *be a unimodular, separable, metrizable, locally compact group. Then the mapping* $U \mapsto U_{\text{ext}}$ *is a bijection of the set of continuous unitary representations of* G *on* E, *onto the set of nondegenerate representations of the involutive Banach algebra* $L^1_{\mathbf{C}}(G)$ *on* E. *Furthermore, in order that a closed vector subspace* F *of* E *should be stable under all the operators* $U(s)$ $(s \in G)$, *it is necessary and sufficient that it should be stable under all the operators* $U(f)$ *for* $f \in \mathscr{L}^1_{\mathbf{C}}(G)$ (*or just for* $f \in \mathscr{K}(G)$).

We have seen in the course of the proof of **(21.1.6)** that, for each $s \in G$ and $x \in E$, the vector $U(s) \cdot x$ is the limit of a sequence $U(u_n) \cdot x$ with $u_n \in \mathscr{K}(G)$. This shows already that the mapping $U \mapsto U_{\text{ext}}$ is injective, and that if a closed subspace F of E is stable under the operators $U(f)$ (where $f \in \mathscr{L}^1_{\mathbf{C}}(G)$ or $f \in \mathscr{K}(G)$), then it is stable under the operators $U(s)$ $(s \in G)$; and the converse follows directly from the definition of $U(\mu)$ **(21.1.4)**. It remains to show that, for each nondegenerate representation $V$ of $L^1_{\mathbf{C}}(G)$ on E, there exists a continuous unitary representation $U$ of G on E such that $V = U_{\text{ext}}$. Let H be the vector subspace of E spanned by the vectors $V(f) \cdot x$, where $f \in \mathscr{L}^1_{\mathbf{C}}(G)$ and $x \in E$; then the hypothesis on $V$ signifies that

H is *dense* in E. Let $s \in G$, and define the sequence of functions $(u_n)$ as in the proof of **(21.1.6)**; then for each $f \in \mathscr{L}^1_{\mathbf{C}}(G)$ we have **(14.11.1)**

$$\lim_{n\to\infty} N_1(u_n * f - \varepsilon_s * f) = 0$$

and consequently **(21.1.5.1)**

$$\lim_{n\to\infty} \| V(u_n)V(f) - V(\varepsilon_s * f)\| = 0.$$

This shows that for each $y \in H$, i.e., each linear combination $\sum_k V(f_k) \cdot x_k$, the sequence $(V(u_n) \cdot y)$ has a limit in E, namely, $\sum_k V(\varepsilon_s * f_k) \cdot x_k$. Let $U(s) \cdot y$ denote this limit. It is clear that the mapping $U(s)\colon H \to E$ so defined is linear and such that for each $f \in \mathscr{L}^1_{\mathbf{C}}(G)$ we have

**(21.1.7.1)**
$$U(s) \circ V(f) = V(\varepsilon_s * f),$$

which shows also that $U(s)$ maps H into itself.

Also, by **(21.1.5.1)**, we have $\|V(u_n)\| \leqq N_1(u_n) = 1$ for all $n$, and therefore $\|U(s) \cdot y\| \leqq \|y\|$ for all $y \in H$; hence $U(s)$ extends uniquely to a continuous operator on E, which we denote also by $U(s)$. Clearly we have $\|U(s)\| \leqq 1$. We have to show that $s \mapsto U(s)$ is a continuous unitary representation of G on E. If $s, t \in G$, then by virtue of **(21.1.7.1)** we have

$$\begin{aligned} U(st) \circ V(f) &= V(\varepsilon_{st} * f) = V(\varepsilon_s * (\varepsilon_t * f)) \\ &= U(s) \circ V(\varepsilon_t * f) = U(s) \circ U(t) \circ V(f), \end{aligned}$$

from which it follows immediately that $U(st) \cdot y = U(s) \cdot (U(t) \cdot y)$ for all $y \in H$ and hence, by continuity, $U(st) = U(s)U(t)$ in $\mathscr{L}(E)$. Next, it follows from **(21.1.7.1)** that $U(e)$ is equal to the identity mapping on H, and therefore also on E. Finally, since $\|U(s) \cdot x\| \leqq \|x\|$ and $\|U(s^{-1}) \cdot x\| \leqq x$, we have also $\|x\| \leqq \|U(s) \cdot x\|$ and therefore $\|U(s) \cdot x\| = \|x\|$ for all $x \in E$, so that $U(s)$ is a unitary operator.

It remains to show that $V = U_{\mathrm{ext}}$. Let $f, g \in \mathscr{L}^1_{\mathbf{C}}(G)$; from the definition of convolution and the Lebesgue–Fubini theorem it follows that for each $h \in \mathscr{L}^\infty_{\mathbf{C}}(G)$ we have

**(21.1.7.2)**
$$\langle h, f * g\rangle = \int f(s)\langle h, \varepsilon_s * g\rangle \, d\beta(s).$$

For each pair of vectors $x, y \in E$, the function $f \mapsto (V(f) \cdot x \mid y)$ is a continuous linear form on $\mathscr{L}^1_{\mathbf{C}}(G)$, hence is of the form $f \mapsto \langle h, f\rangle$ for some $h \in \mathscr{L}^\infty_{\mathbf{C}}(G)$ **(13.17.1)**. Hence, by virtue of **(21.1.7.2)**, we may write

$$\begin{aligned}(V(f)\cdot(V(g)\cdot x)\,|\,y) &= (V(f*g)\cdot x\,|\,y)\\ &= \int f(s)(V(\varepsilon_s * g)\cdot x\,|\,y)\,d\beta(s)\\ &= \int (U(s)\cdot(V(g)\cdot x)\,|\,y)f(s)\,d\beta(s)\\ &= (U(f)\cdot(V(g)\cdot x)\,|\,y),\end{aligned}$$

from which we conclude that $(U(f)\cdot z\,|\,y) = (V(f)\cdot z\,|\,y)$ for all $z \in \mathrm{H}$ and $y \in \mathrm{E}$, and hence that $U(f) = V(f)$ because H is dense in E.

**(21.1.8)** The study of the continuous unitary representations of a unimodular group G is therefore entirely equivalent to that of the nondegenerate representations of $\mathrm{L}^1_{\mathbf{C}}(\mathrm{G})$. Hence we may transfer to the former all the terminology introduced in **(15.5)** for the latter. In particular, two continuous unitary representations $U_1$, $U_2$ of G on spaces $\mathrm{E}_1$, $\mathrm{E}_2$ are said to be *equivalent* if there exists an isomorphism $T$ of the Hilbert space $\mathrm{E}_1$ onto the Hilbert space $\mathrm{E}_2$ such that $U_2(s) = TU_1(s)T^{-1}$ for all $s \in \mathrm{G}$. This is equivalent to saying that $U_2(f) = TU_1(f)T^{-1}$ for all functions $f \in \mathscr{L}^1_{\mathbf{C}}(\mathrm{G})$: in other words, $(U_1)_{\mathrm{ext}}$ and $(U_2)_{\mathrm{ext}}$ are *equivalent* in the sense of **(15.5)**. To say that $U$ is *irreducible* is equivalent to saying, by virtue of **(21.1.7)**, that $U_{\mathrm{ext}}$ is *topologically irreducible*. Finally, if E is the Hilbert sum of a sequence $(\mathrm{F}_n)$ of closed subspaces stable under $U$, then $U$ is said to be the *Hilbert sum* of the subrepresentations corresponding to the $\mathrm{F}_n$.

**(21.1.9)** *Example.* Suppose that G is unimodular. For each $s \in \mathrm{G}$ and each $f \in \mathscr{L}^2_{\mathbf{C}}(\mathrm{G})$, the function $\gamma(s)f = \varepsilon_s * f$ **(14.8.5)** belongs to $\mathscr{L}^2_{\mathbf{C}}(\mathrm{G})$, and we have $\mathrm{N}_2(\varepsilon_s * f) = \mathrm{N}_2(f)$. Hence we may define a unitary operator $R(s)$ on $\mathrm{L}^2_{\mathbf{C}}(\mathrm{G})$ by mapping the class of $f$ to the class of $\varepsilon_s * f$. Further, it follows from **(14.10.6.3)** that $s \mapsto R(s)$ is a continuous unitary representation of G on $\mathrm{L}^2_{\mathbf{C}}(\mathrm{G})$. This representation is called the *regular* (or *left regular*) *representation* of G. It follows from **(14.9.2)** that for each bounded measure $\mu$ on G we have $R(\mu)\cdot\tilde{g} = (\mu * g)^{\sim}$ for all $g \in \mathscr{L}^2_{\mathbf{C}}(\mathrm{G})$, and in particular that $R(f)\cdot\tilde{g} = (f * g)^{\sim}$ for all $f \in \mathscr{L}^1_{\mathbf{C}}(\mathrm{G})$. The representation $R_{\mathrm{ext}}$ is called the *regular* (or *left regular*) *representation* of $\mathrm{L}^1_{\mathbf{C}}(\mathrm{G})$ on $\mathrm{L}^2_{\mathbf{C}}(\mathrm{G})$. It is *injective*, because it follows immediately from regularization **(14.11.1)** that if $f * g$ is negligible for all functions $g \in \mathscr{L}^2_{\mathbf{C}}(\mathrm{G})$, then $f$ is negligible.

**(21.1.10)** Let $\mathrm{E}_1$, $\mathrm{E}_2$ be two Hausdorff topological vector spaces over **C**, and let $U_1$, $U_2$ be continuous linear representations of G on $\mathrm{E}_1$, $\mathrm{E}_2$, respectively **(21.1.1)**. Generalizing the terminology of **(21.1.8)**, we say that $U_1$ and $U_2$ are *equivalent* if there exists an isomorphism $T\colon \mathrm{E}_1 \to \mathrm{E}_2$ of topological vector spaces such that $U_2(s) = TU_1(s)T^{-1}$ for all $s \in G$. When $E_1$, $E_2$ are Hilbert spaces and $U_1$, $U_2$ are continuous unitary representations, it can be shown that this definition is equivalent to that given in **(21.1.8)** (Problem 4). The *direct sum* of two arbitrary continuous linear representations $U_1$, $U_2$ of G is defined to be the continuous linear representation $U$ of G on $E_1 \times E_2$ defined by $U(s) \cdot (x_1, x_2) = (U_1(s) \cdot x_1, U_2(s) \cdot x_2)$. If $E_1$ and $E_2$ are finite-dimensional and $U_1(s)$, $U_2(s)$ are identified with their matrices relative to (arbitrary) bases of $E_1$, $E_2$, respectively, then $U(s)$ is identified with the matrix $\begin{pmatrix} U_1(s) & 0 \\ 0 & U_2(s) \end{pmatrix}$, and we write $U = U_1 \oplus U_2$. The direct sum of a finite number of continuous linear representations of G is defined in the same way. A continuous linear representation of G on a *finite*-dimensional space is said to be *completely reducible* if it is equivalent to a *direct sum of irreducible representations*.

## PROBLEMS

1. Let E be a normed space, G a (separable, metrizable) locally compact group, and $s \mapsto U(s)$ a mapping of G into the group **GL**(E) such that $U(st) = U(s)U(t)$ for all $s, t \in G$. Let A be a dense subset of E such that for each $x \in A$ the mapping $s \mapsto U(s) \cdot x$ is continuous on G.
(a) Show that the function $s \mapsto \|U(s)\|$ is lower semicontinuous on G and that
$$\|U(st)\| \leqq \|U(s)\| \cdot \|U(t)\|$$
for all $s, t \in G$.
(b) Deduce from (a) that for each compact subset K of G the set $\{U(s) : s \in K\}$ is equicontinuous on E (use **(12.16.2)**). Deduce that the mapping $(s, x) \mapsto U(s) \cdot x$ of $G \times E$ into E is continuous.

2. Let E be a separable normed space and D a denumerable dense subset of E; let G be a locally compact group and let $s \mapsto U(s)$ be a mapping of G into **GL**(E) such that $U(st) = U(s)U(t)$ for all $s, t \in G$. Suppose also that for each $x \in D$ the mapping $s \mapsto U(s) \cdot x$ of G into E is *measurable* (relative to a Haar measure on G).
Let V be a symmetric compact neighborhood of $e$ in G. Show that there exists a compact subset K of V, with measure arbitrarily close to that of V, such that the mapping $s \mapsto \|U(s)\|$ is lower semicontinuous on K **(13.9.5)**. Deduce that this mapping is bounded on K (same method as in Problem 1). Show, by using **(14.10.8)**, that there exists a neighborhood $W \subset V$ of $e$ in G such that the mapping $s \mapsto \|U(s)\|$ is bounded on W, and deduce that the mapping $(s, x) \mapsto U(s) \cdot x$ of $G \times E$ into E is continuous.

3. Let G be an infinite (metrizable) compact group, endowed with normalized Haar measure. Show that for each $s \neq e$ in G there exists a function $f \in \mathscr{L}_{\mathbf{C}}^2(G)$ such that $N_2(f) = 1$ and $N_2(\gamma(s)f - f) = \sqrt{2}$. Deduce that the regular representation $s \mapsto R(s)$ of G on $L_{\mathbf{C}}^2(G)$ is not a continuous mapping of G into the Banach algebra $\mathscr{L}(L_{\mathbf{C}}^2(G))$.

4. Let G be a locally compact group, let $E_1$ and $E_2$ be separable complex Hilbert spaces, $U_1$ and $U_2$ continuous unitary representations of G on $E_1$, $E_2$, respectively, and let $T: E_1 \to E_2$ be an isomorphism of *topological vector spaces* such that $U_2(s) = TU_1(s)T^{-1}$ for all $s \in G$.
(a) There exists an isomorphism $T^*: E_2 \to E_1$ of topological vector spaces such that $(T \cdot x_1 | x_2) = (x_1 | T^* \cdot x_2)$ for all $x_1 \in E_1$ and $x_2 \in E_2$. ($T^*$ is the *adjoint* of $T$; cf. Section **15.12**, Problem 1.) The operator $T^* \circ T$ on $E_1$ is self-adjoint, positive, and invertible, and there exists a unique self-adjoint positive invertible operator $A$ such that $A^2 = T^* \circ T$ (**15.11.12**). Show that $A^2 U_1(s) = U_1(s)A^2$ for all $s \in G$, and deduce that $AU_1(s) = U_1(s)A$ for all $s \in G$. (Use the approximation of $t^{1/2}$ by polynomials, together with (**15.11.8.1**).)
(b) Show that $T \circ A^{-1} = S: E_1 \to E_2$ is an isomorphism of *Hilbert spaces*, such that $U_2(s) = SU_1(s)S^{-1}$ for all $s \in G$.

5. (a) Let E be a separable Hilbert space and $A$ an unbounded self-adjoint operator on E. If $U$ is a unitary operator on E that leaves dom($A$) stable and is such that $U \cdot (A \cdot x) = A \cdot (U \cdot x)$ for all $x \in \text{dom}(A)$, show that $U(\text{dom}(A)) = \text{dom}(A)$, and that for each bounded, uniformly measurable function $f$ on **R**, the operator $U$ commutes with the continuous self-adjoint operator $f(A)$ (notation of (**15.12.13**)). In particular, if $A$ is not a homothety, there exists a closed vector subspace F of E, other than E and $\{0\}$, which is stable under $U$.
(b) Let G be a locally compact group and let $s \mapsto U(s)$ be an *irreducible* continuous unitary representation of G on E. Show that if $A$ is an unbounded self-adjoint operator on E, such that dom($A$) is stable under the representation $U$ and such that $U(s) \cdot (A \cdot x) = A \cdot (U(s) \cdot x)$ for all $s \in G$ and all $x \in \text{dom}(A)$, then $A$ is necessarily a homothety. (This is the topological version of *Schur's lemma.*)

6. Let G be a locally compact group and let $U_1$, $U_2$ be continuous unitary representations of G on separable Hilbert spaces $E_1$, $E_2$, respectively. A continuous linear mapping $T: E_1 \to E_2$ is an *intertwining operator* for $U_1$ and $U_2$ if $TU_1(s) = U_2(s)T$ for all $s \in G$. Then $T^*$ (Problem 4) is an intertwining operator for $U_2$ and $U_1$.

Suppose that $U_1$ is *irreducible*. Suppose also that there exists a nonzero unbounded closed operator $T$ from $E_1$ to $E_2$ (Section **15.12**, Problem 1) such that dom($T$) is dense in $E_1$ and stable under $U_1$, and such that $T \cdot (U_1(s) \cdot x) = U_2(s) \cdot (T \cdot x)$ for all $x \in \text{dom}(T)$ and all $s \in G$. Show that dom($T^*$) is dense in $E_2$ and stable under $U_2$, that dom($T^*T$) is dense in $E_1$ and stable under $U_1$, and that $T^*T$ is self-adjoint. (Consider the Hilbert sum of $E_1$ and $E_2$, and the operator $S$ defined on $\text{dom}(T) \oplus E_2$, which is equal to $T$ on dom($T$) and zero on $E_2$.) Deduce from Problem 5 that there exists a constant $c \neq 0$ such that $T^*T = cI$, and hence that $\text{dom}(T) = E_1$ and that $T$ is an isometry of $E_1$ onto a closed subspace of $E_2$. Hence $U_1$ is equivalent to a *subrepresentation* of $U_2$.

7. Let E be a finite-dimensional *real* vector space. If G is a topological group, a *continuous* (*real*) *linear representation* of G on E is any continuous homomorphism of G into **GL(E)**.
(a) Let $F = E_{(\mathbf{C})}$ be the complex vector space obtained from E by extension of scalars; identify E with the (real) subspace of F consisting of all $x \otimes 1$ with $x \in E$. Then every $z \in F$ is uniquely of the form $z = x + iy$ where $x, y \in E$. Define a mapping $J: F \to F$ by $J \cdot (x + iy) = x - iy$, where $x, y \in E$; then $J$ is a semilinear bijection, and $J^2 = I$; also E is the set of $z \in F$ such that $J \cdot z = z$. If $s \mapsto U(s)$ is a continuous (real) linear representation of G on E, the mapping $s \mapsto V(s) = U(s) \otimes 1_{\mathbf{C}}$ is a continuous linear representation of G on F, such that $V(s) \cdot J = J \cdot V(s)$ for all $s \in G$.

(b) Conversely, let F be a finite-dimensional complex vector space, and let $J$ be a semilinear bijection of F onto F such that $J^2 = I$. If $F_{|\mathbf{R}}$ is the real vector space obtained from F by restriction of scalars, then $J$ is an involutory automorphism of $F_{|\mathbf{R}}$. If E is the eigenspace of this automorphism for the eigenvalue 1, then $i$E is the eigenspace for the eigenvalue $-1$, and consequently F may be identified with $E_{(\mathbf{C})}$. Show that if $s \mapsto V(s)$ is a continuous linear representation of G on F such that $V(s) \cdot J = J \cdot V(s)$ for all $s \in G$, then there exists a continuous (real) linear representation $U$ of G on E such that $V$ may be identified with $s \mapsto U(s) \otimes 1_{\mathbf{C}}$.

**8.** Let F be a finite-dimensional left vector space over **H**, the division ring of quaternions. If G is any topological group, a *continuous* (*quaternionic*) *linear representation* of G on F is any continuous homomorphism of G into **GL**(F).
(a) Identify the quaternions of the form $a + bi$ ($a, b \in \mathbf{R}$) with complex numbers, so that every quaternion $a + bi + cj + dk$ is expressed as $(a + bi) + (c + di)j$, and $\mathbf{H} = \mathbf{C} \oplus \mathbf{C}j$ is a left vector space of dimension 2 over **C**. Let $E = F_{|\mathbf{C}}$ be the complex vector space obtained from F by restriction of scalars. If we define $J \cdot z = jz$ for each vector $z \in E$, then we have $J \cdot (\lambda z) = \bar{\lambda}(J \cdot z)$ for all $\lambda \in \mathbf{C}$, so that $J$ is a semilinear bijection of E onto E such that $J^2 = -I$. A quaternionic continuous linear representation $s \mapsto U(s)$ of G on F can be considered as a continuous linear representation of G on E, and we have $U(s) \cdot J = J \cdot U(s)$ for all $s \in G$.
(b) Conversely, let E be a finite-dimensional complex vector space, and let $J$ be a semilinear bijection of E onto E such that $J^2 = -I$. For each vector $z \in E$ and each quaternion $\lambda + \mu j$ (where $\lambda, \mu \in \mathbf{C}$), put $(\lambda + \mu j)z = \lambda z + \mu(J \cdot z)$. This defines on E a structure of left vector space over **H** such that if F denotes this left vector space then E is $F_{|\mathbf{C}}$. If $U$ is a continuous linear representation of G on E such that $U(s) \cdot J = J \cdot U(s)$ for all $s \in G$, then $U$ can be regarded as a quaternionic continuous linear representation of G on F.

**9.** For finite-dimensional real (resp. quaternionic) continuous linear representations of a topological group G, the notions of equivalent representations, direct sum of representations, and irreducible representations are defined exactly as in (21.1.1) and (21.1.10), by replacing the field **C** by **R** (resp. **H**) throughout. If $U$ is a continuous linear representation of G on a finite-dimensional complex vector space, satisfying the condition of Problem 7(b) (resp. 8(b)), we denote by $U^{(\mathbf{R})}$ (resp. $U^{(\mathbf{H})}$) the corresponding real (resp. quaternionic) linear representation.
(a) Let $U$, $V$ be two equivalent complex linear representations of G, so that if E, F are the respective spaces of the representations $U$, $V$, there exists a linear bijection $T$ of E onto F such that $V(s) = TU(s)T^{-1}$ for all $s \in G$. Suppose that there exists a semilinear bijection $J_E$ (resp. $J_F$) of E (resp. F) onto itself such that $J_E^2 = \varepsilon I_E$ and $J_F^2 = \varepsilon I_F$ (where $\varepsilon = \pm 1$) and $U(s)J_E = J_E U(s)$, $V(s)J_F = J_F V(s)$ for all $s \in G$. Show that there exists a linear bijection $S$ of E onto F such that $V(s) = SU(s)S^{-1}$ for all $s \in G$ and also $SJ_E = J_F S$. (Put

$$T' = \frac{1}{2}(T + J_F T J_E^{-1}), \qquad T'' = \frac{1}{2i}(T - J_F T J_E^{-1}),$$

and show that there exists a *real* number $\xi$ such that $T' + \xi T''$ is a bijection.) Deduce that if $\varepsilon = 1$, the representations $U^{(\mathbf{R})}$ and $V^{(\mathbf{R})}$ are equivalent, and that if $\varepsilon = -1$ the representations $U^{(\mathbf{H})}$ and $V^{(\mathbf{H})}$ are equivalent.
(b) Let $U$ be a complex linear representation of G on a (finite-dimensional) complex vector space E, and identify each automorphism $U(s)$ with its matrix relative to a fixed basis of E. In order that $U$ should satisfy the condition of Problem 7(b) (resp. 8(b)), it is necessary and sufficient that there should exist an invertible complex matrix $P$ such that

$U(s) = P\overline{U(s)}P^{-1}$ for all $s \in$ G, and such that $P\bar{P} = \bar{P}P = I$ (resp. $-I$). (For any complex matrix $A = (a_{ij})$, $\bar{A}$ denotes the complex conjugate matrix $(\bar{a}_{ij})$.) In particular, the representation $U$ is equivalent to the complex conjugate representation $s \mapsto \overline{U(s)}$ (denoted by $\bar{U}$).
(c) Conversely, let $U$ be an *irreducible* complex linear representation of G on E that is equivalent to its complex conjugate. Then $U$ satisfies one and only one of the conditions of Problems 7(b) and 8(b); in other words, one of the representations $U^{(\mathbf{R})}$, $U^{(\mathbf{H})}$ is defined, but not the other. (Use (b) and Schur's lemma (A.22.4).) Moreover, whichever of the representations $U^{(\mathbf{R})}$, $U^{(\mathbf{H})}$ is defined is irreducible.

**10.** (a) Let $U$ be a complex linear representation of G on a finite-dimensional vector space E. For each $s \in$ G, $U(s)$ is also an automorphism of the real vector space $E_{|\mathbf{R}}$ obtained from E by restriction of scalars; let $U_{|\mathbf{R}}$ denote the real linear representation so defined. Show that the complex linear representation $s \mapsto U_{|\mathbf{R}}(s) \otimes 1_{\mathbf{C}}$ is equivalent to the direct sum of the representation $U$ and its conjugate $\bar{U}$. (Observe that if $(e_j)$ is a basis of E over C, the vectors $e'_j = \frac{1}{2}(e_j \otimes 1 + (ie_j) \otimes i)$ and $e''_j = \frac{1}{2}(e_j \otimes 1 - (ie_j) \otimes i)$ form a basis of $E_{|\mathbf{R}} \otimes \mathbf{C}$ over C.)
(b) Deduce from (a) that if $U$ is *irreducible* and not equivalent to its conjugate $\bar{U}$, then $U_{|\mathbf{R}}$ is irreducible.
(c) Suppose that $U$ satisfies the condition of Problem 8(b), so that the quaternionic linear representation $U^{(\mathbf{H})}$ is defined. Show that if $U$ is irreducible, then so also is $U_{|\mathbf{R}}$. (Use (a) and observe that if $V$ is an irreducible real linear representation, then $W = V \otimes 1_{\mathbf{C}}$ is irreducible, and $W^{(\mathbf{H})}$ is not defined.)
(d) If $V_1$, $V_2$ are inequivalent irreducible real linear representations of G, show that there exists no irreducible complex linear representation that is equivalent to a subrepresentation of $s \mapsto V_1(s) \otimes 1_{\mathbf{C}}$ and also to a subrepresentation of $s \mapsto V_2(s) \otimes 1_{\mathbf{C}}$. (Use Schur's lemma (A.22.4).)
(e) Deduce from above that the finite-dimensional irreducible real linear representations of G are all obtained (up to equivalence) from the finite-dimensional irreducible complex linear representations $U$ of G, by taking $U^{(\mathbf{R})}$ whenever this is defined, and otherwise taking $U_{|\mathbf{R}}$. Furthermore, if the irreducible complex representations considered are pairwise inequivalent, then the same is true of the irreducible real representations obtained from them.
(f) State and prove the analogous results for irreducible quaternionic linear representations.

**11.** Let $U$, $V$ be two finite-dimensional continuous complex linear representations of G, and let $W(s) = U(s) \otimes V(s)$ (A.10.5). If the representations $U^{(\mathbf{R})}$ and $V^{(\mathbf{R})}$ (resp. $U^{(\mathbf{H})}$ and $V^{(\mathbf{H})}$) are defined, then $W^{(\mathbf{R})}$ is defined; and if $U^{(\mathbf{R})}$ and $V^{(\mathbf{H})}$ are defined, then $W^{(\mathbf{H})}$ is defined. State and prove the analogous results for the representations $s \mapsto \bigwedge^p U(s)$ (A.13.4), and the representations $s \mapsto \mathbf{S}^p U(s)$ defined by symmetric powers (A.17). If $U^{(\mathbf{R})}$ (resp. $U^{(\mathbf{H})}$) is defined, then we have ${}^t(U^{(\mathbf{R})})^{-1} = ({}^tU^{-1})^{(\mathbf{R})}$ (resp. ${}^t(U^{(\mathbf{H})})^{-1} = ({}^tU^{-1})^{(\mathbf{H})}$).

**12.** Let G and H be two topological groups and let $(s, t) \mapsto U((s, t))$ be a continuous linear representation of G × H on a finite-dimensional complex vector space E. Suppose that $U$ is irreducible and that the representations $s \mapsto U((s, e'))$ and $t \mapsto U((e, t))$ of G and H, respectively, on E are completely reducible ($e$, $e'$ being the identity elements of G, H, respectively). Show that there exists an irreducible representation $V$ of G and an irreducible representation $W$ of H such that $U$ is equivalent to the representation

$$(s, t) \mapsto V(s) \otimes W(t).$$

(Use Schur's lemma.)

13. Let G be a separable, metrizable, locally compact group and let $\Delta$ be its modulus (14.3). If $\beta$ is a left Haar measure on G and if for each function $f \in \mathscr{L}_{\mathbf{C}}^{1}(G)$ we put $f^* = \check{\bar{f}} \cdot \Delta^{-1}$, show that the transform of the measure $f \cdot \beta$ under the involution $\mu \mapsto \check{\bar{\mu}}$ of $M_{\mathbf{C}}^{1}(G)$ is $f^* \cdot \beta$. Extend the results of Section 21.1 to nonunimodular locally compact groups.

## 2. THE HILBERT ALGEBRA OF A COMPACT GROUP

(21.2.1) In this section, G denotes a (metrizable) *compact* group and $\beta$ the Haar measure on G with total mass 1 (we recall that compact groups are unimodular (14.3.3)). If $f, g \in \mathscr{L}_{\mathbf{C}}^{2}(G)$, the function $f * g$ is *continuous* on G and satisfies

$$\|f * g\| \leqq N_2(f)N_2(g) \tag{21.2.1.1}$$

by virtue of (14.10.7). It follows that

$$N_2(f * g) \leqq N_2(f)N_2(g), \tag{21.2.1.2}$$

so that $L_{\mathbf{C}}^{2}(G)$ is a separable Banach algebra with respect to convolution and its Hilbert space structure. Also we have $N_2(\check{\bar{f}}) = N_2(f)$ since G is unimodular, and therefore $L_{\mathbf{C}}^{2}(G)$ is a Banach algebra with involution. In fact, it is a *Hilbert* algebra (15.7.5), relative to the scalar product in $L_{\mathbf{C}}^{2}(G)$. For the condition (15.7.5.1) follows immediately from the definition of the involution and of the scalar product, having regard to (14.3.4); and (15.7.5.3) follows from (21.2.1.2). The condition (15.7.5.4) is a consequence of regularization (14.11.1). Finally, condition (15.7.5.2) takes the form

$$\langle f * g, h\rangle = \langle g, \check{\bar{f}} * h\rangle \tag{21.2.1.3}$$

for all $f, g, h \in L_{\mathbf{C}}^{2}(G)$; when $g$ is continuous, this formula is a special case of (14.9.4.1), and for arbitrary $g$ the result follows by continuity, because of (13.11.6) and (21.2.1.2).

(21.2.2) A function $h \in \mathscr{L}_{\mathbf{C}}^{2}(G)$ is said to be *central* if its class in $L_{\mathbf{C}}^{2}(G)$ belongs to the *center* of this algebra. This signifies that for all functions $f \in \mathscr{L}_{\mathbf{C}}^{2}(G)$, the functions $f * h$ and $h * f$ are equal almost everywhere; but they are *continuous* functions, and therefore they are equal (since $\beta$ has support G (14.1.2)). In other words, for all $s \in G$ we must have

$$\int_G f(t^{-1})(h(st) - h(ts))\, d\beta(t) = 0.$$

This is possible only if $h(st) = h(ts)$ for all $t$ in the complement of a negligible set (depending on $s$) **(13.14.4)**; if in addition $h$ is *continuous*, then this negligible set is necessarily empty, again because the support of $\beta$ is the whole of G **(14.1.2)**. Hence *the continuous central functions on* G *are the continuous functions h which satisfy*

$$\textbf{(21.2.2.1)} \qquad h(sts^{-1}) = h(t) \quad \textit{for all} \quad s, t \in G.$$

We remark that the classes of these functions belong also to the *center* of $M_{\mathbf{C}}(G)$; this follows immediately from **(14.8.2)** and **(14.8.4)**.

**(21.2.3)** (Peter–Weyl theorem) *Let* G *be a metrizable compact group. The complete Hilbert algebra* $L^2_{\mathbf{C}}(G)$ *is the Hilbert sum of an at most denumerable family* $(\mathfrak{a}_\rho)_{\rho \in R}$ *of finite-dimensional simple algebras; each* $\mathfrak{a}_\rho$ *is isomorphic to a matrix algebra* $\mathbf{M}_{n_\rho}(\mathbf{C})$ *and is a minimal two-sided ideal in* $L^2_{\mathbf{C}}(G)$. *The elements of* $\mathfrak{a}_\rho$ *are classes of continuous functions on* G; *the identity element of* $\mathfrak{a}_\rho$ *is the class of a continuous function* $u_\rho$ *such that* $\check{u}_\rho = u_\rho$; *and the orthogonal projection of* $L^2_{\mathbf{C}}(G)$ *onto* $\mathfrak{a}_\rho$ **(6.3.1)** *maps the class of a function f to the class of* $f * u_\rho = u_\rho * f$. *Consequently, for all* $f \in \mathscr{L}^2_{\mathbf{C}}(G)$ *we have*

$$\textbf{(21.2.3.1)} \qquad \tilde{f} = \sum_{\rho \in R} (f * u_\rho)^\sim,$$

*the right-hand side being a convergent series in* $L^2_{\mathbf{C}}(G)$, *regardless of the way in which the elements of* R *are arranged as a sequence.*

Since $L^2_{\mathbf{C}}(G)$ is *complete*, it is the Hilbert sum of an at most denumerable family $(\mathfrak{a}_\rho)_{\rho \in R}$ of distinct two-sided ideals that are topologically simple Hilbert algebras and annihilate each other in pairs **(15.8.13)**. Everything therefore reduces to proving that each $\mathfrak{a}_\rho$ is *finite-dimensional.* For each $\mathfrak{a}_\rho$ will then be the Hilbert sum of a finite number of minimal left ideals, each of which is generated by an irreducible self-adjoint idempotent, and the sum of these idempotents will be the identity element of the algebra $\mathfrak{a}_\rho$. If $v$ is a function whose class is this identity element, every element of $\mathfrak{a}_\rho$ will be the class of a function of the form $f * v$, hence *continuous* **(21.2.1)**. The remaining assertions of the theorem then follow from **(15.8.11)**.

In view of **(15.8.15)**, it will be enough to prove the following assertion:

**(21.2.3.2)** *Each closed two-sided ideal* $\mathfrak{b} \neq \{0\}$ *in* $L^2_{\mathbf{C}}(G)$ *contains a nonzero element of the center of* $L^2_{\mathbf{C}}(G)$.

We shall use the following remark:

(21.2.3.3) *For a closed vector subspace* $\mathfrak{b}$ *of* $L^2_{\mathbf{C}}(G)$, *the following conditions are equivalent:*

(a) $\mathfrak{b}$ *is a left ideal in* $L^2_{\mathbf{C}}(G)$;
(b) $\mathfrak{b}$ *is stable under the regular representation of* $L^1_{\mathbf{C}}(G)$ *on* $L^2_{\mathbf{C}}(G)$ (21.1.9);
(c) *for each function* $f$ *whose class is in* $\mathfrak{b}$, *and each* $s \in G$, *the class of* $\varepsilon_s * f = \gamma(s) f$ *lies in* $\mathfrak{b}$.

The equivalence of (b) and (c) is a particular case of (21.1.7), applied to the regular representation. It is clear that (b) implies (a); on the other hand, $\mathscr{L}^2_{\mathbf{C}}(G)$ is dense in $\mathscr{L}^1_{\mathbf{C}}(G)$ (13.11.6) and the mapping $f \mapsto f * g$ of $\mathscr{L}^1_{\mathbf{C}}(G)$ into $\mathscr{L}^2_{\mathbf{C}}(G)$ is continuous for all $g \in \mathscr{L}^2_{\mathbf{C}}(G)$ (14.10.6), whence (a) implies (b).

There is of course an analogous statement for right ideals in $L^2_{\mathbf{C}}(G)$.

We now come to the proof of (21.2.3.2). We shall first show that $\mathfrak{b}$ contains the class of a continuous function $f$, not identically zero. For if $g$ is a nonnegligible function whose class belongs to $\mathfrak{b}$, then the class of $g * \check{g}$ also belongs to $\mathfrak{b}$; but $g * \check{g}$ is continuous (21.2.1) and $(g * \check{g})(e) = (N_2(g))^2 > 0$ (14.10.4). We may therefore take $f = g * \check{g}$. Next consider the function

(21.2.3.4) $$h(t) = \int_G f(sts^{-1})\, d\beta(s).$$

Since the function $(x, y, z) \mapsto f(xyz)$ is uniformly continuous on $G \times G \times G$ (3.16.5), it follows immediately that $h$ is continuous on G, and since $h(e) = f(e) \neq 0$, $h$ is not identically zero. For all $x \in G$ we have

(21.2.3.5) $$h(xtx^{-1}) = \int_G f((sx)t(sx)^{-1})\, d\beta(s) = h(t)$$

because $\beta$ is right-invariant. It remains to show that the class of $h$ belongs to $\mathfrak{b}$. Now $L^2_{\mathbf{C}}(G)$ is the Hilbert sum of $\mathfrak{b}$ and its orthogonal supplement $\mathfrak{b}^\perp$, which is also a two-sided ideal **(15.8.2)**; hence it is enough to show that $(\tilde{h}|\tilde{w}) = 0$ for all $\tilde{w} \in \mathfrak{b}^{\perp}$. We have

$$(\tilde{h}|\tilde{w}) = \int \bar{w}(t)\,d\beta(t) \int f(sts^{-1})\,d\beta(s)$$

$$= \int d\beta(s) \int \bar{w}(t) f(sts^{-1})\,d\beta(t)$$

$$= \int d\beta(s) \int \bar{w}(s^{-1}ts) f(t)\,d\beta(t)$$

by the Lebesgue–Fubini theorem and the left- and right-invariance of $\beta$. Since $\tilde{w} \in \mathfrak{b}^{\perp}$, the class of $\varepsilon_s * w * \varepsilon_{s^{-1}}$ also belongs to $\mathfrak{b}^{\perp}$ by virtue of **(21.2.3.3)**, hence by definition we have $\int \bar{w}(s^{-1}ts) f(t)\,d\beta(t) = 0$, and the proof is complete.

**(21.2.4)** By virtue of **(21.2.3)** it is convenient to *identify* each element of an ideal $\mathfrak{a}_\rho$ with the unique continuous function in the class, and this we shall do from now on.† For each $\rho \in R$, choose once and for all a decomposition of $\mathfrak{a}_\rho$ as the Hilbert sum of $n_\rho$ minimal left ideals $\mathfrak{l}_j = \mathfrak{a}_\rho * m_j$ (also denoted by $\mathfrak{l}_j^{(\rho)}$), pairwise isomorphic and orthogonal, where each $m_j$ $(1 \leqq j \leqq n_\rho)$ is a minimal self-adjoint idempotent, so that $u_\rho = \sum_{j=1}^{n_\rho} m_j$. Also let $(a_j)_{1 \leq j \leq n_\rho}$ be a Hilbert basis of $\mathfrak{l}_1$, such that $a_j \in m_j * \mathfrak{a}_\rho * m_1$. Then from **(15.8.14)** we know that all the numbers $(m_j | m_j)$ are equal to the same number $\gamma > 0$, and that

$$a_j * \check{a}_j = \gamma m_j, \qquad \check{a}_j * a_j = \gamma m_1.$$

Now put, for each pair of indices $j, k$,

$$m_{jk} = \gamma^{-1} a_j * \check{a}_k$$

(so that $m_{jj} = m_j$); then we have

**(21.2.4.1)** $$m_{jk} * a_h = \delta_{kh} a_j$$

where $\delta_{kh}$ is the Kronecker delta. We shall also write $m_{ij}^{(\rho)}$ in place of $m_{ij}$.

† More generally, from now on we shall *identify* each *continuous* function $f$ on a locally compact group G, belonging to one or other of the spaces $\mathscr{L}^1_{\mathbf{C}}(G, \beta)$, $\mathscr{L}^2_{\mathbf{C}}(G, \beta)$, $\mathscr{L}^\infty_{\mathbf{C}}(G, \beta)$ (where $\beta$ is a left or right Haar measure on G), with its class in the corresponding space $L^1_{\mathbf{C}}(G, \beta)$, $L^2_{\mathbf{C}}(G, \beta)$, $L^\infty_{\mathbf{C}}(G, \beta)$. This can cause no confusion because $f$ is the *only* continuous function in its class, since the support of $\beta$ is the whole of G.

**(21.2.5)** *With the notation of* **(21.2.4)**:

(a) *For each index j, the* $m_{ij}$ $(1 \leq i \leq n_\rho)$ *form an orthogonal basis of* $\mathfrak{l}_j$.

(b) $m_{ji} = \check{m}_{ij}$, $m_{ij} * m_{hk} = \delta_{jh} m_{ik}$.

(c) $(m_{ij} | m_{ij}) = n_\rho$, $m_{ij}(e) = n_\rho \delta_{ij}$ *for all pairs* $(i, j)$ (*so that* $\gamma = n_\rho^{-1}$). *The functions* $n_\rho^{-1/2} m_{ij}^{(\rho)}$ $(1 \leq i, j \leq n_\rho, \rho \in R)$ *therefore form a Hilbert basis* **(6.5)** *of the Hilbert space* $L_{\mathbf{C}}^2(G)$.

(d) *Let* $M_\rho(s) = (n_\rho^{-1} m_{ij}(s))$ *for all* $s \in G$; *then the matrices* $M_\rho(s)$ *satisfy the relations*

$$M_\rho(st) = M_\rho(s) M_\rho(t), \qquad M_\rho(s^{-1}) = (M_\rho(s))^*, \tag{21.2.5.1}$$

*so that* $s \mapsto M_\rho(s)$ *is a continuous unitary representation of* G *on* $\mathbf{C}^{n_\rho}$, *relative to the Hermitian scalar product* $\sum_{j=1}^{n_\rho} \xi_j \bar{\eta}_j$.

The assertions in (a) and (b) are immediate consequences of the definitions in **(21.2.4)**, since the $a_j \in \mathfrak{l}_1$ and the $m_j$ are pairwise orthogonal. Since $\mathfrak{a}_\rho$ is a Hilbert algebra, we have

$$(m_{ij} | m_{ij}) = \gamma^{-2}(a_i * \check{a}_j | a_i * \check{a}_j) = \gamma^{-2}(\check{a}_i * a_i | \check{a}_j * a_j) = (m_1 | m_1).$$

To calculate this number we remark that for each index $k$ the function $t \mapsto m_{ik}(st)$ belongs to $\mathfrak{l}_k$ for each $s \in G$ **(21.2.3.3)** and can therefore be written in the form

$$m_{ik}(st) = \sum_{j=1}^{n_\rho} c_{ij}(s) m_{jk}(t). \tag{21.2.5.2}$$

On the other hand,

$$m_{jk}(t) = (m_{j1} * m_{1k})(t) = \int_G m_{j1}(tx) \overline{m_{k1}(x)} \, d\beta(x),$$

hence in particular $m_{jk}(e) = (m_{j1} | m_{k1})$, and by putting $t = e$ in **(21.2.5.2)** we obtain, using the orthogonality properties of the $m_{ij}$,

$$m_{ik}(s) = (m_1 | m_1) c_{ik}(s). \tag{21.2.5.3}$$

Next, putting $s = t^{-1}$ and $i = k = 1$ in **(21.2.5.2)**, we obtain by use of **(14.10.4)**

$$(m_1 | m_1) = m_1(e) = \sum_{j=1}^{n_\rho} c_{1j}(s) \overline{m_{1j}(s)}$$

and therefore, using **(21.2.5.3)**

$$\sum_{j=1}^{n_\rho} m_{1j}(s)\overline{m_{1j}(s)} = (m_1 \mid m_1)^2.$$

Integrating over G, we finally obtain

$$(m_1 \mid m_1)^2 = n_\rho (m_1 \mid m_1)$$

which proves (c); and then the relations **(21.2.5.1)** follow immediately from (b) and **(21.2.5.2)** and **(21.2.5.3)**.

**(21.2.6)** *The center of the Hilbert algebra* $L^2_{\mathbf{C}}(G)$ *is the Hilbert sum of the* 1-*dimensional subspaces* $\mathbf{C}u_\rho$ $(\rho \in R)$. *In particular, if* G *is commutative, all the ideals* $\mathfrak{a}_\rho$ *are of dimension* $n_\rho = 1$.

That the $u_\rho$ belong to the center of $L^2_{\mathbf{C}}(G)$ follows from the facts that $u_\rho$ is the identity element of $\mathfrak{a}_\rho$ and that $\mathfrak{a}_\rho * \mathfrak{a}_{\rho'} = \{0\}$ whenever $\rho \neq \rho'$. Conversely, if the class $\tilde{f}$ of a function $f$ belongs to the center of $L^2_{\mathbf{C}}(G)$, then so also does the class of $f * u_\rho \in \mathfrak{a}_\rho$, hence $f * u_\rho = c_\rho u_\rho$ for some scalar $c_\rho \in \mathbf{C}$; now apply the formula **(21.2.3.1)**.

**(21.2.7)** The classes of the complex *constant* functions form a two-sided ideal of dimension 1 in $L^2_{\mathbf{C}}(G)$ **(14.6.3)**, which is therefore of the form $\mathfrak{a}_{\rho_0}$. It is called the *trivial* ideal. The corresponding linear representation $M_{\rho_0}$ of dimension 1 is such that $M_{\rho_0}(s) = 1$ for all $s \in G$, that is to say, it is the *trivial* linear representation **(21.1.1)**. For each $\rho \neq \rho_0$ in R, we have

**(21.2.7.1)**
$$\int_G m_{ij}^{(\rho)}(s)\, d\beta(s) = 0$$

since the subspaces $\mathfrak{a}_\rho$ and $\mathfrak{a}_{\rho_0}$ are orthogonal.

**(21.2.8)** (i) *If $f$ and $g$ are continuous complex-valued functions on* G, *then*

**(21.2.8.1)**
$$f * g = \sum_{\rho \in R} \left( \sum_{1 \le i,j \le n_\rho} n_\rho^{-1} (g \mid m_{ij}^{(\rho)})(f * m_{ij}^{(\rho)}) \right),$$

*the series on the right being summable for the topology of uniform convergence.*

(ii) *The functions* $m_{ij}^{(\rho)}$ $(\rho \in R,\ 1 \leqq i, j \leqq n_\rho)$ *form a total system in the space of continuous functions on* G, *for the topology of uniform convergence.*

(i) Identifying continuous functions with their classes in $L^2_C(G)$, we may write

$$g = \sum_{\rho \in R,\, 1 \leq i,\, j \leq n_\rho} \frac{1}{n_\rho} (g \mid m_{ij}^{(\rho)}) m_{ij}^{(\rho)}$$

the series on the right being summable *in* $L^2_C(G)$, because the functions $n_\rho^{-1/2} m_{ij}^{(\rho)}$ form a Hilbert basis of this space. Now form the convolution product of both sides with $f$; since $\tilde{u} \mapsto f * u$ is a continuous mapping of $L^2_C(G)$ into $\mathscr{C}_C(G)$ (21.2.1.1), we obtain the formula (21.2.8.1).

(ii) It follows from regularization (14.11.1) that for each continuous function $g$ on G there exists a continuous function $f$ on G such that $\| f * g - g \|$ is arbitrarily small. Now, for each $\rho \in R$, the functions $f * m_{ij}^{(\rho)}$ belong to $\mathfrak{a}_\rho$, and therefore are linear combinations of the $m_{hk}^{(\rho)}$ $(1 \leqq h, k \leqq n_\rho)$ with complex coefficients. This completes the proof.

## PROBLEMS

1. Let E be a finite-dimensional complex vector space, E* its dual, G a topological group, and $U$ a continuous linear representation of G on E. For each pair of vectors $x \in E, x^* \in E^*$, the function $s \mapsto \langle U(s) \cdot x, x^* \rangle$ is continuous on G; it is called the *coefficient of U relative to* $(x, x^*)$ and is denoted by $c_U(x, x^*)$. For all $t \in G$ we have

$$\gamma(t) c_U(x, x^*) = c_U(x, {}^tU(t)^{-1} \cdot x^*), \qquad \delta(t) c_U(x, x^*) = c_U(U(t) \cdot x, x^*).$$

If we identify $U(s)$ with its matrix $(u_{jk}(s))$ relative to a fixed basis of E, then the functions $c_U(x, x^*)$ are linear combinations of the $u_{jk}$. We have

$$c_{{}^tU^{-1}}(x^*, x) = \check{c}_U(x, x^*), \qquad c_{\overline{U}}(x, x^*) = \overline{c_{\overline{U}}(x, x^*)}.$$

(a) Let $\mathscr{V}(U)$ (or $\mathscr{V}_C(U)$) denote the vector subspace of $\mathscr{C}_C(G)$ spanned by the coefficients of the continuous linear representation $U$ of G. If $U_1$, $U_2$ are equivalent, then $\mathscr{V}(U_1) = \mathscr{V}(U_2)$; also $\mathscr{V}({}^tU^{-1}) = \mathscr{V}(U)$ and $\mathscr{V}(\overline{U}) = \overline{\mathscr{V}(U)}$. If $U_1$, $U_2$ are finite-dimensional continuous linear representations of G, then $\mathscr{V}(U_1 \oplus U_2) = \mathscr{V}(U_1) + \mathscr{V}(U_2)$ and $\mathscr{V}(U_1 \otimes U_2) = \mathscr{V}(U_1)\mathscr{V}(U_2)$, the vector subspace of $\mathscr{C}_C(G)$ spanned by the products $c_1 c_2$, where $c_1 \in \mathscr{V}(U_1)$ and $c_2 \in \mathscr{V}(U_2)$. The vector subspace $\mathscr{V}(U)$ has finite dimension $\leqq (\dim U)^2$ and is stable under left and right translations $f \mapsto \gamma(s) f$, $f \mapsto \delta(s) f$ for all $s \in G$. Conversely, if E is a vector subspace of $\mathscr{C}_C(G)$ that is stable under left translations $f \mapsto \gamma(s) f$ and is finite-dimensional, and if we denote by $U(s)$ the endomorphism $f \mapsto \gamma(s) f$ of E, then $U$ is a continuous linear representation of G on E, and $E \subset \mathscr{V}(U)$. A function $f \in \mathscr{C}_C(G)$ is called a *representative function* on G if the vector subspace of $\mathscr{C}_C(G)$ spanned by the left-translates $\gamma(s) f$ of $f$, for all $s \in G$, is finite-dimensional. The representative functions on G form a subalgebra $\mathscr{B}(G)$ (or $\mathscr{B}_C(G)$) of $\mathscr{C}_C(G)$, which is the same as the subalgebra generated by the coefficients of all the finite-dimensional continuous linear representations of G.

(b) Let $U$ be a continuous linear representation of G, of dimension $n < \infty$, and let $U'$ be the continuous linear representation of G on $\mathscr{V}(U)$ defined by $U'(s) \cdot f = \gamma(s) f$. Show that

$U$ is equivalent to a subrepresentation of $U'$. If $U$ is irreducible, $U'$ is the direct sum of $n$ representations equivalent to $U$. Give an example of a reducible representation where this is not the case (cf. **(21.1.3)**). Deduce that if $U$ is irreducible and if $U_1$ is a finite-dimensional continuous linear representation such that $\mathscr{V}(U_1) \subset \mathscr{V}(U)$, then $U_1$ is the direct sum of $m$ representations equivalent to $U$, where $m \leqq n$.
(c) Extend the above definitions and results to finite-dimensional continuous *real* linear representations (Section **21.1**, Problem 7); in place of $\mathscr{V}_{\mathbf{C}}(U)$ and $\mathscr{B}_{\mathbf{C}}(\mathrm{G})$ we have $\mathscr{V}_{\mathbf{R}}(U)$ and $\mathscr{B}_{\mathbf{R}}(\mathrm{G})$.

**2.** Let G be a metrizable compact group.
(a) Show that the algebra $\mathscr{B}_{\mathbf{C}}(\mathrm{G})$ of complex representative functions is the direct sum of the two-sided ideals $\mathfrak{a}_\rho$ ($\rho \in \mathrm{R}$), and that the algebra $\mathscr{B}_{\mathbf{R}}(\mathrm{G})$ consists of the real and imaginary parts of the functions belonging to $\mathscr{B}_{\mathbf{C}}(\mathrm{G})$.
(b) Let M be a subset of $\mathscr{B}_{\mathbf{C}}(\mathrm{G})$. The set H of elements $t \in \mathrm{G}$ such that $\delta(t)f = f$ for all $f \in \mathrm{M}$ is a closed subgroup of G. Show that the set of functions $g \in \mathscr{B}_{\mathbf{C}}(\mathrm{G})$ such that $\gamma(t)g = g$ for all $t \in \mathrm{H}$ is the left ideal $\mathfrak{b}$ of $\mathscr{B}_{\mathbf{C}}(\mathrm{G})$ generated by M. The functions belonging to $\mathfrak{b}$ may be canonically identified with continuous functions on G/H, and $\mathfrak{b}$ may be identified with the intersection of $\mathscr{B}_{\mathbf{C}}(\mathrm{G})$ with $\mathscr{C}_{\mathbf{C}}(\mathrm{G/H})$ (considered as a subalgebra of $\mathscr{C}_{\mathbf{C}}(\mathrm{G})$); also $\mathfrak{b}$ is dense in $\mathscr{C}_{\mathbf{C}}(\mathrm{G/H})$. (Use the Stone–Weierstrass theorem.)
(c) Let K be a closed subgroup of G. Show that every function in $\mathscr{B}_{\mathbf{C}}(\mathrm{K})$ is the restriction to K of a function belonging to $\mathscr{B}_{\mathbf{C}}(\mathrm{G})$. (Consider the set of functions in $\mathscr{B}_{\mathbf{C}}(\mathrm{K})$ that are restrictions to K of functions belonging to $\mathscr{B}_{\mathbf{C}}(\mathrm{G})$ and use (a) above, with G replaced by K.) If $\mathfrak{b}$ is the left ideal in $\mathscr{B}_{\mathbf{C}}(\mathrm{G})$ that is the intersection of $\mathscr{B}_{\mathbf{C}}(\mathrm{G})$ with $\mathscr{C}_{\mathbf{C}}(\mathrm{G/K})$, show that K is equal to the subgroup H of elements $t \in \mathrm{G}$ such that $\gamma(t)f = f$ for all $f \in \mathfrak{b}$. (Observe that a function belonging to $\mathfrak{b}$ that is constant on K is constant on H.)

**3.** Let G be an infinite compact group. With the notation of **(21.2.3)**, if $p$, $q$ are two functions defined on R, with values $> 0$, we write $p = o(q)$ to mean that for each $\varepsilon > 0$ there exists a finite subset J of R such that $p(\rho) \leqq \varepsilon q(\rho)$ for all $\rho \in \mathrm{R} - \mathrm{J}$.
(a) Show that for each function $f \in \mathscr{L}^2_{\mathbf{C}}(\mathrm{G})$, the operator $R(f)$ is a Hilbert–Schmidt operator on $\mathrm{L}^2_{\mathbf{C}}(\mathrm{G})$, and that the mapping $\tilde{f} \mapsto R(f)$ is an isometry of the Hilbert algebra $\mathrm{L}^2_{\mathbf{C}}(\mathrm{G})$ onto a closed subalgebra of the Hilbert algebra $\mathscr{L}_2(\mathrm{L}^2_{\mathbf{C}}(\mathrm{G}))$ **(15.4.8)**. In particular, for all $f, g \in \mathscr{L}^2_{\mathbf{C}}(\mathrm{G})$, the operator $R(f)R(g)$ is *nuclear* (Section **15.11**, Problem 7), and we have

$$\mathrm{Tr}(R(f)R(g)) = \sum_{\rho \in \mathrm{R}} \mathrm{Tr}(R(f * g * u_\rho)) = (f \mid g).$$

(b) We have $\|R(f * u_\rho)\|_2 = \mathrm{N}_2(f * u_\rho) = o(1)$ and $\mathrm{N}_\infty(f * u_\rho) = o(n_\rho)$ for all $f \in \mathscr{L}^2_{\mathbf{C}}(\mathrm{G})$. (Use (a) above and the relation $f * u_\rho = f * u_\rho * u_{\rho}$.)
(c) Give an example of a continuous function on G such that $R(f)$ is not nuclear. (Take $\mathrm{G} = \mathbf{T}$.)
(d) Show that $\|R(m_{ij}^{(\rho)})\| = 1$ and $\|R(m_{ij}^{(\rho)})\|_1 = n_\rho$. (Observe that the eigenvalues of $R(m_{ij}^{(\rho)}) * R(m_{ij}^{(\rho)})$ are known.)
(e) Let $f \in \mathscr{L}^1_{\mathbf{C}}(\mathrm{G})$. Show that $R(f)$ is a compact operator on $\mathrm{L}^2_{\mathbf{C}}(\mathrm{G})$ and that $\|R(f * u_\rho)\| = o(1)$. (Use the fact that $\mathrm{L}^2_{\mathbf{C}}(\mathrm{G})$ is dense in $\mathrm{L}^1_{\mathbf{C}}(\mathrm{G})$, the inequality **(21.1.4.3)**, and (a) above.) Deduce that $\|R(f * u_\rho)\|_2 = \mathrm{N}_2(f * u_\rho) = o(n_\rho)$ and that $\mathrm{N}_\infty(f * u_\rho) = o(n_\rho^2)$.

**4.** Let M be a compact differential manifold and G a compact group acting continuously on M such that, for each $s \in \mathrm{G}$, the mapping $x \mapsto s \cdot x$ is a diffeomorphism of M.

(a) Show that for each real-valued function $f$ belonging to the Banach space $\mathscr{E}^{(1)}(M)$ of $C^1$-functions on M (17.1), there exists a function $u \in \mathscr{B}_{\mathbf{R}}(G)$ (Problem 2) such that, if we put

$$f_u(x) = \int_G u(s) f(s^{-1} \cdot x)\, d\beta(s)$$

(where $\beta$ is a Haar measure on G), $f_u$ is of class $C^1$ and the norm $\| f - f_u \|$ is arbitrarily small in $\mathscr{E}^{(1)}(M)$. (Use regularization, together with Problem 2.) If $f$ is of class $C^r$, where $r$ is a positive integer or $+\infty$, then so also is $f_u$. The set of functions $x \mapsto f_u(t \cdot x)$ as $t$ runs through G is then a finite-dimensional vector space.
(b) Show that there exists an embedding $F: M \to \mathbf{R}^N$ and a continuous homomorphism $\rho$ of G into the orthogonal group $\mathbf{O}(n, \mathbf{R})$ such that $F(s \cdot x) = \rho(s) \cdot F(x)$ for all $s \in G$ and all $x \in M$. (Start with an embedding $x \mapsto (f_1(x), \ldots, f_n(x))$ of M in $\mathbf{R}^n$ (16.25.1). Show first that there exists $u \in \mathscr{B}_{\mathbf{R}}(G)$ such that, if $g_i = (f_i)_u$ (in the notation of (a) above), the mapping $x \mapsto (g_1(x), \ldots, g_n(x))$ is an immersion, not necessarily injective. There exists then a finite open covering $(U_\alpha)$ of M such that the restriction of this immersion to each $U_\alpha$ is an embedding. Next show that there exists $v \in \mathscr{B}_{\mathbf{R}}(G)$ such that, if $h_i = (f_i)_v$, the relations $h_i(x) = h_i(y)$ for $1 \leq i \leq n$ imply that $x$ and $y$ belong to the same $U_\alpha$. Finally consider the finite-dimensional vector space spanned by all the functions $x \mapsto g_i(t \cdot x)$ and $x \mapsto h_i(t \cdot x)$ as $t$ runs through G.)

**5.** Let M be a compact differential manifold and G a compact Lie group acting differentiably on M; let $x$ be a point of M and $S_x$ the stabilizer of $x$ in G.
(a) Show that there exists a submanifold W of M, contained in a neighborhood of $x$ and containing $x$, which is stable under $S_x$ and which is such that $T_x(W)$ is a supplement in $T_x(M)$ to the tangent space $T_x(G \cdot x)$ to the orbit of $x$. (Use Problem 4 above, or Problem 6 of Section **19.1**.)
(b) Let V be a submanifold of G, passing through $e$ and such that the tangent space to V at $e$ is supplementary in $\mathfrak{g}_e = T_e(G)$ to the Lie algebra $T_e(S_x)$ of $S_x$. Show that there exists a relatively compact open neighborhood U of $e$ in V and a relatively compact open neighborhood K of $x$ in W such that the mapping $(s, y) \mapsto s \cdot y$ of $U \times K$ into M is a diffeomorphism onto a neighborhood of $x$ in M, and such that K is stable under $S_x$. Deduce that $s \cdot K \cap K = \emptyset$ for all $s \in US_x$ not belonging to $S_x$.
(c) Deduce from (b) that there exists a relatively compact open neighborhood $K' \subset K$ of $x$ in W having the following properties: (i) $K'$ is stable under $S_x$; (ii) the mapping $(s, y) \mapsto s \cdot y$ of $U \times K'$ into M is a diffeomorphism onto a neighborhood of $x$ in M; (iii) $s \cdot K' \cap K' = \emptyset$ for $s \notin S_x$. (Use Problem 4.) Such a set $K'$ is called a *slice* of M at the point $x$ (for the action of G on M). Show that for all $z \in K'$ we have $S_z \subset S_x$.

**6.** If M is a pure differential manifold and G is a Lie group acting differentiably on M, let L(G, M) denote the set of conjugacy classes in G of the stabilizers of the points of M (two stabilizers being in the same class if they stabilize two points of the same orbit). We shall show that, if G and M are *compact*, the set L(G, M) is *finite*. The proof will be by induction on $\dim(M) = n$.
(a) Show that if the result is true for every differential manifold M of dimension $n - 1$, then $L(G, \mathbf{R}^n)$ is finite for all compact subgroups G of $\mathbf{O}(n)$ (apply the hypothesis to $S_{n-1}$).
(b) There exists a finite number of slices $K_i$ ($1 \leq i \leq r$) of M (Problem 5) relative to points $x_i$ of M, such that M is the union of the sets $G \cdot K_i$. Deduce from (a) that each of the sets $L(S_{x_i}, K_i)$ is finite, and show that $L(G, G \cdot K_i)$ is finite by using Problem 5(c).

7. Let G be a compact Lie group. Show that there are only finitely many conjugacy classes of normalizers of connected Lie groups immersed in G. (Consider the projective space $\mathbf{P}(\Lambda(\mathfrak{g}_e))$ corresponding to the exterior algebra on the vector space $\mathfrak{g}_e$, and the action of G on this compact manifold induced by the adjoint representation of G on $\mathfrak{g}_e$, and apply the result of Problem 6.)

8. Let G be a compact group and $\beta$ the Haar measure on G for which the total mass is 1. In order that a sequence $(x_n)$ of points of G should be *equirepartitioned* relative to the measure $\beta$ (Section 13.4, Problem 7) it is necessary and sufficient that, for each $\boldsymbol{\rho} \neq \boldsymbol{\rho}_0$ in R, we should have

$$\lim_{N \to \infty} \frac{1}{N} \sum_{k=1}^{N} M_{\boldsymbol{\rho}}(x_k) = 0.$$

(Use (21.2.8) and (21.7.1).) In particular, for a point $s \in G$ to be such that the sequence $(s^n)_{n \geq 0}$ is equirepartitioned relative to $\beta$, it is necessary and sufficient that 1 is an eigenvalue of none of the matrices $M_{\boldsymbol{\rho}}(s)$ for $\boldsymbol{\rho} \neq \boldsymbol{\rho}_0$. (This condition implies that G is commutative.)

## 3. CHARACTERS OF A COMPACT GROUP

We retain the hypotheses and notation of (21.2). For each $\boldsymbol{\rho} \in R$ and each $s \in G$, let

(21.3.1)
$$\chi_{\boldsymbol{\rho}}(s) = \frac{1}{n_{\boldsymbol{\rho}}} u_{\boldsymbol{\rho}}(s) = \frac{1}{n_{\boldsymbol{\rho}}} \sum_{j=1}^{n_{\boldsymbol{\rho}}} m_{jj}^{(\boldsymbol{\rho})}(s).$$

The function $\chi_{\boldsymbol{\rho}}$ is called the *character* of the compact group G associated with the minimal two-sided ideal $\mathfrak{a}_{\boldsymbol{\rho}}$.

The character $\chi_{\boldsymbol{\rho}_0}$ associated with $\mathfrak{a}_{\boldsymbol{\rho}_0}$ (21.2.7) is the constant function $\chi_{\boldsymbol{\rho}_0}(s) = 1$ for all $s \in G$. It is called the *trivial character* of G.

The following properties are immediate consequences of (21.2.3) and (21.2.5):

(21.3.2) *Every character $\chi_{\boldsymbol{\rho}}$ is a continuous central function on* G; *in other words*

(21.3.2.1)
$$\chi_{\boldsymbol{\rho}}(sts^{-1}) = \chi_{\boldsymbol{\rho}}(t) \qquad \textit{for all} \quad s, t \in G.$$

*We have*

(21.3.2.2)
$$\chi_{\boldsymbol{\rho}}(s^{-1}) = \overline{\chi_{\boldsymbol{\rho}}(s)} \qquad \textit{for all} \quad s \in G,$$

*and*

(21.3.2.3)
$$\chi_{\boldsymbol{\rho}} * \chi_{\boldsymbol{\rho}} = \frac{1}{n_{\boldsymbol{\rho}}} \chi_{\boldsymbol{\rho}}.$$

*The characters form a Hilbert basis of the center of* $\mathrm{L}^2_{\mathbf{C}}(\mathrm{G})$, *indexed by* R. *In other words,*

(21.3.2.4) $$\int \chi_{\rho'}(s)\overline{\chi_\rho(s)}\, d\beta(s) = \begin{cases} 0 & \text{if } \rho' \neq \rho, \\ 1 & \text{if } \rho' = \rho, \end{cases}$$

*and if f is any central function in* $\mathscr{L}^2_{\mathbf{C}}(\mathrm{G})$, *then*

(21.3.2.5) $$\tilde{f} = \sum_{\rho \in \mathrm{R}} (f \mid \chi_\rho)\chi_\rho = \sum_{\rho \in \mathrm{R}} n_\rho (f * \chi_\rho)$$

*in* $\mathrm{L}^2_{\mathbf{C}}(\mathrm{G})$. *Furthermore, we have*

(21.3.2.6) $$\int \chi_\rho(s)\, d\beta(s) = 0 \qquad \text{for all} \quad \rho \neq \rho_0 .$$

*Finally, for each* $s \in \mathrm{G}$,

(21.3.2.7) $$\chi_\rho(s) = \mathrm{Tr}(M_\rho(s))$$

*and in particular*

(21.3.2.8) $$\chi_\rho(e) = n_\rho .$$

**(21.3.3)** *If f, g are continuous complex-valued central functions on* G, *then*

(21.3.3.1) $$f * g = \sum_{\rho \in \mathrm{R}} (g \mid \chi_\rho)(f * \chi_\rho)$$

*the series on the right being summable for the topology of uniform convergence.*

This follows directly from (21.2.1.1) and the fact that the $\chi_\rho$ form a Hilbert basis of the center of $\mathrm{L}^2_{\mathbf{C}}(\mathrm{G})$ (21.3.2).

**(21.3.4)** *The functions* $\chi_\rho$ ($\rho \in \mathrm{R}$) *form a total system in the space of continuous complex-valued central functions on* G, *for the topology of uniform convergence.*

For each continuous central function $f$, $f * \chi_\rho$ is a scalar multiple of $\chi_\rho$. Taking account of (21.3.3), it is enough to show that for each continuous central function $g$, there exists a continuous *central* function $f$ such that $\| f * g - g \|$ is arbitrarily small. We shall first establish the following topological lemma:

**(21.3.4.1)** (i) *Let* G *be a metrizable topological group and* K *a compact subset of* G. *For each open neighborhood* U *of the identity element e of* G, *there exists a neighborhood* $\mathrm{V} \subset \mathrm{U}$ *of e, such that* $t\mathrm{V}t^{-1} \subset \mathrm{U}$ *for all* $t \in \mathrm{K}$.

(ii) *In a compact metrizable group* G, *there exists a fundamental system of neighborhoods of e that are invariant under all inner automorphisms. If* T *is such a neighborhood, there exists a continuous central function* $h \geqq 0$, *with support contained in* T, *and such that* $\int h(s)\, d\beta(s) = 1$.

(i) Let $U_0$ be a neighborhood of $e$ such that $U_0^3 \subset U$. For each $s \in G$, there exists a neighborhood $V_s$ of $e$ in G such that $sV_s s^{-1} \subset U_0$; by continuity, there is therefore a neighborhood $W_s$ of $s$ such that $tV_s t^{-1} \subset U$ for all $t \in W_s$. There exist a finite number of points $s_j \in K$ $(1 \leqq j \leqq m)$ such that the $W_{s_j}$ cover K; if we put $V = \bigcap_j V_{s_j}$, we shall have $tVt^{-1} \subset U$ for all $t \in K$.

(ii) We may apply (i) with $K = G$. The union T of the $tVt^{-1}$ as $t$ runs through G is then a neighborhood of $e$ contained in U and invariant under all inner automorphisms.

To construct the function $h$, choose a continuous function $f \geqq 0$, with support contained in T, and such that $f(e) > 0$; then let

$$h(t) = c \int_G f(sts^{-1})\, d\beta(s)$$

where $c$ is a suitably chosen positive constant. The proof that $h$ satisfies the required conditions is the same as in **(21.2.3)**.

The proof that for any given continuous central function $g$, the number $\|h * g - g\|$ can be made arbitrarily small by suitable choice of a continuous central function $h$, now follows from the lemma **(21.3.4.1)** and regularization **(14.11.1)**.

**(21.3.5)** (i) *For each element* $s \neq e$ *in* G, *there exists* $\mathfrak{p} \in R$ *such that* $\chi_\mathfrak{p}(s) \neq \chi_\mathfrak{p}(e)$.

(ii) *The intersection of the kernels* $N_\mathfrak{p}$ *of the homomorphisms* $s \mapsto M_\mathfrak{p}(s)$, *as* $\mathfrak{p}$ *runs through* R, *consists of the identity element alone.*

(i) If not, it would follow from **(21.3.4)** that $f(s) = f(e)$ for all continuous central functions $f$ on G, contradicting **(21.3.4.1)**.

(ii) It is enough to remark that $s \in N_\mathfrak{p}$ implies that $\chi_\mathfrak{p}(s) = \chi_\mathfrak{p}(e)$.

**(21.3.6)** *For all characters* $\chi$ *of* G *we have*

**(21.3.6.1)**
$$\chi(s)\chi(t) = \chi(e) \int \chi(usu^{-1}t)\, d\beta(u)$$

*for all* $s, t \in G$.

From the definition of $\chi_\rho$, we have

$$\chi_\rho(usu^{-1}t) = \frac{1}{n_\rho}\sum_i m_{ii}(usu^{-1}t)$$

$$= \frac{1}{n_\rho^4}\sum_{i,j,h,k} m_{ij}(u)m_{jk}(s)m_{hk}(u^{-1})m_{ki}(t)$$

by virtue of (**21.2.5.1**), and therefore

$$\int \chi_\rho(usu^{-1}t)\,d\beta(u) = \frac{1}{n_\rho^4}\sum_{i,j,h,k} m_{jh}(s)m_{ki}(t)\int m_{ij}(u)m_{hk}(u^{-1})\,d\beta(u)$$

$$= \frac{1}{n_\rho^3}\sum_{i,j,h,k} \delta_{jh}\,\delta_{ik}\,m_{jh}(s)m_{ki}(t)$$

$$= \frac{1}{n_\rho^3}\sum_{i,j} m_{jj}(s)m_{ii}(t) = \frac{1}{n_\rho}\chi_\rho(s)\chi_\rho(t),$$

making use of (**21.2.5**).

(**21.3.7**) The mapping that sends the class of a function $f \in \mathscr{L}^2_{\mathbf{C}}(\mathrm{G})$ to the class of its complex conjugate $\bar{f}$ is clearly a semilinear bijection of the algebra $\mathrm{L}^2_{\mathbf{C}}(\mathrm{G})$ onto itself, which is an automorphism of its *ring* structure. This automorphism therefore transforms each minimal two-sided ideal $\mathfrak{a}_\rho$ into another minimal two-sided ideal, which we denote by $\mathfrak{a}_{\bar{\rho}}$. If in general $\bar{X}$ denotes the matrix whose entries are the complex conjugates of those of a matrix $X$, then we have $M_{\bar{\rho}}(s) = \overline{M_\rho(s)}$ for all $s \in \mathrm{G}$, and for the corresponding characters we have

(**21.3.7.1**) $$\chi_{\bar{\rho}} = \overline{\chi_\rho}\,.$$

The relation $\mathfrak{a}_\rho = \mathfrak{a}_{\bar{\rho}}$ is therefore equivalent to the character $\chi_\rho$ taking only *real* values on G.

*Particular Cases: I. Commutative Compact Groups*

(**21.3.8**) Let $f \in \mathscr{L}^2_{\mathbf{C}}(\mathrm{G})$ be nonnegligible and such that, for all $s \in \mathrm{G}$, $f(st) = f(s)f(t)$ for almost all $t \in \mathrm{G}$. This means that in $\mathrm{L}^2_{\mathbf{C}}(\mathrm{G})$ the subspace $\mathbf{C}\cdot\tilde{f}$ is stable under all the mappings $\tilde{g} \mapsto (\varepsilon_s * g)^\sim$, and hence is a minimal closed left ideal of dimension 1 (**21.2.3.3**). This is possible only if this ideal is one of the $\mathfrak{a}_\rho$ *such that* $n_\rho = 1$, and then $f$ is equal almost everywhere to the corresponding *character* $\chi_\rho$. These characters are called the *abelian characters* of G. We have just seen that they are the only continuous homomorphisms of G into $\mathbf{C}^*$; the image of G under such a homomorphism, being a compact subgroup of $\mathbf{C}^*$, is necessarily *contained in* $\mathbf{U}$ (the unit circle in $\mathbf{C}^*$), because $\mathbf{C}^* = \mathbf{U} \times \mathbf{R}_+^*$, and $\mathbf{R}_+^*$ contains no compact subgroup other than $\{1\}$.

If the compact group G is *commutative, every character of* G *is abelian,* because the algebras $\mathfrak{a}_\rho$ are commutative. The classes of the characters of G then form a *Hilbert basis* of $L_{\mathbf{C}}^2(G)$ (21.2.5), and *every* continuous function on G is the uniform limit of a sequence of linear combinations of characters (21.3.4).

(21.3.9) *Every character of the group* $\mathbf{U}^n$ *is of the form*

(21.3.9.1) $$(\zeta_1, \zeta_2, \ldots, \zeta_n) \mapsto \zeta_1^{k_1}\zeta_2^{k_2} \cdots \zeta_n^{k_n}$$

*where* $k_1, \ldots, k_n$ *are integers (positive, negative, or zero). The only character of* $\mathbf{U}^n$ *that takes only real values is the trivial character* ($k_1 = \cdots = k_n = 0$).

The group $\mathbf{U}^n$ is isomorphic to $\mathbf{T}^n$, hence to $\mathbf{R}^n/\mathbf{Z}^n$. If $u$ is a continuous homomorphism of $\mathbf{R}^n/\mathbf{Z}^n$ into $\mathbf{T} = \mathbf{R}/\mathbf{Z}$, and if $\varphi\colon \mathbf{R}^n \to \mathbf{R}^n/\mathbf{Z}^n$ and $\psi\colon \mathbf{R} \to \mathbf{R}/\mathbf{Z}$ are the canonical homomorphisms, then $u \circ \varphi$ is a continuous homomorphism of $\mathbf{R}^n$ into $\mathbf{R}/\mathbf{Z}$, and therefore (16.30.3) factorizes as $\psi \circ v$, where $v$ is a continuous homomorphism of $\mathbf{R}^n$ into $\mathbf{R}$. By restricting $v$ to each of the subgroups $\mathbf{R}\mathbf{e}_j$ of $\mathbf{R}^n$ (where $(\mathbf{e}_j)$ is the canonical basis of $\mathbf{R}^n$) and using (4.1.3), it follows that $v$ is a *linear* mapping of $\mathbf{R}^n$ into $\mathbf{R}$. Moreover, since $u(\varphi(\mathbf{Z}^n)) = \{0\}$, we must have $v(\mathbf{Z}^n) \subset \mathbf{Z}$, and therefore each of the $v(\mathbf{e}_j)$ must be an *integer* $k_j$. This completes the proof.

Observe that if we apply to the group $\mathbf{U}^n$ the theorems (21.3.2) and (21.3.4) we regain, in view of (21.3.9), the facts that the orthogonal system $(\zeta^n)_{n \in \mathbf{Z}}$ is total in $L_{\mathbf{C}}^2(\mathbf{U})$(7.4.3), and that every continuous function on $\mathbf{U}^n$ is a uniform limit of trigonometric polynomials (7.4.2).

*Particular Cases: II. Finite Groups*

(21.3.10) If G is a finite group, the algebras $M_{\mathbf{C}}^1(G)$, $L_{\mathbf{C}}^1(G)$ and $L_{\mathbf{C}}^2(G)$ all coincide with the *group algebra* of G over $\mathbf{C}$, which is also denoted by $\mathbf{C}[G]$ (14.7.4), and all the elements of this algebra may be considered as continuous functions on G. Let $C_1, \ldots, C_r$ denote the *conjugacy classes* of G (with $C_1 = \{e\}$), that is to say, the equivalence classes for the relation:

$$\text{there exists } t \in G \text{ such that } s' = tst^{-1}$$

between elements $s, s' \in G$. It follows from the definition (21.2.2.1) that each central function is constant on each set $C_j$, and from (21.3.4) that the central functions are in this case *linear combinations of characters*. Since the latter are linearly independent, we see that *the number r of conjugacy classes* $C_j$ *is equal to the number of characters of* G *and to the dimension of the center of* C[G].

Let $\boldsymbol{\rho}_1, \boldsymbol{\rho}_2, \ldots, \boldsymbol{\rho}_r$ denote the elements of R, and for brevity let $\chi_{ij}$ denote the value taken by the character $\chi_{\boldsymbol{\rho}_i}$ on the class $C_j$. If $g$ is the order of G and $h_j$ the number of elements in the class $C_j$, the orthogonality relations **(21.3.2.4)** take the form

**(21.3.10.1)** $$\frac{1}{g}\sum_{k=1}^{r} h_k \chi_{ik}\overline{\chi_{jk}} = \delta_{ij} \qquad (1 \leqq i, j \leqq r).$$

In other words, the $r \times r$ matrix

**(21.3.10.2)** $$(h_k^{1/2} g^{-1/2} \chi_{ik})_{1 \leq i,\, k \leq r}$$

is *unitary*. We obtain other orthogonality relations by expressing that the transpose of **(21.3.10.2)** is unitary:

**(21.3.10.3)** $$\sum_{i=1}^{r} \chi_{ik}\overline{\chi_{il}} = 0 \qquad \text{if} \quad k \neq l,$$

**(21.3.10.4)** $$\sum_{i=1}^{r} |\chi_{ik}|^2 = g/h_k .$$

These formulas can also be written as

**(21.3.10.5)** $$\sum_{\boldsymbol{\rho} \in R} \chi_{\boldsymbol{\rho}}(s)\chi_{\boldsymbol{\rho}}(t^{-1}) = 0$$

if $s$, $t$ are not conjugate in G, and

**(21.3.10.6)** $$\sum_{\boldsymbol{\rho} \in R} |\chi_{\boldsymbol{\rho}}(s)|^2 = g/h_k$$

if $s \in C_k$.

Since $e$ is not conjugate to any other element of G, by putting $t = e$ in **(21.3.10.5)** and taking account of **(21.3.2.8)** we obtain

**(21.3.10.7)** $$\sum_{\boldsymbol{\rho} \in R} n_{\boldsymbol{\rho}} \chi_{\boldsymbol{\rho}}(s) = 0 \qquad \text{if} \quad s \neq e.$$

Of course, if we put $s = e$ in (21.3.10.6) we obtain the relation

(21.3.10.8)
$$\sum_{\rho \in R} n_\rho^2 = g,$$

which also follows from the fact that $L_{\mathbf{C}}^2(G)$ is the direct sum of the $\mathfrak{a}_\rho$.

PROBLEMS

1. Let G be a metrizable compact group. Show that the center of the Banach algebra $M_{\mathbf{C}}(G)$ of measures on G is the closure of the center of $L_{\mathbf{C}}^2(G)$, in the vague topology.

2. Show that:

(a) If the representations $M_\rho$ and $\bar{M}_\rho = M_{\bar{\rho}}$ are not equivalent, then $\int \chi_\rho(s^2)\, d\beta(s) = 0$.

(b) If the representation $M_\rho^{(R)}$ is defined (Section 21.1, Problem 9), then $\int \chi_\rho(s^2)\, d\beta(s) = 1$.

(c) If the representation $M_\rho^{(H)}$ is defined (Section 21.1, Problem 9), then $\int \chi_\rho(s^2)\, d\beta(s) = -1$.

(Use the orthogonality relations for characters and observe that if $M_\rho^{(R)}$ (resp. $M_\rho^{(H)}$) is defined, there exists a *unitary* matrix $U$ such that $\overline{M_\rho(s)} = UM_\rho(s)U^{-1}$ for all $s \in G$, and ${}^tU = U$ (resp. ${}^tU = -U$) (cf. Section 21.1, Problem 9).)

3. Let G be a finite group.
(a) Show that for each character $\chi$ of G and each $s \in G$, the complex number $\chi(s)$ is an algebraic integer.† (Consider the eigenvalues of $U(s)$, where $U$ is an irreducible representation of G with character $\chi$, and remark that each element of G has finite order.)
(b) Deduce from (a) that the number of elements of a conjugacy class $C_j$ in G divides the order $g$ of G. Give a direct proof of this fact.
(c) The characteristic functions $e_j$ of the subsets $C_j$ of G form a basis of the center of $\mathbf{C}[G]$ for which the coefficients in the multiplication table are rational integers $\geqq 0$. Deduce that if we put $M_\rho(e_j) = \lambda_j I_n$, then the complex numbers $\lambda_j$ are algebraic integers (remark that the image of the center of $\mathbf{C}[G]$ under $M_\rho$ is a finitely generated $\mathbf{Z}$-module). Show that for each conjugacy class $C_j$ in G the number $n_\rho^{-1} \sum_{s \in C_j} \chi_\rho(s)$ is an algebraic integer, for each $\rho \in R$.
(d) Deduce from (a) and (c) that each $n_\rho$ divides the order $g$ of G (use (21.3.10.1)).

4. Let G be a finite group. For each $s \in G$, let $t(s)$ denote the number of elements $u \in G$ such that $u^2 = s$. Show that

$$t(s) = \sum_{\rho \in R} \nu(\rho)\chi_\rho(s)$$

† See R. Godement, *Algebra*, Hermann (Paris) 1968, p. 560.

where

$$v(\rho) = \frac{1}{g} \sum_{s \in G} \chi_\rho(s^2)$$

(which is equal to 1, 0 or $-1$, by Problem 2).

5. Let G be a finite group. With the notation of (21.3.10), show that each of the numbers $\sum_{k=1}^{r} \chi_{jk}$ is a rational integer $\geqq 0$. (Decompose into irreducible representations the representation of degree $g$ (the order of G) defined on the vector space $\mathbf{C}[G] = \mathbf{C}^G$ by $U(s) \cdot e_t = e_{tst^{-1}}$.)

6. Let G be a finite group.
(a) For each $\rho \in R$ and each $s \in G$, we have $|\chi_\rho(s)| \leqq n_\rho$ (cf. (22.1.3.5)); we have $\chi_\rho(s) = n_\rho$ if and only if $s$ lies in the kernel $N_\rho$ of $M_\rho$; and $|\chi_\rho(s)| = n_\rho$ if and only if the coset of $s$ in $G/N_\rho$ belongs to the center of this group. (Consider the eigenvalues of $M_\rho(s)$.)
(b) Suppose that for some conjugacy class $C_j$ in G, the number $h_j$ of elements of $C_j$ is prime to $n_\rho$. Show that for each $s \in C_j$, either $\chi_\rho(s) = 0$ or $|\chi_\rho(s)| = n_\rho$. (Deduce from Problem 3(c) that the number $\chi_\rho(s)/n_\rho$ is an algebraic integer, and use (a) above.)
(c) Suppose that $C_j \neq \{e\}$ and that the number of elements in $C_j$ is a prime power. Show that there exists $\rho \neq \rho_0$ in R such that $|\chi_\rho(s)| = n_\rho$ for $s \in C_j$. (Use (21.3.2.6) and the fact that $n_\rho \chi_\rho(s)$ is an algebraic integer.) Deduce that G cannot be a noncommutative simple group.

7. Let G be a finite group of order $p^a q^b$, where $p$ and $q$ are prime numbers. Then G is solvable (*Burnside's theorem*). (Argue by induction on the order of G, by considering a Sylow $p$-subgroup of G, which has a nontrivial center.† Consider the number of elements conjugate in G to an element $\neq e$ of this center, and use Problem 6(c).)

## 4. CONTINUOUS UNITARY REPRESENTATIONS OF COMPACT GROUPS

(21.4.1) *Let* G *be a compact group and let V be a continuous unitary representation of* G *on a separable complex Hilbert space* E. *Then (with the notation of* (21.2)):

(i) *For each* $\rho \in R$, *the operator* $V(u_\rho)$ (21.1.4.2) *is an orthogonal projection of* E *onto a closed subspace* $E_\rho$ *of* E, *and* E *is the Hilbert sum of the* $E_\rho$.

(ii) *Each subspace* $E_\rho$ *is stable under V, and the restriction of V to* $E_\rho$ *is the Hilbert sum of a (finite or infinite) sequence of irreducible representations of* G, *each equivalent to* $M_{\bar\rho}$ ((21.2.5) *and* (21.3.7)).

(i) Since $u_\rho * u_\rho = u_\rho$ and $\check{u}_\rho = u_\rho$ (21.2.3), $V(u_\rho)$ is a continuous operator on E which is idempotent and Hermitian (21.1.6), hence (15.5.3.1) is an

† See J.-P. Serre, *Représentations linéaires des groupes finis*, Paris (Hermann), 1967.

orthogonal projection. Further, since $u_\rho * u_{\rho'} = 0$ if $\rho' \neq \rho$, we have $V(u_\rho) \circ V(u_{\rho'}) = 0$, and therefore the images $E_\rho$ of the projections $V(u_\rho)$ are pairwise orthogonal closed subspaces. To show that E is the Hilbert sum of the $E_\rho$, it is enough to show that the sum of the subspaces $E_\rho$ is dense in E. Now, we know that as $f$ runs through $\mathscr{L}^1_{\mathbf{C}}(G)$ and $x$ runs through E, the vectors $V(f) \cdot x$ span a dense subspace of E **(21.1.7)**. Since the continuous functions are dense in $\mathscr{L}^1_{\mathbf{C}}(G)$, and since $\|V(f)\| \leqq N_1(f)$ **(21.1.4.3)**, it follows that the $V(f) \cdot x$ already span a dense subspace of E as $x$ runs through E and $f$ runs through the space $\mathscr{C}_{\mathbf{C}}(G)$ of continuous functions on G. But if $f$ is continuous, then for each $\varepsilon > 0$ there exists a finite linear combination $\sum_{i,j,\rho} c_{ij}^{(\rho)} m_{ij}^{(\rho)}$ such that $\|f - \sum_{i,j,\rho} c_{ij}^{(\rho)} m_{ij}^{(\rho)}\| \leqq \varepsilon$ **(21.2.8)**, and *a fortiori*

$$\left\| V(f) - \sum_{i,j,\rho} c_{ij}^{(\rho)} V(m_{ij}^{(\rho)}) \right\| \leqq N_1\left( f - \sum_{i,j,\rho} c_{ij}^{(\rho)} m_{ij}^{(\rho)} \right) \leqq \varepsilon.$$

Since $m_{ij}^{(\rho)} = u_\rho * m_{ij}^{(\rho)}$, we have $V(m_{ij}^{(\rho)}) = V(u_\rho) V(m_{ij}^{(\rho)})$, and therefore the vector $\sum_{i,j,\rho} c_{ij}^{(\rho)} V(m_{ij}^{(\rho)}) \cdot x$ belongs to the sum of the $E_\rho$. This shows that the sum of the $E_\rho$ is dense in E.

(ii) That each $E_\rho$ is stable under $V$ follows from the fact that the $u_\rho$ belong to the center of the algebra $M_{\mathbf{C}}(G)$. If $V_\rho$ is the restriction of $V$ to $E_\rho$, then $V_\rho(u_{\rho'}) = 0$ for $\rho' \neq \rho$, because $u_{\rho'} * u_\rho = 0$. The restriction of $(V_\rho)_{\text{ext}}$ to the algebra $L^2_{\mathbf{C}}(G)$ may therefore be considered as a nondegenerate representation of the algebra $\mathfrak{a}_\rho$ on $E_\rho$; it follows therefore from **(15.8.16)** that this representation is the Hilbert sum, finite or infinite (according as the dimension of $E_\rho$ is finite or not), of irreducible representations each equivalent to the representation $U_{\mathfrak{l}_1}$, in the notation of **(21.2.4)**. But it follows from the definition of $U_{\mathfrak{l}_1}$ **(15.8.1)** and from **(21.1.9)** that $U_{\mathfrak{l}_1}$ is the restriction of $R_{\text{ext}}$ to $\mathfrak{a}_\rho$. Now we have

$$(\varepsilon_s * m_{i1})(t) = m_{i1}(s^{-1}t) = \frac{1}{n_\rho} \sum_{j=1}^{n_\rho} m_{ij}(s^{-1}) m_{j1}(t)$$

by **(21.2.5)**; this shows that relative to the basis of $\mathfrak{l}_1$ formed by the $n_\rho^{-1} m_{i1}$ $(1 \leqq i \leqq n_\rho)$, the matrix of $R(s)$ is ${}^t M_\rho(s^{-1}) = \overline{M_\rho(s)} = M_{\bar{\rho}}(s)$ by **(21.2.5)**.

**(21.4.1.1)** If G is a *commutative* compact group, every continuous unitary representation of G is therefore a Hilbert sum of *one-dimensional* representations **(21.3.8)**.

**(21.4.2)** With the same notation, if $E_\rho \neq \{0\}$, the irreducible representation $M_{\bar{\rho}}$ is said to be *contained* in the representation $V$; if $E_\rho$ is of finite dimension $d_\rho n_\rho > 0$ (resp. of infinite dimension), then $M_{\bar{\rho}}$ is said to be *contained* $d_\rho$

*times* (resp. *infinitely many times*) in $V$, and $d_\rho$ is called the *multiplicity* of $M_{\bar{\rho}}$ in $V$. The $M_{\bar{\rho}}$ such that $d_\rho > 0$ are also called the *irreducible components* of the representation $V$.

It follows from (21.4.1) that every *irreducible* continuous unitary representation of G is *equivalent to one of the representations* $M_\rho$, and that $M_\rho$ is *contained* $n_\rho$ *times in the regular representation* (21.1.9) *of* G.

(21.4.3) A continuous linear representation $U$ of a *compact* group G on a *finite*-dimensional complex vector space E (21.1.1) may always be considered as a continuous *unitary* representation, because there exists a positive definite Hermitian form on E (in other words, a *scalar product* (6.2)) that is *invariant* under the action $(s, x) \mapsto U(s) \cdot x$ of G on E (20.11.3.3). For *compact* groups there is therefore no loss of generality, where finite-dimensional continuous linear representations are concerned, in restricting consideration to unitary representations. If, for such a representation $U$, we identify $U(s)$ with its matrix relative to a fixed basis of E that is orthonormal with respect to the scalar product referred to above, we have (21.1.2.1)

(21.4.3.1) $$\overline{U(s)} = {}^tU(s)^{-1} = {}^tU(s^{-1}).$$

(21.4.4) *Let $V$ be a continuous unitary representation of a compact group* G *on a vector space* E *of finite dimension d, and suppose that for each* $\boldsymbol{\rho} \in$ R *the irreducible representation $M_\rho$ is contained $d_\rho$ times in $V$, so that* $d = \sum_{\rho \in R} d_\rho n_\rho$. *Then, for all* $s \in$ G, *we have*

(21.4.4.1) $$\mathrm{Tr}(V(s)) = \sum_{\rho \in R} d_\rho \chi_\rho(s).$$

This follows from (21.3.2.7) and the fact that $\mathrm{Tr}(PUP^{-1}) = \mathrm{Tr}(U)$ for any square matrix $U$ and invertible matrix $P$ of the same size.

(21.4.5) *Two finite-dimensional continuous unitary representations $V_1$, $V_2$ of a compact group* G *are equivalent if and only if* $\mathrm{Tr}(V_1(s)) = \mathrm{Tr}(V_2(s))$ *for all* $s \in$ G.

This follows immediately from the formula (21.4.4.1) and the linear independence of characters (21.3.2).

(21.4.6) Let $V'$, $V''$ be continuous linear representations of a topological group G on spaces E′, E″ of *finite* dimensions $d'$, $d''$, respectively. Then it is clear that the mapping

(21.4.6.1) $$V' \otimes V'' : s \mapsto V'(s) \otimes V''(s)$$

is a continuous linear representation of G on the vector space $E' \otimes E''$ of dimension $d'd''$. This representation is called the *tensor product* of $V'$ and $V''$, and we have ((A.10.5) and (A.11.3))

$$\text{(21.4.6.2)} \qquad \operatorname{Tr}(V'(s) \otimes V''(s)) = \operatorname{Tr}(V'(s)) \operatorname{Tr}(V''(s)).$$

In particular, if G is compact we may form the tensor product $M_{\rho'} \otimes M_{\rho''}$ for any two elements $\rho'$, $\rho''$ of R, and then by (21.4.4.1) we have

$$\text{(21.4.6.3)} \qquad \chi_{\rho'} \chi_{\rho''} = \sum_{\rho} c^{\rho}_{\rho'\rho''} \chi_{\rho}$$

where $c^{\rho}_{\rho'\rho''}$ is the number of times the representation $M_{\rho}$ is contained in $M_{\rho'} \otimes M_{\rho''}$, and is therefore a nonnegative *integer*. Since the $\chi_{\rho}$ are linearly independent over **C** and *a fortiori* over **Z**, we see that the *subring* of $\mathscr{C}_{\mathbf{C}}(G)$ *generated by the characters* of G is a **Z**-*algebra*; its identity element is the trivial character, the characters $\chi_{\rho}$ form a *basis over* **Z**, and the multiplication table is given by (21.4.6.3).

(21.4.6.4) For each $\rho \in R$, the *trivial* representation (21.2.7) is contained in $M_{\rho} \otimes M_{\bar{\rho}} = M_{\rho} \otimes \bar{M}_{\rho}$; for if it were not so, then by (21.3.2.6) and (21.4.6.3) we should have

$$\int \chi_{\rho}(s)\chi_{\bar{\rho}}(s)\, d\beta(s) = \int |\chi_{\rho}(s)|^2\, d\beta(s) = 0,$$

which is absurd.

(21.4.7) Since any irreducible representation $V$ of G is equivalent to a representation $M_{\rho}$ for a unique index $\rho$, we shall say (by abuse of language) that $\rho$ is the *class* of the representation $V$, and we write $\rho = \operatorname{cl}(V)$. The class $\rho_0$ of the trivial representation is called the *trivial class.* The class $\bar{\rho}$ is called the *conjugate* of the class $\rho$.

If $V$ is a finite-dimensional continuous unitary representation of G, and if for each $\rho \in R$ the representation $M_{\rho}$ is contained $d_{\rho}$ times in $V$, then the element $\sum_{\rho \in R} d_{\rho} \cdot \rho$ of the **Z**-module $\mathbf{Z}^{(R)}$ of formal linear combinations of elements of R with integer coefficients is called the *class* of the representation $V$, and is written $\operatorname{cl}(V)$. The relation $\operatorname{cl}(V_1) = \operatorname{cl}(V_2)$ therefore signifies that the representations $V_1$ and $V_2$ are *equivalent*, which justifies this terminology. We say also that the class $\rho$ is *contained* $d_{\rho}$ *times* in $\operatorname{cl}(V)$, or that $d_{\rho}$ is the *multiplicity* of $\rho$ in $\operatorname{cl}(V)$.

Conversely, every element $\sum_{\rho \in R} d_{\rho} \cdot \rho$ of $\mathbf{Z}^{(R)}$ in which the coefficients $d_{\rho}$

are *positive or zero* is the class of a linear representation of G, namely the Hilbert sum of a family of $m = \sum_{\rho \in R} d_\rho$ irreducible representations, containing $d_\rho$ representations equal to $M_\rho$ for each $\rho \in R$. It is clear that the bijection $\rho \mapsto \chi_\rho$ extends by linearity to an isomorphism of the Z-module $\mathbf{Z}^{(R)}$ onto the subring of $\mathscr{C}_C(G)$ generated by the characters of G. Transporting the ring structure back to $\mathbf{Z}^{(R)}$ by means of the inverse of this isomorphism, we define on $\mathbf{Z}^{(R)}$ a structure of a *commutative ring*, for which $\rho_0$ is the identity element and the multiplication is given by

(21.4.7.1)
$$\rho'\rho'' = \sum_{\rho} c^{\rho}_{\rho'\rho''} \cdot \rho.$$

For this ring structure we have

(21.4.7.2)
$$\mathrm{cl}(V_1 \otimes V_2) = \mathrm{cl}(V_1) \cdot \mathrm{cl}(V_2)$$

for any two finite-dimensional continuous linear representations $V_1$, $V_2$ of G.

By abuse of language, the ring $\mathbf{Z}^{(R)}$ just defined is called the *ring of classes of continuous linear representations of* G. (The abuse of language lies in the fact that a linear combination of the elements of R with integer coefficients is the class of a representation only if all the coefficients are $\geq 0$.) Also we shall sometimes write R(G) in place of R.

For example, if $G = \mathbf{U}^n$ (isomorphic to $\mathbf{T}^n$), it follows from (21.3.9) that the ring of classes of linear representations of G is isomorphic to the subring $\mathbf{Z}[X_1, \dots, X_n, X_1^{-1}, \dots, X_n^{-1}]$ of the field of rational functions $\mathbf{Q}(X_1, \dots, X_n)$ in $n$ indeterminates over the field $\mathbf{Q}$ of rational numbers.

(21.4.8) With the notation of (21.2.4), the formula (21.2.3.1) may be written as

$$\tilde{f} = \sum_{\rho \in R} \left( \sum_{j=1}^{n_\rho} (f * m_{jj}^{(\rho)})^{\sim} \right).$$

Now we have, by definition,

$$(f * m_{jj}^{(\rho)})(s) = \int f(t) m_{jj}^{(\rho)}(t^{-1}s)\, d\beta(t),$$

and therefore

$$\sum_{j=1}^{n_\rho} (f * m_{jj}^{(\rho)})(s) = n_\rho \cdot \mathrm{Tr}\left( \int f(t) M_\rho(t^{-1}s)\, d\beta(t) \right).$$

Also we have $M_\rho(t^{-1}s) = M_\rho(t^{-1})M_\rho(s)$, and

$$\int f(t)M_\rho(t^{-1})\, d\beta(t) = M_\rho(\check{f})$$

by virtue of (21.1.4.2) and the fact that G, being compact, is unimodular. Hence, for all $f \in \mathscr{L}^2_{\mathbf{C}}(G)$, we obtain the formula

(21.4.8.1) $$\check{f} = \sum_{\rho \in R} n_\rho \cdot \operatorname{Tr}(M_\rho(\check{f})M_\rho(\cdot))^{\sim}$$

where the series on the right converges in $\mathscr{L}^2_{\mathbf{C}}(G)$, no matter how the elements of R are arranged in a sequence. The function $\rho \mapsto M_\rho(\check{f})$, defined on R and taking its values in the space of *all* complex square matrices, is sometimes called the "Fourier transform" of $f$, and the formula (21.4.8.1) is the "Fourier inversion formula for compact groups" (cf. Chapter XXII).

## PROBLEMS

1. Let G, H be two compact groups. Show that the ring $\mathbf{Z}^{(R(G \times H))}$ of classes of continuous linear representations of G × H is isomorphic to the tensor product $\mathbf{Z}^{(R(G))} \otimes \mathbf{Z}^{(R(H))}$ (cf. Section 21.1, Problem 12).

2. Let $P_k(\sigma_1, \sigma_2, \ldots, \sigma_m)$ be the polynomial with rational integer coefficients that expresses the sum $X_1^k + \cdots + X_m^k$ of the $k$th powers of $m$ indeterminates in terms of the elementary symmetric functions $\sigma_h = \sum_{(j_i)} X_{j_1}X_{j_2} \cdots X_{j_h}$ of these indeterminates (the summation is over all strictly increasing sequences $j_1 < j_2 < \cdots < j_h$ of $h \leq m$ indices). Let $U$ be a finite-dimensional linear representation of a compact group G and consider the element of $\mathbf{Z}^{(R(G))}$ given by

(*) $$P_k\left(\operatorname{cl}(U), \operatorname{cl}\left(\bigwedge^2 U\right), \ldots, \operatorname{cl}\left(\bigwedge^m U\right)\right)$$

(Section 21.1, Problem 11). Consider also the canonical homomorphism $\chi$ of $\mathbf{Z}^{(R(G))}$ into $\mathscr{C}(G)$, which maps $\rho \in R(G)$ to $\chi_\rho$. Show that the image under $\chi$ of the element (*) above is equal to the function $s \mapsto \operatorname{Tr}(U(s^k))$.

3. Let G be a locally compact group and let $U$ be a continuous unitary representation of G on a separable Hilbert space E. Let $\mathscr{R}(U)$ denote the algebra of intertwining operators of $U$ with itself (Section 21.1, Problem 6), i.e., the algebra of continuous operators $T \in \mathscr{L}(E)$ such that $TU(s) = U(s)T$ for all $s \in G$.

   The representation $U$ is said to be *primary* if the center of $\mathscr{R}(U)$ consists only of the homotheties of E, and *isotypic* if it is primary and if there exists a nontrivial irreducible subrepresentation of $U$.

(a) For $U$ to be primary, it is necessary and sufficient that the center of $\mathscr{R}(U)$ contain no orthogonal projection other than 0 and $1_E$. (Observe that the center is a closed self-adjoint subalgebra of $\mathscr{L}(E)$, and use the Gelfand–Neumark theorem.)

(b) For $U$ to be isotypic, it is necessary and sufficient that $U$ should be equivalent to a (finite or infinite) Hilbert sum of equivalent irreducible representations. (To show that the condition is necessary, consider a closed subspace F of E that is stable under $U$ and such that the restriction $V$ of $U$ to F is irreducible. If $W$ is the restriction of $U$ to the orthogonal supplement $F^\perp$ of F, which is assumed to be $\neq \{0\}$, deduce from the fact that the projection $P_F$ cannot belong to the center of $\mathscr{R}(U)$ that there exists a nonzero intertwining operator between $V$ and $W$ (Section 21.1, Problem 6), and hence that $W$contains a subrepresentation equivalent to $V$; then use induction. To show that the condition is sufficient, E being now the Hilbert sum of subspaces $F_k$ stable under $U$ and such that the restrictions $U_k$ of $U$ to the $F_k$ are equivalent irreducible representations, consider an orthogonal projection $P \neq 0$ belonging to the center of $\mathscr{R}(U)$; show that there exists at least one index $k$ such that $P \cdot P_{F_k} \neq 0$, and deduce that $P \cdot P_{F_j} \neq 0$ for all indices $j$, and thence that $P = 1_E$.)

(c) If $U$ is equivalent to a Hilbert sum of irreducible representations all equivalent to the same representation $V$, show that the number $n$ (finite or $+\infty$) of these representations is finite if and only if $\mathscr{R}(U)$ is of finite dimension over **C**, and that this dimension is then $n^2$. (Use the topological version of Schur's lemma (Section 21.1, Problem 5).) Furthermore, every subrepresentation $W$ of $U$ is a Hilbert sum of representations equivalent to $V$. (With the notation of (b) above, let $L \subset E$ be the subspace of the representation $W$; there exists at least one index $k$ such that the orthogonal projection of $F_k$ on L is nonzero. Deduce that there exists a nonzero intertwining operator between $U_k$ and $W$, and use Section 21.1, Problem 6, to obtain a subrepresentation of $W$ equivalent to $U_k$; then proceed by induction.)

4. Let G be a unimodular locally compact group. A continuous unitary representation of G on a separable Hilbert space E is said to admit a *discrete decomposition* if it is a Hilbert sum of irreducible representations.

(a) Let R(G) be the set of equivalence classes of irreducible continuous unitary representations of G. Let $U$ be a continuous unitary representation of G on E, and suppose that E is a Hilbert sum of subspaces $E_k$ such that the restriction $U_k$ of $U$ to $E_k$ is irreducible. For each $\boldsymbol{\rho} \in R(G)$, let $M_\rho$ be the Hilbert sum of the $E_k$ such that $U_k$ is in the class $\boldsymbol{\rho}$. The nonzero $M_\rho$ are called the *isotypic components* of E. Show that for every irreducible subrepresentation $V$ of $U$, the space of $V$ is necessarily contained in one of the $M_\rho$, and that $V$ is then of class $\boldsymbol{\rho}$, so that $M_\rho$ may be defined as the smallest closed subspace of E that contains the spaces of all the irreducible subrepresentations of $U$ of class $\boldsymbol{\rho}$ (and is therefore defined independently of the decomposition $(E_k)$ chosen). (Use Problem 3(c) above and Section 21.1, Problem 6.) If the restriction of $U$ to $M_\rho$ is the Hilbert sum of $n_\rho$ representations of class $\boldsymbol{\rho}$, where $n_\rho$ is finite or $+\infty$, this number $n_\rho$ is called the *multiplicity* of $\boldsymbol{\rho}$ in $U$ (or in the class of $U$).

(b) Let $U$ be a continuous unitary representation of G on E that has the following property: for every closed subspace F of E stable under $U$, there exists a closed subspace L of F that is minimal among those that are stable under $U$. Show that $U$ admits a discrete decomposition. (Argue by induction, as in (15.8.10).)

(c) Let $f_n$ be a sequence of continuous functions on G satisfying the conditions of (14.11.2). Let $U$ be a continuous unitary representation of G such that, for each $n$, the operator $U(f_n)$ is compact. Show that $U$ admits a discrete decomposition into irreducible representations and that, for each $\boldsymbol{\rho} \in R(G)$, the multiplicity of $\boldsymbol{\rho}$ in $U$ is *finite*. (Show that the criterion of (b) above is satisfied. If $F \subset E$ is closed and stable under $U$, there exists an integer $n$ such that the restriction of $U(f_n)$ to F is $\neq 0$. Consider an eigenvalue $\lambda \neq 0$ of this restriction, and the corresponding eigenspace M, which is finite-dimensional. For each vector $x \neq 0$ in M, let $P_x$ be the smallest closed subspace of E that contains $x$ and is stable under $U$. If $P_x$ is the Hilbert sum of two $U$-stable subspaces Q and R, show that $P_x \cap M$ is the Hilbert sum of $Q \cap M$ and $R \cap M$, and hence deduce that there exists $x \in M$ for which $P_x$ is minimal. Furthermore, if the subrepresentation of $U$ corresponding to $P_x$ is of class $\mathfrak{p}$, then $n_{\mathfrak{p}}$ is at most equal to the dimension of M.)

**5.** Let G be a unimodular locally compact group, and let $U$ be a continuous unitary representation of G on a separable Hilbert space E. For each pair $(x, y)$ of points of E, the function $s \mapsto (U(s) \cdot x \mid y)$, which is continuous and bounded on G, is called the *coefficient of U relative to* $(x, y)$, and is denoted by $c_U(x, y)$. For each bounded measure $\mu$ on G, we have $c_U(U(\mu) \cdot x, y) = c_U(x, y) * \check{\mu}$ and $c_U(x, U(\mu) \cdot y) = \bar{\mu} * c_U(x, y)$. If $J$ is a semilinear bijection of E into itself such that $(J \cdot x \mid J \cdot y) = \overline{(x \mid y)}$ (we may take $J(e_n) = e_n$, where $(e_n)$ is a Hilbert basis of E), let $\bar{U}$ denote the continuous unitary representation $s \mapsto JU(s)J^{-1}$ of G on E, which is well-defined up to equivalence. Show that $c_{\bar{U}}(x, y) = \overline{c_U(x, y)}$.
(a) Suppose that $U$ is irreducible and that there exist two nonzero vectors $x$, $y$ in E such that the function $c_U(x, y)$ belongs to $L^2_{\mathbf{C}}(G)$. Then $c_U(x, U(\mu) \cdot y)$ belongs to $L^2_{\mathbf{C}}(G)$, for every bounded measure $\mu$ on G. Deduce that the set of $z \in E$ such that $c_U(x, z) \in L^2_{\mathbf{C}}(G)$ is a dense vector subspace F of E, and that the linear mapping $z \mapsto \overline{c_U(x, z)}$ of F into $L^2_{\mathbf{C}}(G)$ is closed (Section **15.12**, Problem 1). Use Section **21.1**, Problem 6, to show that $F = E$ and that $U$ is equivalent to a subrepresentation of the regular representation $R$ of G on $L^2_{\mathbf{C}}(G)$; also that $c_U(x, y)$ belongs to $L^2_{\mathbf{C}}(G)$ for *all* pairs $x$, $y$ in E.
(b) Show that, for each function $f \in \mathscr{K}(G)$, the coefficient $c_R(\tilde{g}, \tilde{f})$ of the regular representation $R$ belongs to $L^2_{\mathbf{C}}(G)$ for each $\tilde{g} \in L^2_{\mathbf{C}}(G)$. Deduce that all the coefficients of an irreducible subrepresentation of $R$ belong to $L^2_{\mathbf{C}}(G)$. An *irreducible* continuous unitary representation of G is said to be *square-integrable* if it is equivalent to a subrepresentation of the regular representation of G.
(c) Show that if at least one irreducible unitary representation $U$ of G is square-integrable, then the center Z of G is necessarily *compact*. (Observe that the function $|c_U(x, y)|$ on $G \times G$ is invariant under left and right translations by elements of Z.)

**6.** Let G be a unimodular locally compact group, $U$ an irreducible unitary representation of G on a Hilbert space E, and assume that $U$ is square-integrable (Problem 5).
(a) For all $x$, $y$, $x'$, $y'$ in E we have

$$(c_U(x, y) \mid c_U(x', y')) = d_U^{-1}(x \mid x')\overline{(y \mid y')}$$

in $L^2_{\mathbf{C}}(G)$, where $d_U$ is a number $> 0$ that depends only on the equivalence class of $U$. (Observe that, as a result of Problem 5, the mapping $S_x$: $z \mapsto \overline{c_U(x, z)}$ is an intertwining operator between $U$ and the regular representation $R$, and consequently $S^*_{x'} S_x$ is a homothety in E, by virtue of Schur's lemma; in other words, there exists a constant $a(x, x')$ such that

$$(\overline{c_U(x, y)} \mid \overline{c_U(x', y')}) = a(x, x')(y \mid y').$$

Show on the other hand that

$$(c_U(x, y) \mid c_U(x', y')) = (c_U(y', x') \mid c_U(y, x))$$

by using the fact that G is unimodular.)

The number $d_U$ is multiplied by $a^{-1}$ when the Haar measure $\beta$ on G is replaced by $a\beta$. When G is compact and $\beta$ is the Haar measure with total mass 1, the number $d_U$ is equal to the *dimension* of the representation $U$.
(b) Deduce from the Banach–Steinhaus theorem that there exists a constant $b > 0$ such that $N_2(c_U(x, y)) \leq b \cdot \|x\| \cdot \|y\|$ for all $x$, $y$ in E.
(c) Let $A$, $B$ be two *nuclear* operators on E (Section **15.11**, Problem 7). Show that

$$\int_G U(s)AU(s)^{-1}\, d\beta(s) = d_U^{-1}\, \mathrm{Tr}(A),$$

$$\int_G \mathrm{Tr}(U(s)AU(s)^{-1}B)\, d\beta(s) = d_U^{-1}\, \mathrm{Tr}(A)\, \mathrm{Tr}(B),$$

$$\int_G \mathrm{Tr}(AU(s)^{-1})\, \mathrm{Tr}(BU(s))\, d\beta(s) = d_U^{-1}\, \mathrm{Tr}(AB).$$

(Observe that there exists a Hilbert basis $(e_n)$ of E and a sequence $(f_n)$ of vectors of norm 1 in E such that, for all $x \in$ E, we have $A \cdot x = \sum_n \lambda_n (x \mid e_n) f_n$, where $\sum_n |\lambda_n| < \infty$, and use (a) and (b) above.)

**7.** Let $U$, $U'$ be two square-integrable irreducible unitary representations of G on separable Hilbert spaces E, E′, respectively. Show that if $U$ and $U'$ are inequivalent, then every coefficient of $U$ is orthogonal in $L^2_{\mathbf{C}}(G)$ to every coefficient of $U'$. (Consider on E′ × E the sesquilinear form $(x', x) \mapsto (c_{U'}(x', a') \mid c_U(x, a))$; show that it is continuous (Problem 6(b)) and that it can be written in the form $(x', x) \mapsto (x' \mid A \cdot x)$, where $A$ is a continuous operator from E to E′; finally prove that $A$ is an intertwining operator of $U$ with $U'$.)

**8.** Given two Hilbert spaces $E_1$ and $E_2$, a continuous operator $T\colon E_1 \to E_2$ is said to be a *Hilbert–Schmidt operator* if the operator on $E_1 \oplus E_2$ that is equal to $T$ on $E_1$ and 0 on $E_2$ is Hilbert–Schmidt (**15.4.8**). The space $\mathscr{L}_2(E_1, E_2) \subset \mathscr{L}_2(E_1 \oplus E_2)$ of Hilbert–Schmidt operators from $E_1$ to $E_2$ is a Hilbert space.

For each $x_1 \in E_1$ and $x_2 \in E_2$, let $u_{x_1, x_2}$ denote the linear mapping $z \mapsto (z \mid x_1)x_2$ of $E_1$ into $E_2$. This mapping belongs to $\mathscr{L}_2(E_1, E_2)$, and we have $\|u_{x_1, x_2}\|_2 = \|x_1\| \cdot \|x_2\|$. If $(a_m)$ (resp. $(b_n)$) is a Hilbert basis of $E_1$ (resp. $E_2$), then the $u_{a_m, b_n}$ form a Hilbert basis of $\mathscr{L}_2(E_1, E_2)$.
(a) Let $G_1$, $G_2$ be two locally compact groups and let $U_1$ (resp. $U_2$) be a continuous unitary representation of $G_1$ (resp. $G_2$) on a separable Hilbert space $E_1$ (resp. $E_2$). For $s_1 \in G_1$, $s_2 \in G_2$, and $T \in \mathscr{L}_2(E_1, E_2)$, show that the mapping $U_2(s_2)TU_1(s_1)^{-1}$, which we denote by $U(s_1, s_2) \cdot T$, belongs to $\mathscr{L}_2(E_1, E_2)$, and that $U(s_1, s_2)$ is a continuous unitary representation of $G_1 \times G_2$ on the Hilbert space $\mathscr{L}_2(E_1, E_2)$.
(b) Suppose that $U_1$ and $U_2$ are irreducible. Show that $U$ is irreducible. (Remark that the closed subspace of $\mathscr{L}_2(E_1, E_2)$ generated by the transforms $U_2(s_2)u_{a,b}$, where $a \neq 0$ in $E_1$ and $b \neq 0$ in $E_2$, contains all the elements $u_{a,y}$ for $y \in E_2$; likewise for the transforms $u_{a,b}U_1(s_1)$.) The restriction of $U$ to the subgroup $G_1 \times \{e_2\}$ of $G_1 \times G_2$ is then an isotypic unitary representation (Problem 3), a Hilbert sum of representations equivalent to $\bar{U}_1$, the multiplicity of the class of $\bar{U}_1$ in this restriction being equal to the dimension of $E_2$. Likewise for the restriction of $U$ to the subgroup $\{e_1\} \times G_2$.

**9.** Let G be a unimodular locally compact group, and let $U$ be a square-integrable irreducible continuous unitary representation of G on a Hilbert space E (Problem 5). Let $M_U$ be the closed vector subspace of $L_{\mathbf{C}}^2(G)$ spanned by the coefficients (Problem 5) of $U$. It is stable under the operators $\gamma(s)$ and $\delta(s)$ for each $s \in G$.

(a) Let $U'$ be another square-integrable irreducible representation of G. Show that if $U'$ is equivalent to $U$, then $M_{U'} = M_U$, so that $M_U$ depends only on the class $\rho$ of $U$, and is therefore also denoted by $M_\rho$. If on the other hand $U'$ is not equivalent to $U$, then the subspaces $M_U$ and $M_{U'}$ are orthogonal. If $(e_j)$ is a Hilbert basis of E, the elements $f_{jk} = d_U^{1/2} c_U(e_j, e_k)$ form a Hilbert basis of $M_U$.

(b) Define a continuous unitary representation $(s, t) \mapsto V(s, t)$ of $G \times G$ on $M_U$ by $V(s, t) \cdot c_U(x, y) = \gamma(s)\,\delta(t) c_U(x, y)$. Show that this representation is equivalent to the continuous unitary representation $(s, t) \mapsto W(s, t)$ of $G \times G$ on the Hilbert space $L_2(E)$ of Hilbert–Schmidt operators on E, defined by $W(s, t) \cdot T = U(s) T U(t)^{-1}$ (Problem 8). Deduce that $V$ is irreducible, and that the restriction to $M_U$ of the regular representation $R$ is a Hilbert sum of irreducible representations equivalent to $\bar{U}$, the multiplicity of $\bar{U}$ in this decomposition (Problem 4) being the dimension of E.

(c) Let $f$ be a function in $\mathscr{L}_{\mathbf{C}}^2(G)$, with compact support, and let $P$ be the orthogonal projection of $L_{\mathbf{C}}^2(G)$ onto the subspace $M_U$. Show that $U(f)$ is a Hilbert–Schmidt operator on E and that $\|U(f)\|_2 \leqq d_U^{1/2} N_2(P \cdot \check{f})$. (Use the basis $(f_{jk})$ of $M_U$ to calculate $N_2(P \cdot \check{f})$.)

(d) Let $L_{\mathbf{C}}^2(G)_d$ be the closed subspace of $L_{\mathbf{C}}^2(G)$ that is the Hilbert sum of the subspaces $M_\rho$, as $\rho$ runs through the set of equivalence classes of square-integrable irreducible representations of G. Show that $L_{\mathbf{C}}^2(G)_d$ contains every closed subspace F of $L_{\mathbf{C}}^2(G)$ that is stable under $\gamma(s)$ (resp. $\delta(s)$) for all $s \in G$ and is minimal with respect to this property among nonzero subspaces. (Let $P$ be the orthogonal projection of $L_{\mathbf{C}}^2(G)$ onto F. If $V$ is the irreducible representation that is the restriction of $R$ to F, calculate the coefficients $c_V(f, P \cdot g)$ for $f \in F$ and $g \in \mathscr{K}(G)$.)

**10.** Let G be a locally compact group and let $U$ be a continuous linear representation of G on a *finite*-dimensional complex vector space. Assume that the coefficients $c_U(x, x^*)$ (Section **21.2**, Problem 1) belong to $\mathscr{L}_{\mathbf{C}}^2(G)$.

(a) Show that there exists on E a (nondegenerate) Hermitian scalar product $\Phi$ that is invariant under $U$ (same method as in **(20.11.3.1)**).

(b) Deduce from (a) that the group G is necessarily compact. (Observe that the coefficients of the matrix of $\Phi$, relative to a basis of E, belong to $\mathscr{L}_{\mathbf{C}}^1(G)$.)

**11.** (a) Let G be a topological group, let $U$ be a continuous linear representation of G on a complex vector space E of dimension $d$, and let $V_0$ be the trivial representation of G on a vector space F of dimension $n$. Let $W$ be the representation $s \mapsto U(s) \otimes V_0(s)$ of G on $E \otimes F$. Show that if $n > d$, there exists no vector $z \in E \otimes F$ such that the vectors $W(s) \cdot z$ $(s \in G)$ generate $E \otimes F$. (Write $z$ in the form $\sum_{j=1}^{d} x_j \otimes y_j$, where the $x_j$ form a basis of E, and the $y_j$ belong to F.)

(b) Let G be a compact group. With the notation of **(21.4.1)**, if $V$ is the Hilbert sum of $q \leqq n_\rho$ representations equal to $M_\rho$, then there exists a totalizing vector $x_0$ in the space E of the representation $V$ (in other words the vectors $V(s) \cdot x_0$ for $s \in G$ span E). (Reduce to the case where E is the sum $\mathfrak{l}_1 + \mathfrak{l}_2 + \cdots + \mathfrak{l}_q$ in $\mathfrak{a}_\rho$ and $V$ is the restriction to E of the regular representation. Show that we may take $x_0 = m_{11} + m_{22} + \cdots + m_{qq}$, by showing that no nonzero vector in E is orthogonal to all the $V(s) \cdot x_0$.)

(c) Let G be a compact group. Show that a continuous unitary representation $V$ of G on a separable Hilbert space E is topologically cyclic if and only if, for each $\rho \in \mathbf{R}$, the multiplicity of $M_\rho$ in $V$ is $\leqq n_\rho$. (To show that the condition is sufficient, we may assume that E is the Hilbert sum of left ideals $\mathfrak{b}_\rho \subset \mathfrak{a}_\rho$, where $\rho$ runs through a subset $\mathbf{R}'$ of $\mathbf{R}$, and

$V$ is the restriction to E of the regular representation. If $x_\rho \in \mathfrak{b}_\rho$ is a totalizer for the restriction of $V$ to $\mathfrak{b}_\rho$, consider a vector $x_0 = \sum_{\rho \in R'} \xi_\rho x_\rho$, where $\sum_{\rho \in R'} \xi_\rho^2 \|x_\rho\|^2 < \infty$ and $\xi_\rho > 0$.)

## 5. INVARIANT BILINEAR FORMS; THE KILLING FORM

**(21.5.1)** From now on in this chapter we shall consider only (real or complex) *Lie groups*. By a *linear representation* of a *real* Lie group G on a *finite-dimensional real or complex* vector space E we shall mean (as in **(16.9.7)**), unless the contrary is expressly stated, a *Lie group homomorphism* (hence of class $C^\infty$) $s \mapsto U(s)$ of G into **GL**(E). (If E is a complex vector space, we consider **GL**(E) as equipped with its underlying structure of a *real* Lie group.) By virtue of **(19.10.2)**, this notion in fact coincides with the notion of continuous linear representation (on a finite-dimensional complex vector space) introduced in **(21.1)**.

If G is a *complex* Lie group, a linear representation of G on a finite-dimensional *complex* vector space E is by definition a homomorphism of complex Lie groups $s \mapsto U(s)$ (hence a *holomorphic* mapping) of G into **GL**(E). One must be careful to distinguish these representations from linear representations of the underlying real Lie group $G_{|\mathbf{R}}$ on E; every linear representation of G on E is also a linear representation of $G_{|\mathbf{R}}$, but the converse is false.

Let E be a finite-dimensional real vector space, and let $E_{(\mathbf{C})} = E \otimes_{\mathbf{R}} \mathbf{C}$ be its complexification. Every endomorphism $P$ of E has a unique extension to an endomorphism $P \otimes 1_{\mathbf{C}}$ of $E_{(\mathbf{C})}$, such that $(P \otimes 1_{\mathbf{C}}) \cdot (x \otimes \zeta) = (P \cdot x) \otimes \zeta$ for all $x \in E$ and all $\zeta \in \mathbf{C}$ **(A.10.6)**. The matrix of $P$ relative to a basis $(e_j)$ of E is the same as the matrix of $P \otimes 1_{\mathbf{C}}$ relative to the basis $(e_j \otimes 1)$ of $E_{(\mathbf{C})}$. It follows immediately that every linear representation $s \mapsto U(s)$ of a real Lie group G on E extends uniquely to a linear representation $s \mapsto U(s) \otimes 1_{\mathbf{C}}$ of G on $E_{(\mathbf{C})}$.

**(21.5.2)** Given any linear representation $s \mapsto U(s)$ of a real (resp. complex) Lie group G on a finite-dimensional real or complex (resp. complex) vector space E, we have a *derived homomorphism* $\mathbf{u} \mapsto U_*(\mathbf{u})$ of the Lie algebra $\mathfrak{g}_e$ of G into the Lie algebra $\mathfrak{gl}(E)$. For each $\mathbf{w} \in \mathfrak{g}_e$, we have **(19.8.9)**

**(21.5.2.1)**
$$U(\exp(\mathbf{w})) = \exp(U_*(\mathbf{w})).$$

If G is a real Lie group and E a real vector space, the derived homomorphism of the representation $U \otimes 1_{\mathbf{C}}$ of G on $E_{(\mathbf{C})}$ **(21.5.1)** is the homomorphism $\mathbf{u} \mapsto U_*(\mathbf{u}) \otimes 1_{\mathbf{C}}$ of $\mathfrak{g}_e$ into $\mathfrak{gl}(E_{(\mathbf{C})}) = \mathfrak{gl}(E) \otimes_{\mathbf{R}} \mathbf{C}$.

We remark also that if F is a finite-dimensional *complex* vector space, $\mathfrak{a}$ a *real* Lie algebra (of finite or infinite dimension) and $\rho: \mathfrak{a} \to \mathfrak{gl}(\mathrm{F})$ a homomorphism of *real* Lie algebras, then the mapping $\rho_{(\mathbf{C})}: \mathbf{u} \otimes \zeta \mapsto \rho(\mathbf{u})\zeta$ is a ($\mathbf{C}$-linear) homomorphism of the complexification $\mathfrak{a}_{(\mathbf{C})} = \mathfrak{a} \otimes_{\mathbf{R}} \mathbf{C}$ of $\mathfrak{a}$ into $\mathfrak{gl}(\mathrm{F})$ that extends $\rho$.

(21.5.3) Let $s \mapsto U(s)$ be a linear representation of a real (resp. complex) Lie group G on a finite-dimensional real or complex (resp. complex) vector space E. Canonically associated with $U$ is the following *bilinear form* on the real (resp. complex) vector space $\mathfrak{g}_e \times \mathfrak{g}_e$:

(21.5.3.1) $$\mathrm{B}_U: (\mathbf{u}, \mathbf{v}) \mapsto \mathrm{Tr}(U_*(\mathbf{u}) \circ U_*(\mathbf{v})).$$

From the symmetry $\mathrm{Tr}(PQ) = \mathrm{Tr}(QP)$ of the trace it follows that the form $\mathrm{B}_U$ is *symmetric*, but it can be degenerate. Furthermore, it is *invariant* under the action $(s, \mathbf{u}) \mapsto \mathrm{Ad}(s) \cdot \mathbf{u}$ of G on $\mathfrak{g}_e$: for by (16.5.4) and (19.2.1.1) we have

$$U_*(\mathrm{Ad}(s) \cdot \mathbf{u}) = \mathrm{Ad}(U(s)) \cdot U_*(\mathbf{u}) = U(s)U_*(\mathbf{u})U(s)^{-1},$$

and the relation

(21.5.3.2) $$\mathrm{B}_U(\mathrm{Ad}(s) \cdot \mathbf{u}, \mathrm{Ad}(s) \cdot \mathbf{v}) = \mathrm{B}_U(\mathbf{u}, \mathbf{v})$$

therefore follows from the symmetry of the trace.

(21.5.4) In general, let $\Phi$ be any $\mathbf{R}$-bilinear mapping of $\mathfrak{g}_e \times \mathfrak{g}_e$ into a real vector space E that is *invariant* under the action $(s, \mathbf{u}) \mapsto \mathrm{Ad}(s) \cdot \mathbf{u}$ of G on $\mathfrak{g}_e$; then, for all $\mathbf{u}, \mathbf{v}, \mathbf{w}$ in $\mathfrak{g}_e$, we have

(21.5.4.1) $$\Phi([\mathbf{w}, \mathbf{u}], \mathbf{v}) + \Phi(\mathbf{u}, [\mathbf{w}, \mathbf{v}]) = 0.$$

For by hypothesis we have, for all $t \in \mathbf{R}$,

$$\Phi(\mathrm{Ad}(\exp(t\mathbf{w})) \cdot \mathbf{u}, \mathrm{Ad}(\exp(t\mathbf{w})) \cdot \mathbf{v}) = \Phi(\mathbf{u}, \mathbf{v});$$

if we now differentiate both sides of this relation with respect to $t$ at $t = 0$, we obtain (21.5.4.1) by use of (8.1.4) and (19.11.2.2).

(21.5.5) More generally, if $\mathfrak{a}$ is a Lie algebra over $\mathbf{R}$ (resp. $\mathbf{C}$) and F is a finite-dimensional vector space over $\mathbf{R}$ (resp. $\mathbf{C}$), then to each Lie algebra homomorphism $\rho: \mathfrak{a} \to \mathfrak{gl}(\mathrm{F})$ we may associate a symmetric bilinear $\mathbf{R}$-form (resp. $\mathbf{C}$-form) on $\mathfrak{a} \times \mathfrak{a}$ by the formula

(21.5.5.1) $$\mathrm{B}_\rho(\mathbf{u}, \mathbf{v}) = \mathrm{Tr}(\rho(\mathbf{u}) \circ \rho(\mathbf{v})).$$

Since $\rho([\mathbf{u}, \mathbf{v}]) = \rho(\mathbf{u}) \circ \rho(\mathbf{v}) - \rho(\mathbf{v}) \circ \rho(\mathbf{u})$, the symmetry property of the trace shows again that we have

(21.5.5.2) $$B_\rho(\mathrm{ad}(\mathbf{w}) \cdot \mathbf{u}, \mathbf{v}) + B_\rho(\mathbf{u}, \mathrm{ad}(\mathbf{w}) \cdot \mathbf{v}) = 0$$

for all $\mathbf{u}, \mathbf{v}, \mathbf{w} \in \mathfrak{a}$.

**(21.5.6)** Consider a finite-dimensional real or complex Lie algebra $\mathfrak{a}$, and its *adjoint representation* $\mathbf{u} \mapsto \mathrm{ad}(\mathbf{u})$, which is a homomorphism of $\mathfrak{a}$ into $\mathfrak{gl}(\mathfrak{a})$. We denote by $B_\mathfrak{a}$ or simply B the symmetric bilinear form corresponding to this homomorphism according to **(21.5.5)**; it is called the *Killing form* of the Lie algebra $\mathfrak{a}$. By **(21.5.5.2)** we have

(21.5.6.1) $$B([\mathbf{w}, \mathbf{u}], \mathbf{v}) + B(\mathbf{u}, [\mathbf{w}, \mathbf{v}]) = 0.$$

If $\sigma$ is any *automorphism* of the Lie algebra $\mathfrak{a}$, we have $\sigma([\mathbf{u}, \mathbf{v}]) = [\sigma(\mathbf{u}), \sigma(\mathbf{v})]$, or equivalently $\sigma \circ \mathrm{ad}(\mathbf{u}) = \mathrm{ad}(\sigma(\mathbf{u})) \circ \sigma$ in $\mathrm{End}(\mathfrak{a})$. From this and the symmetry of the trace we deduce immediately that

(21.5.6.2) $$B(\sigma(\mathbf{u}), \sigma(\mathbf{v})) = B(\mathbf{u}, \mathbf{v}).$$

**(21.5.7)** *If* $\mathfrak{b}$ *is an ideal in a Lie algebra* $\mathfrak{a}$, *the restriction to* $\mathfrak{b} \times \mathfrak{b}$ *of the Killing form* $B_\mathfrak{a}$ *is the Killing form* $B_\mathfrak{b}$.

By hypothesis, for each $\mathbf{x} \in \mathfrak{b}$, we have $\mathrm{ad}(\mathbf{x}) \cdot \mathfrak{a} \subset \mathfrak{b}$; hence, for $\mathbf{x}$ and $\mathbf{y} \in \mathfrak{b}$, if we put $U = \mathrm{ad}(\mathbf{x}) \circ \mathrm{ad}(\mathbf{y})$, we have $U(\mathfrak{a}) \subset \mathfrak{b}$. If we now calculate the trace of $U$ by means of a basis of $\mathfrak{a}$ consisting of a basis of $\mathfrak{b}$ and a basis of a subspace of $\mathfrak{a}$ supplementary to $\mathfrak{b}$, we see that this trace is equal to that of the *restriction* of $U$ to $\mathfrak{b}$.

It should be remarked, however, that there is no simple relation between the Killing form of an arbitrary *Lie subalgebra* of $\mathfrak{a}$, and the restriction to this subalgebra of the Killing form of $\mathfrak{a}$.

**(21.5.8)** If G is a (real or complex) Lie group, $\mathfrak{g}_e$ its Lie algebra, the Killing form of $\mathfrak{g}_e$ is called the *Killing form of* G.

**(21.5.9)** Let G be a connected Lie group with center $\{e\}$, and let $U$ be a linear representation of G on a finite-dimensional vector space, such that the bilinear form $B_U$ is *nondegenerate*. Then there exists a pseudo-Riemannian structure on G whose metric tensor $\boldsymbol{g}$ satisfies $\boldsymbol{g}(e) = B_U$, and which is *invariant under left and right translations by elements of* G **(20.11.8)**.

## 6. SEMISIMPLE LIE GROUPS. CRITERION OF SEMISIMPLICITY FOR A COMPACT LIE GROUP

**(21.6.1)** A finite-dimensional real or complex Lie algebra $\mathfrak{a}$ is said to be *semisimple* if its Killing form **(21.5.6)** is *nondegenerate*. A real or complex Lie group is said to be *semisimple* if its Lie algebra is semisimple.

If $\mathfrak{a}$ is a finite-dimensional real Lie algebra, any basis of $\mathfrak{a}$ over $\mathbf{R}$ can be canonically identified with a basis of its complexification $\mathfrak{a}_{(\mathbf{C})}$ over $\mathbf{C}$. Consequently, the Killing form $\mathrm{B}_{\mathfrak{a}_{(\mathbf{C})}}$ of $\mathfrak{a}_{(\mathbf{C})}$ has the same matrix relative to this basis as does the Killing form $\mathrm{B}_{\mathfrak{a}}$. It follows immediately that *if* $\mathfrak{a}$ *is semisimple, so also is its complexification, and conversely.*

On the other hand, if $\mathfrak{a}$ is a *complex* Lie algebra and $\mathfrak{a}_{|\mathbf{R}}$ the real Lie algebra obtained from $\mathfrak{a}$ by restriction of scalars, then we have $\mathrm{B}_{\mathfrak{a}_{|\mathbf{R}}} = 2\mathscr{R}(\mathrm{B}_{\mathfrak{a}})$. For if $u$ is an endomorphism of a finite-dimensional complex vector space E, and if $u_0$ is the same mapping $u$ considered as an $\mathbf{R}$-linear mapping, then it is easy to verify that $\mathrm{Tr}(u_0) = 2\mathscr{R}(\mathrm{Tr}(u))$ **(16.21.13.1)**. Hence it follows that *if* $\mathfrak{a}$ *is semisimple, so also is* $\mathfrak{a}_{|\mathbf{R}}$: for by taking a basis of $\mathfrak{a}$ that is orthogonal relative to $\mathrm{B}_{\mathfrak{a}}$, we see from the remarks above that $\mathrm{B}_{\mathfrak{a}_{|\mathbf{R}}}$ has signature $(n, n)$ if $n = \dim_{\mathbf{C}}(\mathfrak{a})$, and therefore is nondegenerate.

**(21.6.2)** *Let* $\mathfrak{a}$ *be a real or complex semisimple Lie algebra.*

(i) *The only commutative ideal in* $\mathfrak{a}$ *is the zero ideal.*

(ii) *For each ideal* $\mathfrak{b}$ *in* $\mathfrak{a}$, *the subspace* $\mathfrak{b}^{\perp}$ *of* $\mathfrak{a}$ *orthogonal to* $\mathfrak{b}$ *with respect to the Killing form* $\mathrm{B}_{\mathfrak{a}}$ *is an ideal of* $\mathfrak{a}$, *supplementary to* $\mathfrak{b}$, *and the Lie algebras* $\mathfrak{b}$ *and* $\mathfrak{b}^{\perp}$ *are semisimple.*

(i) Let $\mathfrak{c}$ be a commutative ideal in $\mathfrak{a}$. For each $\mathbf{y} \in \mathfrak{a}$, we have $\mathrm{ad}(\mathbf{y}) \cdot \mathfrak{c} \subset \mathfrak{c}$ and therefore $\mathrm{ad}(\mathbf{x}) \cdot (\mathrm{ad}(\mathbf{y}) \cdot \mathfrak{c}) = \{0\}$ for all $\mathbf{x} \in \mathfrak{c}$. On the other hand, $\mathrm{ad}(\mathbf{x}) \cdot (\mathrm{ad}(\mathbf{y}) \cdot \mathfrak{a}) \subset \mathfrak{c}$, because $\mathbf{x} \in \mathfrak{c}$. If we compute the trace of $U = \mathrm{ad}(\mathbf{x}) \circ \mathrm{ad}(\mathbf{y})$ with the help of a basis of $\mathfrak{a}$ consisting of a basis of $\mathfrak{c}$ and a basis of a subspace supplementary to $\mathfrak{c}$, it follows that we obtain 0: in other words, $\mathrm{B}_{\mathfrak{a}}(\mathbf{x}, \mathbf{y}) = 0$ for all $\mathbf{x} \in \mathfrak{c}$ and $\mathbf{y} \in \mathfrak{a}$. Since $\mathrm{B}_{\mathfrak{a}}$ is nondegenerate, this forces $\mathbf{x} = 0$.

(ii) It follows immediately from **(21.5.6.1)** that if $\mathfrak{b}$ is an ideal in $\mathfrak{a}$, then so also is $\mathfrak{b}^{\perp}$. Hence $\mathfrak{b} \cap \mathfrak{b}^{\perp}$ is an ideal in $\mathfrak{a}$, and we shall show that it is *commutative*. Indeed, if $\mathbf{u}, \mathbf{v}$ are any two elements of $\mathfrak{b} \cap \mathfrak{b}^{\perp}$, then by **(21.5.6.1)** we have $\mathrm{B}_{\mathfrak{a}}(\mathbf{w}, [\mathbf{u}, \mathbf{v}]) = \mathrm{B}_{\mathfrak{a}}([\mathbf{w}, \mathbf{u}], \mathbf{v}) = 0$ for all $\mathbf{w} \in \mathfrak{a}$, because $[\mathbf{w}, \mathbf{u}] \in \mathfrak{b}$ and $\mathbf{v} \in \mathfrak{b}^{\perp}$. Since $\mathrm{B}_{\mathfrak{a}}$ is nondegenerate, it follows that $[\mathbf{u}, \mathbf{v}] = 0$, which proves our assertion. Hence, by virtue of (i) above, we have $\mathfrak{b} \cap \mathfrak{b}^{\perp} = \{0\}$ and therefore $\mathfrak{b} + \mathfrak{b}^{\perp} = \mathfrak{a}$, so that $\mathfrak{b}$ and $\mathfrak{b}^{\perp}$ are supplementary ideals. The restrictions of $\mathrm{B}_{\mathfrak{a}}$ to the nonisotropic subspaces $\mathfrak{b}$ and $\mathfrak{b}^{\perp}$ are therefore nondegenerate, so that $\mathfrak{b}$ and $\mathfrak{b}^{\perp}$ are semisimple Lie algebras, by virtue of **(21.5.7)**.

We shall show later that, conversely, a finite-dimensional real or complex Lie algebra that has no nonzero commutative ideals is semisimple **(21.22.4)**.

From **(21.6.2)** it follows immediately that:

**(21.6.3)** *The center of a semisimple Lie algebra is* $\{0\}$.

In particular **(19.11.9)**, every semisimple Lie algebra over $\mathbf{R}$ (resp. $\mathbf{C}$) is the Lie algebra of a real (resp. complex) semisimple Lie group, and there is a *one-to-one correspondence* between semisimple Lie algebras and *simply connected* semisimple Lie groups (up to isomorphism).

A finite-dimensional real or complex Lie algebra is said to be *simple* if it is noncommutative and if it contains no ideals other than itself and $\{0\}$.

It can be shown that if $\mathfrak{a}$ is a simple Lie algebra over $\mathbf{C}$, then the Lie algebra $\mathfrak{a}_{|\mathbf{R}}$ obtained by restriction of scalars is also simple (Problem 1). On the other hand, if $\mathfrak{g}$ is a simple Lie algebra over $\mathbf{R}$, then the Lie algebra $\mathfrak{g}_{(\mathbf{C})}$ over $\mathbf{C}$ obtained by extension of scalars is semisimple, but not necessarily simple (Problem 1).

**(21.6.4)** *Every semisimple Lie algebra* $\mathfrak{g}$ *is the direct sum of a finite number of ideals* $\mathfrak{g}_i$ $(1 \leqq i \leqq r)$, *each of which is a simple Lie algebra, and which are mutually orthogonal with respect to* $\mathrm{B}_{\mathfrak{g}}$. *Every ideal of* $\mathfrak{g}$ *is the direct sum of a subfamily of* $(\mathfrak{g}_i)_{1 \leq i \leq r}$.

The proof is by induction on the dimension of $\mathfrak{g}$. Let $\mathfrak{a}$ be a nonzero ideal of $\mathfrak{g}$ of smallest possible dimension; by virtue of **(21.6.2)**, $\mathfrak{g}$ is the direct sum of $\mathfrak{a}$ and the ideal $\mathfrak{a}^{\perp}$, which implies that $[\mathfrak{a}, \mathfrak{a}^{\perp}] = \{0\}$. Every ideal in the Lie algebra $\mathfrak{a}$ is therefore also an ideal in $\mathfrak{g}$, and therefore by hypothesis the Lie algebra $\mathfrak{a}$ contains no ideals other than $\mathfrak{a}$ and $\{0\}$. Since moreover $\mathfrak{a}$ is not commutative **(21.6.2)**, it is a *simple* Lie algebra. By applying the inductive hypothesis to the semisimple Lie algebra $\mathfrak{a}^{\perp}$, the first assertion is established. If now $\mathfrak{b}$ is any ideal in $\mathfrak{g}$, then $\mathfrak{b} \cap \mathfrak{g}_i$ is an ideal in $\mathfrak{g}_i$, hence is either $\mathfrak{g}_i$ or $\{0\}$. If $\mathfrak{a}$ is the sum of the $\mathfrak{g}_i$ contained in $\mathfrak{b}$, then $\mathfrak{a}^{\perp}$ is the sum of the remaining $\mathfrak{g}_i$, and we have $\mathfrak{b} = \mathfrak{a} \oplus \mathfrak{c}$, where $\mathfrak{c} = \mathfrak{b} \cap \mathfrak{a}^{\perp}$. Since $\mathfrak{b} \cap \mathfrak{g}_i = \{0\}$ for each $\mathfrak{g}_i \subset \mathfrak{a}^{\perp}$, we have also $[\mathfrak{b}, \mathfrak{g}_i] = \{0\}$ for these $\mathfrak{g}_i$, hence $[\mathfrak{b}, \mathfrak{a}^{\perp}] = \{0\}$ and so *a fortiori* $[\mathfrak{c}, \mathfrak{c}] = \{0\}$. But since the Lie algebra $\mathfrak{a}^{\perp}$ is semisimple, it has no nonzero commutative ideals **(21.6.2)**, so that $\mathfrak{c} = 0$ and therefore $\mathfrak{b} = \mathfrak{a}$.

**(21.6.5)** *Every semisimple Lie algebra* $\mathfrak{g}$ *is equal to its derived algebra* $\mathfrak{D}(\mathfrak{g})$.

This is obvious if $\mathfrak{g}$ is simple, because by definition $\mathfrak{D}(\mathfrak{g})$ cannot be zero. The general case now follows from (21.6.4).

(21.6.5.1) With the notation of the proof of (21.6.4), the ideals $\mathfrak{a}$ and $\mathfrak{a}^{\perp}$ are orthogonal relative to *any invariant* symmetric **R**-bilinear form $\Phi$ on $\mathfrak{g}$ (21.5.4). For, by virtue of (21.6.5), it is enough to show that for all $\mathbf{x}, \mathbf{y} \in \mathfrak{a}$ and $\mathbf{z} \in \mathfrak{a}^{\perp}$ we have $\Phi([\mathbf{x}, \mathbf{y}], \mathbf{z}) = 0$; but by (21.5.4.1) this is equivalent to $\Phi(\mathbf{x}, [\mathbf{y}, \mathbf{z}]) = 0$, and since $\mathbf{y} \in \mathfrak{a}$ and $\mathbf{z} \in \mathfrak{a}^{\perp}$ we have $[\mathbf{y}, \mathbf{z}] \in \mathfrak{a} \cap \mathfrak{a}^{\perp} = \{0\}$.

(21.6.6) A Lie group is said to be *almost simple* if its Lie algebra is *simple*. It follows from (21.6.4) that a *simply connected* semisimple Lie group G is isomorphic to a *product* of *simply connected almost simple* Lie groups $G_j$. The only connected Lie groups immersed in G that are *normal* in G and of positive dimension are the products of subfamilies of the $G_j$; they are closed in G. It follows from (21.6.3) that the *center* of a semisimple Lie group is *discrete*, and from (21.6.5) that the *commutator subgroup* of a semisimple Lie group is an *open* subgroup (19.7.1).

This last result shows in particular that a *connected* semisimple Lie group G is unimodular, since the kernel of the modulus function $s \mapsto \Delta_G(s)$ contains the commutator subgroup of G.

(21.6.7) *Every derivation* (A.18.2) *of a semisimple Lie algebra* $\mathfrak{g}$ *is inner* (A.19.4).

Let $\mathfrak{D} = \mathrm{Der}(\mathfrak{g})$ be the Lie algebra of derivations of $\mathfrak{g}$ (A.19). Since the center of $\mathfrak{g}$ is $\{0\}$ (21.6.3), the image $\mathrm{ad}(\mathfrak{g})$ of $\mathfrak{g}$ under the adjoint representation $\mathbf{x} \mapsto \mathrm{ad}(\mathbf{x})$ is a Lie subalgebra isomorphic to $\mathfrak{g}$, and therefore semisimple; moreover, since $\mathrm{ad}(D\mathbf{u}) = [D, \mathrm{ad}(\mathbf{u})]$ for $\mathbf{u} \in \mathfrak{g}$ and $D \in \mathfrak{D}$ (A.19.4), $\mathrm{ad}(\mathfrak{g})$ is an *ideal* of $\mathfrak{D}$. Consider the subspace $\mathfrak{a}$ of $\mathfrak{D}$ that is orthogonal to $\mathrm{ad}(\mathfrak{g})$ relative to the Killing form $B_{\mathfrak{D}}$ (which *a priori* might be degenerate). Since the restriction of $B_{\mathfrak{D}}$ to the ideal $\mathrm{ad}(\mathfrak{g})$ is the Killing form $B_{\mathrm{ad}(\mathfrak{g})}$ (21.5.6), and since this form is nondegenerate, it follows that the intersection $\mathfrak{a} \cap \mathrm{ad}(\mathfrak{g})$, which is the subspace of $\mathrm{ad}(\mathfrak{g})$ orthogonal to $\mathrm{ad}(\mathfrak{g})$ relative to $B_{\mathrm{ad}(\mathfrak{g})}$, is zero. Also, by (21.5.6.1), $\mathfrak{a}$ is an ideal of $\mathfrak{D}$, and therefore $[\mathfrak{a}, \mathrm{ad}(\mathfrak{g})] \subset \mathfrak{a} \cap \mathrm{ad}(\mathfrak{g}) = \{0\}$. Consequently, for $D \in \mathfrak{a}$ and $\mathbf{u} \in \mathfrak{g}$, we have $\mathrm{ad}(D\mathbf{u}) = [D, \mathrm{ad}\,\mathbf{u}] = 0$, and since the mapping $\mathbf{x} \mapsto \mathrm{ad}(\mathbf{x})$ is injective, it follows that $D\mathbf{u} = 0$, hence $D = 0$ and so $\mathfrak{a} = \{0\}$. This proves that $B_{\mathfrak{D}}$ is nondegenerate and that $\mathfrak{D} = \mathrm{ad}(\mathfrak{g})$.

(21.6.8) *Let* G *be a connected semisimple (real or complex) Lie group. Then the image* $\mathrm{Ad}(G)$ *of* G *under the homomorphism* $s \mapsto \mathrm{Ad}(s)$ *is an open subgroup of the group* $\mathrm{Aut}(\mathfrak{g}_e)$ *of automorphisms of the Lie algebra of* G.

For this image is a connected Lie group immersed in $\mathrm{Aut}(\mathfrak{g}_e)$, whose Lie algebra is $\mathrm{ad}(\mathfrak{g}_e)$ (19.13.9); but the latter is equal to the Lie algebra $\mathrm{Der}(\mathfrak{g}_e)$ of $\mathrm{Aut}(\mathfrak{g}_e)$ by virtue of (21.6.7) and (19.3.8). Hence the result, by (19.7.1).

**(21.6.9)** *Let* G *be a connected (real) Lie group,* C *its center and* $\mathfrak{g}$ *its Lie algebra. The following conditions are equivalent:*

(a) *The quotient group* G/C *is compact.*
(b) G *is isomorphic to a product* $\mathbf{R}^m \times G_1$, *where* $G_1$ *is compact.*
(c) *The Lie group* $\tilde{G}$, *the universal covering of* G, *is isomorphic to a product* $\mathbf{R}^n \times K$, *where* K *is a simply connected semisimple compact group.*
(d) *The Lie algebra* $\mathfrak{g}$ *is the direct sum* $\mathfrak{c} \oplus \mathfrak{D}(\mathfrak{g})$ *of its center* $\mathfrak{c}$ *and its derived algebra* $\mathfrak{D}(\mathfrak{g})$, *and the restriction to* $\mathfrak{D}(\mathfrak{g})$ *of the Killing form* $B_{\mathfrak{g}}$ *is negative definite.*

*When these equivalent conditions are satisfied,* $\mathfrak{D}(\mathfrak{g})$ *is isomorphic to the Lie algebra of* K; *the center* Z *of* K *is finite;* G *is isomorphic to* $\tilde{G}/D$, *where* D *is a discrete subgroup of* $\mathbf{R}^n \times Z$; *the center* C *of* G *is isomorphic to* $(\mathbf{R}^n \times Z)/D$; *and the center of* G/C *consists only of the identity element. The subgroups* $\mathrm{Ad}(\tilde{G})$, $\mathrm{Ad}(G)$, *and* $\mathrm{Ad}(K)$ *of* $\mathrm{Aut}(\mathfrak{g})$ *may be identified with the same (compact) open subgroup of* $\mathrm{Aut}(\mathfrak{D}(\mathfrak{g}))$ *(itself a direct factor of* $\mathrm{Aut}(\mathfrak{g})$*), and are isomorphic to* G/C *and to* K/Z.

*The commutator subgroup* $\mathscr{D}(K)$ *of* K *is equal to* K, *and the commutator subgroup* $\mathscr{D}(G)$ *of* G *may be identified with the group* $K/(D \cap Z)$ *(and is therefore compact).*

Clearly (b) implies (a). We shall first prove that (a) implies (d).

The homomorphism $s \mapsto \mathrm{Ad}(s)$ of G into $\mathrm{Aut}(\mathfrak{g}) \subset \mathbf{GL}(\mathfrak{g})$ has kernel C (19.11.6) and therefore factorizes as $G \to G/C \xrightarrow{v} \mathrm{Aut}(\mathfrak{g})$, where $v$ is an injective homomorphism of Lie groups (16.10.9). If G/C is compact, then so also is its image Ad(G) under $v$, and $v$ is therefore an isomorphism of G/C onto the compact Lie subgroup Ad(G) of $\mathrm{Aut}(\mathfrak{g})$ ((19.10.1) and (16.9.9)). Hence, by (20.11.3.1), there exists a positive definite symmetric bilinear form $\Phi$ on the vector space $\mathfrak{g}$ that is *invariant* under the canonical action of Ad(G) on $\mathfrak{g}$. It is clear that $\mathrm{Ad}(s) \cdot \mathfrak{c} = \{0\}$ for all $s \in G$ (19.11.6); the subspace $\mathfrak{c}^\perp$ of $\mathfrak{g}$, which is the orthogonal supplement of $\mathfrak{c}$ relative to $\Phi$, is therefore also stable under every automorphism $\mathrm{Ad}(s)$ of $\mathfrak{g}$, hence is an *ideal* in $\mathfrak{g}$ (19.11.3). But since Ad(G) may be canonically identified with a closed subgroup of the orthogonal group $\mathbf{O}(\Phi)$, its Lie algebra $\mathrm{ad}(\mathfrak{g})$ is identified with a Lie subalgebra of the Lie algebra $\mathfrak{o}(\Phi)$ of $\mathbf{O}(\Phi)$. Relative to a basis of $\mathfrak{g}$ that is orthonormal with respect to $\Phi$, the matrix $(\alpha_{jk})$ of the endomorphism $\mathrm{ad}(\mathbf{u})$

of $\mathfrak{g}$, where $\mathbf{u}$ is any element of $\mathfrak{g}$, satisfies $\alpha_{kj} = -\alpha_{jk}$ **(19.4.4.3)**. It follows that

$$B_{\mathfrak{g}}(\mathbf{u}, \mathbf{u}) = \operatorname{Tr}((\operatorname{ad}(\mathbf{u}))^2) = \sum_{j,k} \alpha_{jk}\alpha_{kj} = -\sum_{j,k} \alpha_{jk}^2 \leqq 0;$$

and, moreover, that we have $B_{\mathfrak{g}}(\mathbf{u}, \mathbf{u}) = 0$ only if $\operatorname{ad}(\mathbf{u}) = 0$, hence if $\mathbf{u} \in \mathfrak{c}$. The restriction of $B_{\mathfrak{g}}$ to the ideal $\mathfrak{c}^{\perp}$ is therefore negative definite. By virtue of **(21.5.7)**, this shows that the Lie algebra $\mathfrak{c}^{\perp}$ is semisimple, hence equal to its derived algebra **(21.6.5)**; and since $[\mathfrak{c}, \mathfrak{g}] = \{0\}$ by definition, we have also $\mathfrak{D}(\mathfrak{g}) = [\mathfrak{g}, \mathfrak{g}] = \mathfrak{c}^{\perp}$.

Next we shall prove that (d) implies (c). Clearly it is enough to show that if $\mathfrak{k}$ is a semisimple real Lie algebra, such that the Killing form $B_{\mathfrak{k}}$ is negative definite, then a *simply connected* Lie group K whose Lie algebra is isomorphic to $\mathfrak{k}$ **(21.6.3)** is necessarily *compact*. Now, since $B_{\mathfrak{k}}$ is invariant under the adjoint action of K on $\mathfrak{k}$, the subgroup Ad(K) of Aut($\mathfrak{k}$), which is closed **(21.6.8)**, may be identified with a closed subgroup of the orthogonal group $\mathbf{O}(B_{\mathfrak{k}})$, hence is *compact* **(16.11.2)**. On the other hand, the Lie algebra ad($\mathfrak{k}$) of Ad(K) is isomorphic to $\mathfrak{k}$ and therefore has center $\{0\}$ **(21.6.3)**. Hence the center of Ad(K) is discrete, and it follows from Weyl's theorem **(20.22.5)** that the Lie group K, which is the universal covering of Ad(K), is also compact.

We go on to prove the assertions in the second and third paragraphs of **(21.6.9)**. From **(16.30.2.1)** we have $G = \tilde{G}/D$, where D is a discrete subgroup of the center $\mathbf{R}^n \times Z$ of $\tilde{G} = \mathbf{R}^n \times K$. In view of **(21.6.8)** and the fact that every automorphism of $\mathfrak{g}$ leaves $\mathfrak{c}$ and $\mathfrak{D}(\mathfrak{g})$ stable, these assertions (except for those relating to the derived groups) follow from **(20.22.5.1)**. The derived group $\mathscr{D}(K)$ has Lie algebra $\mathfrak{D}(\mathfrak{k}) = \mathfrak{k}$ **(19.12.1)**, and because K is connected it follows that $\mathscr{D}(K) = K$. We deduce that $\mathscr{D}(\tilde{G}) = K$, and since $\mathscr{D}(G)$ is evidently the canonical image of $\mathscr{D}(\tilde{G})$, it is therefore equal to the canonical image of K, which is isomorphic to $K/(D \cap K) = K/(D \cap Z)$ **(12.12.5)**.

Finally, we shall prove that (c) implies (b). Let $p$ be the order of the center Z of K. The projection of the group D on $\mathbf{R}^n$ is a discrete group, because the inverse image in $\mathbf{R}^n \times Z$ of a compact neighborhood of 0 in $\mathbf{R}^n$ is a compact set, and therefore intersects D in a finite set. It follows **(19.7.9.1)** that D is finitely-generated, and hence the set of $z^p$ as $z$ runs through D is a subgroup D′ of $D \cap \mathbf{R}^n$, of finite index in D (and *a fortiori* in $D \cap \mathbf{R}^n$). By **(19.7.9.1)**, the group $\mathbf{R}^n/D'$ is therefore isomorphic to a product $\mathbf{R}^m \times \mathbf{T}^{n-m}$, and hence $\tilde{G}/D'$ is isomorphic to $\mathbf{R}^m \times G'$, where $G' = \mathbf{T}^{n-m} \times K$ is compact. Furthermore, D/D′ is a finite subgroup of the center of $\tilde{G}/D'$, and since $\mathbf{R}^m$ has no finite subgroup other than $\{0\}$, it follows that D/D′ may be identified with a finite subgroup C′ of the center of G′. Hence $\tilde{G}/D$, being isomorphic to $(\tilde{G}/D')/(D/D')$, is isomorphic to $\mathbf{R}^m \times G_1$, where $G_1 = G'/C'$ is compact.

*Remarks*

**(21.6.10)** (i) Since $\{0\}$ is the only compact subgroup of $\mathbf{R}^m$, the subgroup $G_1$ of $G = \mathbf{R}^m \times G_1$ is a *maximal* compact subgroup of G.

(ii) If a real Lie algebra $\mathfrak{g}$ satisfies condition (d) of **(21.6.9)**, it is isomorphic to the Lie algebra of a *compact* connected Lie group, namely the group $\mathbf{T}^n \times K$ (in the notation of **(21.6.9)**). It follows that this condition *characterizes* the Lie algebras of compact connected Lie groups. Since the Lie algebra $\mathfrak{D}(\mathfrak{g})$ is semisimple, it is equal to its derived algebra. The same argument as in **(21.6.5.1)** then shows that $\mathfrak{c}$ and $\mathfrak{D}(\mathfrak{g})$ are orthogonal with respect to *any invariant* **R**-bilinear form on $\mathfrak{g}$.

(iii) The discrete subgroups of the group $\mathbf{R}^n \times Z$ are easily determined (Problem 7), and therefore the structure of compact connected Lie groups is essentially reduced to that of simply connected *semisimple* compact Lie groups.

(iv) It can be shown (Section **21.11**, Problem 12(b)) that, under the conditions of **(21.6.9)**, the group Ad(K) is of *finite* index in Aut(K), and the latter is therefore compact.

(v) In view of **(19.16.4.3)**, a connected Lie group G is unimodular if and only if Ad(G) is unimodular. Since every compact group is unimodular **(14.3.3)**, it follows from **(21.6.9)** that every connected Lie group G, such that the quotient of G by its center is compact, is unimodular.

## PROBLEMS

**1.** (a) Let $\mathfrak{a}$ be a simple Lie algebra over **C**. Show that the Lie algebra $\mathfrak{a}_{|\mathbf{R}}$ over **R** is simple. (Observe that if $\mathfrak{b}$ is an ideal in the semisimple Lie algebra $\mathfrak{a}_{|\mathbf{R}}$, then $[\mathfrak{a}, \mathfrak{b}] = \mathfrak{b}$.)
(b) Let $\mathfrak{a}$ be a simple Lie algebra over **R**. Show that the Lie algebra $\mathfrak{a}_{(\mathbf{C})}$ over **C** is either simple or the direct sum of two isomorphic simple algebras. (Let $c$ be the semilinear bijection of $\mathfrak{a}_{(\mathbf{C})} = \mathfrak{a} \oplus i\mathfrak{a}$ onto itself such that $c(\mathbf{x} + i\mathbf{y}) = \mathbf{x} - i\mathbf{y}$ for all $\mathbf{x}, \mathbf{y} \in \mathfrak{a}$. Show first that if V is a complex vector subspace of $\mathfrak{a}_{(\mathbf{C})}$ such that $c(V) = V$, and if $W = \mathfrak{a} \cap V$, then $V = W \oplus iW$. Deduce that if $\mathfrak{b}$ is an ideal of $\mathfrak{a}_{(\mathbf{C})}$ other than $\mathfrak{a}$, then we have

$$\mathfrak{b} \cap c(\mathfrak{b}) = \{0\}, \qquad \mathfrak{b} \cap \mathfrak{a} = \{0\}, \qquad \mathfrak{a}_{(\mathbf{C})} = \mathfrak{b} \oplus c(\mathfrak{b}),$$

and that $\mathfrak{b}$ is a simple Lie algebra over **C**.)
(c) Let $\mathfrak{a}$ be a simple Lie algebra over **C**. Show that the Lie algebra $(\mathfrak{a}_{|\mathbf{R}})_{(\mathbf{C})}$ over **C** is the direct sum of two simple Lie algebras, each isomorphic to $\mathfrak{a}$. (For each $\mathbf{x} \in \mathfrak{a}$, consider the element $\frac{1}{2}(\mathbf{x} \otimes 1 + (i\mathbf{x}) \otimes i) \in (\mathfrak{a}_{|\mathbf{R}}) \otimes_{\mathbf{R}} \mathbf{C}$.)

**2.** (a) In order that a finite-dimensional real Lie algebra $\mathfrak{g}$ should be the Lie algebra of a compact Lie group, it is necessary and sufficient that for each $\mathbf{u} \in \mathfrak{g}$ the endomorphism $\operatorname{ad}(\mathbf{u}) \otimes 1$ of $\mathfrak{g}_{(\mathbf{C})}$ be diagonalizable and that its eigenvalues be pure imaginary. (Argue as in **(21.6.9)**.)

(b) Deduce from (a) that if $\mathfrak{g}$ is the Lie algebra of a compact group, then every Lie subalgebra $\mathfrak{h}$ of $\mathfrak{g}$ is also the Lie algebra of a compact group; in particular, $\mathfrak{h}$ cannot be solvable unless $\mathfrak{h}$ is commutative.

**3.** Show that the Killing form of a real Lie algebra $\mathfrak{g}$ of positive finite dimension cannot be positive definite. (Use Problem 2, by noting that $\mathrm{ad}(\mathbf{u}) \otimes i$ is a self-adjoint endomorphism relative to the form $B_{\mathfrak{g}(C)}$.)

**4.** (a) Let $\mathfrak{g}$ be a real or complex Lie algebra, $\mathfrak{h}$ a semisimple ideal of $\mathfrak{g}$. Show that $\mathfrak{g}$ is the direct sum of $\mathfrak{h}$ and the centralizer $\mathfrak{z}(\mathfrak{h})$ of $\mathfrak{h}$. (Use **(21.6.7)**.)
(b) Let $\mathfrak{g}$ be the Lie algebra of a compact Lie group and let $\mathfrak{n}$ be an ideal in $\mathfrak{g}$. If $\mathfrak{c}$ is the center of $\mathfrak{g}$, show that

$$\mathfrak{n} = (\mathfrak{n} \cap \mathfrak{c}) \oplus (\mathfrak{n} \cap \mathfrak{D}\mathfrak{g})$$

(consider the Killing form of $\mathfrak{n}$), and deduce that there exists an ideal $\mathfrak{n}'$ in $\mathfrak{g}$ such that $\mathfrak{g} = \mathfrak{n} \oplus \mathfrak{n}'$.
(c) Let $\mathfrak{g}$ be a real Lie algebra and $\mathfrak{n}$ an ideal in $\mathfrak{g}$; suppose that $\mathfrak{n}$ and $\mathfrak{g}/\mathfrak{n}$ are the Lie algebras of compact Lie groups. Show that $\mathfrak{g}$ is the Lie algebra of a compact group if and only if $\mathfrak{g}$ is the direct sum of $\mathfrak{n}$ and another ideal. (Use (b).)

**5.** (a) Let G be a connected Lie group, $\mathfrak{g}$ its Lie algebra, $\mathfrak{h}$ a semisimple subalgebra of $\mathfrak{g}$, and H the connected Lie group immersed in G corresponding to $\mathfrak{h}$. Show that if the center of H is finite, then H is closed in G. (Use Section **19.11**, Problem 4.) (Cf. Section **21.18**, Problem 18.)
(b) Let G be a connected, almost simple, *noncompact* Lie group with finite center. Show that there exists no nontrivial continuous *unitary* linear representation of G on a finite-dimensional complex vector space.

**6.** (a) Let $\mathfrak{g}$ be a finite-dimensional (real or complex) Lie algebra. Show that the sum $\mathfrak{a}$ of all the semisimple ideals of $\mathfrak{g}$ is a semisimple ideal of $\mathfrak{g}$ (and hence is the unique largest semisimple ideal of $\mathfrak{g}$), and deduce that the number of semisimple ideals of $\mathfrak{g}$ is finite.
(b) Use (a) and Section **21.2**, Problem 7, to show that in a compact Lie group G the number of conjugacy classes of connected semisimple Lie subgroups of G is finite.

**7.** Let A be a finite commutative group. Then every discrete subgroup of $\mathbf{R}^n \times A$ is of the form EB (isomorphic to $E \times B$), where B is a subgroup of A, and E is a subgroup of $\mathbf{R}^n \times A$ such that the restriction to E of the projection $\mathbf{R}^n \times A \to \mathbf{R}^n$ is an isomorphism of E onto a discrete subgroup of $\mathbf{R}^n$ (hence isomorphic to $\mathbf{Z}^p$ for some $p \leqq n$).

**8.** Let G be a Lie group for which the number of connected components is finite, and let $G_0$ be the identity component of G. Suppose that $\mathrm{Lie}(G) = \mathrm{Lie}(G_0) = \mathfrak{g}$ is the Lie algebra of a compact group. Show that G is the semidirect product of a maximal compact subgroup K and a normal subgroup V isomorphic to $\mathbf{R}^m$ for some $m$; also that $K \cap G_0$ is the identity component of K, and that $G_0$ is the direct product of $K \cap G_0$ and V. (Use **(21.6.9)** and the fact that the group Ad(G) is compact. By considering a scalar product on $\mathfrak{g}$ that is invariant under Ad(G), we may assume that in the decomposition $G_0 = V \times K_0$ of $G_0$ as the direct product of a subgroup V isomorphic to $\mathbf{R}^m$ and a compact connected group $K_0$, the Lie algebra of V is orthogonal to that of $K_0$ for the scalar product in question, and hence that V is a normal subgroup of G. Then use Section **19.14**, Problem 3.) Under what conditions is the subgroup K (resp. V) above unique?

**9.** Let N be the nilpotent Lie group consisting of all $3 \times 3$ matrices $(x_{ij})$ such that $x_{ij} = 0$ if $i < j$ and $x_{ii} = 1$ for $i = 1, 2, 3$. Let $G_1$ be the closed subgroup of N consisting of the matrices $(x_{ij})$ for which $x_{12}$ and $x_{23}$ are rational integers, and let $H_1$ be the subgroup of $G_1$ consisting of the $(x_{ij})$ for which $x_{12} = x_{23} = 0$ and $x_{13}$ is a rational integer. Show that the Lie group $G_1/H_1$ has infinitely many connected components; its center Z, which is also its commutator subgroup, is compact and connected, and is also the identity component of G, and is the largest compact subgroup of G; but G is not the semidirect product of Z with any other subgroup.

**10.** Let G be a connected Lie group. Define inductively $\mathscr{D}'^0(G) = G$, and $\mathscr{D}'^p(G)$ to be the closure of the commutator subgroup of $\mathscr{D}'^{(p-1)}(G)$, for $p \geqq 1$. Show that if $\mathscr{D}'^p(G)$ is compact, then $\mathscr{D}'^{(p+1)}(G)$ is compact and semisimple, and that $G = \mathscr{D}'^{(p+1)}(G) \cdot H$, where H is the identity component of the centralizer of $\mathscr{D}'^{(p+1)}(G)$ in G. The group $\mathscr{D}'^{(p+1)}(G) \cap H$ is finite and commutative, and $\mathscr{D}'^{(p)}(H)$ is contained in the identity component of the center of H. Show that the connected Lie group $N/H_1 = G_2$ (in the notation of Problem 9) is such that $\mathscr{D}(G_2)$ is compact, but that the Lie algebra of $G_2$ is not the Lie algebra of a compact group.

**11.** (a) Let G be a connected Lie group, $\mathfrak{g}$ its Lie algebra. Show that if the closure of Ad(G) in Aut($\mathfrak{g}$) is compact, then the quotient of G by its center is compact, and consequently Ad(G) is compact. (Observe that there exists an Ad(G)-invariant scalar product on $\mathfrak{g}$.)
(b) In order that a connected Lie group G should be such that the quotient of G by its center is compact, it is necessary and sufficient that for each neighborhood U of $e$ in G there should exist a neighborhood $V \subset U$ of $e$ such that $xVx^{-1} = V$ for all $x \in G$. (The condition is necessary by **(21.3.4.1)**. To show that it is sufficient, use (a) above, by proving that the closure of Ad(G) in End($\mathfrak{g}$) is contained in Aut($\mathfrak{g}$).)

**12.** Let G be a nondiscrete, almost simple Lie group, and $G_0$ its identity component. Show that each normal subgroup N of G either contains $G_0$ or is contained in the centralizer $\mathscr{Z}(G_0)$ of $G_0$, which is the largest discrete normal subgroup of G. In particular, if G is compact, then $\mathscr{Z}(G_0)$ is finite, and there are only finitely many elements $s \in G$ such that Ad($s$) is the identity mapping.

**13.** Let G be an *almost simple* compact Lie group of dimension $n \geqq 1$. For each $s \in G$, each integer $m \geqq 1$ and each neighborhood V of $e$ in G, let M($s$, $m$, V) denote the set of elements of G of the form

$$(x_1(y_1, s)x_1^{-1})(x_2(y_2, s)x_2^{-1}) \cdots (x_m(y_m, s)x_m^{-1})$$

where $x_1, \dots, x_m, y_1, \dots, y_m$ belong to V.
(a) Show that if $s \in G$ is such that Ad($s$) is not the identity mapping of the Lie algebra $\mathfrak{g}$ of G, and if $m \geqq n$, then for each neighborhood V of $e$ the set M($s$, $m$, V) is a neighborhood of $e$. (There exists a vector $\mathbf{a} \in \mathfrak{g}$ such that $\mathbf{b} = \mathrm{Ad}(s) \cdot \mathbf{a} - \mathbf{a} \neq 0$. Show that there exist elements $x_1, \dots, x_m$ in V such that the sequence $(\mathrm{Ad}(x_j) \cdot \mathbf{b})_{1 \leq j \leq m}$ contains a basis of $\mathfrak{g}$. Then consider the mapping

$$(z_1, \dots, z_m, y_1, \dots, y_m) \mapsto (z_1(y_1, s)z_1^{-1}) \cdots (z_m(y_m, s)z_m^{-1})$$

of $G^{2m}$ into G, and its tangent linear mapping at the point $(x_1, \dots, x_m, e, \dots, e)$.)
(b) Let U be a neighborhood of $e$ in G and let $m$ be an integer $\geqq 1$. Show that there exists an element $s \in G$ such that Ad($s$) is not the identity mapping and such that

M($s$, $m$, G) $\subset$ U. (Argue by contradiction, using the compactness of G and Problem 12.)
(c) Let G, G′ be almost simple compact Lie groups and let $\varphi$: G $\rightarrow$ G′ be an (*a priori* not necessarily continuous) isomorphism of abstract groups. Show that $\varphi$ is in fact an isomorphism of Lie groups. (Apply (b) to G′ and (a) to G.)

**14.** Let G be a compact connected Lie group of dimension $n$, and let $(\mathbf{u} \mid \mathbf{v})$ be a scalar product on the Lie algebra $\mathfrak{g}_e$, invariant under the operators Ad($s$) for all $s \in$ G (**20.11.3.1**); also let $\|\mathbf{u}\|^2 = (\mathbf{u} \mid \mathbf{u})$. This scalar product induces canonically a Riemannian metric tensor $\boldsymbol{g}$ on G, invariant under left and right translations (**20.11.8**), and for which the geodesic trajectories are the left-translates of the one-parameter subgroups.
(a) Let $t \mapsto x(t) = \exp(t\mathbf{u})$ be a geodesic passing through $e$, and let $y \in$ G. Put $z(t) = x(t)yx(-t)$. Show that

$$z'(t) = -x(t)y \cdot ((1_{\mathfrak{g}_e} - \mathrm{Ad}(y^{-1})) \cdot \mathbf{u}) \cdot x(-t)$$

(Use (**16.9.9**) and the relations $x'(t) = x(t) \cdot \mathbf{u} = \mathbf{u} \cdot x(t)$ (**19.11.2.2**)). Deduce that

$$\|z'(t)\|_{\boldsymbol{g}} = \|(1_{\mathfrak{g}_e} - \mathrm{Ad}(y^{-1})) \cdot \mathbf{u}\|.$$

(b) By means of the scalar product $(\mathbf{u} \mid \mathbf{v})$, the group Ad(G) may be identified with a subgroup of $\mathbf{O}(n) \subset \mathbf{U}(n)$. Consider on $\mathbf{U}(n)$ the function $s \mapsto \theta(s)$ defined in Section **16.11**, Problem 1. For each $x \in$ G put $\delta(x) = \theta(\mathrm{Ad}(x))$; then we have $0 \leqq \delta(x) \leqq \pi$, and

$$\delta(x^{-1}) = \delta(x), \qquad \delta(yxy^{-1}) = \delta(x), \qquad \delta(xy) \leqq \delta(x) + \delta(y)$$

for all $x$, $y \in$ G, and $\delta(xz) = \delta(x)$ for all $z$ in the center of G.

Let $d(x, y)$ be the Riemannian distance on G defined by the metric tensor $\boldsymbol{g}$ (**20.16.3**). Show that for any two points $x$, $y \in$ G we have

$$d(e, (x, y)) \leqq (2 \sin \tfrac{1}{2}\, \delta(y)) \cdot d(e, x).$$

(Join $e$ to $x$ by a geodesic arc of length $d(e, x)$ (**20.18.5**), and then use (a) above and the definition of $\theta(s)$ in Section **16.11**, Problem 1.)

**15.** Let G be an almost simple connected Lie group, N an *arbitrary* normal subgroup of G.
(a) Consider the Lie subalgebra $\mathfrak{n}_e$ of $\mathfrak{g}_e = \mathrm{Lie}(G)$ associated with N by the procedure of Section **19.11**, Problem 7(b). Show that if N $\neq$ G, we have $\mathfrak{n}_e = \{0\}$.
(b) Show that if N $\neq$ G, then N must be contained in the center C of G (and consequently G/C is a *simple* group). (If $x \in$ N, apply Section **19.11**, Problem 7(c) to the mapping $y \mapsto yxy^{-1}x^{-1}$ of G into N.)

## 7. MAXIMAL TORI IN COMPACT CONNECTED LIE GROUPS

(**21.7.1**) A compact, connected, *commutative* Lie group is necessarily isomorphic to a group $\mathbf{T}^n$ (**19.7.9.2**). For brevity's sake, such a group will be called an *n-dimensional torus*.

In a *compact* Lie group G, a *connected closed* commutative subgroup T is a Lie subgroup of G (**19.10.1**), hence is a *torus*. We say that T is a *maximal torus* in G if there exists no torus in G that properly contains T.

**(21.7.2)** *A connected Lie group* H *immersed in a compact Lie group* G *is a maximal torus of* G *if and only if its Lie algebra* $\mathfrak{h}_e$ *is a maximal commutative subalgebra of the Lie algebra* $\mathfrak{g}_e$ *of* G.

In view of the canonical one-to-one correspondence between Lie subalgebras of $\mathfrak{g}_e$ and connected Lie groups immersed in G **(19.7.4)**, it is enough to show that if $\mathfrak{h}_e$ is a maximal commutative Lie subalgebra of $\mathfrak{g}_e$, then the corresponding subgroup H is necessarily *closed* in G. If this were not the case, its closure $\bar{H} = H'$ in G would be a compact group (hence a Lie subgroup **(19.10.1)**), connected **(3.19.2)** and commutative **(12.8.5)**; consequently its Lie algebra $\mathfrak{h}'_e$ would be commutative and would contain $\mathfrak{h}_e$ properly: contradiction.

**(21.7.2.1)** The condition in **(21.7.2)** may also be put in the following equivalent form: the commutative subalgebra $\mathfrak{h}_e$ is *equal to its centralizer* $\mathfrak{z}(\mathfrak{h}_e)$ in $\mathfrak{g}_e$. For it is clear that $\mathfrak{h}_e$ is maximal if this condition is satisfied; and, conversely, if $\mathfrak{h}_e$ is commutative and $\mathbf{u} \in \mathfrak{z}(\mathfrak{h}_e)$, the vector subspace $\mathfrak{h}_e + \mathbf{R}\mathbf{u}$ of $\mathfrak{g}_e$ is a commutative Lie subalgebra, and therefore if $\mathfrak{h}_e$ is maximal we must have $\mathbf{u} \in \mathfrak{h}_e$, and hence $\mathfrak{z}(\mathfrak{h}_e) = \mathfrak{h}_e$.

**(21.7.3)** *Every connected commutative Lie group* H *immersed in a compact Lie group* G *is contained in a maximal torus of* G.

The Lie algebra $\mathfrak{h}_e$ of H is commutative, hence is contained in a maximal commutative Lie subalgebra of $\mathfrak{g}_e$ (for example, a commutative subalgebra whose dimension is maximal among those which contain $\mathfrak{h}_e$). The result now follows from **(21.7.2)** and **(19.7.4)**.

**(21.7.4)** *Every compact connected Lie group* G *is the union of its maximal tori.*

Since **(21.7.3)** may be applied to the one-parameter subgroups of G, the result to be proved is equivalent to the assertion that the exponential mapping $\exp_G$ is *surjective*. Now, there exists on G a Riemannian structure for which the one-parameter subgroups are the geodesic trajectories passing through *e* **(20.11.8)**. Since G is compact and connected, the proposition is therefore a consequence of the Hopf–Rinow theorem **(20.18.5)**.

**(21.7.5)** The importance of the tori in a compact Lie group is that one knows explicitly all their *linear representations* **(21.3.8)**. By virtue of **(19.7.2)** and **(19.8.7.2)**, the Lie algebra of the commutative real Lie group $(\mathbf{C}^*)^n$ may be canonically identified with the real vector space $\mathbf{C}^n$, and the exponential mapping is

(21.7.5.1) $$(z_1, \dots, z_n) \mapsto (e^{z_1}, \dots, e^{z_n}).$$

The Lie algebra of the subgroup $\mathbf{U}^n$ of $(\mathbf{C}^*)^n$ is therefore the subspace $i\mathbf{R}^n$ of $\mathbf{C}^n$, and the exponential mapping of $i\mathbf{R}^n$ into $\mathbf{U}^n$ is the restriction of **(21.7.5.1)** to $i\mathbf{R}^n$; its *kernel* is therefore the *discrete subgroup* $2\pi i\mathbf{Z}^n$ of $i\mathbf{R}^n$. Every *character* $\chi$ of $\mathbf{U}^n$, being a homomorphism of $\mathbf{U}^n$ into $\mathbf{U}$, has therefore a *derived homomorphism*, which is an $\mathbf{R}$-*linear* mapping $\alpha: i\mathbf{R}^n \to i\mathbf{R}$ such that, by virtue of **(21.5.2.1)**,

(21.7.5.2) $$\chi(e^{i\xi_1}, \dots, e^{i\xi_n}) = e^{\alpha(i\xi_1, \dots, i\xi_n)}$$

for all $(\xi_1, \dots, \xi_n) \in \mathbf{R}^n$. This implies that we must have

$$\alpha(2\pi i m_1, \dots, 2\pi i m_n) \in 2\pi i\mathbf{Z}$$

for all $(m_1, \dots, m_n) \in \mathbf{Z}^n$. Conversely, if this condition is satisfied, the mapping $(i\xi_1, \dots, i\xi_n) \mapsto e^{\alpha(i\xi_1, \dots, i\xi_n)}$ factorizes as

$$(i\xi_1, \dots, i\xi_n) \mapsto (e^{i\xi_1}, \dots, e^{i\xi_n}) \overset{\chi}{\mapsto} e^{\alpha(i\xi_1, \dots, i\xi_n)}$$

where $\chi$ is a character of $\mathbf{U}^n$.

By transport of structure, it therefore follows that if T is an *n-dimensional torus* and t its Lie algebra, the exponential mapping $\exp_{\mathrm{T}}$ is a *homomorphism of Lie groups* from t (regarded as an additive group) to T, the kernel $\Gamma_{\mathrm{T}}$ of which is a *lattice* in t, that is to say, a free $\mathbf{Z}$-module that spans the real vector space t. The *characters* of T are the continuous mappings $\chi$ of T into $\mathbf{U}$ such that

(21.7.5.3) $$\chi(\exp(\mathbf{u})) = e^{2\pi i\varphi(\mathbf{u})}$$

for all $\mathbf{u} \in$ t, where $\varphi \in$ t* is an $\mathbf{R}$-linear form on the vector space t such that $\varphi(\mathbf{u})$ is an *integer* for all $\mathbf{u} \in \Gamma_{\mathrm{T}}$. These linear forms constitute a *lattice* $\Gamma_{\mathrm{T}}^*$ in the real vector space t*, called the *dual* of the lattice $\Gamma_{\mathrm{T}}$ **(22.14.6)**. The elements of the lattice $2\pi i\Gamma_{\mathrm{T}}^*$ in the complexification $\mathrm{t}^*_{(\mathbf{C})}$ of t* are called the *weights* of T; they are therefore $\mathbf{R}$-linear mappings of t into $i\mathbf{R} \subset \mathbf{C}$, namely the *derived homomorphisms* of the characters of T.

If now $V: \mathrm{T} \to \mathbf{GL}(\mathrm{E})$ is a *linear representation* of T on a *complex* vector space E of finite dimension $m$, it leaves invariant a scalar product **(6.2)** on E **(21.4.3)**; and E is the *Hilbert sum*, relative to this scalar product, of subspaces $\mathrm{E}_k$ $(1 \leqq k \leqq m)$ of complex dimension 1, such that for all $x \in \mathrm{E}_k$ we have

(21.7.5.4) $$V(s) \cdot x = \chi_k(s)x$$

where $\chi_k$ is a character of T (21.4.4). Bearing in mind (21.7.5.3) and (21.5.2.1), we see therefore that the derived homomorphism $V_*: \mathfrak{t} \to \mathfrak{gl}(E) = \text{End}(E)$ is such that

(21.7.5.5) $$V_*(\mathbf{u}) \cdot x = \alpha_k(\mathbf{u})x$$

for all $\mathbf{u} \in \mathfrak{t}$ and $x \in E_k$ $(1 \leqq k \leqq m)$, where $\alpha_k$ is a *weight* of T. We remark that the $\alpha_k$ are not *a priori* necessarily distinct, for distinct values of the index $k$.

(21.7.5.6) *There exists* $\mathbf{u}_0 \in \mathfrak{t}$ *such that* $\text{Ker}(V_*(\mathbf{u}_0))$ *is the intersection of all the kernels* $\text{Ker}(V_*(\mathbf{u}))$ *in* $\mathfrak{t}$, *as* $\mathbf{u}$ *runs through* $\mathfrak{t}$.

Since $\mathfrak{t}$ is not the union of any finite number of hyperplanes (12.16.1), there exists an element $\mathbf{u}_0 \in \mathfrak{t}$ such that $\alpha_k(\mathbf{u}_0) \neq 0$ for *all* the indices $k$ such that the linear form $\alpha_k$ is not identically zero. This clearly proves the proposition (A.4.17).

(21.7.6) The study of the structure of a compact connected Lie group G and of its linear representations rests entirely on the consideration of the *restrictions to the tori in* G (and especially to the *maximal* tori) of the *linear representations* of G (on complex vector spaces). Let $\mathfrak{g}$ be the Lie algebra of G. Up to the end of Section 21.12, we shall study from this point of view the extension of the *adjoint representation* of G to the complex vector space $\mathfrak{g}_{(\mathbf{C})} = \mathfrak{g} \otimes_{\mathbf{R}} \mathbf{C}$, that is to say (21.5.1) the homomorphism

(21.7.6.1) $$s \mapsto \text{Ad}(s) \otimes 1_{\mathbf{C}}$$

of G into $\mathbf{GL}(\mathfrak{g}_{(\mathbf{C})})$. If we consider the restriction of this homomorphism to a *torus* T in G, its derived homomorphism is the restriction, to the Lie algebra $\mathfrak{t}$ of T, of the homomorphism

(21.7.6.2) $$\mathbf{u} \mapsto \text{ad}(\mathbf{u}) \otimes 1_{\mathbf{C}}$$

of $\mathfrak{g}$ into $\mathfrak{gl}(\mathfrak{g}_{(\mathbf{C})})$ (19.11.2). Applying (21.7.5.6) to this restriction, we obtain:

(21.7.6.3) *If* $\mathfrak{t}$ *is the Lie algebra of a torus* T *in the compact Lie group* G, *there exists a vector* $\mathbf{u}_0 \in \mathfrak{t}$ *such that* $\mathfrak{z}(\mathfrak{t}) = \mathfrak{z}(\mathbf{u}_0)$ *in* $\mathfrak{g}$.

We shall use this result to prove the fundamental theorem on the *conjugacy of maximal tori*:

(21.7.7) *Let* G *be a compact connected Lie group,* T *a maximal torus in* G, *and* A *a torus in* G. *Then there exists an element* $s \in G$ *such that* $sAs^{-1} \subset T$ (*which implies that* $sAs^{-1} = T$ *if* A *is a* maximal *torus*).

Let $\mathfrak{g}$, $\mathfrak{t}$, $\mathfrak{a}$ be the Lie algebras of G, T, A, respectively. Since all three groups are connected, it follows from (19.7.4) and (19.2.1.1) that it is enough to prove the following proposition:

(21.7.7.1) *There exists* $s \in G$ *such that* $\mathrm{Ad}(s) \cdot \mathfrak{a} \subset \mathfrak{t}$.

By virtue of (21.7.6.3), there exists a vector $\mathbf{u} \in \mathfrak{a}$ and a vector $\mathbf{v} \in \mathfrak{t}$ such that $\mathfrak{z}(\mathfrak{a}) = \mathfrak{z}(\mathbf{u})$ and $\mathfrak{z}(\mathfrak{t}) = \mathfrak{z}(\mathbf{v})$. Consider a scalar product $(\mathbf{x} \mid \mathbf{y})$ on $\mathfrak{g}$ that is *invariant* under the action $(s, \mathbf{x}) \mapsto \mathrm{Ad}(s) \cdot \mathbf{x}$ of G on $\mathfrak{g}$ (20.11.3.2); for this scalar product and the corresponding norm $\|\mathbf{x}\| = (\mathbf{x} \mid \mathbf{x})^{1/2}$, the function $s \mapsto \|\mathrm{Ad}(s) \cdot \mathbf{u} - \mathbf{v}\|^2$ is continuous on the compact group G, hence attains its *minimum* at some $s_0 \in G$ (3.17.10). We shall show that $\mathrm{Ad}(s_0) \cdot \mathbf{u} \in \mathfrak{t}$; this will imply, by virtue of the commutativity of $\mathfrak{t}$ and by transport of structure, that

$$\mathfrak{t} \subset \mathfrak{z}(\mathrm{Ad}(s_0) \cdot \mathbf{u}) = \mathrm{Ad}(s_0) \cdot \mathfrak{z}(\mathbf{u}) = \mathrm{Ad}(s_0) \cdot \mathfrak{z}(\mathfrak{a}) = \mathfrak{z}(\mathrm{Ad}(s_0) \cdot \mathfrak{a})$$

and consequently that the vector subspace $\mathfrak{t} + \mathrm{Ad}(s_0) \cdot \mathfrak{a}$ of $\mathfrak{g}$ is a *commutative* Lie subalgebra. But since $\mathfrak{t}$ is maximal among such subalgebras, it follows that $\mathrm{Ad}(s_0) \cdot \mathfrak{a} \subset \mathfrak{t}$, which will establish (21.7.7.1).

By replacing $\mathbf{u}$ by $\mathrm{Ad}(s_0) \cdot \mathbf{u}$, and $\mathfrak{a}$ by $\mathrm{Ad}(s_0) \cdot \mathfrak{a}$, we may assume (because of the invariance of the scalar product) that $s_0 = e$. Let us express that, for each $\mathbf{x} \in \mathfrak{g}$, the derivative of the function of a real variable

$$t \mapsto \|\mathrm{Ad}(\exp(t\mathbf{x})) \cdot \mathbf{u} - \mathbf{v}\|^2$$

vanishes at $t = 0$. Since the derivative of $t \mapsto \mathrm{Ad}(\exp(t\mathbf{x})) \cdot \mathbf{u}$ at $t = 0$ is $[\mathbf{x}, \mathbf{u}]$ (19.11.2), we obtain by use of (21.5.4.1)

$$0 = 2([\mathbf{x}, \mathbf{u}] \mid \mathbf{u} - \mathbf{v}) = 2(\mathbf{x} \mid [\mathbf{u}, \mathbf{u} - \mathbf{v}])$$

for all $\mathbf{x} \in \mathfrak{g}$; in other words, $[\mathbf{u}, \mathbf{v}] = 0$ and therefore $\mathbf{u} \in \mathfrak{z}(\mathbf{v}) = \mathfrak{z}(\mathfrak{t})$.

Q.E.D.

(21.7.7.2) In particular, any two maximal tori of G have the *same dimension*; this dimension is called the *rank* of the compact connected Lie group G or of its Lie algebra $\mathfrak{g}$. If G is a compact semisimple Lie group, $\mathfrak{g} = \bigoplus_j \mathfrak{g}_j$ the decomposition of its Lie algebra as a direct sum of *simple* algebras (21.6.4), and $\mathfrak{t}_j$ a maximal commutative subalgebra of $\mathfrak{g}_j$, then it is immediately verified that $\mathfrak{t} = \bigoplus_j \mathfrak{t}_j$ is equal to its centralizer in $\mathfrak{g}$, so that the rank of $\mathfrak{g}$ is the *sum* of the ranks of the $\mathfrak{g}_j$.

**(21.7.8)** *Let* G *be a compact connected Lie group and* T *a maximal torus in* G. *For each* $x \in G$, *there exists* $s \in G$ *such that* $sxs^{-1} \in T$.

For $x$ lies in some torus in G **(21.7.4)**.

**(21.7.9)** *Let* G *be a compact connected Lie group. For each torus* A *in* G, *the centralizer* $\mathscr{Z}(A)$ *of* A *in* G *is connected. Moreover, for each* $s \in \mathscr{Z}(A)$, $A \cup \{s\}$ *is contained in a maximal torus of* G.

Let $s \in \mathscr{Z}(A)$; the centralizer $\mathscr{Z}(s)$ of $s$ is a Lie subgroup of G, and its identity component H is a compact connected subgroup of G. There exists a maximal torus T of G containing $s$ **(21.7.8)**, and by definition we have $T \subset H$ and $A \subset H$; moreover, T is clearly a maximal torus *of* H. By **(21.7.7)**, therefore, there exists an element $h \in H$ such that $A \subset T' = hTh^{-1}$, and therefore T′ is contained in the identity component of $\mathscr{Z}(A)$. But we have $hsh^{-1} = s$ by definition of H, hence $s$ lies in the identity component of $\mathscr{Z}(A)$; and T′ is a maximal torus of G containing A and $s$. Q.E.D.

In particular:

**(21.7.10)** *In a compact connected Lie group* G, *every maximal torus is equal to its centralizer, and is therefore maximal in the set of* all *commutative subgroups of* G.

It should be noted, however, that the set of commutative subgroups of G in general contains maximal elements that are *not* tori (Problem 1).

**(21.7.11)** *The center of a compact connected Lie group* G *is the intersection of the maximal tori of* G.

Since G is the union of its maximal tori **(21.7.4)**, their intersection is contained in the center C of G. Conversely, if T is any maximal torus of G, then CT is a commutative subgroup of G containing T, hence is equal to T by **(21.7.10)**, so that $C \subset T$.

**(21.7.12)** *For each element* $s$ *of a compact connected Lie group* G, *the identity component of* $\mathscr{Z}(s)$ *is the union of the maximal tori of* G *that contain* $s$.

Clearly this identity component $(\mathscr{Z}(s))_0$ contains every maximal torus that contains $s$. Conversely, if $x \in (\mathscr{Z}(s))_0$, then $x$ belongs to some maximal torus A of the compact connected Lie group $(\mathscr{Z}(s))_0$ **(21.7.4)**; but since A is a torus in G, there exists a maximal torus T of G containing both A and $s$, because by definition $s \in \mathscr{Z}(A)$ **(21.7.9)**.

We remark that the centralizer of an arbitrary element $s \in G$ is not necessarily connected (Problem 1).

**(21.7.13)** An element $s$ in a compact connected Lie group G is said to be *regular* if it belongs to *only one* maximal torus of G, and *singular* if it belongs to more than one. Likewise, an element $\mathbf{u}$ of the Lie algebra $\mathfrak{g}_e$ of G is said to be *regular* if it belongs to only one maximal commutative Lie subalgebra, and *singular* if it belongs to more than one (see **(21.8.4.2)**).

It follows from **(21.7.12)** that an element $s \in G$ is regular if and only if *the identity component of* $\mathscr{Z}(s)$ *is a maximal torus.* An equivalent condition is that *the dimension of* $\mathscr{Z}(s)$ *is equal to the rank of* G; for $\mathscr{Z}(s)$ contains at least one maximal torus T, and if the dimension of $\mathscr{Z}(s)$ is equal to that of T, then T must be open in $\mathscr{Z}(s)$ **(16.8.3.3)**, hence is the identity component of $\mathscr{Z}(s)$. Likewise, in order that $\mathbf{u} \in \mathfrak{g}_e$ should be regular, it is necessary and sufficient that *the centralizer* $\mathfrak{z}(\mathbf{u})$ *should be a maximal commutative subalgebra.* For this condition is clearly sufficient; and conversely if $\mathbf{u}$ is regular and $\mathfrak{t}$ is a maximal commutative subalgebra that contains $\mathbf{u}$, then $\mathfrak{z}(\mathbf{u})$ cannot contain any element $\mathbf{v} \notin \mathfrak{t}$, otherwise the commutative subalgebra generated by $\mathbf{u}$ and $\mathbf{v}$ would be contained in a maximal commutative subalgebra distinct from $\mathfrak{t}$.

**(21.7.14)** *Let* G *be a compact connected Lie group. If* $\mathbf{u}$ *is a regular element of the Lie algebra of* G, *then the centralizer* $\mathscr{Z}(\mathbf{R}\mathbf{u})$ *in* G **(19.11.3)** *is a maximal torus of* G.

For if A is the torus in G that is the closure of the one-parameter subgroup corresponding to $\mathbf{R}\mathbf{u}$, we have $\mathscr{Z}(A) = \mathscr{Z}(\mathbf{R}\mathbf{u})$ (**(19.11.6)** and **(12.8.6)**). Since $\mathscr{Z}(A)$ is connected **(21.7.9)**, its Lie algebra is the centralizer of $\mathbf{R}\mathbf{u}$ **(19.11.6)**, and the proposition follows.

**(21.7.15)** *Let* G *be a compact connected Lie group. Every maximal torus* T *of* G *is open (hence of finite index) in its normalizer* $\mathscr{N}(T)$ *in* G.

Let H be the identity component of the Lie group $\mathscr{N}(T)$, and for each $s \in H$ let $\sigma_s$ denote the automorphism $x \mapsto sxs^{-1}$ of T. The argument of **(19.14.4)** shows that the mapping $s \mapsto \sigma_s$ is a Lie group homomorphism of H into the group Aut(T). But Aut(T) is discrete **(19.13.6)** and H is connected, hence $\sigma_s$ is the identity mapping for each $s \in H$ **(3.19.7)**. In other words, H is contained in the centralizer of T, hence H = T **(21.7.10)**.

**(21.7.16)** With the notation of **(21.7.15)**, the finite quotient group $W(G, T) = \mathscr{N}(T)/T$ is called the *Weyl group* of the compact connected Lie group G, relative to the maximal torus T. For each $s \in G$, we have $\mathcal{N}(sTs^{-1}) = s\mathcal{N}(T)s^{-1}$ by transport of structure, hence $W(G, sTs^{-1})$ is isomorphic to W(G, T). By virtue of the conjugacy theorem (21.7.7), the Weyl groups corresponding to the various maximal tori of G are all isomorphic; any one of these Weyl groups W(G, T) is called the *Weyl group of* G, and denoted by W(G), or simply W. The group W(G, T) acts differentiably on T in a canonical way: every element $w \in W(G, T)$ is the coset of an element $s \in \mathcal{N}(T)$, and we define $w \cdot t = sts^{-1}$, which is independent of the choice of representative $s \in w$, because T is commutative. Since $\mathcal{Z}(T) = T$ (21.7.10), it follows that W(G, T) acts *faithfully* on T.

(21.7.17) *Let* G *be a compact connected Lie group and* T *a maximal torus in* G. *If two elements* $t_1$, $t_2$ *of* T *are conjugate in* G, *there exists an element* $w \in W(G, T)$ *such that* $t_2 = w \cdot t_1$.

We have $t_2 = st_1 s^{-1}$ for some $s \in G$, and we have to show that $t_2 = nt_1 n^{-1}$ for some $n \in \mathcal{N}(T)$. The torus T is contained in $\mathcal{Z}(t_1)$ and $\mathcal{Z}(t_2)$, hence $\mathcal{Z}(t_2)$ also contains the torus $sTs^{-1}$. Hence, if H is the identity component of $\mathcal{Z}(t_2)$, both T and $sTs^{-1}$ are maximal tori in the compact connected Lie group H, and therefore there exists an element $h \in H$ such that $sTs^{-1} = hTh^{-1}$ (21.7.7). It follows that $h^{-1}sT(h^{-1}s)^{-1} = T$, so that $n = h^{-1}s \in \mathcal{N}(T)$; and since $h \in \mathcal{Z}(t_2)$, we have $h^{-1}(st_1 s^{-1})h = t_2$, that is to say, $nt_1 n^{-1} = t_2$.

(21.7.18) *Let* G *be a compact connected Lie group and* C′ *a closed subgroup of the center* C *of* G. *Then, if* T *is a maximal torus of* G, *the quotient group* T/C′ *is a maximal torus of* G/C′; *also* $\mathcal{N}(T/C') = \mathcal{N}(T)/C'$, *and the Weyl group* W(G/C′, T/C′) *is isomorphic to* W(G, T).

Recall that C is contained in T (21.7.11). If $\pi: G \to G/C'$ is the canonical homomorphism, and if $\pi(s)$ belongs to the normalizer of $\pi(T) = T/C'$, we have $\pi(sts^{-1}) \in \pi(T)$ for all $t \in T$, and therefore $sts^{-1} \in C'T = T$, whence $s \in \mathcal{N}(T)$ and thus $\mathcal{N}(T/C') = \mathcal{N}(T)/C'$. It is clear that $\pi(T)$ is a compact connected commutative subgroup of G/C′, hence is a torus. Let U be a compact connected commutative subgroup of G/C′ containing $\pi(T) = T/C'$; then $U \subset \mathcal{N}(T/C') = \mathcal{N}(T)/C'$, hence $\pi^{-1}(U)$ is a compact subgroup of G such that $T \subset \pi^{-1}(U) \subset \mathcal{N}(T)$. By (21.7.15), T is open in $\pi^{-1}(U)$, hence T/C′ is the identity component of $U = \pi^{-1}(U)/C'$. Since U is connected, we have $T/C' = U$, and hence T/C′ is a maximal torus in G/C′. Finally,

$$W(G/C', T/C') = \mathcal{N}(T/C')/(T/C') = (\mathcal{N}(T)/C')/(T/C') = \mathcal{N}(T)/T$$
$$= W(G, T)$$

up to canonical isomorphism ((12.12.2) and (19.10.2)).

## PROBLEMS

1. Let G be the rotation group $\mathbf{SO}(4k, \mathbf{R})$, where $k \geqq 1$. Let V be a vector subspace of dimension $2k$ in $\mathbf{R}^{4k}$, $V^{\perp}$ its orthogonal supplement, and let $s$ be the automorphism of $\mathbf{R}^{4k}$ such that $s(x) = x$ for $x \in V$ and $s(x) = -x$ for $x \in V^{\perp}$; then $s \in G$. Let H (resp. H′) be a hyperplane in V (resp. $V^{\perp}$), and let $t$ be the automorphism of $\mathbf{R}^{4k}$ whose restriction to V (resp. $V^{\perp}$) is the orthogonal reflection with respect to H (resp. H′). We have $t \in G$ and $ts = st$. Let S be the centralizer of $s$ in G, and $S_0$ the identity component of S. Show that $\det(u \mid V) > 0$ for all $u \in S_0$ and hence that $t \notin S_0$. Deduce that S is not connected, and that the commutative subgroup generated by $s$ and $t$ is not contained in any maximal torus of G.

2. Let G be a Lie group, $\mathfrak{g}$ its Lie algebra, and $\Phi$ a G-invariant symmetric **R**-bilinear form on $\mathfrak{g} \times \mathfrak{g}$ (**21.5.4**). Let K be a compact subgroup of G. Let $\mathbf{x}, \mathbf{y}$ be elements of $\mathfrak{g}$; show that there exists an element $t \in K$ such that $\Phi(\mathbf{u}, [\mathrm{Ad}(t) \cdot \mathbf{x}, \mathbf{y}]) = 0$ for all $\mathbf{u} \in \mathfrak{k} = \mathrm{Lie}(K)$. (Argue as in (**21.7.7.1**).)

3. Let G be a compact Lie group, $\mathfrak{g}$ its Lie algebra, and $\mathfrak{m}$ a vector subspace of $\mathfrak{g}$ such that $[[\mathbf{u}, \mathbf{v}], \mathbf{w}] \in \mathfrak{m}$ for all $\mathbf{u}, \mathbf{v}, \mathbf{w}$ in $\mathfrak{m}$. Let G′ be the identity component of the normalizer $\mathcal{N}(\mathfrak{m})$ of $\mathfrak{m}$ in G (**19.11.3**), and let $\mathfrak{t}$ be a maximal element of the set of commutative subalgebras of $\mathfrak{g}$ contained in $\mathfrak{m}$. Show that for each commutative subalgebra $\mathfrak{a}$ of $\mathfrak{g}$ contained in $\mathfrak{m}$, there exists an element $s \in G'$ such that $\mathrm{Ad}(s) \cdot \mathfrak{a} \subset \mathfrak{t}$. (Argue as in (**21.7.7**).)

4. Let G be a Lie group, $s$ an element of G, and $\mathfrak{g}$ the Lie algebra of G. Let $\mathfrak{n}$ be the union of the kernels of the endomorphisms $(\mathrm{Ad}(s) - 1_{\mathfrak{g}})^k$ of the vector space $\mathfrak{g}$, for all integers $k \geqq 1$. Show that $\mathfrak{n}$ is a Lie subalgebra of $\mathfrak{g}$. Let N be the connected Lie group immersed in G corresponding to the Lie algebra $\mathfrak{n}$. Show that the mapping $(s, t) \mapsto sxts^{-1}$ of $G \times N$ into G is a submersion (**16.7.1**) at the point $(e, e)$. (The vector space $\mathfrak{g}$ is the direct sum of $\mathfrak{n}$ and a subspace $\mathfrak{m}$ such that the restriction of $\mathrm{Ad}(s) - 1_{\mathfrak{g}}$ to $\mathfrak{m}$ is an automorphism of this subspace (**11.4.1**). Calculate the tangent linear mapping at the point $\xi = 0$ of the mapping

$$\xi \mapsto \exp(\xi\mathbf{u})x \exp(\xi\mathbf{v}) \exp(-\xi\mathbf{u})$$

of **R** into G, for $\mathbf{u} \in \mathfrak{n}$ and $\mathbf{v} \in \mathfrak{m}$.)

5. Give a proof of (**21.7.4**) without using the Riemannian structure of G, but using instead (**21.7.7**) and proceeding by induction on $n = \dim(G)$. (Given a maximal torus T of G, show that the union E of the conjugates of G is open in G. For this it is enough to show that E is a neighborhood of any point $s \in T$; distinguish two cases, according as $s$ belongs or does not belong to the center of G. In the second case consider the identity component of the centralizer of $s$ in G, and use the inductive hypothesis and Problem 4.)

6. Let G be a compact connected Lie group, T and T′ two maximal tori in G. Let A (resp. A′) be a subset of T (resp. T′) and let $\sigma$ be an automorphism of G such that $\sigma(A) = A'$. Show that there exists $s \in G$ such that $s\sigma(a)s^{-1} = \sigma(a)$ for all $a \in A$ and such that $s\sigma(T)s^{-1} = T'$. (Consider the maximal torus $T'' = \sigma(T)$ and the identity component of the centralizer of A′ in G.)

   In particular, there exists an inner automorphism of G that transforms T into T′ and fixes the elements of $T \cap T'$.

7. Let G be a compact connected Lie group and N a closed normal subgroup of G. Show that if there exists a maximal torus T of G such that $T \cap N = \{e\}$, then $N = \{e\}$. (Consider a maximal torus S of N, and use **(16.30.2.2)**.)

8. Let G be a compact connected Lie group.
(a) If H is a connected closed subgroup of G, every maximal torus of H is of the form $T \cap H$, where T is a maximal torus of G.
The Weyl group $W(H, T \cap H)$ is isomorphic to the quotient $F/F'$, where F is the subgroup of W(G, T) that leaves $T \cap H$ stable as a whole, and $F'$ is the normal subgroup of F that fixes $T \cap H$ elementwise.
(b) If N is a closed normal subgroup of G, every maximal torus of G/N is of the form TN/N, where T is a maximal torus of G (use Section **21.6**, Problem 4). If in addition N is discrete, every maximal torus of G is the inverse image of a maximal torus of G/N under the canonical homomorphism (use **(16.30.2.2)**).

9. Let G be a compact connected Lie group, T a maximal torus in G, and N its normalizer in G. Then G (resp. N) acts differentiably on G (resp. T) by inner automorphisms; let E and F be the respective orbit spaces **(12.10)** and $\pi: G \to E$, $\pi': T \to F$ the canonical mappings. If $j: T \to G$ is the canonical injection, there exists one and only one continuous mapping $f: F \to E$ such that $\pi \circ j = f \circ \pi'$. Show that $f$ is a homeomorphism of F onto E.

10. Let G be a compact connected Lie group and T a maximal torus of G. Show that the manifold G/T is simply connected, and that if G′ is a compact connected Lie group, locally isomorphic to G, and if T′ is a maximal torus of G′, then G/T and G′/T′ are diffeomorphic. (Reduce to the case where G and G′ are semisimple, and use Section **16.30**, Problem 11.)

## 8. ROOTS AND ALMOST SIMPLE SUBGROUPS OF RANK ONE

Throughout this section, G denotes a compact connected Lie group, T a maximal torus of G, $\mathfrak{g}$ (resp. $\mathfrak{t}$) the Lie algebra of G (resp. T), and $W = W(G, T)$ the Weyl group of G relative to T. Since W may be identified with a group of automorphisms of T, it acts linearly and faithfully on $\mathfrak{t}$ via the derived automorphisms. To be precise, if $s$ is a representative of $w \in W$ in $\mathscr{N}(T)$, we have $w \cdot \mathbf{u} = \mathrm{Ad}(s) \cdot \mathbf{u}$ for all $\mathbf{u} \in \mathfrak{t}$. Since $w$ is now an automorphism of the real vector space $\mathfrak{t}$, its contragredient ${}^t w^{-1}$ is an automorphism of the dual $\mathfrak{t}^*$ of $\mathfrak{t}$; we shall write $w \cdot \lambda$ in place of ${}^t w^{-1}(\lambda)$ for $\lambda \in \mathfrak{t}^*$, so that we have $\langle w \cdot \mathbf{u}, w \cdot \lambda \rangle = \langle \mathbf{u}, \lambda \rangle$ for all $\mathbf{u} \in \mathfrak{t}$ and $\lambda \in \mathfrak{t}^*$, and W acts linearly on the vector space $\mathfrak{t}^*$ by the rule $(w, \lambda) \mapsto w \cdot \lambda$.

**(21.8.1)** Consider again the linear representation $s \mapsto \mathrm{Ad}(s) \otimes 1_{\mathbf{C}}$ of T on the complex vector space $\mathfrak{g}_{(\mathbf{C})} = \mathfrak{g} \otimes_{\mathbf{R}} \mathbf{C} = \mathfrak{g} \oplus i\mathfrak{g}$, and the derived homomorphism $\mathbf{u} \mapsto \mathrm{ad}(\mathbf{u}) \otimes 1_{\mathbf{C}}$ of $\mathfrak{t}$ into $\mathfrak{gl}(\mathfrak{g}_{(\mathbf{C})})$. It follows from **(21.7.5)** that there is a *finite* set $\mathbf{S} \subset 2\pi i \Gamma_T^*$ of nonzero *weights* of T (also denoted by $\mathbf{S}(G, T)$ or $\mathbf{S}(G)$) such that the complex vector space $\mathfrak{g}_{(\mathbf{C})}$ is the *direct sum* of a vector subspace $\mathfrak{g}_0 \supset \mathfrak{t}_{(\mathbf{C})}$ and vector subspaces $\mathfrak{g}_\alpha$, each of the latter being nonzero and corresponding to a weight $\alpha \in \mathbf{S}$. Furthermore, for each $\mathbf{u} \in \mathfrak{t}$ we have

(21.8.1.1) $$[\mathbf{u}, \mathbf{x}] = 0$$

for all $\mathbf{x} \in \mathfrak{g}_0$, and

(21.8.1.2) $$[\mathbf{u}, \mathbf{x}] = \alpha(\mathbf{u})\mathbf{x}$$

for all $\alpha \in \mathbf{S}$ and all $\mathbf{x} \in \mathfrak{g}_\alpha$.

In fact, $\mathfrak{g}_0 = \mathfrak{t} \oplus i\mathfrak{t} = \mathfrak{t}_{(\mathbf{C})}$, the *complexification* of $\mathfrak{t}$, because if $\mathbf{x} = \mathbf{y} + i\mathbf{z}$ with $\mathbf{y}, \mathbf{z} \in \mathfrak{g}$, the relation $[\mathbf{u}, \mathbf{x}] = 0$ for $\mathbf{u} \in \mathfrak{t}$ implies that $[\mathbf{u}, \mathbf{y}] = 0$ and $[\mathbf{u}, \mathbf{z}] = 0$ for all $\mathbf{u} \in \mathfrak{t}$, and therefore $\mathbf{y} \in \mathfrak{t}$ and $\mathbf{z} \in \mathfrak{t}$, since $\mathfrak{t}$ is its own centralizer in $\mathfrak{g}$ (21.7.2.1).

The weights $\alpha \in \mathbf{S}$ are called the *roots* of G relative to T, or of $\mathfrak{g}$ relative to $\mathfrak{t}$.

(21.8.1.3) The roots $\alpha \in \mathbf{S}$ are the *only* **R**-linear mappings $\lambda$ of $\mathfrak{t}$ into **C** that are not identically zero and are such that for some $\mathbf{x}_0 \neq 0$ in $\mathfrak{g}_{(\mathbf{C})}$, we have $[\mathbf{u}, \mathbf{x}_0] = \lambda(\mathbf{u})\mathbf{x}_0$ for *all* $\mathbf{u} \in \mathfrak{t}$ (A.24.4).

(21.8.2) Let $c$ be the semilinear bijection of the complex vector space $\mathfrak{g}_{(\mathbf{C})}$ onto itself defined by $c(\mathbf{y} + i\mathbf{z}) = \mathbf{y} - i\mathbf{z}$ for $\mathbf{y}, \mathbf{z} \in \mathfrak{g}$. The real subspace $\mathfrak{g}$ (resp. $i\mathfrak{g}$) is therefore the set of all $\mathbf{x} \in \mathfrak{g}_{(\mathbf{C})}$ such that $c(\mathbf{x}) = \mathbf{x}$ (resp. $c(\mathbf{x}) = -\mathbf{x}$). It is clear that $[\mathbf{u}, c(\mathbf{x})] = c([\mathbf{u}, \mathbf{x}])$ for all $\mathbf{x} \in \mathfrak{g}_{(\mathbf{C})}$ and all $\mathbf{u} \in \mathfrak{t}$; hence it follows from (21.8.1.2) that, for all $\mathbf{x} \in \mathfrak{g}_\alpha$ and all $\mathbf{u} \in \mathfrak{t}$, we have $[\mathbf{u}, c(\mathbf{x})] = \overline{\alpha(\mathbf{u})}c(\mathbf{x}) = -\alpha(\mathbf{u})c(\mathbf{x})$, since $\alpha(\mathbf{u})$ is pure imaginary. Consequently, *if $\alpha$ is a root, so also is $-\alpha$, and we have $\mathfrak{g}_{-\alpha} = c(\mathfrak{g}_\alpha)$.*

(21.8.3) Consider a root $\alpha = 2\pi i\varphi$, where $\varphi$ is an **R**-linear form on $\mathfrak{t}$. For each $\mathbf{x}_\alpha \in \mathfrak{g}_\alpha$, the two elements

(21.8.3.1) $$\mathbf{y}_\alpha = \mathbf{x}_\alpha + c(\mathbf{x}_\alpha), \qquad \mathbf{z}_\alpha = i(\mathbf{x}_\alpha - c(\mathbf{x}_\alpha))$$

belong to $\mathfrak{g}$, and as $\mathbf{x}_\alpha$ runs through a **C**-basis of $\mathfrak{g}_\alpha$, the $\mathbf{y}_\alpha$ and $\mathbf{z}_\alpha$ form an **R**-basis of $(\mathfrak{g}_\alpha \oplus \mathfrak{g}_{-\alpha}) \cap \mathfrak{g}$. For each $\mathbf{u} \in \mathfrak{t}$, we have

(21.8.3.2) $$[\mathbf{u}, \mathbf{y}_\alpha] = 2\pi\varphi(\mathbf{u})\mathbf{z}_\alpha, \qquad [\mathbf{u}, \mathbf{z}_\alpha] = -2\pi\varphi(\mathbf{u})\mathbf{y}_\alpha.$$

Since the *center* $\mathfrak{c}$ of $\mathfrak{g}$ is contained in $\mathfrak{t}$, it follows from these formulas that $\mathfrak{c}$ *is the set of vectors* $\mathbf{u} \in \mathfrak{t}$ *such that* $\alpha(\mathbf{u}) = 0$ *for all* $\alpha \in \mathbf{S}$.

Moreover, since $(\mathrm{Ad}(\exp(\mathbf{u})) \otimes 1_{\mathbf{C}}) \cdot \mathbf{x}_\alpha = e^{2\pi i \varphi(\mathbf{u})}\mathbf{x}_\alpha$ (21.7.5.1), we deduce from (21.8.3.1) that

(21.8.3.3)

$$\mathrm{Ad}(\exp(\mathbf{u})) \cdot \mathbf{y}_\alpha = \cos(2\pi\varphi(\mathbf{u}))\mathbf{y}_\alpha + \sin(2\pi\varphi(\mathbf{u}))\mathbf{z}_\alpha ,$$
$$\mathrm{Ad}(\exp(\mathbf{u})) \cdot \mathbf{z}_\alpha = -\sin(2\pi\varphi(\mathbf{u}))\mathbf{y}_\alpha + \cos(2\pi\varphi(\mathbf{u}))\mathbf{z}_\alpha .$$

From these formulas we obtain a characterization of the *regular* elements of G (resp. $\mathfrak{g}$) (21.7.13) that are *contained in* T (resp. $\mathfrak{t}$):

(21.8.4) *In order that an element* $\mathbf{u} \in \mathfrak{t}$ *be regular, it is necessary and sufficient that it belong to none of the hyperplanes* $\mathfrak{u}_\alpha = \mathfrak{u}_{-\alpha} = \alpha^{-1}(0)$ *in* $\mathfrak{t}$, *where* $\alpha \in \mathbf{S}$.

*In order that an element* $s \in \mathrm{T}$ *be regular, it is necessary and sufficient that it belong to none of the subgroups* $\mathrm{U}_\alpha = \mathrm{U}_{-\alpha} = \chi_\alpha^{-1}(1)$ *in* T, *where* $\chi_\alpha$ *is the character of* T *defined by* $\chi_\alpha(\exp(\mathbf{u})) = e^{\alpha(\mathbf{u})}$ *for* $\mathbf{u} \in \mathfrak{t}$ *and* $\alpha \in \mathbf{S}$.

If $\mathbf{u}$ belongs to none of the $\mathfrak{u}_\alpha$, it follows from (21.8.3.2) that the image of $\mathfrak{g}$ under $\mathrm{ad}(\mathbf{u})$ is the sum of the subspaces $\mathfrak{g} \cap (\mathfrak{g}_\alpha \oplus \mathfrak{g}_{-\alpha})$, where $\alpha \in \mathbf{S}$; hence its kernel has the same dimension as $\mathfrak{t}$ (A.4.17), and since this kernel contains $\mathfrak{t}$ it coincides with $\mathfrak{t}$, and therefore $\mathbf{u}$ is regular (21.7.13). Conversely, if $\mathbf{u} \in \mathfrak{u}_\alpha$, the kernel of $\mathrm{ad}(\mathbf{u})$ contains $\mathfrak{t} \oplus ((\mathfrak{g}_\alpha \oplus \mathfrak{g}_{-\alpha}) \cap \mathfrak{g})$, hence $\mathbf{u}$ is singular (21.7.13).

Likewise, if $s$ lies in none of the subgroups $\mathrm{U}_\alpha$, then the set of vectors in $\mathfrak{g}_{(\mathbf{C})}$ that are fixed by $\mathrm{Ad}(s) \otimes 1_{\mathbf{C}}$ is precisely $\mathfrak{t}_{(\mathbf{C})}$, because in each of the $\mathfrak{g}_\alpha$ the restriction of $\mathrm{Ad}(s) \otimes 1_{\mathbf{C}}$ is multiplication by $\chi_\alpha(s)$. Since $\mathrm{Ad}(s)$ fixes all the elements of the Lie algebra of $\mathscr{Z}(s)$, this Lie algebra is equal to $\mathfrak{t}$, and therefore $s$ is regular (21.7.13). Conversely, if $s \in \mathrm{U}_\alpha$, then $\mathrm{Ad}(s)$ fixes the vectors belonging to $(\mathfrak{g}_\alpha \oplus \mathfrak{g}_{-\alpha}) \cap \mathfrak{g}$, and therefore (19.11.2.3) the Lie algebra of $\mathscr{Z}(s)$ has dimension strictly larger than the rank of G, and $s$ is singular.

More precisely, this proof shows that the Lie algebra of $\mathscr{Z}(s)$ is the *direct sum* of $\mathfrak{t}$ and the subspaces $(\mathfrak{g}_\alpha \oplus \mathfrak{g}_{-\alpha}) \cap \mathfrak{g}$ for all roots $\alpha$ such that $s \in \mathrm{U}_\alpha$. If the number of these roots is $2k$ (it is an even number because $\mathrm{U}_{-\alpha} = \mathrm{U}_\alpha$), we have therefore

(21.8.4.1)
$$\dim(\mathscr{Z}(s)) \geqq l + 2k$$

where $l$ is the *rank* of G. (We shall see later (21.10.3) that in fact the two sides of (21.8.4.1) are equal.)

For each $\alpha \in \mathbf{S}$, the group $U_\alpha$ is a *compact subgroup* of T, of dimension $l-1$, and $\mathfrak{u}_\alpha$ is its Lie algebra. It should be remarked that $U_\alpha$ need not be connected (Problem 2).

(21.8.4.2) *In order that* $\mathbf{u} \in \mathfrak{t}$ *should be such that* $\exp(\mathbf{u})$ *is a singular element of* G, *it is necessary and sufficient that* $\alpha(\mathbf{u})$ *should be an integer multiple of* $2\pi i$ *for some root* $\alpha \in \mathbf{S}$. *In order that* $\mathbf{u}$ *should be such that* $\exp(\mathbf{u})$ *lies in the center of* G, *it is necessary and sufficient that* $\alpha(\mathbf{u})$ *should be an integer multiple of* $2\pi i$ *for all roots* $\alpha \in \mathbf{S}$.

This follows from (21.8.3.3).

(21.8.4.3) For each $\mathbf{u} \in \mathfrak{t}$, the set consisting of 0 and the $\alpha(\mathbf{u})$, $\alpha \in \mathbf{S}$, is the set of eigenvalues of the endomorphism $\mathrm{ad}(\mathbf{u}) \otimes 1_{\mathbf{C}}$ of $\mathfrak{g}_{(\mathbf{C})}$. It follows therefore from (19.16.6) that the set of elements $\mathbf{u} \in \mathfrak{t}$ such that the tangent linear mapping $T_{\mathbf{u}}(\exp_G)$ *is not bijective* is the set of $\mathbf{u} \in \mathfrak{t}$ such that *at least one of the numbers* $\alpha(\mathbf{u})$, $\alpha \in \mathbf{S}$, is a nonzero integer multiple of $2\pi i$.

(21.8.5) *For each root* $\alpha \in \mathbf{S}$ *and each element* $\mathbf{x}_\alpha \neq 0$ *in* $\mathfrak{g}_\alpha$, *the two elements* $\mathbf{y}_\alpha$, $\mathbf{z}_\alpha$ *of* $\mathfrak{g}$ *defined in* (21.8.3.1), *and the element* $\mathbf{h}_\alpha^0 = [\mathbf{y}_\alpha, \mathbf{z}_\alpha]$ (*which belongs to* $\mathfrak{t}$), *form a basis of the Lie algebra of a connected closed subgroup* $K_\alpha$ *of* G, *locally isomorphic to* SU(2, C). *Furthermore,* $-i\alpha(\mathbf{h}_\alpha^0) > 0$.

The Jacobi identity shows that, for each $\mathbf{u} \in \mathfrak{t}$,

$$[\mathbf{u}, \mathbf{h}_\alpha^0] = [\mathbf{u}, [\mathbf{y}_\alpha, \mathbf{z}_\alpha]] = [\mathbf{y}_\alpha, [\mathbf{u}, \mathbf{z}_\alpha]] - [\mathbf{z}_\alpha, [\mathbf{u}, \mathbf{y}_\alpha]] = 0$$

by virtue of (21.8.3.2). Since $\mathfrak{t}$ is its own centralizer in $\mathfrak{g}$, it follows that $\mathbf{h}_\alpha^0 \in \mathfrak{t}$.

Now let $(\mathbf{u} \mid \mathbf{v})$ be a G-*invariant scalar product* on $\mathfrak{g}$ (21.4.3). Bearing in mind (21.5.4.1) and (21.8.3.2), we have

(21.8.5.1) $$(\mathbf{u} \mid \mathbf{h}_\alpha^0) = (\mathbf{u} \mid [\mathbf{y}_\alpha, \mathbf{z}_\alpha]) = ([\mathbf{u}, \mathbf{y}_\alpha] \mid \mathbf{z}_\alpha) = 2\pi\varphi(\mathbf{u})(\mathbf{z}_\alpha \mid \mathbf{z}_\alpha).$$

Since $\mathbf{z}_\alpha \neq 0$, we have $(\mathbf{z}_\alpha \mid \mathbf{z}_\alpha) \neq 0$; since the linear form $\varphi = (2\pi i)^{-1}\alpha$ is not identically zero on $\mathfrak{t}$, we have $\mathbf{h}_\alpha^0 \neq 0$; and therefore from (21.8.5.1) we obtain

(21.8.5.2) $$i(\mathbf{h}_\alpha^0 \mid \mathbf{h}_\alpha^0) = \alpha(\mathbf{h}_\alpha^0)(\mathbf{z}_\alpha \mid \mathbf{z}_\alpha),$$

so that $\alpha(\mathbf{h}_\alpha^0) \neq 0$, and $-i\alpha(\mathbf{h}_\alpha^0) > 0$.

We have therefore shown that the three vectors $\mathbf{h}_\alpha^0$, $\mathbf{y}_\alpha$, $\mathbf{z}_\alpha$ form a basis of a Lie subalgebra $\mathfrak{k}_\alpha$ of $\mathfrak{g}$, for which the multiplication table is as follows:

(21.8.5.3)
$$[\mathbf{h}_\alpha^0, \mathbf{y}_\alpha] = -i\alpha(\mathbf{h}_\alpha^0)\mathbf{z}_\alpha, \quad [\mathbf{h}_\alpha^0, \mathbf{z}_\alpha] = i\alpha(\mathbf{h}_\alpha^0)\mathbf{y}_\alpha, \quad [\mathbf{y}_\alpha, \mathbf{z}_\alpha] = \mathbf{h}_\alpha^0.$$

We now recall that the compact connected Lie group $\mathbf{SU}(2, \mathbf{C}) = \mathbf{SU}(2)$ (16.11.3) has as Lie algebra the Lie subalgebra $\mathfrak{su}(2)$ (over $\mathbf{R}$) of $\mathfrak{gl}(2, \mathbf{C}) = \mathbf{M}_2(\mathbf{C})$ consisting of the *antihermitian matrices* $S$ (i.e., matrices $S$ such that ${}^t\bar{S} + S = 0$) *with trace* 0 ((19.4.3.2) and (19.7.1.1)). It is immediately seen that the three matrices

(21.8.5.4)
$$iH = \begin{pmatrix} i & 0 \\ 0 & -i \end{pmatrix}, \qquad Y = \begin{pmatrix} 0 & i \\ i & 0 \end{pmatrix}, \qquad Z = \begin{pmatrix} 0 & -1 \\ 1 & 0 \end{pmatrix}$$

form an $\mathbf{R}$-basis of this Lie algebra, and that the multiplication table is

(21.8.5.5)
$$[iH, Y] = 2Z, \quad [iH, Z] = -2Y, \quad [Y, Z] = 2iH.$$

If we now observe that by (21.8.5.2) the number

$$a_\alpha = -i\alpha(\mathbf{h}_\alpha^0)$$

is strictly positive, it is clear that we may define an *isomorphism* $\sigma$ of $\mathfrak{su}(2)$ onto $\mathfrak{k}_\alpha$ by the formulas

$$\sigma(iH) = \frac{2}{a_\alpha}\mathbf{h}_\alpha^0, \qquad \sigma(Y) = \frac{2}{a_\alpha^{1/2}}\mathbf{y}_\alpha, \qquad \sigma(Z) = \frac{2}{a_\alpha^{1/2}}\mathbf{z}_\alpha.$$

It is immediately verified that no matrix $\neq \lambda I$ in $\mathbf{M}_2(\mathbf{C})$ commutes with each of $H$, $Y$, and $Z$, and therefore the center of $\mathfrak{su}(2)$ is $\{0\}$, so that the compact group $\mathbf{SU}(2)$ is semisimple, with discrete and therefore finite center (in fact, it is easily shown that the center consists of $\pm I$). Moreover, $\mathbf{SU}(2)$ is simply connected (16.30.6), hence the groups locally isomorphic to $\mathbf{SU}(2)$ are compact (in fact, there are up to isomorphism only two of them, namely $\mathbf{SU}(2)$ itself and its quotient by its center, which is isomorphic to $\mathbf{SO}(3)$). The connected Lie group $\mathrm{K}_\alpha$ immersed in G that has $\mathfrak{k}_\alpha$ as its Lie algebra is therefore compact with respect to its proper topology, hence is closed in G (3.17.2).

It follows immediately from **(21.8.5.5)** that $\mathbf{R}iH$ is a *maximal* commutative subalgebra of $\mathfrak{su}(2)$, and consequently **(21.7.7.2)** $\mathbf{SU}(2)$ and the $K_\alpha$ are *almost simple groups of rank* 1.

**(21.8.6)** If $v$ is any automorphism of G, it is clear by transport of structure that $v(T)$ is a maximal torus of G. The derived automorphism $v_*$ of $\mathfrak{g}$ transforms $\mathfrak{t}$ into the Lie algebra of $v(T)$, and the contragredient ${}^t v_*^{-1}$ of this automorphism therefore transforms the lattice of weights of T into that of $v(T)$, and the set **S** of roots of G relative to T into the set of roots of G relative to $v(T)$.

In particular, taking $v = \mathrm{Int}(s)$, where $s \in \mathcal{N}(T)$, these remarks show that *the Weyl group* $W = W(G, T)$ *leaves invariant (globally) the lattice of weights of* T *and the set* **S** *of roots of* G *relative to* T. For each $w \in W$ we have, evidently,

**(21.8.6.1)**
$$(w \otimes 1_{\mathbf{C}})(\mathfrak{g}_\alpha) = \mathfrak{g}_{w \cdot \alpha}.$$

Furthermore:

**(21.8.7)** *There exists an element* $r_\alpha \in K_\alpha$ *belonging to the normalizer* $\mathcal{N}(T)$, *whose coset* $s_\alpha$ *modulo* T *in the Weyl group* W *acts on* $\mathfrak{t}$ *as the orthogonal reflection in the hyperplane* $\mathfrak{u}_\alpha$ (relative to any G-invariant scalar product on $\mathfrak{g}$), *mapping* $\mathbf{h}_\alpha^0$ *to* $-\mathbf{h}_\alpha^0$.

We shall show that we may take $r_\alpha = \exp(\xi \mathbf{y}_\alpha)$ for a suitably chosen real number $\xi$. We have **(19.11.2.2)**

$$\mathrm{Ad}(r_\alpha) = \exp(\xi\, \mathrm{ad}(\mathbf{y}_\alpha)) = \sum_{n=0}^{\infty} \frac{\xi^n}{n!} (\mathrm{ad}(\mathbf{y}_\alpha))^n$$

in the algebra $\mathrm{End}(\mathfrak{g})$. Now, for each $\mathbf{u} \in \mathfrak{u}_\alpha$, we have

$$\mathrm{ad}(\mathbf{y}_\alpha) \cdot \mathbf{u} = -[\mathbf{u}, \mathbf{y}_\alpha] = -i\alpha(\mathbf{u})\mathbf{z}_\alpha = 0$$

by definition, hence $\mathrm{Ad}(r_\alpha) \cdot \mathbf{u} = 0$; also, if we put $a_\alpha = -i\alpha(\mathbf{h}_\alpha^0) > 0$ as above, it follows from the formulas **(21.8.5.3)** that

$$\mathrm{Ad}(r_\alpha) \cdot \mathbf{h}_\alpha^0 = \cos(-a_\alpha \xi)\mathbf{h}_\alpha^0 + \sin(-a_\alpha \xi)\mathbf{z}_\alpha,$$

and therefore, if we put $\xi = \pi/a_\alpha$, we obtain $\mathrm{Ad}(r_\alpha) \cdot \mathbf{h}_\alpha^0 = -\mathbf{h}_\alpha^0$. Hence, for this value of $\xi$, the automorphism $\mathrm{Ad}(r_\alpha)$ leaves invariant $\mathfrak{t} = \mathfrak{u}_\alpha \oplus \mathbf{R}\mathbf{h}_\alpha^0$ so that $r_\alpha \in \mathcal{N}(T)$ **(19.11.4)**; and since $\mathbf{h}_\alpha^0$ is orthogonal to $\mathfrak{u}_\alpha$, relative to any G-invariant scalar product on $\mathfrak{g}$ by virtue of **(21.8.5.1)**, it follows that the restriction of $\mathrm{Ad}(r_\alpha)$ to $\mathfrak{t}$ is the orthogonal reflection in the hyperplane $\mathfrak{u}_\alpha$.

**(21.8.8)** The definition of the roots involves only the Lie algebra $\mathfrak{g}$ and a maximal commutative subalgebra $\mathfrak{t}$ of $\mathfrak{g}$ **(21.8.1)**. Since any one of these subalgebras can be transformed into any other by an automorphism of $\mathfrak{g}$ **(21.7.7)**, it follows that the set of roots, up to automorphisms of $\mathfrak{g}$, *depends only on* $\mathfrak{g}$ and is therefore the same for any two *locally isomorphic* compact connected Lie groups. Moreover, if we canonically decompose $\mathfrak{g}$ as the direct sum $\mathfrak{c} \oplus \mathfrak{D}(\mathfrak{g})$ of its center $\mathfrak{c}$ and its derived algebra $\mathfrak{D}(\mathfrak{g})$ **(21.6.9)**, we have $\mathfrak{t} = \mathfrak{c} \oplus \mathfrak{t}'$, where $\mathfrak{t}'$ is a maximal commutative subalgebra of $\mathfrak{D}(\mathfrak{g})$. Also $\mathfrak{g}_{(\mathbf{C})} = \mathfrak{c}_{(\mathbf{C})} \oplus (\mathfrak{D}(\mathfrak{g}))_{(\mathbf{C})}$, and it is immediate that $\mathfrak{c}_{(\mathbf{C})}$ is the center and $(\mathfrak{D}(\mathfrak{g}))_{(\mathbf{C})}$ the derived algebra of $\mathfrak{g}_{(\mathbf{C})}$; and by decomposing an element $\mathbf{x} \in \mathfrak{g}_\alpha$ relative to this direct sum, it follows from **(21.8.1.2)** that the component of $\mathbf{x}$ in $\mathfrak{c}_{(\mathbf{C})}$ is zero, so that the $\mathfrak{g}_\alpha$ are contained in $(\mathfrak{D}(\mathfrak{g}))_{(\mathbf{C})}$. Finally, for each root $\alpha \in \mathbf{S}$, we have $\alpha(\mathbf{u}) = 0$ for all $\mathbf{u} \in \mathfrak{c}$, so that $\alpha$ may be identified with its restriction to $\mathfrak{t}'$. The result of these considerations is that the set of roots of $\mathfrak{g}$ relative to $\mathfrak{t}$ may be *identified* with the set of roots of $\mathfrak{D}(\mathfrak{g})$ relative to $\mathfrak{t}'$, so that in the study of the roots of a compact connected Lie group G we may assume without loss of generality that G is semisimple (or even semisimple and simply connected).

## PROBLEMS

1. Every quaternion of norm 1 can be written uniquely as $x + yj$, where $x$ and $y$ are complex numbers such that $|x|^2 + |y|^2 = 1$. Show that the mapping

$$x + yj \mapsto \begin{pmatrix} x & y \\ -\bar{y} & \bar{x} \end{pmatrix}$$

defines an isomorphism of the Lie group $\mathbf{U}(1, \mathbf{H})$ onto the Lie group $\mathbf{SU}(2, \mathbf{C})$.

2. With the notation of **(21.8.4)** and **(21.8.5)**, show that the group $U_\alpha$ has two connected components if $K_\alpha$ is isomorphic to $\mathbf{SU}(2, \mathbf{C})$, and is connected if $K_\alpha$ is isomorphic to $\mathbf{SO}(3, \mathbf{R})$.

3. With the notation of **(21.8.1)**, show that for *every* symmetric $\mathbf{C}$-bilinear form $\Phi$ on $\mathfrak{g}_{(\mathbf{C})} \times \mathfrak{g}_{(\mathbf{C})}$, the spaces $\mathfrak{g}_\alpha$ and $\mathfrak{g}_\beta$ are orthogonal relative to $\Phi$ if $\alpha + \beta \neq 0$, and $\mathfrak{g}_0 = \mathfrak{t}_{(\mathbf{C})}$ is orthogonal to $\mathfrak{g}_\alpha$ for each $\alpha \in \mathbf{S}$.

4. Show that a singular element of a compact connected Lie group lies in infinitely many maximal tori.

## 9. LINEAR REPRESENTATIONS OF SU(2)

The study of the roots of a compact connected Lie group G is based on the existence of the subgroups $K_\alpha$ of G, locally isomorphic to **SU**(2) (21.8.5). In this section we shall show that it is possible to describe *explicitly* all the linear representations of **SU**(2). This result, applied to the restrictions to the subgroups $K_\alpha$ of the linear representation $s \mapsto \operatorname{Ad}(s) \otimes 1_{\mathbf{C}}$ of G on $\mathfrak{g}_{(\mathbf{C})}$, will enable us in Section 20.10 to derive certain properties of the set **S** of roots of G, on the basis of which it is possible to describe *explicitly* all possible sets of roots.

(21.9.1) We recall that if G is a connected, simply connected, real Lie group and E is a finite-dimensional *complex* vector space, the mapping $V \mapsto V_*$ is a bijection of the set of linear representations of G on E, onto the set of (**R**-linear) homomorphisms of the Lie algebra $\mathfrak{g}_e$ of G into the Lie algebra $\mathfrak{gl}(E) = \operatorname{End}(E)$ (19.7.6). By virtue of (21.5.2.1) and the connectedness of G, a complex vector subspace F of E is *stable* under $V$ if and only if it is stable under $V_*$.

Furthermore, from any **R**-linear homomorphism of Lie algebras

$$\rho : \mathfrak{g}_e \to \mathfrak{gl}(E),$$

we obtain a **C**-linear homomorphism $\rho \otimes 1_{\mathbf{C}} : \mathfrak{g}_e \otimes_{\mathbf{R}} \mathbf{C} \to \mathfrak{gl}(E)$ of the complexification of $\mathfrak{g}_e$ into $\mathfrak{gl}(E)$, and we obtain in this way a *bijection* of the set of **R**-homomorphisms of $\mathfrak{g}_e$ into $\mathfrak{gl}(E)$ onto the set of **C**-homomorphisms of $\mathfrak{g}_e \otimes_{\mathbf{R}} \mathbf{C}$ into $\mathfrak{gl}(E)$, the inverse bijection being the restriction $\rho' \mapsto \rho' | \mathfrak{g}_e$ (21.5.2); equivalently, a complex vector subspace F of E is *stable* under $\rho$ if and only if it is stable under $\rho \otimes 1_{\mathbf{C}}$.

Finally, if $\mathfrak{g}$ is the Lie algebra of a (real or complex) Lie group, and $\rho$ is a Lie algebra homomorphism of $\mathfrak{g}$ into $\mathfrak{gl}(E) = \operatorname{End}(E)$, then (19.6.4) $\rho$ extends uniquely to a homomorphism of associative algebras $U(\rho)$: $U(\mathfrak{g}) \to \operatorname{End}(E)$, which makes E into a *left module* over the enveloping algebra $U(\mathfrak{g})$ of $\mathfrak{g}$; to say that a complex vector subspace F of E is *stable* under $\rho$ therefore signifies that it is a $U(\mathfrak{g})$-*submodule* of E.

From these remarks it follows that a linear representation $V$ of a compact Lie group G on a finite-dimensional complex vector space E is *irreducible* if and only if E is a *simple* $U(\mathfrak{g}_e \otimes_{\mathbf{R}} \mathbf{C})$-module, where the module structure on E is defined by $U(V_* \otimes 1_{\mathbf{C}})$.

Moreover, for *every* linear representation $V$ of the compact group G on a finite-dimensional complex vector space E, the module E is a *direct sum* of *simple* $U(\mathfrak{g}_e \otimes_{\mathbf{R}} \mathbf{C})$-submodules. Among these simple submodules there may appear *trivial* $U(\mathfrak{g}_e \otimes_{\mathbf{R}} \mathbf{C})$-modules, corresponding to the trivial representation of G (21.1.1): in such a submodule, the product of any element of the module by any element of $\mathfrak{g}_e \otimes_{\mathbf{R}} \mathbf{C}$ is *zero*.

(21.9.2) The three matrices $Y, Z, H$ (21.8.5.4) are clearly linearly independent *over* **C** in $\mathbf{M}_2(\mathbf{C})$, and therefore form a *basis* of the complexification of $\mathfrak{su}(2)$. Another, more convenient basis consists of the three matrices

(21.9.2.1)
$$H = \begin{pmatrix} 1 & 0 \\ 0 & -1 \end{pmatrix},$$
$$X^+ = -\tfrac{1}{2}(Z + iY) = \begin{pmatrix} 0 & 1 \\ 0 & 0 \end{pmatrix},$$
$$X^- = \tfrac{1}{2}(Z - iY) = \begin{pmatrix} 0 & 0 \\ 1 & 0 \end{pmatrix},$$

from which it is clear that the complexification of $\mathfrak{su}(2)$ is $\mathfrak{sl}(2, \mathbf{C})$ (19.7.1.1). The multiplication table for this basis is

(21.9.2.2)
$$[H, X^+] = 2X^+,$$
$$[H, X^-] = -2X^-,$$
$$[X^+, X^-] = H.$$

If $\rho: \mathfrak{sl}(2, \mathbf{C}) \to \mathfrak{gl}(E)$ is a **C**-homomorphism of $\mathfrak{sl}(2, \mathbf{C})$ into the Lie algebra $\mathfrak{gl}(E)$, where E is a finite-dimensional complex vector space, we denote by $(P, x) \mapsto P \cdot x$ the corresponding $U(\mathfrak{sl}(2, \mathbf{C}))$-module structure on E (21.9.1) (so that $P$ is a sum of *noncommutative* monomials in $H$, $X^+$, and $X^-$ with complex coefficients).

(21.9.3) *For each integer $m \geq 0$ there exists an irreducible linear representation of* **SU**(2) *on a complex vector space* $L_m$ *of dimension* $m + 1$, *which has a basis* $(e_j)_{0 \leq j \leq m}$ *for which the* $U(\mathfrak{sl}(2, \mathbf{C}))$*-module structure of* $L_m$ *is defined by the formulas*

(21.9.3.1)
$$H \cdot e_j = (m - 2j)e_j,$$
$$X^+ \cdot e_j = (m - j + 1)e_{j-1},$$
$$X^- \cdot e_j = (j + 1)e_{j+1}$$

(*with the convention that* $e_{-1} = e_{m+1} = 0$). *Every irreducible linear representation of* **SU**(2) *is equivalent to one of these representations.*

Consider a linear representation $V$, not necessarily irreducible, of $\mathbf{SU}(2)$ on a nonzero finite-dimensional complex vector space E. For each $\lambda \in \mathbf{C}$, let $E_\lambda$ denote the subspace of all vectors $x$ in E such that $H \cdot x = \lambda x$.

The set P of complex numbers $\lambda$ such that $E_\lambda \neq \{0\}$ is the finite set of *eigenvalues* of the endomorphism $V_*(H)$ of E. It follows from **(21.7.5)**, applied to the maximal torus of $\mathbf{SU}(2)$ for which $\mathbf{R}iH$ is the Lie algebra, that E is the *direct sum* of the $E_\lambda$ for $\lambda \in P$.

**(21.9.3.2)** *If* $x \in E_\lambda$, *then* $X^+ \cdot x \in E_{\lambda+2}$ *and* $X^- \cdot x \in E_{\lambda-2}$.

For by virtue of **(21.9.2.2)** we have

$$H \cdot (X^+ \cdot x) = [H, X^+] \cdot x + X^+ \cdot (H \cdot x) = (\lambda + 2)X^+ \cdot x,$$

$$H \cdot (X^- \cdot x) = [H, X^-] \cdot x + X^- \cdot (H \cdot x) = (\lambda - 2)X^- \cdot x.$$

An element $x \in E$ is said to be *primitive for the eigenvalue* $\lambda \in P$ if $x \neq 0$ and if

**(21.9.3.3)**
$$H \cdot x = \lambda x, \qquad X^+ \cdot x = 0.$$

We then have:

**(21.9.3.4)** *There exists a primitive element in* E.

Let $z$ be a nonzero element of some $E_\lambda$. Not all of the vectors $(X^+)^n \cdot z$ ($n \geqq 0$) can be different from 0, otherwise by virtue of **(21.9.3.2)** they would all belong to *distinct* eigenspaces $E_\mu$, which is absurd. If $k \geqq 0$ is the smallest integer such that $(X^+)^{k+1} \cdot z = 0$, the vector $x = (X^+)^k \cdot z$ is primitive for the eigenvalue $\lambda + 2k$.

**(21.9.3.5)** *Let* M *be a (not necessarily finite-dimensional) complex vector space,* $(P, z) \to P \cdot z$ *a* $U(\mathfrak{sl}(2, \mathbf{C}))$*-module structure on* M, *and let* $x \neq 0$ *be an element of* M *satisfying* **(21.9.3.3)**. *For each integer* $j \geqq 0$, *put* $x_j = (X^-)^j \cdot x/j!$

(i) *For each integer* $j \geqq 0$, *we have*

**(21.9.3.6)**
$$H \cdot x_j = (\lambda - 2j)x_j,$$
$$X^- \cdot x_j = (j + 1)x_{j+1},$$
$$X^+ \cdot x_j = (\lambda - j + 1)x_{j-1}$$

*with the convention that* $x_{-1} = 0$.

(ii) *If* M *is finite-dimensional, then* $\lambda$ *is an integer* $m \geqq 0$, *the subspace* F *of* M *spanned by the* $x_j$ *is a simple* U($\mathfrak{sl}$(2, **C**))-*submodule, and the corresponding irreducible linear representation of* **SU**(2) *is equivalent to that defined by* (21.9.3.1).

The first of the formulas (21.9.3.6) follows from the calculation in (21.9.3.2), and the second follows from the definition of the $x_j$. As to the third, it is true by hypothesis when $j = 0$, and we proceed by induction on $j \geqq 1$. By use of (21.9.2.2) and the first of the formulas (21.9.3.6), we obtain

$$\begin{aligned} jX^+ \cdot x_j &= (X^+X^-) \cdot x_{j-1} \\ &= [X^+, X^-] \cdot x_{j-1} + X^- \cdot (X^+ \cdot x_{j-1}) \\ &= H \cdot x_{j-1} + (\lambda - j + 2)X^- \cdot x_{j-2} \\ &= (\lambda - 2j + 2 + (j-1)(\lambda - j + 2))x_{j-1} \\ &= j(\lambda - j + 1)x_{j-1}. \end{aligned}$$

The $x_j$ belong to eigenspaces of $H$ corresponding to *distinct* eigenvalues of $H$ in M, hence the nonzero $x_j$ are *linearly independent*. Consequently, if M is finite-dimensional, there exists a smallest integer $m \geqq 0$ such that $x_{m+1} = 0$; by (21.9.3.6), we then have $0 = X^+ \cdot x_{m+1} = (\lambda - m)x_m$, whence $\lambda = m$. The subspace F spanned by the $m + 1$ vectors $x_j$ $(0 \leqq j \leqq m)$ then has these vectors as a *basis*, because they are all nonzero and hence linearly independent. The formulas (21.9.3.6) show that F is a U($\mathfrak{sl}$(2, **C**))-submodule of M. To show that F is *simple*, we remark that if F′ is a nonzero submodule of F, it is a vector subspace of F that is stable under the endomorphism $z \mapsto H \cdot z$, hence is a direct sum of a certain number of 1-dimensional eigenspaces $\mathbf{C}x_j$ of this endomorphism (A.24.3); in other words, there exists at least one index $j \in [0, m]$ such that $x_j \in$ F′. But then, since $m - j + 1 \neq 0$ for all $j$ such that $0 \leqq j \leqq m$, the third of the formulas (21.9.3.6) shows that $x_0 \in$ F′, and the second that $x_k \in$ F′ for $0 \leqq k \leqq m$, so that finally F′ = F.

It remains to show that for *each* integer $m \geqq 0$, the formulas (21.9.3.1) effectively define a homomorphism of $\mathfrak{sl}$(2, **C**) into $\mathfrak{gl}$($\mathrm{L}_m$) for which $\mathrm{L}_m$ is a simple U($\mathfrak{sl}$(2, **C**))-module. The first point amounts to the verification that

$$\begin{aligned} H \cdot (X^+ \cdot e_j) - X^+ \cdot (H \cdot e_j) &= 2X^+ \cdot e_j, \\ H \cdot (X^- \cdot e_j) - X^- \cdot (H \cdot e_j) &= -2X^- \cdot e_j, \\ X^+ \cdot (X^- \cdot e_j) - X^- \cdot (X^+ \cdot e_j) &= H \cdot e_j, \end{aligned}$$

which is a simple calculation. Finally, the proof above that F is simple shows that $\mathrm{L}_m$ is simple, and the proof of (21.9.3) is therefore complete.

**(21.9.4)** Let $V$ be an arbitrary continuous linear representation of $\mathbf{SU}(2)$ on a finite-dimensional complex vector space E, and let $\rho = V_*$ be the derived homomorphism. Then E is the direct sum of $r$ simple $\mathrm{U}(\mathfrak{sl}(2, \mathbf{C}))$-submodules $E_1, \ldots, E_r$, with $E_k$ isomorphic to $L_{m_k}$, say. From **(21.9.3)** we deduce immediately the following simple rule for determining the integer $r$ and the integers $m_k$: $r$ is the dimension of the *kernel* N of the endomorphism $\rho(X^+)$ of E; this kernel N is stable under the endomorphism $\rho(H)$, and the *eigenvalues* (each counted according to its multiplicity) of this endomorphism of N are precisely the numbers $m_k$.

Another characterization of the number $r$, which follows from the first of the formulas **(21.9.3.1)**, is that it is *the sum of the multiplicities of the eigenvalues* 0 *and* 1 *of* $\rho(H)$.

## PROBLEMS

1. In the almost simple compact Lie group $\mathbf{SU}(2) = \mathrm{G}$, the matrices $r(t) = \begin{pmatrix} e^{it} & 0 \\ 0 & e^{-it} \end{pmatrix}$, where $t \in \mathrm{I} = [-\pi, \pi] \subset \mathbf{R}$, form a maximal torus T of G. For each central function $f$ on G, put $f^0(t) = f(r(t))$. For $f$ to belong to $\mathscr{L}^1(\mathrm{G})$ it is necessary and sufficient that $f^0(t) \sin^2 t$ belong to $\mathscr{L}^1(\mathrm{I})$, and we have

$$\int_{\mathrm{G}} f(s)\, dm_{\mathrm{G}}(s) = \frac{1}{\pi} \int_{-\pi}^{\pi} f^0(t) \sin^2 t\, dt$$

**(21.15.4.2)**.

(a) Let $\chi_m$ denote the character of the irreducible representation of G on $\mathrm{L}_m$. Then

$$\chi_m^0(t) = \frac{\sin(m+1)t}{\sin t} = \sum_{j=0}^{m} e^{i(m-2j)t}.$$

For each $f \in \mathscr{L}^1(\mathrm{G})$ we have $(f \mid \chi_m) = 2 \sum_{j=0}^{m} c_{m-2j}$, where $c_k$ is the coefficient of $e^{kit}$ in the Fourier series of the function $f^0(t) \sin^2 t$.

(b) Let $p$ be a real number such that $0 < p < 1$, and let $a_n = (n+1)^p - (n-1)^p$ for each integer $n \geqq 1$. Then the series

$$g_p(t) = 2 + \sum_{n=1}^{\infty} a_n \cos nt$$

converges for each $t$ that is not an integer multiple of $2\pi$, and its sum belongs to $\mathscr{L}^1(\mathrm{I})$ (Section **22.19**, Problem 10(c)). Hence there exists a central function $f_p \in \mathscr{L}^1(\mathrm{G})$ such that $f_p^0(t) \sin^2 t = g_p(t)$. Deduce (by taking $p$ close to 1) that the majorations of Section **21.2**, Problem 3(e) cannot be improved by replacing the exponents of $n_\rho$ by smaller exponents.

(c) There exists a continuous central function $h$ on G such that $h^0(t) = \sum_{n=2}^{\infty} n^{-2} \cos n^3 t$.

With the notation of Section 21.2, Problem 3, show that the family of numbers $N_\infty(h * u_\rho)$ is unbounded.

2. With the notation of Problem 1, show that

$$\chi_m \chi_n = \chi_{m+n} + \chi_{m+n-2} + \cdots + \chi_{|m-n|}.$$

## 10. PROPERTIES OF THE ROOTS OF A COMPACT SEMISIMPLE GROUP

(21.10.1) Let K be a *compact, connected, semisimple* Lie group, $\mathfrak{k}$ its Lie algebra, and $\mathfrak{t}$ a maximal commutative subalgebra of $\mathfrak{k}$. Let $\mathfrak{g}$ denote the complexified Lie algebra $\mathfrak{k}_{(\mathbf{C})}$, $\mathfrak{h}$ its commutative subalgebra $\mathfrak{t}_{(\mathbf{C})}$, and $\mathbf{S}$ the set of roots of $\mathfrak{k}$ relative to $\mathfrak{t}$. By abuse of notation, for each root $\alpha \in \mathbf{S}$ we shall denote again by $\alpha$ its extension $\alpha \otimes 1_{\mathbf{C}}$ to a **C**-linear form on $\mathfrak{h}$; these linear forms are called the *roots* of $\mathfrak{g}$ relative to $\mathfrak{h}$. The complex Lie algebra $\mathfrak{g}$, its commutative subalgebra $\mathfrak{h}$, and the finite set $\mathbf{S} \subset \mathfrak{h}^* - \{0\}$ of linear forms have the following properties:

(A) There exists a decomposition of $\mathfrak{g}$ as a direct sum of nonzero complex vector subspaces:

(21.10.1.1) $$\mathfrak{g} = \mathfrak{h} \oplus \left( \bigoplus_{\alpha \in \mathbf{S}} \mathfrak{g}_\alpha \right)$$

such that for each $\mathbf{h} \in \mathfrak{h}$ and each $\mathbf{x} \in \mathfrak{g}_\alpha$, we have

(21.10.1.2) $$[\mathbf{h}, \mathbf{x}] = \alpha(\mathbf{h})\mathbf{x}.$$

(B) There exists a *nondegenerate symmetric* **C**-*bilinear form* $\Phi$ on the vector space $\mathfrak{g}$ such that

(21.10.1.3) $$\Phi([\mathbf{x}, \mathbf{y}], \mathbf{z}) + \Phi(\mathbf{y}, [\mathbf{x}, \mathbf{z}]) = 0$$

for all $\mathbf{x}, \mathbf{y}, \mathbf{z} \in \mathfrak{g}$; also, for each root $\alpha \in \mathbf{S}$, there exists $\mathbf{h}_\alpha^0 \in \mathfrak{h}$ such that

(21.10.1.4) $$\alpha(\mathbf{h}) = \Phi(\mathbf{h}, \mathbf{h}_\alpha^0)$$

for all $\mathbf{h} \in \mathfrak{h}$, and

(21.10.1.5) $$\alpha(\mathbf{h}_\alpha^0) \neq 0.$$

(C) The center of $\mathfrak{g}$ is $\{0\}$.

Property (A) has in effect been established in **(21.8.1)**; property (C) follows from the hypothesis that K is semisimple **(21.6.3)**. As to (B), take any K-invariant **R**-bilinear scalar product on $\mathfrak{k}$ **(21.4.3)**, and extend it to a symmetric **C**-bilinear form $\Phi$ on $\mathfrak{k}_{(\mathbf{C})} = \mathfrak{g}$: this is always possible, and in one way only (note that $\Phi$ is *not* a *hermitian* scalar product). The existence of the element $\mathbf{h}_\alpha^0$ satisfying **(21.10.1.4)** and **(21.10.1.5)** then follows from **(21.8.5)** and **(21.8.5.1)**.

In view of later applications (Section **21.20**), in this and the following section *we shall not make use* (unless otherwise stated) of the fact that $\mathfrak{g}$ and $\mathfrak{h}$ arise by complexification of the Lie algebras of a compact semisimple group and one of its maximal tori; we shall use *only* the properties (A), (B), (C) listed above.

**(21.10.2)** (i) *If $\alpha$, $\beta$ are two roots in* **S** *such that $\alpha + \beta \neq 0$, then $\mathfrak{g}_\alpha$ and $\mathfrak{g}_\beta$ are orthogonal relative to $\Phi$. In particular, $\mathfrak{g}_\alpha$ is totally isotropic, for each $\alpha \in$* **S**.

(ii) *For each $\alpha \in$* **S**, *we have $-\alpha \in$* **S**, *and $\mathfrak{g}_\alpha$ and $\mathfrak{g}_{-\alpha}$ are totally isotropic subspaces of the same dimension. Each of the subspaces $\mathfrak{h}$, $\mathfrak{g}_\alpha \oplus \mathfrak{g}_{-\alpha}$ is nonisotropic.*

(iii) *For each $\alpha \in$* **S**, *and each $\mathbf{x} \in \mathfrak{g}_\alpha$, $\mathbf{y} \in \mathfrak{g}_{-\alpha}$, we have*

**(21.10.2.1)** $$[\mathbf{x}, \mathbf{y}] = \Phi(\mathbf{x}, \mathbf{y})\mathbf{h}_\alpha^0.$$

(i) If $\mathbf{x} \in \mathfrak{g}_\alpha$, $\mathbf{y} \in \mathfrak{g}_\beta$, $\mathbf{h} \in \mathfrak{h}$, then by **(21.10.1.3)** and **(21.10.1.2)** we have $\Phi([\mathbf{h}, \mathbf{x}], \mathbf{y}) + \Phi(\mathbf{x}, [\mathbf{h}, \mathbf{y}]) = 0$, and hence $(\alpha(\mathbf{h}) + \beta(\mathbf{h}))\Phi(\mathbf{x}, \mathbf{y}) = 0$. Since $\alpha + \beta \neq 0$, there exists $\mathbf{h} \in \mathfrak{h}$ such that $\alpha(\mathbf{h}) + \beta(\mathbf{h}) \neq 0$, and therefore $\Phi(\mathbf{x}, \mathbf{y}) = 0$.

(ii) The same proof shows that $\mathfrak{h}$ is orthogonal to $\mathfrak{g}_\alpha$ for each $\alpha \in$ **S**; furthermore, if the roots $\alpha$, $\beta$ are such that $\beta \neq \alpha$ and $\beta \neq -\alpha$, then $\mathfrak{g}_\beta$ and $\mathfrak{g}_{-\beta}$ are orthogonal to $\mathfrak{g}_\alpha \oplus \mathfrak{g}_{-\alpha}$. In other words, $\mathfrak{g}$ is the direct sum of $\mathfrak{h}$ and the *distinct* subspaces $\mathfrak{g}_\alpha \oplus \mathfrak{g}_{-\alpha}$ ($\alpha \in$ **S**), which are mutually orthogonal; each of these subspaces is therefore nonisotropic, and since each $\mathfrak{g}_\alpha$ is totally isotropic by (i), the assertions in (ii) are consequences of the elementary properties of nondegenerate symmetric **C**-bilinear forms.

(iii) Since the restriction of $\Phi$ to $\mathfrak{h} \times \mathfrak{h}$ is nondegenerate, by (ii), the element $\mathbf{h}_\alpha^0$ satisfying **(21.10.1.4)** for all $\mathbf{h} \in \mathfrak{h}$ is *unique*. If $\mathbf{x} \in \mathfrak{g}_\alpha$ and $\mathbf{y} \in \mathfrak{g}_{-\alpha}$, then by **(21.10.1.2)** and **(21.10.1.3)** we have, for all $\mathbf{h} \in \mathfrak{h}$,

$$\Phi(\mathbf{h}, [\mathbf{x}, \mathbf{y}]) = \Phi([\mathbf{h}, \mathbf{x}], \mathbf{y}) = \alpha(\mathbf{h})\Phi(\mathbf{x}, \mathbf{y}) = \Phi(\mathbf{h}, \Phi(\mathbf{x}, \mathbf{y})\mathbf{h}_\alpha^0)$$

and **(21.10.2.1)** follows from this, since the restriction of $\Phi$ to $\mathfrak{h} \times \mathfrak{h}$ is nondegenerate.

(21.10.2.2) With the convention that $\mathfrak{g}_\lambda = \{0\}$ for $\lambda \in \mathfrak{h}^*$ whenever $\lambda$ is not a root, we have $[\mathfrak{g}_\alpha, \mathfrak{g}_\beta] \subset \mathfrak{g}_{\alpha+\beta}$ for any two roots $\alpha$, $\beta$ (cf. (21.10.5)). For it follows from the Jacobi identity that

$$[\mathbf{h}, [\mathbf{x}, \mathbf{y}]] = [[\mathbf{h}, \mathbf{x}], \mathbf{y}] + [\mathbf{x}, [\mathbf{h}, \mathbf{y}]]$$
$$= (\alpha(\mathbf{h}) + \beta(\mathbf{h}))[\mathbf{x}, \mathbf{y}]$$

for all $\mathbf{h} \in \mathfrak{h}$, $\mathbf{x} \in \mathfrak{g}_\alpha$, $\mathbf{y} \in \mathfrak{g}_\beta$, and the result therefore follows from (21.8.1.3).

(21.10.3) (i) *For each $\alpha \in \mathbf{S}$ there exists one and only one element $\mathbf{h}_\alpha \in \mathfrak{h}$ belonging to $[\mathfrak{g}_\alpha, \mathfrak{g}_{-\alpha}]$ and such that*

(21.10.3.1) $$\alpha(\mathbf{h}_\alpha) = 2.$$

*The spaces $\mathfrak{g}_\alpha$ and $\mathfrak{h}_\alpha = [\mathfrak{g}_\alpha, \mathfrak{g}_{-\alpha}] \subset \mathfrak{h}$ are one-dimensional. For each nonzero $\mathbf{x}_\alpha \in \mathfrak{g}_\alpha$ there exists one and only one element $\mathbf{x}_{-\alpha} \in \mathfrak{g}_{-\alpha}$ such that*

(21.10.3.2) $$[\mathbf{x}_\alpha, \mathbf{x}_{-\alpha}] = \mathbf{h}_\alpha$$

*and we have*

(21.10.3.3) $$[\mathbf{h}_\alpha, \mathbf{x}_\alpha] = 2\mathbf{x}_\alpha, \qquad [\mathbf{h}_\alpha, \mathbf{x}_{-\alpha}] = -2\mathbf{x}_{-\alpha},$$

*so that the subspace $\mathfrak{s}_\alpha = \mathbf{C}\mathbf{h}_\alpha \oplus \mathbf{C}\mathbf{x}_\alpha \oplus \mathbf{C}\mathbf{x}_{-\alpha}$ of $\mathfrak{g}$ is a Lie subalgebra of $\mathfrak{g}$ isomorphic to $\mathfrak{sl}(2, \mathbf{C})$.*

(ii) *The set $\mathbf{S}$ of roots has the following properties:*

($S_1$) *$\mathbf{S}$ spans the vector space $\mathfrak{h}^*$.*
($S_2$) *For each $\alpha \in \mathbf{S}$, the linear mapping $\sigma_\alpha$: $\lambda \mapsto \lambda - \lambda(\mathbf{h}_\alpha)\alpha$ is an involutory bijection of $\mathfrak{h}^*$ onto itself, not equal to the identity mapping, and $\mathbf{S}$ is stable under $\sigma_\alpha$.*
($S_3$) *For each pair $\alpha, \beta$ in $\mathbf{S}$, the number $\beta(\mathbf{h}_\alpha)$ is an integer (positive, negative, or zero).*
($S_4$) *For each $\alpha \in \mathbf{S}$, the only element of the form $t\alpha$ (where $t \in \mathbf{C}$ and $t \neq 1$) belonging to $\mathbf{S}$ is $-\alpha$.*

(i) Since the restriction of $\Phi$ to $\mathfrak{g}_\alpha \oplus \mathfrak{g}_{-\alpha}$ is nondegenerate (21.10.2(ii)), it follows from (21.10.2.1) that the vector space $[\mathfrak{g}_\alpha, \mathfrak{g}_{-\alpha}]$ is one-dimensional. It is clear that the only element $\mathbf{h}_\alpha$ in this space that satisfies (21.10.3.1) is $\mathbf{h}_\alpha = 2(\alpha(\mathbf{h}_\alpha^0))^{-1}\mathbf{h}_\alpha^0$. For each $\mathbf{x}_\alpha \neq 0$ in $\mathfrak{g}_\alpha$, by virtue of (21.10.2(ii)) and the elementary properties of nondegenerate symmetric bilinear forms, there exists an element $\mathbf{y} \in \mathfrak{g}_{-\alpha}$ such that $\Phi(\mathbf{x}_\alpha, \mathbf{y}) \neq 0$. By multiplying $\mathbf{y}$ by a suitable nonzero scalar and using (21.10.2.1), we obtain an element

$\mathbf{x}_{-\alpha} \in \mathfrak{g}_{-\alpha}$ satisfying (21.10.3.2), and then the relations (21.10.3.3) follow from (21.10.3.1). For each $\mathbf{x}_\alpha \neq 0$ in $\mathfrak{g}_\alpha$ and each $\mathbf{x}_{-\alpha} \in \mathfrak{g}_{-\alpha}$ satisfying (21.10.3.2), the subspace $\mathfrak{s}_\alpha = \mathbf{C}\mathbf{h}_\alpha \oplus \mathbf{C}\mathbf{x}_\alpha \oplus \mathbf{C}\mathbf{x}_{-\alpha}$ of $\mathfrak{g}$ is therefore a *Lie subalgebra isomorphic to* $\mathfrak{sl}(2, \mathbf{C})$ (21.9.2). Next, suppose that $\dim(\mathfrak{g}_\alpha) > 1$, and hence also $\dim(\mathfrak{g}_{-\alpha}) > 1$. Then the hyperplane in $\mathfrak{g}$ with equation $\Phi(\mathbf{x}_\alpha, \mathbf{y}) = 0$ would have nonzero intersection with $\mathfrak{g}_{-\alpha}$; in other words, there would exist a vector $\mathbf{y} \neq 0$ in $\mathfrak{g}_{-\alpha}$ such that $[\mathbf{x}_\alpha, \mathbf{y}] = 0$ and $[\mathbf{h}_\alpha, \mathbf{y}] = -2\mathbf{y}$, by virtue of (21.10.2.1). But this would mean that, for the homomorphism $\mathbf{u} \mapsto \operatorname{ad}(\mathbf{u})$ of $\mathfrak{s}_\alpha$ into $\mathfrak{gl}(\mathfrak{g})$, $\mathbf{y}$ was a *primitive* element (21.9.3) for the eigenvalue $-2$ of $\operatorname{ad}(\mathbf{h}_\alpha)$, contrary to (21.9.3.5). The uniqueness of $\mathbf{x}_{-\alpha} \in \mathfrak{g}_{-\alpha}$ satisfying (21.10.3.2) is now evident.

We shall henceforth identify the Lie algebra $\mathfrak{s}_\alpha$ with $\mathfrak{sl}(2, \mathbf{C})$, by identifying $\mathbf{h}_\alpha$, $\mathbf{x}_\alpha$, and $\mathbf{x}_{-\alpha}$ with $H$, $X^+$, and $X^-$, respectively (21.9.2.1).

(ii) If the vector subspace of $\mathfrak{h}^*$ spanned by $\mathbf{S}$ were not the whole of $\mathfrak{h}^*$, there would exist $\mathbf{h} \neq 0$ in $\mathfrak{h}$ such that $\alpha(\mathbf{h}) = 0$ for all roots $\alpha \in \mathbf{S}$; hence $\mathbf{h}$ would belong to the center of $\mathfrak{g}$, contrary to hypothesis. This establishes $(S_1)$.

Let $\alpha, \beta \in \mathbf{S}$ and let $\mathbf{y}$ be a nonzero element of $\mathfrak{g}_\beta$. Then $[\mathbf{h}_\alpha, \mathbf{y}] = \beta(\mathbf{h}_\alpha)\mathbf{y}$, so that $\beta(\mathbf{h}_\alpha)$ is an eigenvalue of $\operatorname{ad}(\mathbf{h}_\alpha)$ in $\mathfrak{g}$. By virtue of the identification of $\mathfrak{s}_\alpha$ with $\mathfrak{sl}(2, \mathbf{C})$, it follows from (21.9.3) that $\beta(\mathbf{h}_\alpha)$ is an *integer* $p$, which proves $(S_3)$.

It follows from the relation $\alpha(\mathbf{h}_\alpha) = 2$ that the linear mapping $\sigma_\alpha$: $\lambda \mapsto \lambda - \lambda(\mathbf{h}_\alpha)\alpha$ of $\mathfrak{h}^*$ into itself is such that $\sigma_\alpha^2 = 1$. With the identification of $\mathfrak{s}_\alpha$ with $\mathfrak{sl}(2, \mathbf{C})$, the $\mathrm{U}(\mathfrak{sl}(2, \mathbf{C}))$-submodule of $\mathfrak{g}$ generated by the element $\mathbf{y}$ above may be identified with one of the modules $\mathrm{L}_m$ (21.9.3), the element $\mathbf{y}$ being identified with $e_j$, where $p = m - 2j$ (A.24.4). Now define $\mathbf{z} = (\operatorname{ad}(\mathbf{x}_{-\alpha}))^p \cdot \mathbf{y}$ if $p \geqq 0$, and $\mathbf{z} = (\operatorname{ad}(\mathbf{x}_\alpha))^{-p} \cdot \mathbf{y}$ if $p \leqq 0$. By virtue of the formulas (21.9.3.1), in all cases $\mathbf{z}$ is a nonzero multiple of $e_{j+p} = e_{m-j}$, and we have $\mathbf{z} \in \mathfrak{g}_{\beta - p\alpha}$ by (21.10.2.2). This shows that $\beta - p\alpha \in \mathbf{S}$, and proves $(S_2)$.

If $\alpha$ and $\beta = t\alpha$ both belong to $\mathbf{S}$, it follows from $(S_3)$ and (21.10.3.1) that $2t \in \mathbf{Z}$. Since $\alpha = t^{-1}\beta$, we may assume that $0 < |t| \leqq 1$, and then the only possible values of $t$ are $\pm\frac{1}{2}$, $\pm 1$. Suppose that $\beta \in \mathbf{S}$ is such that $2\beta \in \mathbf{S}$, and let $\mathbf{y}$ be a nonzero element of $\mathfrak{g}_{2\beta}$; then $[\mathbf{h}_\beta, \mathbf{y}] = 2\beta(\mathbf{h}_\beta)\mathbf{y} = 4\mathbf{y} \neq 0$. Now $3\beta = 2\beta + \beta$ is not a root, hence from (21.10.2.2) $\operatorname{ad}(\mathbf{x}_\beta) \cdot \mathbf{y} = 0$. But since $\mathbf{h}_\beta = [\mathbf{x}_\beta, \mathbf{x}_{-\beta}]$, we have

$$\begin{aligned}\operatorname{ad}(\mathbf{h}_\beta) \cdot \mathbf{y} &= \operatorname{ad}(\mathbf{x}_\beta)(\operatorname{ad}(\mathbf{x}_{-\beta}) \cdot \mathbf{y}) - \operatorname{ad}(\mathbf{x}_{-\beta}) \cdot (\operatorname{ad}(\mathbf{x}_\beta) \cdot \mathbf{y}) \\ &= \operatorname{ad}(\mathbf{x}_\beta)(\operatorname{ad}(\mathbf{x}_{-\beta}) \cdot \mathbf{y}).\end{aligned}$$

By (21.10.2.2) again, $\operatorname{ad}(\mathbf{x}_{-\beta}) \cdot \mathbf{y} \in \mathfrak{g}_\beta$; since $\mathfrak{g}_\beta$ is one-dimensional, it follows that $\operatorname{ad}(\mathbf{x}_{-\beta}) \cdot \mathbf{y}$ is a scalar multiple of $\mathbf{x}_\beta$, so that $\operatorname{ad}(\mathbf{x}_\beta) \cdot (\operatorname{ad}(\mathbf{x}_{-\beta}) \cdot \mathbf{y}) = 0$. This contradicts the previous calculation that $\operatorname{ad}(\mathbf{h}_\beta) \cdot \mathbf{y} = 4\mathbf{y} \neq 0$, and thereby proves $(S_4)$.

**(21.10.4)** *With the notation of* **(21.10.3)**, *let* $\alpha$, $\beta$ *be two roots that are not proportional to each other, and let* $p$ *(resp.* $q$*) be the largest integer* $\geqq 0$ *such that* $\beta - p\alpha$ *(resp.* $\beta + q\alpha$*) is a root. Then* $\beta + k\alpha$ *is a root for all integers* $k$ *such that* $-p \leqq k \leqq q$*; also* $\beta(\mathbf{h}_\alpha) = p - q$, *and* $\mathrm{ad}(\mathbf{x}_\alpha)$ *is a bijection of* $\mathfrak{g}_{\beta+k\alpha}$ *onto* $\mathfrak{g}_{\beta+(k+1)\alpha}$ *for* $-p \leqq k \leqq q - 1$.

Let E denote the vector subspace of $\mathfrak{g}$ that is the direct sum of the $\mathfrak{g}_{\beta+k\alpha}$ for all integers $k \in \mathbf{Z}$ such that $\beta + k\alpha$ is a root. With the identification **(21.10.3)** of the subalgebra $\mathfrak{s}_\alpha$ of $\mathfrak{g}$ with $\mathfrak{sl}(2, \mathbf{C})$, it follows from **(21.10.2.2)** that E is an $\mathrm{U}(\mathfrak{sl}(2, \mathbf{C}))$-module. Since the $\mathfrak{g}_{\beta+k\alpha}$ such that $\beta + k\alpha$ is a root are one-dimensional, and since all the numbers $\beta(\mathbf{h}_\alpha) + k\alpha(\mathbf{h}_\alpha) = \beta(\mathbf{h}_\alpha) + 2k$ are all distinct and of the *same parity*, it follows immediately from **(21.9.4)** that E is *simple*, hence isomorphic to $\mathrm{L}_m$ for some integer $m \geqq 0$. Hence E is the direct sum of $m + 1$ subspaces $\mathfrak{g}_{\beta+k\alpha}$, with $a \leqq k \leqq b$, where $a$ and $b$ are rational integers such that $b - a = m$, $\beta + k\alpha \in \mathbf{S}$ for $a \leqq k \leqq b$, and $\beta(\mathbf{h}_\alpha) + 2a = -m$, $\beta(\mathbf{h}_\alpha) + 2b = m$. Since the interval $[a, b]$ of $\mathbf{Z}$ contains 0, we have $a = -p \leqq 0$, $b = q \geqq 0$, and $\beta(\mathbf{h}_\alpha) = p - q$. Finally, the last assertion of the proposition follows from the second of the formulas **(21.9.3.1)**.

**(21.10.5)** *If* $\alpha$, $\beta$ *are two roots, then*

**(21.10.5.1)** $$[\mathfrak{g}_\alpha, \mathfrak{g}_\beta] = \{0\} \quad \text{if} \quad \alpha + \beta \text{ is not a root,}$$

**(21.10.5.2)** $$[\mathfrak{g}_\alpha, \mathfrak{g}_\beta] = \mathfrak{g}_{\alpha+\beta} \quad \text{if} \quad \alpha + \beta \text{ is a root.}$$

The first assertion has already been proved **(21.10.2.2)**. If $\alpha + \beta$ is a root, then with the notation of **(21.10.4)** we have $q \geqq 1$, and $\mathrm{ad}(\mathbf{x}_\alpha)$ is a bijection of $\mathfrak{g}_\beta$ onto $\mathfrak{g}_{\alpha+\beta}$, by **(21.10.4)**.

**(21.10.6)** When $\mathfrak{g}$, $\mathfrak{h}$ and $\mathbf{S}$ arise from a compact connected semisimple group K and a maximal torus of K, as at the beginning of this section, we can say more about the properties of the elements $\mathbf{h}_\alpha$, $\mathbf{x}_\alpha$, and $\mathbf{x}_{-\alpha}$ of **(21.10.3)**. Starting with an element $\mathbf{x}'_\alpha \neq 0$ in $\mathfrak{g}_\alpha$, we have $c(\mathbf{x}'_\alpha) \in \mathfrak{g}_{-\alpha}$ **(21.8.2)**; writing $\mathbf{y}_\alpha = \mathbf{x}'_\alpha + c(\mathbf{x}'_\alpha)$, $\mathbf{z}_\alpha = i(\mathbf{x}'_\alpha - c(\mathbf{x}'_\alpha))$ as in **(21.8.3)**, we obtain the formulas **(21.8.5.3)**, with $-i\alpha(\mathbf{h}^0_\alpha) = a_\alpha > 0$, from which we deduce

**(21.10.6.1)** $$[2\mathbf{x}'_\alpha, -2c(\mathbf{x}'_\alpha)] = a_\alpha \mathbf{h}_\alpha$$

where $\mathbf{h}_\alpha = -2ia_\alpha^{-1}\mathbf{h}^0_\alpha \in i\mathfrak{t}$ satisfies **(21.10.3.1)**. It follows that the vectors

$\mathbf{x}_\alpha = 2a_\alpha^{-1/2}\mathbf{x}'_\alpha$, $\mathbf{x}_{-\alpha} = -2a_\alpha^{-1/2}c(\mathbf{x}'_\alpha)$ satisfy the relation **(21.10.3.2)**, and are such that

**(21.10.6.2)** $$\mathbf{x}_{-\alpha} = -c(\mathbf{x}_\alpha).$$

By virtue of **(21.10.5)**, we may write

**(21.10.6.3)** $$[\mathbf{x}_\alpha, \mathbf{x}_\beta] = \mathrm{N}_{\alpha,\beta}\,\mathbf{x}_{\alpha+\beta}$$

for all pairs of roots $\alpha$, $\beta$ such that $\alpha + \beta$ is a root. Since $[c(\mathbf{x}_\alpha), c(\mathbf{x}_\beta)] = c([\mathbf{x}_\alpha, \mathbf{x}_\beta])$, it follows from **(21.10.6.2)** that

**(21.10.6.4)** $$\mathrm{N}_{-\alpha,-\beta} = -\overline{\mathrm{N}}_{\alpha,\beta}$$

if $\alpha + \beta \in \mathbf{S}$. It may be shown **(21.20.7)** that it is possible to choose the $\mathbf{h}_\alpha$, $\mathbf{x}_\alpha$, $\mathbf{x}_{-\alpha}$ such that the numbers $\mathrm{N}_{\alpha,\beta}$ are *real.*

A basis (over **C**) of $\mathfrak{g} = \mathfrak{k}_{(\mathbf{C})}$ consisting of elements $\mathbf{x}_\alpha$ satisfying the conditions of **(21.10.3)** and also **(21.10.6.4)** for which the $\mathrm{N}_{\alpha,\beta}$ are *real*, together with an **R**-basis of $i\mathfrak{t}$, is called a *Weyl basis* of $\mathfrak{g}$ (cf. Section **21.20**).

We remark also that the linear mapping $\lambda \mapsto s_\alpha \cdot \lambda$ of $\mathfrak{h}^*$ onto itself, defined by the element $s_\alpha$ of the Weyl group constructed in **(21.8.7)**, is the same as the mapping $\sigma_\alpha : \lambda \mapsto \lambda - \lambda(\mathbf{h}_\alpha)\alpha$, which features in $(\mathrm{S}_2)$ of **(21.10.3)**: for it follows immediately from **(21.8.7)** that $(s_\alpha \cdot \lambda)(\mathbf{u}) = \lambda(\mathbf{u})$ for $\mathbf{u} \in \mathfrak{u}_\alpha$, and $(s_\alpha \cdot \lambda)(\mathbf{h}_\alpha) = -\lambda(\mathbf{h}_\alpha)$.

## PROBLEMS

1. Let G be a compact connected Lie group and G′ a connected closed subgroup of G; let $\mathfrak{g}$, $\mathfrak{g}'$ be the Lie algebras of G, G′; let T be a maximal torus of G such that $\mathrm{T}' = \mathrm{T} \cap \mathrm{G}'$ is a maximal torus of G′ (Section 21.7, Problem 8), and let $\mathfrak{t}$, $\mathfrak{t}'$ be the Lie algebras of T, T′.

   Show that every root of G′ relative to T′ is the restriction to $\mathfrak{t}'$ of at least one root of G relative to T. (Observe that $\mathfrak{g}'_{(\mathbf{C})}$ is stable under $\mathrm{Ad}(t)$ for all $t \in \mathrm{T}'$, and that $\mathfrak{g}_{(\mathbf{C})}$ is the direct sum of $\mathfrak{t}_{(\mathbf{C})}$ and the $\mathfrak{g}_{\alpha'}$, where $\alpha'$ runs through the set of restrictions to $\mathfrak{t}'$ of the roots of G relative to T, and $\mathfrak{g}_{\alpha'}$ denotes the sum of the $\mathfrak{g}_\alpha$ for the roots $\alpha$ whose restriction to $\mathfrak{t}'$ is $\alpha'$.)

2. With the notation of Problem 1, assume that $\mathrm{T}' = \mathrm{T}$, so that G′ has the *same rank* as G. Then every root of G′ relative to T is also a root of G relative to T, i.e., $\mathbf{S}(\mathrm{G}') \subset \mathbf{S}(\mathrm{G})$.
   (a) Suppose that G is the product of almost simple compact groups $\mathrm{G}_i$ ($1 \leqq i \leqq r$), T being

the product of maximal tori $T_i \subset G_i$. Show that $G'$ is necessarily a product of connected closed subgroups $G'_i \supset T_i$ of $G_i$. (If $G'_i$ is the projection of $G'$ in $G_i$, use the fact that every $x \in G'$ is of the form $yty^{-1}$, where $t \in T$ and $y \in G'$, and deduce that the projection $x_i$ of $x$ in $G'_i$ belongs to $G'$.)

(b) The Lie algebra $\mathfrak{g}_{(\mathbf{C})}$ is the direct sum of $\mathfrak{g}'_{(\mathbf{C})}$ and the $\mathfrak{g}_\alpha$ for the roots $\alpha \in \mathbf{S}(G) - \mathbf{S}(G')$. Show that the subgroup of G consisting of the elements $s \in G$ such that the restriction of $\mathrm{Ad}(s)$ to $\mathfrak{g}_\alpha$ is the identity mapping, for each $\alpha \in \mathbf{S}(G) - \mathbf{S}(G')$, is the largest normal subgroup of G contained in $G'$. (Consider the homogeneous space $G/G'$.)

(c) Let D(G) denote the union of the hyperplanes in $\mathfrak{t}$ described by the equations $\alpha(\mathbf{u}) = 2\pi i n$, where $\alpha \in \mathbf{S}(G)$ and $n \in \mathbf{Z}$; D(G) is also the inverse image under $\exp_G$ of the set of singular elements of T (**21.8.4.2**). Define D(G′) in the same way. In general, if $\Delta$ is the union of a family of hyperplanes in $\mathfrak{t}$ consisting of a finite number of families of parallel hyperplanes, a point of $\Delta$ is called *special* if it lies in a hyperplane of *each* of the parallel families. The special points of D(G) form the inverse image under $\exp_G$ of the *center* of G.

Deduce from (b) that if G is *almost simple* and if $\Delta$ is the union of the hyperplanes contained in D(G) but not in D(G′), then every special point of $\Delta$ is also a special point of D(G) (i.e., its image under $\exp_G$ lies in the center of G). Deduce that if $G' \neq G$, then $\dim(G') \leq \dim(G) - 2\,\mathrm{rank}(G)$ (where $\mathrm{rank}(G) = \dim(T)$).

3. With the notation of (**21.8.4**), show that the union of the conjugates of a subgroup $U_\alpha$ in G is the continuous image of a compact manifold of dimension $\dim(G) - 3$. (Use the fact that $\dim(\mathscr{Z}(U_\alpha)) = \dim(T) + 2$.) Deduce that the set of regular points of G (**21.7.13**) is a dense open subset of G (cf. (**16.23.2**)).

## 11. BASES OF A ROOT SYSTEM

**(21.11.1)** Let F be a complex vector space of finite dimension $n$. A finite subset $\mathbf{S}$ of F that does not contain 0 is called a *reduced root system in* F if it satisfies the conditions $(S_1)$, $(S_2)$, $(S_3)$, and $(S_4)$ of (**21.10.3**), with $\mathfrak{h}^*$ replaced by F and the **C**-linear forms $\lambda \mapsto \lambda(\mathbf{h}_\alpha)$ replaced by **C**-linear forms $v_\alpha$ on F, so that $\sigma_\alpha(\lambda) = \lambda - v_\alpha(\lambda)\alpha$.

In this terminology, we have proved in (**21.10.3**) that the set $\mathbf{S}$ of roots of $\mathfrak{g}$ relative to $\mathfrak{h}$ (or of K relative to T, if we had started with a compact connected semisimple group K and a maximal torus T of K) is a reduced root system in $\mathfrak{h}^*$. Conversely, it can be shown that *every* reduced root system is (up to isomorphisms of complex vector spaces) the set of roots of a compact connected semisimple Lie group K, whose Lie algebra is determined up to isomorphism by the root system. Moreover, it is possible to describe explicitly *all* reduced root systems (and hence *all* compact connected Lie groups). We shall not give the proofs of these facts, for which we refer to [79] and [85]; our purpose in this section is to deduce from the definition some properties of reduced root systems that are useful in the theory of compact connected Lie groups.

(21.11.2) *Let* **S** *be a reduced root system in a vector space* F *of dimension n over* **C** (21.11.1).

(i) *The vector subspace* $F_0$ *over* **R** *spanned by* **S** *has dimension n, and the real vector subspace of the dual* $F^*$ *of* F *spanned by the forms* $v_\alpha$ *is of dimension n, and may be identified with the dual* $F_0^*$ *of* $F_0$.

(ii) *There exists a scalar product* $(\lambda \mid \mu)$ *on* $F_0$, *with respect to which the* **R**-*linear mappings* $\sigma_\alpha\colon \lambda \mapsto \lambda - v_\alpha(\lambda)\alpha$ *of* $F_0$ *into itself are orthogonal reflections in hyperplanes, such that* $\sigma_\alpha(\alpha) = -\alpha$, *and the group* $W_{\mathbf{S}}$ *of orthogonal transformations of* $F_0$ *generated by the* $\sigma_\alpha$ *is finite.*

The restrictions $v_\alpha^0$ of the linear forms $v_\alpha$ to $F_0$ are *real*-valued, because by hypothesis the numbers $v_\alpha(\beta)$ ($\alpha, \beta \in \mathbf{S}$) are *integers*, hence $F_0$ is stable under the mappings $\lambda \mapsto \lambda - v_\alpha(\lambda)\alpha$. Since **S** spans $F_0$, any endomorphism of $F_0$ that fixes each element of **S** is the identity mapping; consequently the restriction mapping $w \mapsto w \mid \mathbf{S}$ of $W_{\mathbf{S}}$ into the group of all permutations of **S** is injective, and therefore $W_{\mathbf{S}}$ is *finite*. Hence there exists a scalar product $(\lambda \mid \mu)$ on $F_0$ that is *invariant under* $W_{\mathbf{S}}$ (20.11.3.1); each element of $W_{\mathbf{S}}$ is therefore an orthogonal transformation relative to this scalar product. In particular, since $\sigma_\alpha$ is an orthogonal transformation that is not the identity and that fixes the points of the hyperplane $M_\alpha$ in $F_0$ given by the equation $v_\alpha^0(\lambda) = 0$, it is necessarily the orthogonal reflection in this hyperplane $M_\alpha$. Next, by expressing $\sigma_\alpha^2$ as the identity, we obtain $v_\alpha^0(\lambda)(v_\alpha^0(\alpha) - 2)\alpha = 0$, and since $v_\alpha^0$ is not identically zero on $F_0$ (because $F_0$ spans F), we have $v_\alpha^0(\alpha) = 2$ and $\sigma_\alpha(\alpha) = -\alpha$; this implies that $\alpha$ is orthogonal to $M_\alpha$, and consequently $\sigma_\alpha$ is the reflection

(21.11.2.1) $$\sigma_\alpha\colon \lambda \mapsto \lambda - \frac{2(\alpha \mid \lambda)}{(\alpha \mid \alpha)}\alpha,$$

which shows that $v_\alpha^0(\lambda) = 2(\alpha \mid \lambda)/(\alpha \mid \alpha)$. If $j\colon F_0 \to F_0^*$ is the bijective linear mapping canonically associated with the scalar product, so that the image of $\mu \in F_0$ under $j$ is the linear form $\lambda \mapsto (\mu \mid \lambda)$ on $F_0$, then we have $j(2\alpha/(\alpha \mid \alpha)) = v_\alpha^0$. Since **S** spans $F_0$, the linear forms $v_\alpha^0$ ($\alpha \in \mathbf{S}$) span $F_0^*$. It remains to be shown that the dimension of $F_0$ over **R** cannot exceed $n$; if we had $n + 1$ elements $\alpha_k$ ($1 \leqq k \leqq n + 1$) of **S** linearly independent over **R**, there would exist $n + 1$ *complex* numbers $c_k$, not all zero, such that $\sum_{k=1}^{n+1} c_k \alpha_k = 0$, and therefore $\sum_{k=1}^{n+1} c_k v_\beta(\alpha_k) = 0$ for all $\beta \in \mathbf{S}$. Now, the numbers $v_\beta(\alpha_k)$ are real, and therefore this system of linear equations in the unknowns $c_k$ has a *nontrivial* solution $(c_k^0)$ consisting of *real* numbers, because it has a nontrivial solution consisting of complex numbers. Since the $v_\alpha^0$ span $F_0^*$, we should therefore have $\sum_{k=1}^{n+1} c_k^0 \alpha_k = 0$, contrary to hypothesis.

**(21.11.3)** *With the notation of* **(21.11.2)**, *the numbers*

**(21.11.3.1)** $$n(\beta, \alpha) = \frac{2(\beta\,|\,\alpha)}{(\alpha\,|\,\alpha)} = v_\alpha(\beta) \qquad (\alpha, \beta \in \mathbf{S})$$

*are integers such that*

**(21.11.3.2)** $$0 \leqq n(\beta, \alpha)n(\alpha, \beta) \leqq 4,$$

*and we have* $n(\beta, \alpha)n(\alpha, \beta) = 4$ *only when* $\beta = \pm\alpha$. *If* $\alpha$, $\beta$ *are distinct and* $n(\beta, \alpha) > 0$, *then* $\alpha - \beta$ *is a root.*

With the notation of **(21.11.1)**, we have seen in the proof of **(21.11.2)** that

$$v_\alpha(\beta) = \frac{2(\beta\,|\,\alpha)}{(\alpha\,|\,\alpha)} = n(\beta, \alpha),$$

and therefore $n(\beta, \alpha)$ is an integer, by virtue of $(S_3)$. The inequality **(21.11.3.2)** is a direct consequence of the Cauchy–Schwarz inequality, which also shows that the equality $(\beta\,|\,\alpha)^2 = (\alpha\,|\,\alpha)(\beta\,|\,\beta)$ holds only when $\beta = t\alpha$ with $t \in \mathbf{R}$, and by virtue of $(S_4)$ this implies that $\beta = \pm\alpha$. If $n(\beta, \alpha) > 0$, we cannot have $\beta = -\alpha$ and therefore, if $\alpha$ and $\beta$ are distinct, the product $n(\beta, \alpha)n(\alpha, \beta)$ can take only the values 1, 2, 3; consequently one of the numbers $n(\beta, \alpha)$, $n(\alpha, \beta)$ is equal to 1. Interchanging $\alpha$ and $\beta$ if necessary (which replaces $\alpha - \beta$ by its negative $\beta - \alpha$), we may assume that $n(\alpha, \beta) = 1$, and then $\sigma_\beta(\alpha) = \alpha - \beta$ is a root, by virtue of $(S_2)$.

**(21.11.3.3)** Since we have

**(21.11.3.4)** $$\sigma_\alpha(\beta) = \beta - n(\beta, \alpha)\alpha$$

for each pair of roots $\alpha$, $\beta$, and since **S** spans F, it follows that the reflections $\sigma_\alpha$ are *uniquely* determined by the integers $n(\beta, \alpha)$. Hence the same is true of the linear forms $v_\alpha$, which are therefore independent of the choice of invariant scalar product $(\lambda\,|\,\mu)$.

**(21.11.4)** We shall now change notation, and henceforth denote by E the *real* vector space $F_0^*$, so that its dual E* (the space of **R**-linear forms on E) is canonically identified with the *real* vector space spanned by **S**.

**(21.11.5)** *Let* **S** *be a reduced root system and* E* *the real vector space spanned by* **S**. *There exists a subset* **B** *of* **S** *that is a basis of* E* *over* **R** *and is such that for each root* $\beta \in \mathbf{S}$, *the coefficients* $m_{\beta\alpha}$ *in the expression*

**(21.11.5.1)**
$$\beta = \sum_{\alpha \in \mathbf{B}} m_{\beta\alpha}\alpha$$

*are* integers, all of the same sign.

Such a subset **B** of **S** is called a *basis* of the reduced root system **S**.

Since E is not the union of any finite set of hyperplanes, there exists $\mathbf{x} \in \mathrm{E}$ such that $\alpha(\mathbf{x}) \neq 0$ for all $\alpha \in \mathbf{S}$. Let $\mathbf{S}_{\mathbf{x}}^{+}$ denote the set of roots $\alpha \in \mathbf{S}$ such that $\alpha(\mathbf{x}) > 0$, so that $\mathbf{S} = \mathbf{S}_{\mathbf{x}}^{+} \cup (-\mathbf{S}_{\mathbf{x}}^{+})$, and $\mathbf{S}_{\mathbf{x}}^{+} \cap (-\mathbf{S}_{\mathbf{x}}^{+}) = \varnothing$. A root $\alpha \in \mathbf{S}_{\mathbf{x}}^{+}$ will be called *decomposable* if there exist two roots $\beta, \gamma$ *in* $\mathbf{S}_{\mathbf{x}}^{+}$ such that $\alpha = \beta + \gamma$, and *indecomposable* otherwise. We shall prove **(21.11.5)** in the following more precise form:

**(21.11.5.2)** *For each* $\mathbf{x} \in \mathrm{E}$ *such that* $\alpha(\mathbf{x}) \neq 0$ *for each root* $\alpha \in \mathbf{S}$, *the set* $\mathbf{B}_{\mathbf{x}}$ *of indecomposable elements of* $\mathbf{S}_{\mathbf{x}}^{+}$ *is a basis of* **S**. *Conversely, if* **B** *is a basis of* **S**, *then* $\mathbf{B} = \mathbf{B}_{\mathbf{x}}$ *for each* $\mathbf{x} \in \mathrm{E}$ *such that* $\alpha(\mathbf{x}) > 0$ *for all roots* $\alpha \in \mathbf{B}$.

We shall first show that each root belonging to $\mathbf{S}_{\mathbf{x}}^{+}$ is a linear combination of elements of $\mathbf{B}_{\mathbf{x}}$ with coefficients that are *integers* $\geqq 0$. Suppose then that this is not the case, and let $\mathrm{I} \subset \mathbf{S}_{\mathbf{x}}^{+}$ be the nonempty set of roots that do *not* have this property. Then there exists a root $\alpha \in \mathrm{I}$ for which $\alpha(\mathbf{x}) > 0$ takes the smallest possible value; since $\mathbf{B}_{\mathbf{x}} \cap \mathrm{I} = \varnothing$ by definition, we have $\alpha \notin \mathbf{B}_{\mathbf{x}}$, hence there exist $\beta, \gamma \in \mathbf{S}_{\mathbf{x}}^{+}$ such that $\alpha = \beta + \gamma$. It follows that $\alpha(\mathbf{x}) = \beta(\mathbf{x}) + \gamma(\mathbf{x})$ and $\beta(\mathbf{x}) > 0$, $\gamma(\mathbf{x}) > 0$, so that

$$\beta(\mathbf{x}) < \alpha(\mathbf{x}) \quad \text{and} \quad \gamma(\mathbf{x}) < \alpha(\mathbf{x}),$$

and therefore $\beta \notin \mathrm{I}$ and $\gamma \notin \mathrm{I}$. But then $\alpha = \beta + \gamma \notin \mathrm{I}$, by the definition of I, and we have arrived at a contradiction.

Next, we shall prove that

**(21.11.5.3)** *If* $\alpha, \beta$ *are distinct elements of* $\mathbf{B}_{\mathbf{x}}$, *then* $(\alpha \mid \beta) \leqq 0$.

For otherwise it would follow from **(21.11.3)** that $\gamma = \alpha - \beta$ was a root, and therefore either $\gamma \in \mathbf{S}_{\mathbf{x}}^{+}$ and $\alpha = \beta + \gamma$ would be decomposable, or else $-\gamma \in \mathbf{S}_{\mathbf{x}}^{+}$ and $\beta = \alpha + (-\gamma)$ would be decomposable.

Now suppose that a subset A of E* and an element $\mathbf{x}$ of E are such that $\lambda(\mathbf{x}) > 0$ for all $\lambda \in A$, and $(\lambda \mid \mu) \leqq 0$ for any two *distinct* elements $\lambda$, $\mu$ of A. Then the elements of A are linearly independent over **R**. For otherwise we should have two disjoint nonempty subsets A′, A″ of A and a relation

(21.11.5.4)
$$\sum_{\lambda \in A'} a_\lambda \lambda = \sum_{\mu \in A''} b_\mu \mu = v$$

in which the $a_\lambda$ are all $\geqq 0$, the $b_\mu$ all $\geqq 0$, and at least one of the $a_\lambda$ or the $b_\mu$ is nonzero. But then it follows from the hypotheses and from **(21.11.5.4)** that

$$(v \mid v) = \sum_{(\lambda, \mu) \in A' \times A''} a_\lambda b_\mu (\lambda \mid \mu) \leqq 0$$

and therefore $v = 0$. Consequently $0 = v(\mathbf{x}) = \sum_{\lambda \in A'} a_\lambda \lambda(\mathbf{x})$, and since $a_\lambda \geqq 0$ and $\lambda(\mathbf{x}) > 0$ for all $\lambda \in A'$, we must have $a_\lambda = 0$ for all $\lambda \in A'$; similarly $b_\mu = 0$ for all $\mu \in A''$, and we have a contradiction.

We have therefore now proved by these considerations that $\mathbf{B}_\mathbf{x}$ is a *basis* of $\mathbf{S}$. Conversely, if $\mathbf{B}$ is any basis of $\mathbf{S}$, then $\mathbf{B}$ is a basis of the vector space E*, hence there exists in the dual space E an element $\mathbf{x}$ such that $\alpha(\mathbf{x}) > 0$ for all $\alpha \in \mathbf{B}$. Consider any one of the elements $\mathbf{x} \in E$ having this property. Let $\mathbf{S}^+$ be the set of roots that are linear combinations of elements of $\mathbf{B}$ with coefficients that are all *integers* $\geqq 0$; clearly $\mathbf{S}^+ \subset \mathbf{S}_\mathbf{x}^+$, $-\mathbf{S}^+ \subset -\mathbf{S}_\mathbf{x}^+$, and since by hypothesis $\mathbf{S} = \mathbf{S}^+ \cup (-\mathbf{S}^+) = \mathbf{S}_\mathbf{x}^+ \cup (-\mathbf{S}_\mathbf{x}^+)$, it follows that $\mathbf{S}^+ = \mathbf{S}_\mathbf{x}^+$. If for some root $\alpha \in \mathbf{B}$ we had $\alpha = \beta + \gamma$ with $\beta, \gamma \in \mathbf{S}_\mathbf{x}^+$, it would follow that $\alpha = \sum_{\lambda \in \mathbf{B}} (m_{\beta\lambda} + m_{\gamma\lambda})\lambda$, where the coefficients $m_{\beta\lambda}$ and $m_{\gamma\lambda}$ are integers $\geqq 0$ and at least one of the $m_{\beta\lambda}$ (resp. $m_{\gamma\lambda}$) is $> 0$; consequently $\sum_{\lambda \in \mathbf{B}} (m_{\beta\lambda} + m_{\gamma\lambda}) \geqq 2$, whereas this sum must be equal to 1, because $\mathbf{B}$ is a basis of E*. Hence $\mathbf{B} \subset \mathbf{B}_\mathbf{x}$, and since both $\mathbf{B}$ and $\mathbf{B}_\mathbf{x}$ are bases of E*, we have $\mathbf{B} = \mathbf{B}_\mathbf{x}$.

If $\mathbf{B} = \mathbf{B}_\mathbf{x}$, the set $\mathbf{S}^+ = \mathbf{S}_\mathbf{x}^+$ (resp. $-\mathbf{S}^+$) is called the set of *positive* (resp. *negative*) roots, relative to $\mathbf{B}$; it is the set of roots $\beta \in \mathbf{S}$ such that in **(21.11.5.1)** all the integers $m_{\beta\alpha}$ are $\geqq 0$ (resp. $\leqq 0$).

**(21.11.5.5)** With the notation of **(21.11.1)**, if $\mathbf{S}$ is a reduced root system in F, the set $\mathbf{S}^\vee$ of linear forms $v_\alpha$ is a *reduced root system* in the dual space F*, and is called the *dual* of S. For $\mathbf{S}^\vee$ does not contain 0, because $\sigma_\alpha \neq 1$ for all $\alpha \in \mathbf{S}$, and it spans F*, by **(21.11.2)**. The transpose ${}^t\sigma_\alpha$ is the linear mapping $u \mapsto u - u(\alpha)v_\alpha$ of F* into itself, which is an involutory bijection. Furthermore, if $\sigma_\alpha(\beta) = \gamma \in \mathbf{S}$, where $\alpha$ and $\beta$ are roots, then we have $\sigma_\gamma = \sigma_\alpha \sigma_\beta \sigma_\alpha^{-1}$;

writing this relation in the form $\sigma_\alpha \sigma_\beta = \sigma_\gamma \sigma_\alpha$, we obtain $v_\beta = v_\gamma - v_\gamma(\alpha)v_\alpha$, that is to say, ${}^t\sigma_\alpha(v_\gamma) = v_\beta$, so that the set $\mathbf{S}^\vee$ satisfies $(S_2)$. The verification of $(S_3)$ is immediate, and $(S_4)$ follows from the fact that if $\alpha$ and $t\alpha$ are roots, where $t \in \mathbf{R}$, then $\sigma_\alpha = \sigma_{t\alpha}$, and conversely. It is clear that, if we identify $F^{**}$ with F, we have $(\mathbf{S}^\vee)^\vee = \mathbf{S}$.

If now $\mathbf{B}$ is a basis of $\mathbf{S}$, the set $\mathbf{B}^\vee$ of elements $v_\alpha$, where $\alpha \in \mathbf{B}$, is a *basis* of $\mathbf{S}^\vee$. For, using the bijection $j$ defined in (21.11.2), we may identify $\mathbf{S}^\vee$ with the set $\mathbf{S}'$ of elements $\alpha' = 2\alpha/(\alpha|\alpha)$ of F, where $\alpha \in \mathbf{S}$, and $\mathbf{B}^\vee$ with the set $\mathbf{B}'$ of elements $\alpha'$ with $\alpha \in \mathbf{B}$. We have $\mathbf{B} = \mathbf{B}_\mathbf{x}$ for some $\mathbf{x} \in F^*$, by (21.11.5.2), and since the relations $\alpha(\mathbf{x}) > 0$ and $\alpha'(\mathbf{x}) > 0$ are equivalent, it follows that $\mathbf{S}_\mathbf{x}'^+$ is the set of $\alpha'$ for which $\alpha \in \mathbf{S}_\mathbf{x}^+$. Now if three roots $\alpha, \beta, \gamma \in \mathbf{S}_\mathbf{x}^+$ are such that $\alpha' = \beta' + \gamma'$, then we have $\alpha = t_1\beta + t_2\gamma$ with $t_1 > 0$ and $t_2 > 0$, and therefore (since $\beta'$ and $\gamma'$ are not proportional to each other) there are at least two nonzero components of $\alpha$ with respect to the basis $\mathbf{B}$, in other words $\alpha \notin \mathbf{B}$. This shows also that $\mathbf{B}'$ is contained in the set $\mathbf{B}_\mathbf{x}'$ of indecomposable elements of $\mathbf{S}_\mathbf{x}'^+$; these two sets have the same number of elements, hence $\mathbf{B}' = \mathbf{B}_\mathbf{x}'$ and therefore $\mathbf{B}'$ is a basis of $\mathbf{S}'$ (21.11.5.2).

(21.11.6) *Let* $\mathbf{B}$ *be a basis of the reduced root system* $\mathbf{S}$. *For each root* $\alpha \in \mathbf{B}$, *the reflection* $\sigma_\alpha$ (21.11.2.1) *leaves invariant the set* $\mathbf{S}^+ - \{\alpha\}$ *of positive roots (relative to* $\mathbf{B}$*) other than* $\alpha$, *and transforms* $\alpha$ *into* $-\alpha$.

Let $\beta$ be an element of $\mathbf{S}^+$, other than $\alpha$; we have $\beta = \sum_{\lambda \in \mathbf{B}} m_{\beta\lambda}\lambda$, with coefficients $m_{\beta\lambda}$ that are integers $\geqq 0$. There exists $\gamma \neq \alpha$ in $\mathbf{B}$ such that $m_{\beta\gamma} > 0$, otherwise we should have $\beta = m_{\beta\alpha}\alpha$ and therefore $\beta = \alpha$ by virtue of $(S_4)$. This being so, if $\beta' = \sigma_\alpha(\beta) = \beta - n(\beta, \alpha)\alpha$, the coefficients $m_{\beta'\lambda}$ in the decomposition $\beta' = \sum_{\lambda \in \mathbf{B}} m_{\beta'\lambda}\lambda$ are all integers of the same sign, and by definition we have $m_{\beta'\gamma} = m_{\beta\gamma} > 0$; hence $\beta' \in \mathbf{S}^+$.

(21.11.7) *Let* $\mathbf{B}$ *be a basis of* $\mathbf{S}$, *and let*

(21.11.7.1) $$\delta = \frac{1}{2}\sum_{\lambda \in \mathbf{S}^+} \lambda$$

*be half the sum of the positive roots (relative to* $\mathbf{B}$*). Then we have*

(21.11.7.2) $$\sigma_\alpha(\delta) = \delta - \alpha$$

*for all roots* $\alpha \in \mathbf{B}$. *(In other words,* $v_\alpha(\delta) = 1$ *for all* $\alpha \in \mathbf{B}$.)

For if $\delta_\alpha$ is half the sum of the roots $\mu \in \mathbf{S}^+ - \{\alpha\}$, then it follows from **(21.11.6)** that $\sigma_\alpha(\delta_\alpha) = \delta_\alpha$, and therefore, since $\delta = \delta_\alpha + \frac{1}{2}\alpha$, that $\sigma_\alpha(\delta) = \delta_\alpha - \frac{1}{2}\alpha = \delta - \alpha$.

**(21.11.8)** *Let* **S** *be a reduced root system,* **B** *a basis of* **S**, *and* $\mathrm{W}_\mathbf{S}$ *the finite group generated by the orthogonal reflections* $\sigma_\alpha$ **(21.11.2.1)** *for all* $\alpha \in \mathbf{S}$.

(i) *For each* $\mathbf{x} \in \mathrm{E}$, *there exists* $w \in \mathrm{W}_\mathbf{S}$ *such that* $\alpha(w \cdot \mathbf{x}) \geqq 0$ *for all* $\alpha \in \mathbf{B}$.

(ii) *For each basis* $\mathbf{B}'$ *of* **S**, *there exists* $w \in \mathrm{W}_\mathbf{S}$ *such that* $w(\mathbf{B}') = \mathbf{B}$.

(iii) *For each root* $\beta \in \mathbf{S}$, *there exists* $w \in \mathrm{W}_\mathbf{S}$ *such that* $w(\beta) \in \mathbf{B}$.

(iv) *The group* $\mathrm{W}_\mathbf{S}$ *is generated by the reflections* $\sigma_\alpha$ *for* $\alpha \in \mathbf{B}$.

(Here $w \cdot \mathbf{x}$ is by definition equal to ${}^t w^{-1}(\mathbf{x})$.)

Let $\mathrm{W}_\mathbf{B}$ denote the subgroup of $\mathrm{W}_\mathbf{S}$ generated by the $\sigma_\alpha, \alpha \in \mathbf{B}$. We shall prove (i) by showing, more precisely, that there exists an element $w \in \mathrm{W}_\mathbf{B}$ such that $\alpha(w \cdot \mathbf{x}) \geqq 0$ for all $\alpha \in \mathbf{B}$. If $\delta$ denotes half the sum of the positive roots relative to **B**, choose $w \in \mathrm{W}_\mathbf{B}$ so that the number $\delta(w \cdot \mathbf{x})$ is as large as possible. For $\alpha \in \mathbf{B}$ we have then $\delta(w \cdot \mathbf{x}) \geqq \delta((\sigma_\alpha w) \cdot \mathbf{x})$. But since $\sigma_\alpha^{-1} = \sigma_\alpha$, we have $\delta((\sigma_\alpha w) \cdot \mathbf{x}) = (\sigma_\alpha(\delta))(w \cdot \mathbf{x}) = \delta(w \cdot \mathbf{x}) - \alpha(w \cdot \mathbf{x})$ by virtue of **(21.11.7.2)**, whence $\alpha(w \cdot \mathbf{x}) \geqq 0$.

Likewise, we shall prove (ii) by showing, more precisely, that there exists an element $w \in \mathrm{W}_\mathbf{B}$ such that $w(\mathbf{B}') = \mathbf{B}$. Since $\mathbf{B}'$ is a basis of $\mathrm{E}^*$, there exists $\mathbf{x}' \in \mathrm{E}$ such that $\alpha'(\mathbf{x}') > 0$ for all $\alpha' \in \mathbf{B}'$; from the definition of a basis of **S**, it follows **(21.11.5)** that $\lambda(\mathbf{x}') \neq 0$ for all $\lambda \in \mathbf{S}$. By virtue of (i), there exists $w \in \mathrm{W}_\mathbf{B}$ such that $\alpha(w \cdot \mathbf{x}') \geqq 0$ for all $\alpha \in \mathbf{B}$, that is to say, such that

$$(w^{-1}(\alpha))(\mathbf{x}') \geqq 0;$$

and since $\lambda(\mathbf{x}') \neq 0$ for all $\lambda \in \mathbf{S}$, we have $(w^{-1}(\alpha))(\mathbf{x}') > 0$ or equivalently $\alpha(w \cdot \mathbf{x}') > 0$ for all $\alpha \in \mathbf{B}$, which as above implies that $\lambda(w \cdot \mathbf{x}') \neq 0$ for all $\lambda \in \mathbf{S}$. Hence, with the notation of **(21.11.5)**, we have $\mathbf{B} = \mathbf{B}_{w \cdot \mathbf{x}'}$ and $\mathbf{B}' = \mathbf{B}_{\mathbf{x}'}$, by **(21.11.5.2)**; by transport of structure, it follows that $\mathbf{B} = w(\mathbf{B}')$.

We shall now prove (iii), again by showing that there exists an element $w \in \mathrm{W}_\mathbf{B}$ such that $w(\beta) \in \mathbf{B}$. Let L be the hyperplane in E given by the equation $\beta(\mathbf{x}) = 0$. Since L is not the union of any finite number of subspaces of codimension 2, it follows from $(\mathrm{S}_4)$ that there exists $\mathbf{x}_0 \in \mathrm{L}$ such that $\gamma(\mathbf{x}_0) \neq 0$ for all roots $\gamma \neq \pm\beta$. Hence there exists a number $\varepsilon > 0$ and a point $\mathbf{x} \in \mathrm{E}$ sufficiently close to $\mathbf{x}_0$ so that $\beta(\mathbf{x}) = \varepsilon$ and $|\gamma(x)| > \varepsilon$ for all roots $\gamma$ other than $\pm\beta$. With the notation of **(21.11.5)**, we have therefore $\beta \in \mathbf{B}_\mathbf{x}$, by the definition of $\mathbf{B}_\mathbf{x}$; hence, by virtue of (ii), there exists an element $w \in \mathrm{W}_\mathbf{B}$ such that $w(\mathbf{B}_\mathbf{x}) = \mathbf{B}$, and therefore $w(\beta) \in \mathbf{B}$.

Finally, to establish (iv), it is enough to show that $\sigma_\beta \in W_{\mathbf{B}}$ for each root $\beta \in \mathbf{S}$. But by virtue of (iii) there exists $w \in W_{\mathbf{B}}$ such that $w(\beta) \in \mathbf{B}$, and since $\sigma_{w(\beta)} = w\sigma_\beta w^{-1}$, we have $\sigma_\beta = w^{-1}\sigma_{w(\beta)} w \in W_{\mathbf{B}}$. Q.E.D.

**(21.11.9)** When $\mathfrak{g}$, $\mathfrak{h}$ and $\mathbf{S}$ arise from a compact connected semisimple Lie group K and a maximal torus T of K, as in (21.10.1), the complex vector spaces F, F* defined in (21.11.2) are, respectively, $\mathfrak{h}^* = \mathfrak{t}^* \oplus i\mathfrak{t}^*$ and $\mathfrak{h} = \mathfrak{t} \oplus i\mathfrak{t}$, and the real vector spaces $E^* = F_0$ and $F_0^*$ are, respectively, $i\mathfrak{t}^*$ and $i\mathfrak{t}$. If we choose a K-invariant scalar product $(\mathbf{x} | \mathbf{y})$ on $\mathfrak{k}$ (20.11.3.1), we obtain from it canonically an $\mathbf{R}$-isomorphism $j$ of $\mathfrak{t}$ onto $\mathfrak{t}^*$, for which $j(\mathbf{x})$ (where $\mathbf{x} \in \mathfrak{t}$) is the linear form $\mathbf{y} \mapsto (\mathbf{x} | \mathbf{y})$; and then by transport of structure a scalar product $(\lambda | \mu)$ on E*, by defining $(\lambda | \mu) = (j^{-1}(i\lambda) | j^{-1}(i\mu))$. It is clear that this scalar product is invariant under the *Weyl group* W of K with respect to T, acting faithfully on E* (21.8.6). We have already remarked (21.10.6) that the reflections $\sigma_\alpha$ corresponding to the roots $\alpha \in \mathbf{S}$ (21.11.2.1) are precisely the elements $s_\alpha$ of the Weyl group defined in (21.8.7). In other words, with the notation of (21.11.8), we have $W_{\mathbf{S}} \subset W$. In fact:

**(21.11.10)** *Under the conditions of* (21.11.9), *we have* $W_{\mathbf{S}} = W$.

Let $x$ be an element of the normalizer $\mathcal{N}(T)$ of T in K, and let $w \in W = \mathcal{N}(T)/T$ be the corresponding element of the Weyl group. Clearly, if $\mathbf{B}$ is a basis of $\mathbf{S}$, so also is $w(\mathbf{B})$, by transport of structure; since $W_{\mathbf{S}}$ acts transitively on the set of bases of $\mathbf{S}$ (21.11.8(ii)), it follows that by multiplying $w$ by a suitable element of $W_{\mathbf{S}}$ we may assume that $w(\mathbf{B}) = \mathbf{B}$. Let $\mathbf{u} \in E = i\mathfrak{t}$ be an element such that $\alpha(\mathbf{u}) > 0$ for each root $\alpha \in \mathbf{B}$ (21.11.5.2). Since $w$ permutes the roots in $\mathbf{B}$, it follows that $(w^{-1} \cdot \alpha)(\mathbf{u}) > 0$ for all $\alpha \in \mathbf{B}$, in other words $\alpha(w \cdot \mathbf{u}) > 0$. Let $m$ be the order of $w$ in W, and let $\mathbf{z} = m^{-1} \sum_{k=0}^{m-1} w^k \cdot \mathbf{u} \in E$; then we have $w \cdot \mathbf{z} = \mathbf{z}$, and $\alpha(\mathbf{z}) > 0$ for all $\alpha \in \mathbf{B}$; this implies, as we have seen (21.11.5), that $\beta(\mathbf{z}) \neq 0$ for *all* roots $\beta \in \mathbf{S}$. Hence $i\mathbf{z} \in \mathfrak{t}$ is *regular* (21.8.4); and since $\operatorname{Ad}(x) \cdot i\mathbf{z} = i\mathbf{z}$, it follows from (21.7.14) that $x \in T$ and therefore that $w$ is the identity. Q.E.D.

The proof just given also shows that the relation $w(\mathbf{B}) = \mathbf{B}$ implies that $w = 1$; in other words:

**(21.11.10.1)** *The Weyl group of* K *relative to* T *acts simply transitively on the set of bases of the root system of* K *relative to* T.

*Remarks*

**(21.11.11)** (i) Under the conditions of **(21.11.9)**, the reflections $\sigma_\alpha$ defined in **(21.11.2.1)** are the same as the reflections $\lambda \mapsto s_\alpha \cdot \lambda$ **(21.10.6)** and therefore may be expressed in the form

**(21.11.11.1)**
$$\lambda \mapsto \lambda - \lambda(\mathbf{h}_\alpha)\alpha$$

where $\mathbf{h}_\alpha$ is as defined in **(21.10.3.2)**. We have therefore

**(21.11.11.2)**
$$\lambda(\mathbf{h}_\alpha) = \frac{2(\alpha \mid \lambda)}{(\alpha \mid \alpha)}$$

for all $\lambda \in it^*$, and consequently

**(21.11.11.3)**
$$n(\beta, \alpha) = \beta(\mathbf{h}_\alpha)$$

for all $\alpha, \beta \in \mathbf{S}$.

The integers $n(\alpha, \beta)$, for the elements $\alpha$, $\beta$ of a *basis* of $\mathbf{S}$, are called the *Cartan integers* of $\mathbf{S}$ (or of the Lie algebra $\mathfrak{k}$ or $\mathfrak{g}$, or of the group K). They are independent of the basis chosen, because any basis can be transformed into any other basis by the action of the Weyl group.

The vectors $\mathbf{h}_\alpha \in it$ form a reduced root system $\mathbf{S}^\vee$, the *dual* of $\mathbf{S}$ **(21.11.5.5)**; the Weyl group of $\mathbf{S}^\vee$ may be canonically identified with W.

(ii) Under the conditions of **(21.11.9)**, if $\alpha$ and $\beta$ are roots such that $\beta \neq \pm\alpha$, and if $k$ is an integer such that $\beta + j\alpha$ is a root for $j = 0, 1, \ldots, k$, then we have $k \leqq 3$. For by replacing $\beta$ by $\beta - p\alpha$ for some $p > 0$ if necessary, we may assume that $\beta - \alpha$ is not a root; it follows then from **(21.10.4)** that $k \leqq -\beta(\mathbf{h}_\alpha)$, and the assertion is a consequence of **(21.11.3)**.

(iii) Under the conditions of **(21.11.9)**, if $\mathbf{B}$ is a basis of $\mathbf{S}$ and if $\alpha$, $\beta$ are two roots belonging to $\mathbf{B}$, such that $(\alpha \mid \beta) = 0$, then $\alpha + \beta$ is not a root. For we have $\beta(\mathbf{h}_\alpha) = 0$, which, in the notation of **(21.10.4)**, implies that $p = q$; hence if $q \geqq 1$ we should have also $p \geqq 1$, and then $\beta - \alpha$ would be a root, contrary to the definition **(21.11.5)** of a basis of a root system.

(iv) Again under the conditions of **(21.11.9)**, let $\mathfrak{k} = \bigoplus_j \mathfrak{k}_j$ be the decomposition of $\mathfrak{k}$ as a direct sum of *simple* algebras **(21.6.4)**. From **(21.7.7.2)**, if $\mathfrak{t}_j$ is a maximal commutative subalgebra of $\mathfrak{k}_j$, then $\mathfrak{t} = \bigoplus_j \mathfrak{t}_j$ is a maximal commutative subalgebra of $\mathfrak{t}$. It then follows directly from the definitions (21.8.1) and from the fact that $[\mathfrak{k}_j, \mathfrak{k}_h] = 0$ for $j \neq h$, that if $\mathbf{S}_j$ is the root system of $\mathfrak{k}_j$ relative to $\mathfrak{t}_j$, then the *union* $\mathbf{S}$ of the $\mathbf{S}_j$ is the root system of $\mathfrak{k}$ relative to $\mathfrak{t}$. Note that if $\alpha \in \mathbf{S}_j$ and $\beta \in \mathbf{S}_h$, where $j \neq h$, then $n(\alpha, \beta) = 0$. Finally, it is clear from the definition (21.11.5) that if $\mathbf{B}_j$ is a *basis* of the root system $\mathbf{S}_j$, then the union $\mathbf{B}$ of the $\mathbf{B}_j$ is a basis of $\mathbf{S}$.

## PROBLEMS

1. (a) With the notation of (21.11.3), let $\alpha$ and $\beta$ be two roots in $\mathbf{S}$ such that $\alpha \neq \pm\beta$, and let $\theta$ be the angle (between 0 and $\pi$) between the two vectors $\alpha$, $\beta$ (relative to the scalar product $(\lambda \mid \mu)$). Show that if we write $\|\lambda\| = (\lambda \mid \lambda)^{1/2}$, the following cases exhaust all the possibilities, for $\|\beta\| \geq \|\alpha\|$:

(i) $n(\alpha, \beta) = 0, \quad n(\beta, \alpha) = 0, \quad \theta = \frac{1}{2}\pi.$
(ii) $n(\alpha, \beta) = 1, \quad n(\beta, \alpha) = 1, \quad \theta = \frac{1}{3}\pi, \quad \|\beta\| = \|\alpha\|.$
(iii) $n(\alpha, \beta) = -1, \quad n(\beta, \alpha) = -1, \quad \theta = \frac{2}{3}\pi, \quad \|\beta\| = \|\alpha\|.$
(iv) $n(\alpha, \beta) = 1, \quad n(\beta, \alpha) = 2, \quad \theta = \frac{1}{4}\pi, \quad \|\beta\| = \sqrt{2}\,\|\alpha\|.$
(v) $n(\alpha, \beta) = -1, \quad n(\beta, \alpha) = -2, \quad \theta = \frac{3}{4}\pi, \quad \|\beta\| = \sqrt{2}\,\|\alpha\|.$
(vi) $n(\alpha, \beta) = 1, \quad n(\beta, \alpha) = 3, \quad \theta = \frac{1}{6}\pi, \quad \|\beta\| = \sqrt{3}\,\|\alpha\|.$
(vii) $n(\alpha, \beta) = -1, \quad n(\beta, \alpha) = -3, \quad \theta = \frac{5}{6}\pi, \quad \|\beta\| = \sqrt{3}\,\|\alpha\|.$

(b) If $p$, $q$ are the integers defined in (21.10.4) and if $\alpha + \beta$ is a root, show that

$$\frac{(\beta + \alpha \mid \beta + \alpha)}{(\beta \mid \beta)} = \frac{p+1}{q}.$$

(Consider the various possibilities.)
(c) Show that if $\|\alpha\| = \|\beta\|$ and if $\mathbf{S}$ is irreducible (Problem 10), there exists $w \in \mathbf{W_S}$ such that $w(\alpha) = \beta$. (Observe that by replacing $\alpha$ by one of its transforms under $\mathbf{W_S}$, we may assume that $(\alpha \mid \beta) \neq 0$, and then $n(\alpha, \beta) = n(\beta, \alpha)$, and we may also assume that $n(\alpha, \beta) > 0$. Consider the subgroup of $\mathbf{W_S}$ generated by $\sigma_\alpha$ and $\sigma_\beta$, and use (a) above.)

2. Show that the only reduced root systems in a two-dimensional vector space over $\mathbf{R}$ are the following:

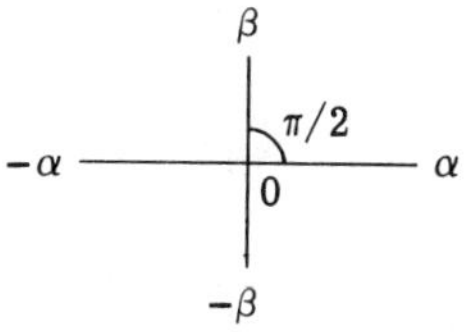


$A_1 \times A_1$

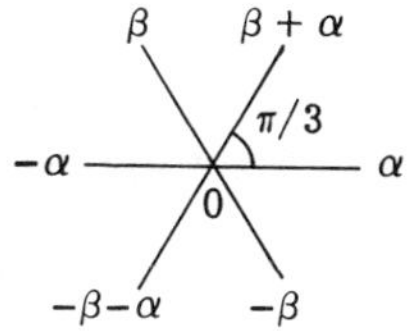


$A_2$

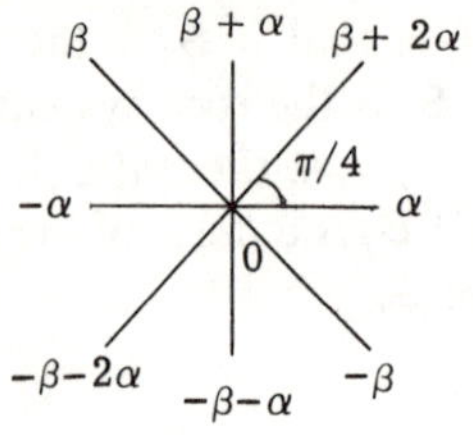


B$_2$

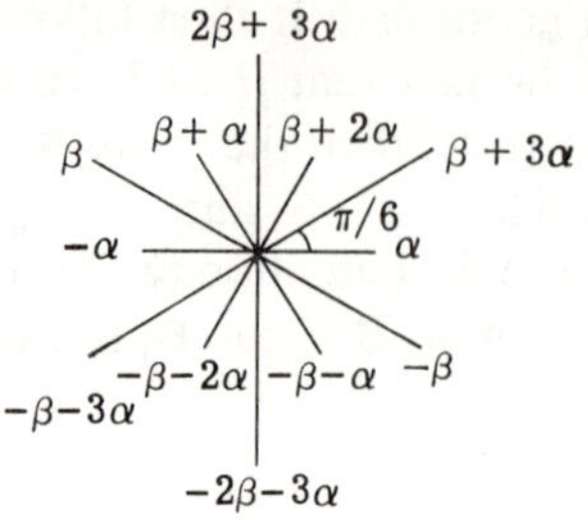


G$_2$

for each of which $(\alpha, \beta)$ is a basis. (Use Problem 1 and (21.10.4).)

3. Let $\alpha_1, \alpha_2, \ldots, \alpha_r$ be linearly independent roots in a reduced root system **S**.
(a) Suppose that a root $\alpha \in \mathbf{S}$ is of the form

$$\alpha = c_1\alpha_1 + c_2\alpha_2 + \cdots + c_p\alpha_p - c_{p+1}\alpha_{p+1} - \cdots - c_r\alpha_r$$

where the $c_j$ are real numbers $\geqq 0$. Show that there exists at least one integer $j \in [1, r]$ such that $\alpha - \alpha_j$ is a root if $j \leqq p$, and such that $\alpha + \alpha_j$ is a root if $j > p$. (Assume the contrary, and show that it leads to $(\alpha | \alpha) \leqq 0$.)
(b) Suppose now that the $c_j$ are *integers* $\geqq 0$. Show that there exists a sequence of indices $(j_k)_{1 \leq k \leq s}$ between 1 and $r$, and a sequence $(\varepsilon_k)_{1 \leq k \leq s}$ of numbers equal to $\pm 1$, such that the linear combinations

$$\varepsilon_1\alpha_{j_1}, \quad \varepsilon_1\alpha_{j_1} + \varepsilon_2\alpha_{j_2}, \ldots, \varepsilon_1\alpha_{j_1} + \varepsilon_2\alpha_{j_2} + \cdots + \varepsilon_s\alpha_{j_s},$$

are roots, the last one being equal to $\alpha$.
(c) In particular, if **B** is a basis of **S** and $\alpha$ is a positive root of **S** (relative to **B**), then there exists a sequence $(\alpha_1, \alpha_2, \ldots, \alpha_s)$ of roots belonging to **B** such that $\alpha_1$, $\alpha_1 + \alpha_2$, $\alpha_1 + \alpha_2 + \alpha_3, \ldots, \alpha_1 + \alpha_2 + \cdots + \alpha_s$ are roots, the last one being equal to $\alpha$.

4. Let G be a compact connected Lie group and T a maximal torus of G; let **S**(G) be the corresponding root system and **S**′ a subset of **S**(G). Show that for there to exist a connected closed subgroup G′ of G *containing* T, such that **S**′ is the root system **S**(G′) of G′ relative to T, it is necessary and sufficient that the following conditions should be satisfied:

(i) There exists a subset **B**′ of **S**′, consisting of linearly independent roots, and such that every element of **S**′ is a linear combination of elements of **B**′ with rational integer coefficients.

(ii) Every linear combination of elements of **S**′ with rational integer coefficients that belongs to **S**(G) belongs to **S**′.

(To show that condition (ii) is necessary, use Problem 3(b) together with (21.10.5.2). To show that the conditions are sufficient, consider in the Lie algebra $\mathfrak{t}$ of T the union $\Delta$ of the hyperplanes given by the equations $\alpha(\mathbf{u}) = 2\pi i n$, where $\alpha \in \mathbf{S}'$ and $n \in \mathbf{Z}$, and the set P of special points of $\Delta$ (Section 21.10, Problem 2). Show that the identity component G′ of the centralizer of P in G (i.e., the subgroup of elements $s \in$ G such that $\mathrm{Ad}(s) \cdot \mathbf{z} = \mathbf{z}$ for all $\mathbf{z} \in$ P) has the required properties, by showing that there exists no root $\alpha \in \mathbf{S}(\mathrm{G}')$ that does not belong to **S**′ and is such that $(2\pi i)^{-1}\alpha(\mathbf{z})$ is an integer for all $\mathbf{z} \in$ P: consider in turn the cases where $\alpha$ is linearly independent of **B**′, and where it is linearly dependent on **B**′.)

A connected closed subgroup $G' \neq G$, with rank equal to that of G, is necessarily the identity component of the centralizer in G of its center. If there exists no connected closed subgroup $G''$ of G such that $G' \subset G'' \subset G$, with $G''$ distinct from $G'$ and G, then $G'$ is also the identity component of the centralizer in G of any element of its center that does not belong to the center of G.

Show that $G'$ is also the identity component of the normalizer of its center in G. (Note that the group of automorphisms of a compact commutative Lie group is discrete.) In order that $G'$ should be the identity component of the centralizer of an element of its center, it is necessary and sufficient that there should exist a special point of $D(G')$ that is not contained in $D(G)$.

5. Let **S**, **S**$'$ be reduced root systems in real vector spaces $E^*$, $E'^*$, respectively, and let **B**, **B**$'$ be bases of **S**, **S**$'$, respectively. Suppose that there exists a bijection $\varphi$ of **B** onto **B**$'$ such that $n(\varphi(\alpha), \varphi(\beta)) = n(\alpha, \beta)$ for all pairs $\alpha, \beta \in$ **B**. Show that there then exists a unique linear bijection $f$ of $E^*$ onto $E'^*$ that extends $\varphi$ and maps **S** onto **S**$'$. (Consider the reflections $\sigma_\alpha$ and $\sigma_{\varphi(\alpha)}$.)

6. Show that if **B** is a basis of a reduced root system **S**, then **B** is the only basis of **S** that consists of positive roots relative to **B**.

7. Let G be a compact connected group, $G_1$ a connected closed subgroup of G, and T a maximal torus of G such that $T_1 = T \cap G_1$ is a maximal torus of $G_1$; let $\mathfrak{g}, \mathfrak{g}_1, \mathfrak{t}, \mathfrak{t}_1$ be the Lie algebras of G, $G_1$, T, $T_1$, respectively. For each root $\lambda$ of $G_1$ relative to $T_1$, let $R(\lambda)$ denote the set of roots $\alpha \in$ **S**(G) whose restriction to $\mathfrak{t}_1$ is equal to $\lambda$ (Section **21.10**, Problem 1).
(a) If $\mu$ is a root of $G_1$ that is the transform of $\lambda$ under an element of the Weyl group of $G_1$, show that $R(\mu)$ is the transform of $R(\lambda)$ under an element of the Weyl group of G (Section **21.7**, Problem 8).
(b) For each root $\lambda \in$ **S**$(G_1)$, let $K_\lambda$ be the corresponding almost simple subgroup of $G_1$ of rank 1 (**21.8.5**). Show that there exists a connected closed subgroup $G_\lambda$ of G containing T, whose root system **S**$(G_\lambda)$ consists of the integral linear combinations of the roots belonging to $R(\lambda)$ that are roots of G (Problem 4), and that $K_\lambda$ is contained in $G_\lambda$.
(c) The subgroup $G_1$ is said to be *nice* if it is contained in no connected closed subgroup $G' \neq G$ of rank *equal* to the rank of G. Show that the center of $G_1$ is then the intersection of $G_1$ with the center of G. (Consider the identity component of the centralizer in G of an element of the center of $G_1$.) In particular, if G is semisimple, every nice subgroup of G is semisimple.
(d) If $G_1$ is a nice subgroup of G, and if $G_2$ is a connected closed subgroup of G, containing $G_1$ and distinct from G or $G_1$, show that the ranks of G, $G_1$, and $G_2$ are all distinct. (Show that if the ranks of $G_1$ and $G_2$ were equal, then $G_1$ and $G_2$ would have the same center; then use Problem 4 to obtain a contradiction.)
(e) Let **B**$(G_1)$ be a basis of the root system **S**$(G_1)$, and let L be the union of the sets $R(\rho)$ for $\rho \in$ **B**$(G_1)$. Show that for each root $\lambda \in$ **S**$(G_1)$, the roots $\alpha \in R(\lambda)$ are integral linear combinations of the roots belonging to L. (Observe that there exists an element $w \in W(G_1)$ such that $w \cdot \lambda = \rho$ belongs to **B**$(G_1)$, and that $w$ is a product of reflections $s_\gamma$ with $\gamma \in$ **B**$(G_1)$; on the other hand, $s_\gamma$ is the restriction to $\mathfrak{t}_1$ of a product of reflections $s_\alpha \in W(G_\gamma) \subset W(G)$ (see (b) above), and for each root $\beta \in$ **S**(G), the vector $s_\alpha(\beta) - \beta$ is an integral linear combination of roots belonging to $R(\gamma)$.) In particular, if $G_1$ is nice, every root in **S**(G) is an integral linear combination of roots belonging to L. (Consider the

connected closed subgroup G′ of G containing T, whose root system $\mathbf{S}(G')$ consists of the integral linear combinations of roots belonging to the union of the sets $R(\lambda)$, for $\lambda \in \mathbf{S}(G_1)$.)

8. Let G be a compact connected semisimple Lie group, $\mathbf{B} = \mathbf{B}(G)$ a basis of the root system $\mathbf{S}(G)$ of G relative to a maximal torus T, with Lie algebra t. A *diagonal* of G (relative to $\mathbf{B}$) is by definition a line in t defined by a system of linear equations of the form

$$\beta_1(\mathbf{x}) = \beta_2(\mathbf{x}) = \cdots = \beta_k(\mathbf{x}) = 0, \qquad \beta_{k+1}(\mathbf{x}) = \cdots = \beta_l(\mathbf{x}),$$

where $\mathbf{B} = \{\beta_1, \beta_2, \ldots, \beta_l\}$. A diagonal is *principal* if $k = 0$ (or, equivalently, if it contains a regular element of t). A diagonal always contains a special point of D(G), other than the origin (Section 21.10, Problem 2).

Under the general hypotheses of Problem 7, let $R_1$ be a principal diagonal of $G_1$, containing points $\mathbf{u} \in t$ such that, for each of the roots $\beta_j$ $(1 \leqq j \leqq l)$ of $\mathbf{B}(G)$, we have $-i\beta_j(\mathbf{u}) \geqq 0$. Let $\mathbf{B}(G_1) = \{\rho_1, \ldots, \rho_h\}$ be a basis of $\mathbf{S}(G_1)$ such that $\rho_1(\mathbf{x}) = \cdots = \rho_h(\mathbf{x})$ are the equations defining $R_1$. Show that if $G_1$ is a nice subgroup of G, every root belonging to the union of the $R(\rho_j)$ $(1 \leqq j \leqq h)$ is of the form $\beta_{k+j} + m_1\beta_1 + \cdots + m_k\beta_k$ where $\beta_1, \ldots, \beta_k$ are the roots of the basis $\mathbf{B}(G)$ that vanish on $R_1$, the index $j$ satisfies $1 \leqq j \leqq l - k$, and the $m_r$ $(1 \leqq r \leqq k)$ are integers $\geqq 0$. (Observe that the point $\mathbf{z} \in R_1$ at which all the roots $\rho_j$ take the value $2\pi i$ is a special point of D(G), and use Problem 7(c).) Deduce that $R_1$ is a diagonal of G.

If $R_1$ is a *principal* diagonal of G, then $k = 0$ and the $R(\rho_j)$ form a *partition* of $\mathbf{B}(G)$. When this is so, the restriction to $t_1$ of a root $\alpha$ of G is never identically zero, and there exists a root $\rho$ of $G_1$ such that $\alpha$ is a positive integral linear combination of the roots belonging to $R(\rho)$. (Observe that the trace on $t_1$ of the hyperplane given the equation $\alpha(\mathbf{z}) = 0$ must be one of the hyperplanes $\rho_j(\mathbf{x}) = 0$ $(1 \leqq j \leqq h)$, by expressing $\alpha$ (resp. its restriction to $t_1$) as a linear combination of the $\beta_j$ (resp. the $\rho_j$).) Deduce that the restriction of $\alpha$ to $t_1$ is a scalar multiple of a root belonging to $\mathbf{S}(G_1)$. If $G_1$ is a nice subgroup of G and $R_1$ is a *principal* diagonal of G, then $G_1$ is said to be a *principal nice* subgroup of G.

9. Let $G_1$ be a *principal nice* subgroup of G. With the notation of Problem 8, let $(\lambda \mid \mu)$ be the scalar product on $it_1^*$ induced by that on $it^*$.

(a) Show that if $(\rho_1 \mid \rho_2) = 0$, then $(\beta' \mid \beta'') = 0$ for all $\beta' \in R(\rho_1)$ and $\beta'' \in R(\rho_2)$. (Observe that, by virtue of Problem 8, the restriction of $\beta' + \beta''$ to $t_1$ cannot belong to $\mathbf{S}(G_1)$, and therefore $\beta' + \beta''$ is not a root of G.) Deduce that if G is almost simple then so also is $G_1$.

(b) Let $\rho_1, \rho_2$ be two roots belonging to $\mathbf{B}(G_1)$ such that $(\rho_1 \mid \rho_2) \neq 0$ (and hence $(\rho_1 \mid \rho_2) < 0$), so that $\rho = \rho_1 + \rho_2$ is a root of $G_1$. We may assume (Problem 2) that the reflection $s_{\rho_2}$ interchanges $\rho$ and $\rho_1$; $s_{\rho_2}$ is the restriction to $t_1$ of an element $w \in W(G_{\rho_2})$ (Problem 7(b)), and $w$ interchanges $R(\rho)$ and $R(\rho_1)$. If $R(\rho_1) = \{\alpha_1, \ldots, \alpha_n\}$, show that for each $j = 1, 2, \ldots, n$ there exists a root $\gamma_j \in R(\rho_2)$ such that $w \cdot \alpha_j = \alpha_j + \gamma_j$, and that if $j \neq h$ the vector $w \cdot \alpha_j - \alpha_h$ does not belong to $R(\rho_2)$. (Consider the restrictions of $\alpha_j$ and $w \cdot \alpha_j$ to $t_1$.) Deduce that $(\alpha_j \mid \gamma) = 0$ for all $\gamma \neq \gamma_j$ in $R(\rho_2)$.

(c) With the hypotheses and notation of (b), show that for each root $\gamma \in R(\rho_2)$ there exists at least one root $\alpha_j \in R(\rho_1)$ such that $\gamma = \gamma_j$, so that the number of elements of $R(\rho_2)$ is $\leqq n$. (Use (b) above.) Furthermore, if $\gamma$ and $\gamma'$ are distinct elements of $R(\rho_2)$, then $(\gamma \mid \gamma') = 0$. (Express that $(w \cdot \alpha_j \mid w \cdot \alpha_h) = (\alpha_j \mid \alpha_h)$.) In particular, if $\|\rho_1\| = \|\rho_2\|$, then $R(\rho_1)$ and $R(\rho_2)$ have the same number of elements, and we have $(\alpha_j \mid \gamma_j) < 0$, and $(\alpha_j \mid \alpha_h) = (\gamma_j \mid \gamma_h) = (\alpha_j \mid \gamma_h) = 0$ for $j \neq h$.

**10.** (a) Let $\mathbf{S}$ be a reduced root system and $\mathbf{B}$ a basis of $\mathbf{S}$. Suppose that there exists a partition $(\mathbf{B}_1, \mathbf{B}_2)$ of $\mathbf{B}$ such that $(\lambda \mid \mu) = 0$ for each $\lambda \in \mathbf{B}_1$ and each $\mu \in \mathbf{B}_2$. Show that every positive root (relative to $\mathbf{B}$) is a linear combination (with integral coefficients $\geqq 0$) of roots that all belong either to $\mathbf{B}_1$ or to $\mathbf{B}_2$. (Use Problem 3(c) to argue by induction, proceeding as in (21.11.11(iii)).) Deduce that $\mathbf{S}$ admits a partition $(\mathbf{S}_1, \mathbf{S}_2)$, where $\mathbf{S}_1$ (resp. $\mathbf{S}_2$) is a reduced root system with basis $\mathbf{B}_1$ (resp. $\mathbf{B}_2$).
(b) A reduced root system $\mathbf{S}$ is said to be *irreducible* if the action of the Weyl group $W_{\mathbf{S}}$ on the vector space spanned by $\mathbf{S}$ is irreducible. If G is a compact semisimple Lie group, show that the root system $\mathbf{S}(G)$ is irreducible if and only if G is almost simple. (Use (a) above.)
(c) Let $\mathbf{S}$ be an irreducible reduced root system spanning a real vector space $E^*$, and let $(\lambda \mid \mu)$ be a scalar product on $E^*$ that is invariant under the Weyl group $W_{\mathbf{S}}$. Show that every $W_{\mathbf{S}}$-invariant bilinear form on $E^* \times E^*$ is a scalar multiple of $(\lambda \mid \mu)$. (Any bilinear form $\Phi$ on $E^* \times E^*$ can be written uniquely as $(\lambda, \mu) \mapsto (u(\lambda) \mid \mu)$, where $u$ is an endomorphism of $E^*$. If $\Phi$ is $W_{\mathbf{S}}$-invariant, then $u$ commutes with all elements of $W_{\mathbf{S}}$, and in particular with the reflections $\sigma_\alpha$. Deduce that $u$ leaves fixed the lines $\mathbf{R}\alpha$, and hence that there exists a scalar $c \in \mathbf{R}$ such that the kernel of $u - c \cdot 1$ is nonzero, hence is equal to $E^*$ by virtue of the hypothesis of irreducibility.)

**11.** Let $\mathbf{S}$ be a reduced root system.
(a) Show that there exists, on the real vector space $E^*$ spanned by $\mathbf{S}$, a unique scalar product—called the *canonical* scalar product—satisfying the relation

$$(*) \qquad (\lambda \mid \mu) = \sum_{\alpha \in \mathbf{S}} (\alpha \mid \lambda)(\alpha \mid \mu)$$

for all $\lambda, \mu \in E^*$. (Use Problem 10 to reduce to the case where $\mathbf{S}$ is irreducible, and consider the bilinear form

$$(\lambda, \mu) \mapsto \sum_{\alpha \in \mathbf{S}} (\alpha \mid \lambda)(\alpha \mid \mu)$$

where $(\lambda \mid \mu)$ is any $W_{\mathbf{S}}$-invariant scalar product on $E^*$.)
(b) Show that for the canonical scalar product we have

$$\sum_{\alpha \in \mathbf{S}} (\alpha \mid \alpha) = \dim(E^*).$$

(Observe that if $M$ is the matrix with entries $(\lambda \mid \mu)$ where $\lambda, \mu \in \mathbf{S}$, then $M^2 = M$, and the rank of $M$ is equal to $\dim(E^*)$.)
(c) If $\mathbf{S}$ is irreducible, show that there exists a constant $\gamma(\mathbf{S})$ such that

$$\sum_{\alpha \in \mathbf{S}} \frac{(\alpha \mid \lambda)(\alpha \mid \mu)}{(\alpha \mid \alpha)^2} = \gamma(\mathbf{S})(\lambda \mid \mu).$$

(Consider the bilinear form $(\lambda, \mu) \mapsto \sum_{\alpha \in \mathbf{S}} \lambda(\mathbf{h}_\alpha)\mu(\mathbf{h}_\alpha)$.)

**12.** (a) Let G be a compact connected semisimple Lie group $\neq \{e\}$, and let T be a maximal torus of G. If $v$ is an automorphism of G such that $v(T) = T$, and such that ${}^t v_*^{-1}$ fixes each of the roots of G relative to T, show that the automorphism $u = v_* \otimes 1$ of $\mathfrak{g}_{(\mathbf{C})}$ is such that, with the notation of (21.10.3),

$$u(\mathbf{x}_\alpha) = v_\alpha \, \mathbf{x}_\alpha$$

for all roots $\alpha \in \mathbf{S}(G)$, where the $v_\alpha$ are complex numbers of absolute value 1 and satisfy the relations

$$(*) \qquad v_\alpha v_{-\alpha} = 1, \qquad v_\alpha v_\beta = v_{\alpha+\beta} \quad \text{if} \quad \alpha + \beta \in \mathbf{S}(G).$$

Show that there exists an element $\mathbf{u} \in \mathfrak{t}$ such that $e^{\alpha(\mathbf{u})} = v_\alpha$ for each $\alpha \in \mathbf{S}(G)$. (Observe that, in view of the relations $(*)$, it is enough that $e^{\alpha(\mathbf{u})} = v_\alpha$ for each root $\alpha$ in a *basis* of $\mathbf{S}(G)$.)

(b) Deduce from (a) that the Lie group Aut(G)/Int(G) is finite. (Use the conjugacy theorem.)

(c) Deduce from (b) that for any compact connected Lie group G, the Lie group Aut(G)/Int(G) is discrete. (Use **(19.13.3)**.)

(d) Show that if G is semisimple and if $v \in \text{Aut}(G)$ is such that $v(T) = T$, then there exists $x \in G$ such that $\text{Int}(x) \circ v$ fixes each point of some one-parameter subgroup of G. (Argue as in **(21.11.10)**.)

**13.** Let $\mathbf{S}$ be a reduced root system.

(a) Let $\mathbf{B}$ be a basis of $\mathbf{S}$. Two roots $\alpha, \beta \in \mathbf{B}$ are said to be *linked* if $(\alpha|\beta) \neq 0$, or equivalently if the Cartan integers $n(\alpha, \beta)$ and $n(\beta, \alpha)$ are $\leqq -1$. Show that there exists no sequence $(\alpha_1, \ldots, \alpha_r)$ of distinct elements of $\mathbf{B}$ such that $\alpha_i$ and $\alpha_{i+1}$ are linked for

$$1 \leqq i \leqq r - 1,$$

and such that $\alpha_r$ and $\alpha_1$ are also linked. (If $\lambda = \alpha_1 + \alpha_2 + \cdots + \alpha_r$, show that the hypothesis would imply that $(\lambda|\lambda) \leqq 0$, by using **(21.11.5.3)**.)

(b) Deduce from (a) that there exists a root $\alpha \in \mathbf{B}$ that is linked to *at most one* other root in $\mathbf{B}$. (Consider a sequence of maximum length $(\alpha_1, \ldots, \alpha_r)$ of roots in $\mathbf{B}$ such that $\alpha_i$ is linked to $\alpha_{i+1}$ for $1 \leqq i \leqq r - 1$.)

(c) Suppose that $\mathbf{B}$ has at least two elements. Show that there exists a partition of $\mathbf{B}$ into two nonempty subsets $\mathbf{B}'$, $\mathbf{B}''$ such that no two roots of $\mathbf{B}'$ or of $\mathbf{B}''$ are linked. (Prove that this result is true not only for $\mathbf{B}$ but more generally for any subset F of $\mathbf{B}$ containing at least two elements, by induction on the number of elements $m$ of F and by observing that F also has the property (b) above.)

(d) Let $\alpha_1, \ldots, \alpha_r$ be distinct elements of $\mathbf{B}$. Show that in the Weyl group $W_{\mathbf{S}}$ all the products $\sigma_{\alpha_{\pi(1)}} \sigma_{\alpha_{\pi(2)}} \cdots \sigma_{\alpha_{\pi(r)}}$, where $\pi$ is any permutation of $\{1, 2, \ldots, r\}$, are *conjugate*. (Observe that if $\alpha_i$ and $\alpha_j$ are not linked, then $\sigma_{\alpha_i}$ and $\sigma_{\alpha_j}$ commute. Then show more generally that if $\alpha_j \mapsto u(\alpha_j)$ is a mapping into any group $\Gamma$ such that $u(\alpha_i)$ and $u(\alpha_j)$ commute whenever $\alpha_i$ and $\alpha_j$ are not linked, all the products $u(\alpha_{\pi(1)}) \cdots u(\alpha_{\pi(r)})$ are conjugate in $\Gamma$. Prove this result by induction on $r$, by using (b) above and reducing to the case where $\alpha_{\pi(r)}$ is linked to at most one other $\alpha_j$, and then to the case where this $\alpha_j$ (if it exists) is $\alpha_{\pi(r-1)}$; in this case, observe that the inductive hypothesis may be applied to the elements $\alpha_{\pi(1)}, \ldots, \alpha_{\pi(r-1)}$ and the mapping $u'$ that coincides with $u$ on $\alpha_{\pi(1)}, \ldots, \alpha_{\pi(r-2)}$ and is such that $u'(\alpha_{\pi(r-1)}) = u(\alpha_{\pi(r-1)})u(\alpha_{\pi(r)})$.)

**14.** Let $\mathbf{S}$ be a reduced root system spanning a vector space E*, and let $\mathbf{B} = \{\beta_1, \ldots, \beta_l\}$ be a basis of $\mathbf{S}$. For every permutation $\pi$ of $\{1, 2, \ldots, l\}$, the product $\sigma_{\beta_{\pi(1)}} \cdots \sigma_{\beta_{\pi(l)}}$ is called a *Coxeter element* of the Weyl group $W_{\mathbf{S}}$.

(a) The conjugacy class in $W_{\mathbf{S}}$ of a Coxeter element depends neither on the permutation $\pi$ nor on the choice of basis $\mathbf{B}$ (cf. Problem 13(d)).

(b) Suppose that $\mathbf{B}$ is numbered in such a way that $\beta_1, \ldots, \beta_r$ are pairwise orthogonal

and $\beta_{r+1}, \ldots, \beta_l$ are pairwise orthogonal (Problem 13(c)). Put $\sigma' = \sigma_{\beta_1}\sigma_{\beta_2} \cdots \sigma_{\beta_r}$ and $\sigma'' = \sigma_{\beta_{r+1}} \cdots \sigma_{\beta_l}$, so that $\sigma = \sigma'\sigma''$ is a Coxeter element; $\sigma'$ is the orthogonal symmetry with respect to the subspace V′ of E* orthogonal to $\beta_1, \beta_2, \ldots, \beta_r$, and $\sigma''$ is the orthogonal symmetry with respect to the subspace V″ of E* orthogonal to $\beta_{r+1}, \ldots, \beta_l$, so that $E^* = V' \oplus V''$. Deduce that no eigenvalue of $\sigma$ is equal to 1. (Observe that if $\sigma'(\lambda) = \sigma''(\lambda)$, then $\lambda - \sigma'(\lambda) = \lambda - \sigma''(\lambda)$ is orthogonal to both V′ and V″.)
(c) The order $h$ of $\sigma$ in $W_{\mathbf{S}}$ is called the *Coxeter number* of **S**. The characteristic polynomial P(T) of $\sigma$ may be written as

$$P(T) = \prod_{j=1}^{l} \left(T - \exp\frac{2\pi i m_j}{h}\right),$$

where $m_1, \ldots, m_l$ are integers such that $0 \leqq m_1 \leqq m_2 \leqq \cdots \leqq m_l < h$. Show that $m_j + m_{l+1-j} = h$ for $1 \leqq j \leqq l$, and that $m_1 > 0$. (Use the fact that the coefficients of P(T) are real.)

**15.** Let F be a real vector space of dimension $l$ and let U′, U″ be two supplementary subspaces of respective dimensions $r$ and $l - r$. Suppose that F is equipped with a scalar product $(x \mid y)$. Let $e_1, \ldots, e_r$ be an orthonormal basis of U′, and $e_{r+1}, \ldots, e_l$ an orthonormal basis of U″. Relative to the basis $(e_1, \ldots, e_l)$ of F, the matrix of the bilinear form $(x \mid y)$ is $A = ((e_j \mid e_k))_{1 \leq j, k \leq l}$; it is positive definite.
(a) Suppose that $(e_j \mid e_{r+k}) \leqq 0$ for each pair of indices $j, k$ such that $1 \leqq j \leqq r$ and $1 \leqq k \leqq l - r$. If $z' \in U'$ and $z'' \in U''$ are two nonzero vectors and if $\theta$ is the angle between them, show that the smallest value of $1 + \cos\theta$ is the smallest eigenvalue $\lambda_0$ of the matrix $A$ (15.11.7).
(b) Suppose in addition that there exists no partition of $\{1, 2, \ldots, l\}$ into two subsets I, I′ such that $(e_i \mid e_j) = 0$ for all $i \in I$ and $j \in I'$. Show that the eigenspace $N(\lambda_0)$ of the self-adjoint operator $u$ on $\mathbf{R}^l$ defined by the matrix $A$ is one-dimensional, spanned by a vector all of whose components are positive. (If $(\xi_j)_{1 \leq j \leq l}$ is a vector in $N(\lambda_0)$, i.e., a vector orthogonal to $\mathbf{R}^l$ relative to the positive semidefinite quadratic form with matrix $A - \lambda_0 I = (a_{ij})$ on $\mathbf{R}^l$, observe that the vector $(|\xi_j|)$ also belongs to $N(\lambda_0)$, by showing that $\sum_{i,j} a_{ij}|\xi_i| \cdot |\xi_j| \leqq \sum a_{ij}\xi_i\xi_j$, and then that if I is the set of indices $i$ such that $\xi_i \neq 0$, we have $a_{ji}|\xi_i| \leqq 0$ for $i \in I$ and $j \notin I$, and $\sum_i |a_{ji}|\xi_i = 0$ for $j \notin I$. Conclude that either $I = \emptyset$ or $I = \{1, 2, \ldots, l\}$.)
(c) Under the hypotheses of (b), deduce that there exists a unique vector

$$z' = \xi_1 e_1 + \cdots + \xi_r e_r$$

in U′ and a unique vector $z'' = \xi_{r+1}e_{r+1} + \cdots + \xi_l e_l$ in U″ such that $\|z'\| = \|z''\| = 1$, the $\xi_j$ are all positive, and the angle $\theta$ between $z'$ and $z''$ satisfies $1 + \cos\theta = \lambda_0$. Show that the line $\mathbf{R}z''$ (resp. $\mathbf{R}z'$) is the orthogonal projection of $\mathbf{R}z'$ (resp. $\mathbf{R}z''$) on U″ (resp. U′). (Use the minimal property of $\cos\theta$.)
(d) Let V′ (resp. V″) be the orthogonal supplement of U′ (resp. U″) in F. If $(f_j)_{1 \leq j \leq l}$ is the basis of F such that $(e_j \mid f_k) = \delta_{jk}$ (Kronecker delta), then $f_1, \ldots, f_r$ form a basis of V″ and $f_{r+1}, \ldots, f_l$ a basis of V′; also $e_j$ is the orthogonal projection of $f_j$ on U′ ($1 \leqq j \leqq r$), and $e_{r+j}$ the orthogonal projection of $f_{r+j}$ on U″ ($1 \leqq j \leqq l - r$). Let $x'$ (resp. $x''$) be the vector in V″ (resp. V′) whose projection on U′ (resp. U″) is $z'$ (resp. $z''$). The plane $P = \mathbf{R}z' + \mathbf{R}z''$ is also equal to $\mathbf{R}x' + \mathbf{R}x''$. If $s'$ (resp. $s''$) is the reflection in V′ (resp. V″), then P is stable under $s'$ and $s''$, and $s' \mid P$, $s'' \mid P$ are the reflections in the lines $\mathbf{R}x'$, $\mathbf{R}x''$, respectively. If C is the set of vectors $y \in F$ such that $(e_j \mid y) > 0$ for $1 \leqq j \leqq l$, then we have $x' \in \bar{C}$, $x'' \in \bar{C}$, and $P \cap C$ is the set of linear combinations $ax' + bx''$ with $a > 0$ and $b > 0$.

16. With the hypotheses and notation of Problem 14, suppose in addition that **S** is *irreducible* (Problem 10). Let C be the set of $\lambda \in E^*$ such that $(\beta_j|\lambda) > 0$ for $1 \leqq j \leqq l$ (21.14.4). For each element $w \in W_{\mathbf{S}}$, either $w(C) = C$ or $w(C) \cap C = \varnothing$ (21.11.10.1).
(a) By using (21.11.5.3) and Problem 15, show that there exist $\gamma' \in V'$ and $\gamma'' \in V''$ such that $\sigma'$ and $\sigma''$ leave invariant the plane $P = \mathbf{R}\gamma' + \mathbf{R}\gamma''$, and that $\sigma'|P$, $\sigma''|P$ are the reflections in the lines $\mathbf{R}\gamma'$, $\mathbf{R}\gamma''$, respectively; also $P \cap C$ is the set of linear combinations of $\gamma'$ and $\gamma''$ with positive coefficients. Show that the restrictions to P of the transformations belonging to the subgroup W′ of $W_{\mathbf{S}}$ generated by $\sigma'$ and $\sigma''$ form a group of order $2h$; deduce that $\sigma|P$ is a rotation through $2\pi/h$, and that $m_1 = 1$.
(b) Show that the only roots in **S** that are orthogonal to $\gamma'$ (resp. $\gamma''$) are $\beta_1, \ldots, \beta_r$ (resp. $\beta_{r+1}, \ldots, \beta_l$). (Observe that if $\alpha$ is orthogonal to $\gamma'$, then $\alpha$ is a linear combination of $\beta_1, \ldots, \beta_r$, by using (21.11.5); observe also that $\sigma_{\beta_j}(\alpha)$ is a linear combination of $\beta_1, \ldots, \beta_r$, for $1 \leqq j \leqq r$.) Deduce that the intersection of P with a hyperplane orthogonal to a root of **S** is necessarily the transform of $\mathbf{R}\gamma'$ or $\mathbf{R}\gamma''$ by an element of W′, and that the number of roots in **S** is $hl$.
(c) Let $\xi$ be an eigenvector of $\sigma \otimes 1$ in $P \otimes_{\mathbf{R}} C$ corresponding to the eigenvalue $\exp(2\pi i/h)$. Show that $(\xi|\alpha) \neq 0$ for each $\alpha \in \mathbf{S}$. (Use (b) above and observe that $\xi$ cannot be of the form $c\eta$ with $\eta \in P$ and $c \in C$.)
(d) A root $\theta \in S$ is said to be *pivotal* (with respect to $\sigma$) if $\theta > 0$ and $\sigma(\theta) < 0$. Show that the pivotal roots are $\theta_k = \sigma_{\beta_l}\sigma_{\beta_{l-1}} \cdots \sigma_{\beta_{k+1}}(\beta_k)$ $(1 \leqq k \leqq l)$ (use (21.11.6)) and deduce that the $\theta_k$ form a basis of $E^*$ (use (21.11.2.1)). If $\varpi_k$ $(1 \leqq k \leqq l)$ are the elements of $\bar{C}$ such that $(\varpi_k|\beta_j) = \delta_{jk}$, then $(1 - \sigma^{-1}) \cdot \varpi_k = \theta_k$. Deduce that two distinct pivotal roots $\theta_j$, $\theta_k$ cannot be in the same orbit in **S** under the action of the cyclic subgroup of $W_{\mathbf{S}}$ generated by $\sigma$ (use (21.14.7.1)). Show that every orbit in **S** under this subgroup contains a pivotal root (consider the sum of the elements of the orbit), and deduce that there are exactly $l$ orbits, each with $h$ elements.
(e) Show that, for each root $\alpha \in \mathbf{S}$, we have

$$\sum_{\beta \in \mathbf{S}} \frac{(\alpha|\beta)^2}{(\beta|\beta)} = h \cdot (\alpha|\alpha).$$

(Use Problem 10.)
(f) Show that for each positive integer $m$ less than $h$ and prime to $h$, the complex number $\exp(2\pi i m/h)$ is an eigenvalue of $\sigma$. (Observe that the matrix of $\sigma$ relative to the basis

$$\{\beta_1, \beta_2, \ldots, \beta_l\}$$

has integer entries, and use that fact that cyclotomic polynomials are irreducible over **Q**.)

17. Let G be a compact Lie group, *not necessarily connected*, and let $G_0$ be its identity component. Suppose that there exists an element $a \in G$ whose centralizer is *discrete*.
(a) Show that the connected component of $a$ in G consists of conjugates of $a$. (Show that the mapping $x \mapsto xax^{-1}$ of $G_0$ into G is a local diffeomorphism.)
(b) Deduce from (a) that $G_0$ is *commutative*. (If not, there would exist $z \in G_0$ such that Int$(za)$ leaves invariant a maximal torus of $G_0$ and the corresponding root system; now use Problem 12(d).)

18. (a) Let G be a compact connected semisimple Lie group $\neq \{e\}$. Show that for each automorphism $v$ of G, the subgroup of fixed points of $v$ is not discrete. (Apply Problem 17 to the compact Lie group Aut(G), using Problem 12 and the fact that the center of G is discrete.)

(b) Let G be a noncommutative compact connected Lie group, and let $v$ be an automorphism of G. Show that the identity component U of the subgroup of fixed points of $v$ is such that every maximal torus of U is contained in a unique maximal torus of G, and therefore contains regular elements of G. (Let V be a maximal torus of U, and let Z be the identity component of the centralizer of V in G. Show that $\mathrm{U} \cap \mathscr{D}(\mathrm{Z})$ is discrete, and use (a) to deduce that $\mathscr{D}(\mathrm{Z}) = \{e\}$.) Hence show that there exists a maximal torus T of G that is globally invariant under $v$, and a basis of the root system of G relative to T that is globally invariant under $v$. (Consider a regular element of G that belongs to U and is arbitrarily close to $e$.)

**19.** Let G be a compact connected *semisimple* Lie group $\neq \{e\}$, let $v$ be an automorphism of G, let F be the closed subgroup of fixed points of $v$, and let $\mathrm{F}_0$ be the identity component of F (so that $\mathrm{F}_0 \neq \{e\}$, by virtue of Problem 18(a)). Then every connected component of F contains regular elements of G (**21.7.13**). The proof is as follows:

(a) Let $x \in \mathrm{F}$ be a singular element of G. Then the identity component $\mathscr{Z}(x)_0$ of the centralizer of $x$ is not commutative (**21.8.4**). Show that a maximal torus S of the identity component of $\mathrm{F} \cap \mathscr{D}(\mathscr{Z}(x)_0)$ has dimension $\geqq 1$. Let U be the identity component of the intersection $\mathscr{Z}(\mathrm{S}) \cap \mathscr{Z}(x)$ of the centralizers of S and $x$, which is globally invariant under $v$, and has rank equal to the rank of G (**21.7.9**).

(b) Show that U is a maximal torus, by proving that its derived group $\mathscr{D}(\mathrm{U}) = \{e\}$. For otherwise $\mathscr{D}(\mathrm{U}) \cap \mathrm{F}$ would contain a torus S′ of dimension $\geqq 1$, and SS′ would be a torus in $\mathscr{Z}(x) \cap \mathrm{F}$ containing S properly.

(c) Show that it is not possible that all the elements of $x\mathrm{S} = \mathrm{S}x$ should be singular in G (cf. (**21.8.4.1**)).

## 12. EXAMPLES: THE CLASSICAL COMPACT GROUPS

### (21.12.1) *The groups* $\mathbf{U}(n)$ $(= \mathbf{U}(n, \mathbf{C}))$ *and* $\mathbf{SU}(n)$ $(n \geqq 2)$.

We shall show that the group T of diagonal matrices $\operatorname{diag}(\zeta_1, \ldots, \zeta_n)$, where each $\zeta_j \in \mathbf{U}$, the unit circle in $\mathbf{C}$ (so that T is isomorphic to $\mathbf{U}^n$) is a maximal torus in $\mathbf{U}(n)$. The Lie algebra $\mathfrak{u}(n)$ of $\mathbf{U}(n)$ is the Lie subalgebra of $\mathfrak{gl}(n, \mathbf{C}) = \mathbf{M}_n(\mathbf{C})$ consisting of the *antihermitian* matrices, i.e., the matrices $X$ such that ${}^t\bar{X} = -X$ (**19.4.3.3**); it has an $\mathbf{R}$-basis consisting of the $n$ matrices $iE_{rr}$ $(1 \leqq r \leqq n)$ and the $n^2 - n$ matrices $E_{rs} - E_{sr}$ and $i(E_{rs} + E_{sr})$

$$(1 \leqq r < s \leqq n),$$

so that the complexification of $\mathfrak{u}(n)$ is $\mathfrak{gl}(n, \mathbf{C})$. The $n$ matrices $iE_{rr}$ $(1 \leqq r \leqq n)$ form a basis of the Lie algebra $\mathfrak{t}$ of T. Since $[iE_{rr}, E_{rs}] = iE_{rs}$, $[iE_{rr}, E_{sr}] = -iE_{sr}$, and $[iE_{rr}, E_{pq}] = 0$ when $p \neq r$ and $q \neq r$ (**19.4.2.2**), we may write, for any $\mathbf{h} \in \mathfrak{t}$,

(21.12.1.1) $$[\mathbf{h}, E_{rs}] = (\varepsilon_r(\mathbf{h}) - \varepsilon_s(\mathbf{h}))E_{rs}$$

for $1 \leqq r, s \leqq n$ and $r \neq s$, where $\varepsilon_r(iE_{ss}) = i\delta_{rs}$ (Kronecker delta). This shows immediately that $\mathfrak{t}$ is a maximal commutative subalgebra of $\mathfrak{u}(n)$, and the center $\mathfrak{c}$ of $\mathfrak{u}(n)$ is the subalgebra $\mathbf{R}iI$ consisting of the pure imaginary multiples of the unit matrix; $\mathfrak{u}(n)$ is the direct sum of $\mathfrak{c}$ and the Lie algebra $\mathfrak{su}(n)$ of the group $\mathbf{SU}(n)$, consisting of the matrices in $\mathfrak{u}(n)$ with zero trace, and $\mathfrak{t}$ is the direct sum of $\mathfrak{c}$ and the maximal commutative subalgebra $\mathfrak{t}'$ of $\mathfrak{su}(n)$ generated by the matrices $i(E_{rr} - E_{ss})$. It follows from **(21.12.1.1)** that the roots of $\mathfrak{u}(n)$ relative to $\mathfrak{t}$ are the $\mathbf{R}$-linear mappings of $\mathfrak{t}$ into $i\mathbf{R}$

**(21.12.1.2)** $$\alpha_{rs} = \varepsilon_r - \varepsilon_s \qquad (1 \leqq r, s \leqq n, r \neq s).$$

The roots of $\mathfrak{su}(n)$ relative to $\mathfrak{t}'$ are obtained by identifying the $\alpha_{rs}$ with their restrictions to $\mathfrak{t}'$ **(21.8.8)**. If we write $\mathfrak{g} = \mathfrak{sl}(n, \mathbf{C})$, the complexification of $\mathfrak{su}(n)$, and $\mathfrak{h} = \mathfrak{t}_{(\mathbf{C})}$, then with the notation of **(21.8.1)** we have $\mathfrak{g}_{\alpha_{rs}} = \mathbf{C}E_{rs}$, and the element $\mathbf{h}_{\alpha_{rs}} \in \mathfrak{h}$ **(21.10.3)** is $E_{rr} - E_{ss}$. It is immediately verified from the definition **(21.11.5)** that the set of roots

**(21.12.1.3)** $$\beta_r = \alpha_{r, r+1} = \varepsilon_r - \varepsilon_{r+1} \qquad (1 \leqq r \leqq n-1)$$

is a *basis* of the root system, for which the positive roots are

**(21.12.1.4)** $$\varepsilon_r - \varepsilon_s = \beta_r + \beta_{r+1} + \cdots + \beta_{s-1} \qquad (1 \leqq r < s \leqq n).$$

It follows also from these facts and from **(21.11.11.3)** that the Cartan integers are given by

**(21.12.1.5)** $$\begin{aligned} n(\beta_r, \beta_s) &= 0 \qquad \text{if } |r-s| \geqq 2, \\ n(\beta_{r+1}, \beta_r) &= n(\beta_r, \beta_{r+1}) = -1 \qquad (1 \leqq r \leqq n-2). \end{aligned}$$

It follows that the group $\mathbf{SU}(n)$ is *almost simple* **(21.6.6)**. Indeed, from **(21.12.1.5)** it is clear that the interval $[1, n-1]$ of $\mathbf{N}$ cannot be partitioned into two nonempty subsets A, B such that $n(\beta_r, \beta_s) = 0$ whenever $r \in$ A and $s \in$ B; for if $r \in$ A, then we must also have $r + 1 \in$ A if $r \leqq n-2$, and $r - 1 \in$ A if $r \geqq 2$. The assertion therefore follows from **(21.11.11)**.

Finally, it follows from **(21.12.1.5)** that the reflection $\sigma_{\beta_r}$ **(21.11.11)**, where $1 \leqq r \leqq n-1$, interchanges $\varepsilon_r$ and $\varepsilon_{r+1}$ and fixes the other $\varepsilon_s$; hence the Weyl group may be identified with the *symmetric group* $\mathfrak{S}_n$ of all permutations of $[1, n]$.

**(21.12.2)** *The group* $\mathbf{U}(n, \mathbf{H})$ $(n \geqq 2)$.

We shall identify the division ring of quaternions

$$\mathbf{H} = \mathbf{R} \oplus \mathbf{R}i \oplus \mathbf{R}j \oplus \mathbf{R}k$$

with $\mathbf{C} \oplus \mathbf{C}j$, so that $jz = \bar{z}j$ for $z \in \mathbf{C}$, and the conjugate of the quaternion $x + yj$ $(x, y \in \mathbf{C})$ is $\bar{x} - yj$. The group $\mathbf{U}(n, \mathbf{C})$ is then identified with the subgroup of $\mathbf{U}(n, \mathbf{H})$ consisting of the unitary matrices with entries in the subfield $\mathbf{C}$ of $\mathbf{H}$. We shall show that the maximal torus T of $\mathbf{U}(n, \mathbf{C})$ defined in (21.12.1) is also a *maximal torus* of $\mathbf{U}(n, \mathbf{H})$. The Lie algebra $\mathfrak{u}(n, \mathbf{H})$ of $\mathbf{U}(n, \mathbf{H})$ consists of the *antihermitian* matrices $X \in \mathbf{M}_n(\mathbf{H})$, i.e., the matrices $X$ satisfying ${}^t\bar{X} = -X$ (19.4.3.2). Observe now that any matrix $X \in \mathbf{M}_n(\mathbf{H})$ can be written uniquely in the form $U + Vj$, where

$$U, V \in \mathbf{M}_n(\mathbf{C});$$

the relation ${}^t\bar{X} = -X$ is equivalent to the two relations

(21.12.2.1) $$ {}^t\bar{U} = -U, \qquad {}^tV = V.$$

Next, it is easily verified that the mapping $\varphi$: $\mathbf{M}_n(\mathbf{H}) \to \mathbf{M}_{2n}(\mathbf{C})$ defined by

(21.12.2.2) $$\varphi(U + Vj) = \begin{pmatrix} U & V \\ -\bar{V} & \bar{U} \end{pmatrix}$$

is an *injective homomorphism of* $\mathbf{R}$*-algebras*. Since $\mathfrak{gl}(n, \mathbf{H})$ (resp. $\mathfrak{gl}(2n, \mathbf{C})$) is just $\mathbf{M}_n(\mathbf{H})$ (resp. $\mathbf{M}_{2n}(\mathbf{C})$) with Lie algebra multiplication $[X, Y] = XY - YX$, it follows that $\varphi$ is also an injective homomorphism of the *real* Lie algebra $\mathfrak{gl}(n, \mathbf{H})$ into the *real* Lie algebra $\mathfrak{gl}(2n, \mathbf{C})$.

We shall from now on identify $\mathfrak{u}(n, \mathbf{H})$ with its image under $\varphi$. Then the Lie algebra $\mathfrak{t}$ of T has an $\mathbf{R}$-basis consisting of the matrices $i(E_{rr} - E_{n+r,n+r})$ for $1 \leqq r \leqq n$, and the Lie algebra $\mathfrak{u}(n, \mathbf{H})$ has an $\mathbf{R}$-basis consisting of this basis of $\mathfrak{t}$ and the matrices

$$\begin{aligned} &E_{rs} - E_{sr} - E_{n+s,n+r} + E_{n+r,n+s} && (1 \leqq r < s \leqq n), \\ &i(E_{rs} + E_{sr} - E_{n+s,n+r} - E_{n+r,n+s}) && (1 \leqq r < s \leqq n), \\ &E_{n+r,s} + E_{n+s,r} - E_{r,n+s} - E_{s,n+r} && (1 \leqq r \leqq s \leqq n), \\ &i(E_{n+r,s} + E_{n+s,r} + E_{r,n+s} + E_{s,n+r}) && (1 \leqq r \leqq s \leqq n). \end{aligned}$$

It follows as in (21.12.1) that the *complexification* of $\mathfrak{u}(n, \mathbf{H})$ may be identified with the Lie subalgebra of $\mathfrak{gl}(2n, \mathbf{C})$ consisting of the matrices

(21.12.2.3) $$Y = \begin{pmatrix} U & W \\ V & -{}^tU \end{pmatrix}$$

such that ${}^tV = V$ and ${}^tW = W$, the matrix $U \in \mathbf{M}_n(\mathbf{C})$ being arbitrary. It is immediately verified that these matrices $Y$ are those which satisfy the relation ${}^tY \cdot J + J \cdot Y = 0$, where

(21.12.2.4) $$J = \begin{pmatrix} 0 & I_n \\ -I_n & 0 \end{pmatrix},$$

and consequently the complexification of $\mathfrak{u}(n, \mathbf{H})$ may be identified, by virtue of **(19.4.3.3)**, with the Lie algebra $\mathfrak{sp}(2n, \mathbf{C})$ of the complex symplectic group $\mathbf{Sp}(2n, \mathbf{C})$, consisting of the matrices $Z \in \mathbf{M}_{2n}(\mathbf{C})$ that satisfy the relation

(21.12.2.5) $${}^tZ \cdot J \cdot Z = J.$$

A basis of $\mathfrak{sp}(2n, \mathbf{C})$ over $\mathbf{C}$ is therefore formed by the $n$ matrices $E_{rr} - E_{n+r,\,n+r}$ $(1 \leqq r \leqq n)$, which form a basis of the complexification $\mathfrak{h} = \mathfrak{t}_{(\mathbf{C})}$ of $\mathfrak{t}$, and the $2n^2$ matrices

$$E_{n+r,\,s} + E_{n+s,\,r}, \qquad E_{r,\,n+s} + E_{s,\,n+r} \quad (1 \leqq r \leqq s \leqq n),$$
$$E_{rs} - E_{n+s,\,n+r} \quad (1 \leqq r, s \leqq n,\ r \neq s).$$

Let $\varepsilon_r$ $(1 \leqq r \leqq n)$ be the $\mathbf{C}$-linear forms on $\mathfrak{h}$ defined by the equations $\varepsilon_r(E_{ss} - E_{n+s,\,n+s}) = \delta_{rs}$ $(1 \leqq s \leqq n)$. Then we have

(21.12.2.6)
$$\left.\begin{aligned} [\mathbf{h}, E_{n+r,\,s} + E_{n+s,\,r}] &= -(\varepsilon_r(\mathbf{h}) + \varepsilon_s(\mathbf{h}))(E_{n+r,\,s} + E_{n+s,\,r}) \\ [\mathbf{h}, E_{r,\,n+s} + E_{s,\,n+r}] &= (\varepsilon_r(\mathbf{h}) + \varepsilon_s(\mathbf{h}))(E_{r,\,n+s} + E_{s,\,n+r}) \end{aligned}\right\} \quad (1 \leqq r \leqq s \leqq n)$$
$$[\mathbf{h}, E_{rs} - E_{n+s,\,n+r}] = (\varepsilon_r(\mathbf{h}) - \varepsilon_s(\mathbf{h}))(E_{rs} - E_{n+s,\,n+r}) \quad (1 \leqq r, s \leqq n,\ r \neq s)$$

for all $\mathbf{h} \in \mathfrak{h}$. This shows on the one hand that $\mathfrak{h}$ is a maximal commutative subalgebra of $\mathfrak{sp}(2n, \mathbf{C})$ (and hence that $\mathfrak{t}$ is a maximal commutative subalgebra of $\mathfrak{u}(n, \mathbf{H})$), and on the other hand that the roots of $\mathfrak{sp}(2n, \mathbf{C})$ relative to $\mathfrak{h}$ (or of $\mathfrak{u}(n, \mathbf{H})$ relative to $\mathfrak{t}$) are the $2n^2$ linear forms

(21.12.2.7) $$\pm 2\varepsilon_r \quad (1 \leqq r \leqq n), \qquad \pm\varepsilon_r \pm \varepsilon_s \quad (1 \leqq r < s \leqq n).$$

If we put

(21.12.2.8) $$\beta_r = \varepsilon_r - \varepsilon_{r+1} \quad (1 \leqq r \leqq n-1), \qquad \beta_n = 2\varepsilon_n$$

then we have the formulas

$$\begin{aligned}
\varepsilon_r - \varepsilon_s &= \beta_r + \beta_{r+1} + \cdots + \beta_{s-1} && (r < s),\\
\varepsilon_r + \varepsilon_s &= \beta_r + \cdots + \beta_{s-1} + 2\beta_s + 2\beta_{s+1} + \cdots + 2\beta_{n-1} + \beta_n && (r < s \leqq n-1),\\
\varepsilon_r + \varepsilon_n &= \beta_r + \cdots + \beta_{n-1} + \beta_n && (r \leqq n-1),\\
2\varepsilon_r &= 2\beta_r + 2\beta_{r+1} + \cdots + 2\beta_{n-1} + \beta_n && (r \leqq n-1),
\end{aligned}$$

which show that the $n$ roots **(21.12.2.8)** form a *basis* of the root system; the positive roots corresponding to this basis are

**(21.12.2.9)** $$2\varepsilon_r \quad (1 \leqq r \leqq n), \qquad \varepsilon_r \pm \varepsilon_s \quad (1 \leqq r < s \leqq n).$$

Also we have

$$\begin{aligned}
\mathbf{h}_{\beta_r} &= E_{rr} - E_{n+r,n+r} - E_{r+1,r+1} + E_{n+r+1,n+r+1} && (1 \leqq r \leqq n-1),\\
\mathbf{h}_{\beta_n} &= E_{nn} - E_{2n,2n}
\end{aligned}$$

and therefore the Cartan integers are

**(21.12.2.10)** $$\begin{aligned}
n(\beta_r, \beta_s) &= 0 \quad \text{if} \quad |r-s| \geqq 2 && (1 \leqq r \leqq n-1),\\
n(\beta_r, \beta_{r+1}) &= n(\beta_{r+1}, \beta_r) = -1 && (1 \leqq r \leqq n-2),\\
n(\beta_{n-1}, \beta_n) &= -1, \quad n(\beta_n, \beta_{n-1}) = -2, &&\\
n(\beta_n, \beta_r) &= 0 && (1 \leqq r \leqq n-2).
\end{aligned}$$

The same proof as in **(21.12.1)** shows that $\mathbf{U}(n, \mathbf{H})$ is *almost simple.* Moreover, it is easily checked with the help of **(21.12.2.10)** that the reflection $\sigma_{\beta_r}$ **(21.11.11)** interchanges $\varepsilon_r$ and $\varepsilon_{r+1}$, and fixes the other $\varepsilon_s$, for

$$1 \leqq r \leqq n-1,$$

while $\sigma_{\beta_n}$ transforms $\varepsilon_n$ into $-\varepsilon_n$ and fixes the other $\varepsilon_s$. It follows that the Weyl group is the *semidirect product* **(19.14)** of the commutative group $\{-1, 1\}^n$ and the symmetric group $\mathfrak{S}_n$, for the action $(\pi, u) \mapsto \pi \cdot u$ of $\mathfrak{S}_n$ on $\{-1, 1\}^n$ defined by $\pi \cdot u = \pi \cdot (u_1, \ldots, u_n) = (u_{\pi^{-1}(1)}, \ldots, u_{\pi^{-1}(n)})$. Its order is therefore $2^n \cdot n!$.

When $n = 1$, the formulas above show that $\mathfrak{sp}(2, \mathbf{C})$ is isomorphic to $\mathfrak{sl}(2, \mathbf{C})$, hence $\mathbf{U}(1, \mathbf{H})$ is isomorphic to $\mathbf{SU}(2)$. This can also be proved directly (Section **21.8**, Problem 1).

**(21.12.3)** *The group* $\mathbf{SO}(2n)$ $(n \geqq 2)$.

The mapping that takes each endomorphism $u$ of $\mathbf{C}^n$ to the same $u$, considered as an endomorphism of $\mathbf{R}^{2n}$ **(16.21.13.1)**, is an injective homomorphism of the $\mathbf{R}$-algebra $\mathbf{M}_n(\mathbf{C})$ into the $\mathbf{R}$-algebra $\mathbf{M}_{2n}(\mathbf{R})$, under which the image of a matrix $(z_{jk}) \in \mathbf{M}_n(\mathbf{C})$, with $z_{jk} = x_{jk} + iy_{jk}$, is the matrix $(Z_{jk}) \in \mathbf{M}_{2n}(\mathbf{R})$, where

$$Z_{jk} = \begin{pmatrix} x_{jk} & -y_{jk} \\ y_{jk} & x_{jk} \end{pmatrix}$$

The image of the group $\mathbf{U}(n)$ under this homomorphism is a subgroup of $\mathbf{SO}(2n)$, which we shall identify with $\mathbf{U}(n)$. We shall show that the maximal torus T of $\mathbf{U}(n)$ defined in **(21.12.1)** is a maximal torus of $\mathbf{SO}(2n)$. The Lie algebra $\mathfrak{so}(2n)$ of $\mathbf{SO}(2n)$ consists of all real skew-symmetric matrices $X$ **(19.4.3.2)**, and its complexification is therefore the Lie algebra of all complex $2n \times 2n$ skew-symmetric matrices. By virtue of **(19.4.3.3)**, this is the Lie algebra $\mathfrak{so}(2n, \mathbf{C})$ of the *complex special orthogonal group* $\mathbf{SO}(2n, \mathbf{C})$, consisting of the complex matrices $Z \in \mathbf{M}_{2n}(\mathbf{C})$ of determinant 1, such that ${}^tZ \cdot Z = I_{2n}$. A basis over $\mathbf{C}$ of $\mathfrak{so}(2n, \mathbf{C})$ is formed by the $n$ matrices

$$H_r = i(E_{2r-1,\,2r} - E_{2r,\,2r-1}) \qquad (1 \leqq r \leqq n),$$

which form a basis of the complexification $\mathfrak{h} = \mathfrak{t}_{(\mathbf{C})}$ of the Lie algebra $\mathfrak{t}$ of T, and by the following $2n(n-1)$ matrices:

$$\bar{M}_{rs} - M_{sr}, \qquad \bar{N}_{rs} - N_{sr}$$

where $1 \leqq r, s \leqq n$, $r \neq s$, and

**(21.12.3.1)**
$$\begin{aligned} M_{rs} &= E_{2r-1,\,2s-1} + E_{2r,\,2s} + i(E_{2r,\,2s-1} - E_{2r-1,\,2s}) = {}^t\bar{M}_{sr}, \\ N_{rs} &= E_{2r-1,\,2s-1} - E_{2r,\,2s} - i(E_{2r,\,2s-1} + E_{2r-1,\,2s}) = {}^tN_{sr}. \end{aligned}$$

Let $\varepsilon_r$ $(1 \leqq r \leqq n)$ be the $\mathbf{C}$-linear forms on $\mathfrak{h}$ defined by the equations $\varepsilon_r(H_s) = \delta_{rs}$ $(1 \leqq s \leqq n)$. Then we have

**(21.12.3.2)**
$$\begin{aligned} [\mathbf{h}, \bar{M}_{rs} - M_{sr}] &= (\varepsilon_r(\mathbf{h}) - \varepsilon_s(\mathbf{h}))(\bar{M}_{rs} - M_{sr}), \\ [\mathbf{h}, \bar{N}_{rs} - N_{sr}] &= (\varepsilon_r(\mathbf{h}) + \varepsilon_s(\mathbf{h}))(\bar{N}_{rs} - N_{sr}), \end{aligned}$$

for $1 \leqq r, s \leqq n$, $r \neq s$, and all $\mathbf{h} \in \mathfrak{h}$. To verify these formulas it is enough to observe that $[H_r, \bar{M}_{rs}] = \bar{M}_{rs}$, $[H_r, N_{rs}] = N_{rs}$, and to use the relations **(21.12.3.1)** together with $\bar{H}_r = -H_r = {}^tH_r$. This shows firstly that $\mathfrak{h}$ is a maximal commutative subalgebra of $\mathfrak{so}(2n, \mathbf{C})$ (and hence that $\mathfrak{t}$ is a maximal commutative subalgebra of $\mathfrak{so}(2n, \mathbf{R})$), and secondly that the roots of $\mathfrak{so}(2n, \mathbf{C})$ relative to $\mathfrak{h}$ (or of $\mathfrak{so}(2n, \mathbf{R})$ relative to $\mathfrak{t}$) are the $2n(n-1)$ linear forms

**(21.12.3.3)** $$\pm\varepsilon_r \pm \varepsilon_s \qquad (1 \leqq r < s \leqq n).$$

If we put

(21.12.3.4) $\beta_r = \varepsilon_r - \varepsilon_{r+1} \quad (1 \leqq r \leqq n-1), \qquad \beta_n = \varepsilon_{n-1} + \varepsilon_n$,

then we have the formulas (for $n \geqq 4$)

$$\begin{aligned}
\varepsilon_r - \varepsilon_s &= \beta_r + \beta_{r+1} + \cdots + \beta_{s-1} && (r < s),\\
\varepsilon_r + \varepsilon_s &= \beta_r + \cdots + \beta_{s-1} + 2\beta_s + 2\beta_{s+1} + \cdots + 2\beta_{n-2} + \beta_{n-1} + \beta_n \\
& && (r < s \leqq n-2),\\
\varepsilon_r + \varepsilon_{n-1} &= \beta_r + \beta_{r+1} + \cdots + \beta_n && (r \leqq n-2),\\
\varepsilon_r + \varepsilon_n &= \beta_r + \beta_{r+1} + \cdots + \beta_{n-2} + \beta_n && (r \leqq n-2).
\end{aligned}$$

which show that the $n$ roots **(21.12.3.4)** form a *basis* of the root system; the positive roots corresponding to this basis are

(21.12.3.5) $\varepsilon_r \pm \varepsilon_s \quad (1 \leqq r < s \leqq n)$.

Also we have

$$\begin{aligned}
\mathbf{h}_{\beta_r} &= H_r - H_{r+1} \qquad (1 \leqq r \leqq n-1),\\
\mathbf{h}_{\beta_n} &= H_{n-1} + H_n,
\end{aligned}$$

from which we obtain the values of the Cartan integers for the basis **(21.12.3.4)**:

(21.12.3.6)
$$\begin{aligned}
n(\beta_r, \beta_s) &= n(\beta_s, \beta_r) && \text{for all } r, s,\\
n(\beta_r, \beta_s) &= 0 && (1 \leqq r \leqq n-3,\ s \geqq r+2),\\
n(\beta_r, \beta_{r+1}) &= -1 && (1 \leqq r \leqq n-2),\\
n(\beta_{n-2}, \beta_n) &= -1, \\
n(\beta_{n-1}, \beta_n) &= 0.
\end{aligned}$$

From these formulas, a proof analogous to that of **(21.12.1)** shows that $\mathbf{SO}(2n)$ is *almost simple* if $n \geqq 4$. Furthermore, the reflection $\sigma_{\beta_r}$ interchanges $\varepsilon_r$ and $\varepsilon_{r+1}$ and fixes the remaining $\varepsilon_s$, for $1 \leqq r \leqq n-1$, while $\sigma_{\beta_n}$ transforms $\varepsilon_{n-1}$ into $-\varepsilon_n$, $\varepsilon_n$ into $-\varepsilon_{n-1}$, and fixes the other $\varepsilon_s$. From this it follows that if E is the subgroup of the multiplicative group $\{-1, 1\}^n$ consisting of the elements $(u_1, \ldots, u_n)$ such that $u_1 u_2 \cdots u_n = 1$ (a subgroup that is isomorphic to $\{-1, 1\}^{n-1}$), the Weyl group is isomorphic to the *semidirect product* of E with the symmetric group $\mathfrak{S}_n$, the action being the same as in **(21.12.2)**. The order of the Weyl group is therefore $2^{n-1} \cdot n!$.

For $n = 3$, it may be shown that $\mathfrak{so}(6)$ is isomorphic to $\mathfrak{su}(4)$; for $n = 2$, that $\mathfrak{so}(4)$ is isomorphic to the direct sum $\mathfrak{su}(2) \oplus \mathfrak{su}(2)$ (Problems 1 and 2).

**(21.12.4)** *The group* $\mathbf{SO}(2n+1)$ $(n \geqq 2)$.

The mapping $\mathbf{M}_{2n}(\mathbf{R}) \to \mathbf{M}_{2n+1}(\mathbf{R})$ defined by

$$U \mapsto \begin{pmatrix} U & 0 \\ 0 & 1 \end{pmatrix}$$

is injective and maps $\mathbf{SO}(2n)$ onto a subgroup of $\mathbf{SO}(2n+1)$; we shall identify $\mathbf{SO}(2n)$ with this subgroup. Then the maximal torus T of $\mathbf{SO}(2n)$ defined in **(21.12.3)** is also a maximal torus of $\mathbf{SO}(2n+1)$. For the complexification of the Lie algebra $\mathfrak{so}(2n+1)$ of $\mathbf{SO}(2n+1)$ is again the Lie algebra $\mathfrak{so}(2n+1, \mathbf{C})$ of complex skew-symmetric matrices (and is the Lie algebra of the *complex special orthogonal group* $\mathbf{SO}(2n+1, \mathbf{C})$, defined as in **(21.12.3)**). With the notation of **(21.12.3)**, a basis of $\mathfrak{so}(2n+1, \mathbf{C})$ may be obtained from the basis of $\mathfrak{so}(2n, \mathbf{C})$ defined in **(21.12.3)** by adjoining the $2n$ matrices

$$P_r^{\pm} = E_{2r-1,\,2n+1} - E_{2n+1,\,2r-1} \pm i(E_{2r,\,2n+1} - E_{2n+1,\,2r}) \qquad (1 \leqq r \leqq n)$$

for which we have

**(21.12.4.1)**
$$[\mathbf{h}, P_r^{\pm}] = \mp \varepsilon_r(\mathbf{h}) P_r^{\pm}$$

for all $\mathbf{h} \in \mathfrak{h}$. This proves our assertion and shows that the roots of the

$$\mathfrak{so}(2n+1, \mathbf{C})$$

relative to $\mathfrak{h}$ (or of $\mathfrak{so}(2n+1)$ relative to $\mathfrak{t}$) are the $2n^2$ linear forms

**(21.12.4.2)** $\pm\varepsilon_r$ $(1 \leqq r \leqq n)$, $\quad \pm\varepsilon_r \pm \varepsilon_s$ $(1 \leqq r < s \leqq n)$.

If we put

**(21.12.4.3)** $\beta_r = \varepsilon_r - \varepsilon_{r+1}$ $(1 \leqq r \leqq n-1)$, $\quad \beta_n = \varepsilon_n$,

then we have

$$\varepsilon_r = \beta_r + \beta_{r+1} + \cdots + \beta_n \qquad (1 \leqq r \leqq n),$$
$$\varepsilon_r - \varepsilon_s = \beta_r + \beta_{r+1} + \cdots + \beta_{s-1} \qquad (1 \leqq r < s \leqq n),$$
$$\varepsilon_r + \varepsilon_s = \beta_r + \cdots + \beta_{s-1} + 2\beta_s + 2\beta_{s+1} + \cdots + 2\beta_n \qquad (1 \leqq r < s \leqq n),$$

showing that the $n$ roots **(21.12.4.3)** form a *basis* of the root system, the positive roots for this basis being

**(21.12.4.4)** $\varepsilon_r \quad (1 \leqq r \leqq n), \qquad \varepsilon_r \pm \varepsilon_s \quad (1 \leqq r < s \leqq n).$

Furthermore, we have

$$\mathbf{h}_{\beta_r} = H_r - H_{r+1} \qquad (1 \leqq r \leqq n-1),$$
$$\mathbf{h}_{\beta_n} = 2H_n,$$

and therefore the Cartan integers are

**(21.12.4.5)**
$$\begin{aligned} & n(\beta_r, \beta_s) = 0 \quad \text{if} \quad |r-s| \geqq 2 && (1 \leqq r \leqq n-1), \\ & n(\beta_r, \beta_{r+1}) = n(\beta_{r+1}, \beta_r) = -1 && (1 \leqq r \leqq n-2), \\ & n(\beta_{n-1}, \beta_n) = -2, \qquad n(\beta_n, \beta_{n-1}) = -1, && \\ & n(\beta_n, \beta_r) = 0 && (1 \leqq r \leqq n-2). \end{aligned}$$

One shows as in **(21.12.2)** that $\mathbf{SO}(2n+1)$ is almost simple and that the Weyl group is isomorphic to the semidirect product of $\{-1, 1\}^n$ and $\mathfrak{S}_n$.

For $n = 1$, the formulas above and in **(21.12.1)** show that $\mathfrak{so}(3)$ is isomorphic to $\mathfrak{su}(2)$; also, using **(21.12.2)**, that $\mathfrak{so}(5)$ is isomorphic to $\mathfrak{u}(2, \mathbf{H})$ (Problem 1).

**(21.12.5)** In the four cases studied above, the scalar product on $it^*$ defined by $(\varepsilon_j | \varepsilon_k) = \delta_{jk}$ (Kronecker delta) is invariant under the action of the Weyl group.

**(21.12.6)** The complex Lie algebras $\mathfrak{sl}(n, \mathbf{C})$, $\mathfrak{so}(2n+1, \mathbf{C})$, $\mathfrak{sp}(2n, \mathbf{C})$, and $\mathfrak{so}(2n, \mathbf{C})$ are denoted, respectively, by $\mathrm{A}_{n-1}$, $\mathrm{B}_n$, $\mathrm{C}_n$, $\mathrm{D}_n$ ($n \geqq 2$). We have seen that, with the exception of $\mathfrak{so}(4, \mathbf{C})$, they are *simple.* It can be shown that, apart from these Lie algebras, there are (up to isomorphism) only five other complex simple Lie algebras, of dimensions 14, 52, 78, 133, and 248 [85]. They are known as the *exceptional* complex simple Lie algebras.

An almost simple compact Lie group that is locally isomorphic to one of the groups of the four types studied in this section is often called a *classical* almost simple compact group.

## PROBLEMS

1. (a) Identify the exterior square $\bigwedge^2(\mathbf{C}^4)$ with the vector space $\mathbf{C}^6$, by identifying the basis

$$e_1 \wedge e_2, \quad e_2 \wedge e_3, \quad e_3 \wedge e_1, \quad e_3 \wedge e_4, \quad e_1 \wedge e_4, \quad e_2 \wedge e_4$$

of $\bigwedge^2 (\mathbf{C}^4)$ (where $(e_1, e_2, e_3, e_4)$ is the canonical basis of $\mathbf{C}^4$) with the canonical basis

$$e'_1, e'_2, e'_3, e'_4, e'_5, e'_6$$

of $\mathbf{C}^6$. If two bivectors $x, y \in \bigwedge^2 (\mathbf{C}^4)$ are identified in this way with the vectors $\sum_{j=1}^{6} \xi^j e'_j$, $\sum_{j=1}^{6} \eta^j e'_j$, then the 4-vector $x \wedge y$ is equal to $\mathrm{B}(x, y) e_1 \wedge e_2 \wedge e_3 \wedge e_4$, where

$$\mathrm{B}(x, y) = \xi^1 \eta^4 + \eta^1 \xi^4 + \xi^2 \eta^5 + \eta^2 \xi^5 + \xi^3 \eta^6 + \eta^3 \xi^6.$$

Consider the mapping that transforms each $u \in \mathbf{SL}(4, \mathbf{C})$ into $\bigwedge^2 u$. Show that $u \mapsto \bigwedge^2 u$ is a surjective homomorphism of the Lie group $\mathbf{SL}(4, \mathbf{C})$ onto a group isomorphic to the orthogonal group $\mathbf{SO}(6, \mathbf{C})$, with kernel $\{I, -I\}$, so that $\mathbf{SL}(4, \mathbf{C})$ is a two-sheeted covering of $\mathbf{SO}(6, \mathbf{C})$. (To show that the homomorphism is surjective, consider the dimensions of the two groups.)

(b) Show that the restriction of $u \mapsto \bigwedge^2 u$ to $\mathbf{SU}(4)$ is a surjective homomorphism onto a group isomorphic to $\mathbf{SO}(6, \mathbf{R}) = \mathbf{SO}(6)$. (Observe that $\mathbf{SU}(4)$ is the subgroup of all $u \in \mathbf{SL}(4, \mathbf{C})$ having the following property: if $\psi$ is the antilinear mapping (relative to complex conjugation) of $\mathbf{C}^4$ onto its dual space $(\mathbf{C}^4)^*$ such that $\psi(e_j) = e_j^*$ for $1 \leqq j \leqq 4$, where $(e_j^*)$ is the basis dual to $(e_j)$, then $\psi \circ u = {}^t u^{-1} \circ \psi$. Express that $(\bigwedge^2 \psi) \circ (\bigwedge^2 u) = (\bigwedge^2 ({}^t u^{-1})) \circ (\bigwedge^2 \psi)$.)

(c) Show that the subgroup of $\mathbf{SL}(4, \mathbf{C})$ consisting of the linear mappings $u$ such that $\bigwedge^2 u$ fixes the coordinate $\xi^6$ of each bivector $x \in \bigwedge^2 (\mathbf{C}^4)$ may be identified with the symplectic group $\mathbf{Sp}(4, \mathbf{C})$, and its image under $u \mapsto \bigwedge^2 u$ with the orthogonal group $\mathbf{SO}(5, \mathbf{C})$. By restricting to $\mathbf{U}(2, \mathbf{H}) \subset \mathbf{Sp}(4, \mathbf{C})$, we obtain a homomorphism $\mathbf{U}(2, \mathbf{H}) \to \mathbf{SO}(5)$, which makes $\mathbf{U}(2, \mathbf{H})$ a double covering of $\mathbf{SO}(5)$.

2. For each pair of matrices $(U_1, U_2)$ in $\mathbf{SL}(2, \mathbf{C})$, let $\varphi(U_1, U_2)$ denote the automorphism $X \mapsto U_1 \cdot X \cdot {}^t U_2$ of the vector space $\mathbf{M}_2(\mathbf{C})$, identified with $\mathbf{C}^4$ by identifying the matrix $\begin{pmatrix} \xi^1 & \xi^2 \\ -\xi^3 & \xi^4 \end{pmatrix}$ with the vector $(\xi^1, \xi^2, \xi^3, \xi^4)$. Show that $\varphi$ is a surjective homomorphism of $\mathbf{SL}(2, \mathbf{C}) \times \mathbf{SL}(2, \mathbf{C})$ onto a group isomorphic to $\mathbf{SO}(4, \mathbf{C})$, with kernel consisting of $(I, I)$ and $(-I, -I)$, so that $\mathbf{SL}(2, \mathbf{C}) \times \mathbf{SL}(2, \mathbf{C})$ is a double covering of $\mathbf{SO}(4, \mathbf{C})$. Show also that the restriction of $\varphi$ to $\mathbf{SU}(2) \times \mathbf{SU}(2)$ is a surjective homomorphism of this group onto a group isomorphic to $\mathbf{SO}(4)$. (Same method as in Problem 1.)

3. Show that apart from the isomorphisms

$$\mathrm{B}_2 \simeq \mathrm{C}_2, \qquad \mathrm{D}_2 \simeq \mathrm{A}_1 \oplus \mathrm{A}_1, \qquad \mathrm{D}_3 \simeq \mathrm{A}_3,$$

which follow from Problems 1 and 2 (or which may be obtained directly from the explicit descriptions of these Lie algebras), there exist no other isomorphisms among the algebras $\mathrm{A}_{n-1}, \mathrm{B}_n, \mathrm{C}_n, \mathrm{D}_n$ for $n \geqq 2$. (Consider the systems of Cartan integers of these Lie algebras.)

4. Let G be an almost simple compact connected Lie group. Show that if $\mathrm{G}_1$ is a principal nice subgroup of G (Section **21.11**, Problem 9), the Lie algebra of $\mathrm{G}_1$ cannot be isomorphic to $\mathfrak{su}(n+1)$ or $\mathfrak{so}(2n)$ for $n \geqq 2$. (Observe that for these Lie algebras, all the roots have the same length.)

5. Let G be a classical compact group. Suppose that G has a principal nice subgroup $\mathrm{G}_1$ (Section **21.11**, Problem 9) that is also a classical group, of rank $\geqq 3$. The complexified Lie

algebra of $G_1$ is necessarily of type $B_n$ or $C_n$ (Problem 4), hence there exists a basis $\{\rho_1, \dots, \rho_h\}$ of $\mathbf{S}(G_1)$ such that $h \geqq 3$, $\|\rho_1\| = \|\rho_2\| = \dots = \|\rho_{h-1}\|$, $(\rho_j|\rho_{j+1}) < 0$, and $(\rho_j|\rho_k) = 0$ if $|k - j| > 1$. Then the sets $R(\rho_1), \dots, R(\rho_{h-1})$ all have the same number $n$ of elements (Section **21.11**, Problem 9).

(a) Show that $n \leqq 2$ (use Section **21.11**, Problem 9(a) and (c).)

(b) Show that if $n = 2$, the inclusion $G_1 \subset G$ is one of the two canonical inclusions $\mathbf{SO}(2h+1) \subset \mathbf{SU}(2h+1)$, $\mathbf{U}(h, \mathbf{H}) \subset \mathbf{SU}(2h)$. (Use Section **21.11**, Problem 9(b).)

(c) Show that if $n = 1$ the inclusion $G_1 \subset G$ is the canonical inclusion

$$\mathbf{SO}(2h+1) \subset \mathbf{SO}(2h+2).$$

(Same method.)

**6.** In the Lie algebra $\mathfrak{so}(7)$, the three roots

$$\gamma_1 = \varepsilon_1, \qquad \gamma_2 = \varepsilon_2 - \varepsilon_1, \qquad \gamma_3 = -\varepsilon_2 - \varepsilon_3,$$

form a basis of the root system. Let $\mathfrak{t}_1$ be the plane in $\mathfrak{t}$ on which the linear form $\gamma_1 - \gamma_3 = \varepsilon_1 + \varepsilon_2 + \varepsilon_3$ vanishes. Show that the restrictions to $\mathfrak{t}_1$ of the roots of $\mathfrak{so}(7)$ form a reduced root system of type $G_2$ (Section **21.11**, Problem 2). Show that there exists a principal nice subgroup of $\mathbf{SO}(7)$ whose root system relative to $\mathfrak{t}_1$ consists of these restrictions.

## 13. LINEAR REPRESENTATIONS OF COMPACT CONNECTED LIE GROUPS

We recall that, until the end of this chapter, by a *linear representation* of a compact Lie group is meant a *continuous* (or, equivalently, $C^\infty$ (**19.10.12**)) linear representation on a *finite-dimensional* complex vector space.

**(21.13.1)** *Every compact Lie group* G *has a faithful linear representation.*

In other words, bearing in mind (**20.11.3.1**), there exists an integer $N > 0$ such that G is *isomorphic to a Lie subgroup of* $\mathbf{U}(N)$.

Consider the set of irreducible representations $s \mapsto M_\rho(s)$ of G ($\rho \in R$) (**21.2.5**). It is enough to show that there exists a *finite* subset J of R such that the kernels $N_\rho$ of the homomorphisms $s \mapsto M_\rho(s)$ for $\rho \in J$ intersect only in $e$, for the Hilbert sum of the representations $s \mapsto M_\rho(s)$ for $\rho \in J$ will then be faithful. Now there exists an open neighborhood V of $e$ in G that contains *no subgroup of* G other than $\{e\}$. To see that this is so, let W be an open neighborhood of 0 in the Lie algebra $\mathfrak{g}_e$ of G, such that $\exp_G$ is a diffeomorphism of W onto an open neighborhood of $e$ in G (**19.8.6**). We may assume that, relative to some norm that defines the topology of $\mathfrak{g}_e$, the open neighborhood W is defined by $\|\mathbf{x}\| < a$. Then the neighborhood $V = \exp(\frac{1}{2}W)$ of $e$ in G has the required property: for if $\mathbf{x} \neq 0$ belongs to $\frac{1}{2}W$, there exists a smallest integer $p > 0$ such that $(p+1)\|\mathbf{x}\| > \frac{1}{2}a$, and necessarily

$(p+1)\mathbf{x} \in W$; if now there were a subgroup $H \neq \{e\}$ of G contained in V and such that $s = \exp(\mathbf{x}) \in H$, then we should have

$$s^{p+1} = \exp((p+1)\mathbf{x}) \in H,$$

contradicting the fact that $(p+1)\mathbf{x} \in W$ and $(p+1)\mathbf{x} \notin \frac{1}{2}W$.

Now the intersection of the kernels $N_\rho$, for all $\rho \in R$, consists only of $e$ **(21.3.5)**, and hence the intersection of the closed sets $N_\rho \cap (G - V)$ $(\rho \in R)$ is empty. Since G is compact, it follows from the Borel–Lebesgue axiom that there exists a finite subset J of R such that the intersection of the sets $N_\rho \cap (G - V)$ for $\rho \in J$ is empty. The set $\bigcap_{\rho \in J} N_\rho$ is then a subgroup of G contained in V, hence consists only of $e$ by the construction of V, and the proof is complete.

**(21.13.1.1)** We remark that this proof shows in fact that every compact metrizable group G, in which there exists a neighborhood of $e$ containing no subgroup other than $\{e\}$, is isomorphic to a subgroup of a unitary group $\mathbf{U}(N)$, hence is a *Lie group* (cf. Section **19.8**, Problem 9).

**(21.13.2)** *Let* G *be a compact Lie group and* $U$ *a faithful linear representation. Then every irreducible linear representation of* G *is contained* **(21.4.2)** *in a tensor product of a certain number of linear representations equal to* $U$ *and a certain number of linear representations equal to its conjugate* $\overline{U}$ **(21.4.3)**.

Put $\mathrm{cl}(U) = \sum_{\rho \in J} d_\rho \cdot \rho$ **(21.4.7)**, where $d_\rho > 0$ for all $\rho$ belonging to a finite subset J of R. Then $\mathrm{cl}(\overline{U}) = \sum_{\rho \in J} d_\rho \cdot \bar{\rho}$.

Suppose that there exists $\rho' \in R$ such that the proposition is false for the representation $M_{\rho'}$. This means that the subring of $\mathbf{Z}^{(R)}$ generated by the classes $\rho \in J$ and their conjugates $\bar{\rho}$ is contained in a $\mathbf{Z}$-module of the form $\mathbf{Z}^{(R')}$, where $R' \subset R$ and $\rho' \notin R'$. It follows from the Peter–Weyl theorem **(21.2.3)** that $\chi_{\rho'}$ is orthogonal to all the functions $m_{ij}^{(\rho)}$, $\rho \in R'$. Consequently, if we put $U(s)^{\otimes m} \otimes \overline{U}(s)^{\otimes n} = (p_{hk}^{(m,n)}(s))$ for each pair of integers $m \geqq 0$, $n \geqq 0$, such that $m + n \geqq 1$, the function $\chi_{\rho'}$ is orthogonal to all the functions $p_{hk}^{(m,n)}$. Moreover, since the trivial representation is contained in $U \otimes \overline{U}$ **(21.4.6.4)**, the class $\rho'$ cannot be the class of the trivial representation, and therefore $\chi_{\rho'}$ is orthogonal also to the constant functions **(21.3.2.6)**. But by the definition of the tensor product of matrices, among the functions $p_{hk}^{(m,n)}$ there appear all the *monomials* with respect to continuous functions that are elements of the matrix $U$ or of $\overline{U}$. By hypothesis, these functions *separate* the points of G, hence the complex vector subspace of $\mathscr{C}_{\mathbf{C}}(G)$ spanned by the constants and the $p_{hk}^{(m,n)}$ is *dense*, by the Stone–Weierstrass theorem **(7.3.1)**. Since the continuous function $\chi_{\rho'}$ is not identically zero, we arrive at a contradiction **(13.14.4)**, which proves **(21.13.2)**.

**(21.13.3)** *Let* G *be a compact Lie group and* H *a closed subgroup of* G. *Then every irreducible linear representation of* H *is contained in the restriction to* H *of a linear representation of* G.

Let $U$ be a faithful linear representation of G **(21.13.1)**. Clearly its restriction $V$ to H is faithful, hence every irreducible linear representation of H is contained in some representation of the form $V^{\otimes m} \otimes \overline{V}^{\otimes n}$ **(21.13.2)**; since this representation is obviously the restriction to H of $U^{\otimes m} \otimes \overline{U}^{\otimes n}$, the proposition is proved.

**(21.13.4)** Let G be a *compact connected* Lie group and T a maximal torus in G. (This notation will be in force up to the end of Section **21.15**.)

As we have already remarked **(21.7.6)**, the study of the linear representations of G is based on the study of their *restrictions* to T. In the first place, *a linear representation of* G *is uniquely determined, up to equivalence, by its restriction to* T. Clearly it is enough to consider irreducible representations, and since up to equivalence such a representation is entirely determined by its character **(21.4.5)**, it is enough to show that if two characters $\chi'$, $\chi''$ have the same restriction to T, then they are equal. We shall in fact prove a more precise result: for this purpose, we remark that if $f$ is a *continuous central function* on G **(21.2.2)**, its restriction to T is a continuous function which, by definition **(21.2.2.1)**, is *invariant under the Weyl group* W of G relative to T.

**(21.13.5)** *The mapping that sends each continuous central function on* G *to its restriction to* T *is an isomorphism of the complex vector space of continuous central functions on* G, *onto the complex vector space* $\mathscr{C}_{\mathbf{C}}(\mathrm{T})^{\mathrm{W}}$ *of continuous complex functions on* T *that are invariant under the Weyl group* W.

The fact that the mapping $f \mapsto f\,|\mathrm{T}$ is injective is immediately obvious. For each $x \in \mathrm{G}$ is of the form $sts^{-1}$ for some $s \in \mathrm{G}$ and $t \in \mathrm{T}$ **(21.7.8)**, hence $f(x) = f(t)$ because $f$ is central. To show that $f \mapsto f\,|\mathrm{T}$ is surjective, suppose we are given a function $g \in \mathscr{C}_{\mathbf{C}}(\mathrm{T})^{\mathrm{W}}$; let us first show that we may *define* a function $f$ on G by the condition $f(sts^{-1}) = g(t)$ for all $t \in \mathrm{T}$ and $s \in \mathrm{G}$. For this purpose, we must verify that if $t_1$, $t_2$ are two elements of T that are conjugate *in* G, then $g(t_1) = g(t_2)$; but by virtue of **(21.7.17)**, there exists $w \in \mathrm{W}$ such that $t_2 = w \cdot t_1$, and the result follows from the W-invariance of $g$. It remains to be shown that the function $f$, so defined, is *continuous* (it is a central function by definition).

If $f$ were not continuous, there would exist a sequence $(x_n)$ of points of G, converging to a limit $x \in G$, and such that $f(x_n)$ does not converge to $f(x)$. We may write $x_n = s_n t_n s_n^{-1}$, where $t_n \in T$ and $s_n \in G$, and because both G and T are compact, we may assume, by passing to a subsequence of $(x_n)$, that $(t_n)$ has a limit $t \in T$, and $(s_n)$ a limit $s \in G$. But then $x = sts^{-1}$; we have $f(x_n) = g(t_n)$ and $f(x) = g(t)$, and the hypothesis on $(x_n)$ contradicts the continuity of $g$.

**(21.13.6)** We recall **(21.7.5)** that the characters of the maximal torus T are the functions $\xi$ with values in **U**, such that $\xi(\exp(\mathbf{u})) = e^{p(\mathbf{u})}$ for all $\mathbf{u} \in \mathfrak{t}$, where $p$ is a *weight* of T. The weights of T are **R**-linear functions on T, with values in $i\mathbf{R}$, which take values belonging to $2\pi i\mathbf{Z}$ at the points of the *lattice* $\Gamma_T$, the kernel of $\exp_T = (\exp_G)|T$. These functions form a lattice $2\pi i\Gamma_T^*$, which we denote by P(G, T) or P(G) (or simply P) and call the *weight lattice* of G (with respect to T). If $\mathbf{u}_1$, $\mathbf{u}_2$ are two points of $\mathfrak{t}$ such that $\exp(\mathbf{u}_1) = \exp(\mathbf{u}_2)$, we have therefore $e^{p(\mathbf{u}_1)} = e^{p(\mathbf{u}_2)}$; this leads us to write $e^p$ (or $s \mapsto e^{p(s)}$) by abuse of notation, for the character $\xi$ corresponding to the weight $p$, whenever there is no risk of confusion.

Consider a character $\chi$ of G. If $U$ is an irreducible representation of G with character $\chi$, the restriction of $U$ to T is a Hilbert sum of one-dimensional representations, and the restriction of $\chi$ to T may therefore be written uniquely in the form

**(21.13.6.1)**
$$\sum_{p \in P} n(p)e^p$$

where each $n(p)$ is an integer $\geqq 0$, and is zero for all but a finite number of values of $p \in P$; it is the *multiplicity* **(21.4.2)** of the representation $s \mapsto e^{p(s)} \cdot 1$ in the representation $U|T$. This number $n(p)$ is called the *multiplicity of the weight* $p$ in the character $\chi$ (or the representation $U$), and we shall say that $p$ *is contained in* $\chi$ (or *is a weight of* $U$) if $n(p) > 0$.

For each element $w$ of the Weyl group W, we have $n(w \cdot p) = n(p)$ **(21.13.5)**. This leads us to consider functions of the form **(21.13.6.1)** in which the integers $n(p)$ *are of arbitrary sign* and satisfy the relations $n(w \cdot p) = n(p)$ for all $w \in W$. It is clear that these functions form a *free* **Z**-*module*, having as a basis the sums

**(21.13.6.2)**
$$S(\Pi) = \sum_{p \in \Pi} e^p$$

where $\Pi$ runs through the set P/W of *orbits* of W in P.

Since the characters $e^p$ of T are linearly independent **(21.3.2)** and since, for any two weights $p', p'' \in P$, we have $e^{p'} \cdot e^{p''} = e^{p'+p''}$, the set of all linear combinations $\sum_{p \in P} n(p)e^p$, with *arbitrary* integers $n(p) \in \mathbf{Z}$, may be identified with the *algebra* $\mathbf{Z}[P]$ *of the additive group* P *over* $\mathbf{Z}$. The $\mathbf{Z}$-module having as basis the $S(\Pi)$, for all $\Pi \in P/W$, is therefore the *subalgebra* $\mathbf{Z}[P]^W$ *of* W-*invariant elements of* $\mathbf{Z}[P]$.

It follows therefore from (21.13.5) that the $\mathbf{Z}$-algebra generated by the characters of G, which may be canonically identified (21.4.7) with the ring $\mathbf{Z}^{(R(G))}$ of *classes of linear representations* of G, is canonically isomorphic to a *subalgebra of* $\mathbf{Z}[P]^W$. In general, the basis elements $S(\Pi)$ of $\mathbf{Z}[P]^W$ are *not* the restrictions of characters of G, as can be seen already from the example of the group $\mathbf{SU}(2)$, for which we know explicitly all the irreducible representations (21.9.3) and the Weyl group, consisting of two elements (21.12.1). We shall nevertheless show that the canonical homomorphism of $\mathbf{Z}^{(R(G))}$ into $\mathbf{Z}[P]^W$ is always *bijective* (21.15.5).

(21.13.7) Let $V$ be a linear representation of G, and suppose that the restriction to T of the function $s \mapsto \operatorname{Tr}(V(s))$ is of the form $S(\Pi)$ for some orbit $\Pi \in P/W$. Then it follows immediately from (21.4.4) and (21.13.6) that the representation $V$ is *irreducible* and that $S(\Pi)$ is the restriction to T of its character.

## PROBLEMS

1. Let G be a compact subgroup of $\mathbf{GL}(n, \mathbf{R})$. Show that if A and B are two compact G-stable subsets of $\mathbf{R}^n$ with no common point, there exists a polynomial $P \in \mathbf{R}[T_1, \ldots, T_n]$ such that $|P(x)| \leq \frac{1}{3}$ for all $x \in A$, $|P(x) - 1| \leq \frac{1}{3}$ for all $x \in B$, and $P(s \cdot x) = P(x)$ for all $s \in G$. (Apply the Weierstrass–Stone theorem and integration with respect to a Haar measure on G.)

2. Deduce from Problem 1 that if G is a compact subgroup of $\mathbf{GL}(n, \mathbf{R})$ there exists a family of polynomials $P_\alpha \in \mathbf{R}[T_{11}, \ldots, T_{nn}]$ in $n^2$ indeterminates, such that G is the set of matrices $s \in \mathbf{GL}(n, \mathbf{R}) \subset \mathbf{R}^{n^2}$ such that $P_\alpha(s) = 0$ for all $\alpha$.

3. Let G be a compact Lie group and H a closed subgroup of G. Show that there exists a neighborhood U of H such that there is no subgroup K of G contained in U that contains H properly. (Use (16.14.2) and argue as in (21.13.1) for the case $H = \{e\}$.)

4. Let G be a compact Lie group and H a closed subgroup of G. Show that if $H \neq G$ there exists at least one irreducible representation of G, other than the trivial representation, whose restriction to H contains the trivial representation. (Assume that the result is false and show, by use of (21.3.4) and (21.2.5), that for all continuous functions $f$ on G we should have $\int_G f\, dm_G = \int_H f\, dm_H$, where $m_G$ and $m_H$ are the normalized Haar measures on G, H, respectively; use this to obtain a contradiction.)

5. Let G be a compact Lie group and H a closed subgroup of G. Show that there exists a continuous linear representation $U$ of G on a finite-dimensional complex vector space E such that H is the stabilizer of some point of E (for the action of G on E defined by $U$). (We may assume that $H \neq G$. For each closed subgroup F that properly contains H, let $V_F$ be an irreducible representation of F, other than the trivial representation, whose restriction to H contains the trivial representation (Problem 4), and let $U_F$ be a linear representation of G whose restriction to F contains $V_F$. Let $H_F \supset H$ be the stabilizer of a point $\neq 0$ in the space of $V_F$. Show that the intersection of the subgroups $H_F$ is equal to H, and observe that this intersection is also the intersection of a finite number of the $H_F$, by using Problem 3.)

## 14. ANTI-INVARIANT ELEMENTS

We shall first study in more detail the structure of the algebras $\mathbf{Z}[P]$ and $\mathbf{Z}[P]^W$, by using the properties of root systems. We shall require the following lemma:

(21.14.1) *Let $u$, $v$ be two linearly independent elements of* P. *If an element* $\Phi \in \mathbf{Z}[P]$ *is divisible by* $1 - e^u$ *and by* $1 - e^v$, *then it is divisible by the product* $(1 - e^u)(1 - e^v)$.

The $\mathbf{Z}$-module P is isomorphic to $\mathbf{Z}^r$ for some $r > 0$. If $(j_1, \dots, j_r)$ are the coordinates of $u$ with respect to a $\mathbf{Z}$-basis of P, and if $d > 0$ is the highest common factor of the $j_k$ $(1 \leqq k \leqq r)$, we may write $u = du_1$, where the coordinates of $u_1$ are relatively coprime. The elementary theory of free $\mathbf{Z}$-modules (A.26.6) shows that there exists a basis $(u_1, \dots, u_r)$ of P containing $u_1$. The projection of $v$ on $\mathbf{Z}u_2 \oplus \mathbf{Z}u_3 \oplus \cdots \oplus \mathbf{Z}u_r$ is nonzero, by hypothesis; by applying the same argument to this projection, we may assume that $u_2$, $\dots$, $u_r$ have been chosen so that $v = mu_2 - nu_1$, where $m, n \in \mathbf{Z}$ and $m \neq 0$. Since the ring $\mathbf{Z}[P]$ is isomorphic to the ring

$$A = \mathbf{Z}[X_1, \dots, X_r, X_1^{-1}, \dots, X_r^{-1}]$$

(21.4.7), it follows that we are reduced to showing that if an element $\Phi$ of this ring is divisible by $X_1^d - 1$ and by $X_2^m - X_1^n$, then it is divisible by their product. Furthermore, since the $X_k$ are invertible in A, we may assume that $m > 0$, and since we have $\Phi = (X_1^d - 1)\Phi_1$ with $\Phi_1 \in A$, we may also assume that $\Phi_1$ is a polynomial in $X_2$ with coefficients in the ring

$$B = \mathbf{Z}[X_1, X_3, \dots, X_r, X_1^{-1}, X_3^{-1}, \dots, X_r^{-1}].$$

The Euclidean algorithm then enables us to write

$$\Phi_1 = (X_2^m - X_1^n)\Phi_2 + (\Psi_1 X_2^{m-1} + \cdots + \Psi_{m-1}),$$

where $\Phi_2 \in A$ and the $\Psi_j$ belong to B. By hypothesis, the product

(21.14.1.1) $$(X_1^d - 1)(\Psi_1 X_2^{m-1} + \cdots + \Psi_{m-1})$$

is divisible by $X_2^m - X_1^n$. If the $\Psi_j$ were not all zero, we should be able to substitute for $X_1, X_3, \ldots, X_r$ nonzero complex numbers $z_1, z_3, \ldots, z_r$ such that $z_1^d \neq 1$ and such that the value of at least one of the coefficients $\Psi_j(z_1, z_3, \ldots, z_r)$ were $\neq 0$. Under this substitution, **(21.14.1.1)** would become a nonzero polynomial of degree $\leqq m-1$ in $X_2$ with complex coefficients, divisible by $X_2^m - z_1^n$; and this is absurd.

We remark that the lemma **(21.14.1)** applies equally to the ring $\mathbf{Z}[cP]$, where $c$ is any nonzero real number.

**(21.14.2)** If $\mathfrak{g} = \mathfrak{c} \oplus \mathfrak{D}(\mathfrak{g})$ is the canonical decomposition of the Lie algebra $\mathfrak{g}$ of G as the direct sum of its center and its derived algebra **(21.6.9)**, the Lie algebra $\mathfrak{t}$ of T takes the form $\mathfrak{t} = \mathfrak{c} \oplus \mathfrak{t}'$, where $\mathfrak{t}'$ is a maximal commutative subalgebra of $\mathfrak{D}(\mathfrak{g})$. We have seen **(21.8.8)** that the root system $\mathbf{S} \subset i\mathfrak{t}'^*$ of $\mathfrak{D}(\mathfrak{g})$ relative to $\mathfrak{t}'$ may be identified with a finite subset of the lattice of weights P(G) ($\mathfrak{t}'^*$ being identified with the annihilator of $\mathfrak{c}$ in $\mathfrak{t}^*$). We shall suppose that a *basis* $\mathbf{B} = \{\beta_1, \ldots, \beta_l\}$ of $\mathbf{S}$ **(21.11.5)** has been chosen. The elements $\mathbf{h}_\alpha$ of $i\mathfrak{t}'$, for $\alpha \in \mathbf{S}$, form a reduced root system $\mathbf{S}^\vee$, the *dual* of $\mathbf{S}$ **(21.11.11)**. For simplicity we shall put $\mathbf{h}_j = \mathbf{h}_{\beta_j}$; we recall **(21.11.5.5)** that the $\mathbf{h}_j$ form a *basis* $\mathbf{B}^\vee$ of the root system $\mathbf{S}^\vee$, and also a basis of the real vector space $i\mathfrak{t}'$.

**(21.14.3)** *The weight lattice* $P = P(G)$ *is contained in the set* $P(\mathfrak{g})$ *of* $\mathbf{C}$*-linear forms* $\lambda$ *on* $\mathfrak{t}_{(\mathbf{C})}$ *such that* $\lambda(\mathbf{h}_j) \in \mathbf{Z}$ *for* $1 \leqq j \leqq l$. (Since each $\mathbf{h}_\alpha \in \mathbf{S}^\vee$ is a linear combination of the $\mathbf{h}_j$ with integer coefficients, this condition is equivalent to requiring that $\lambda(\mathbf{h}_\alpha)$ should be an integer for *all* roots $\alpha \in \mathbf{S}$.)

For each $p \in P(G)$, $e^p$ is a character of T. By virtue of **(21.13.3)**, there exists a linear representation $U$ of G on a vector space E such that for each $\mathbf{h} \in \mathfrak{t}_{(\mathbf{C})}$ the complex number $p(\mathbf{h})$ is an eigenvalue of the endomorphism $U_*(\mathbf{h})$ of E (we identify $U_*$ with its extension $U_* \otimes 1_{\mathbf{C}}$ to $\mathfrak{g}_{(\mathbf{C})}$). With the notation of **(21.10.3)**, we may apply **(21.9.3)** to the restriction of $U_*$ to each subalgebra $\mathfrak{s}_\alpha \subset \mathfrak{g}_{(\mathbf{C})}$ isomorphic to $\mathfrak{sl}(2, \mathbf{C})$, and conclude that $p(\mathbf{h}_\alpha)$ is an integer for each $\alpha \in \mathbf{S}$.

Since the dual $\mathfrak{t}_{(\mathbf{C})}^*$ of $\mathfrak{t}_{(\mathbf{C})}$ may be identified with $\mathfrak{c}_{(\mathbf{C})}^* \oplus \mathfrak{t}'^*_{(\mathbf{C})}$ ($\mathfrak{c}^*$ being identified with the annihilator of $\mathfrak{t}'$ in $\mathfrak{t}^*$), $P(\mathfrak{g})$ may be identified with $\mathfrak{c}_{(\mathbf{C})}^* \oplus P(\mathfrak{D}(\mathfrak{g}))$, where $P(\mathfrak{D}(\mathfrak{g})) \subset i\mathfrak{t}'^*$ is the lattice *dual* **(21.7.5)** to the lattice in $i\mathfrak{t}'$ generated by the $\mathbf{h}_j$.

**(21.14.4)** In the real vector space $it^* = ic^* \oplus it'^*$, the set C (or C(g)) of linear forms $\lambda$ such that $\lambda(\mathbf{h}_j) > 0$ $(1 \leqq j \leqq l)$ is called the *Weyl chamber* relative to the basis **B** of **S**. Since the $\mathbf{h}_j$ form a basis of the real vector space $it'$, the closure $\bar{C}$ of C in $it^*$ is the set of linear forms $\lambda$ such that $\lambda(\mathbf{h}_j) \geqq 0$ for $1 \leqq j \leqq l$. We have $C(\mathfrak{g}) = ic^* + C(\mathfrak{D}(\mathfrak{g}))$ and $\overline{C(\mathfrak{g})} = ic^* + \overline{C(\mathfrak{D}(\mathfrak{g}))}$.

**(21.14.5)** Let L be the set of linear forms $\lambda \in it^*$ that can be written $\lambda = \gamma + \sum_{j=1}^{l} c_j \beta_j$, where $\gamma \in ic^*$ and the $c_j$ are real numbers $\geqq 0$, *not all zero*. If we put $L_0 = L \cup \{0\}$, it is clear that $L_0 + L_0 \subset L_0$, $aL_0 \subset L_0$ for all real $a > 0$, and $L_0 \cap (-L_0) = \{0\}$. The relation $\mu - \lambda \in L_0$ is therefore a (partial) *ordering* on $it^*$, which we denote by $\lambda \leqq \mu$. The relation $\lambda \leqq \mu$ is equivalent to $\lambda + \nu \leqq \mu + \nu$ for all $\nu \in it^*$, and to $a\lambda \leqq a\mu$ for all real $a > 0$; and the relation $\lambda > 0$ is equivalent to $\lambda \in L$.

The positive roots (relative to the basis **B**) in the sense of **(21.11.5)** are therefore exactly those which are $> 0$ in the ordering just defined. This justifies the terminology.

**(21.14.6)** (i) *The Weyl chamber* C *is contained in the set* L *of forms* $> 0$. *For any* W-*invariant scalar product* $(\lambda | \mu)$ *on* $it^*$, *we have* $(\lambda | \mu) > 0$ *for all pairs of forms* $\lambda$, $\mu$ *such that* $\lambda \in C$ *and* $\mu > 0$.

(ii) *The Weyl chamber* C (resp. *its closure* $\bar{C}$) *is the set of linear forms* $\lambda \in it^*$ *such that* $w \cdot \lambda < \lambda$ (resp. $w \cdot \lambda \leqq \lambda$) *for all* $w \neq 1$ *in the Weyl group* W.

(i) By virtue of **(21.11.11.2)**, the Weyl chamber C may also be defined as the set of $\lambda \in it^*$ such that $(\lambda | \beta_j) > 0$ for $1 \leqq j \leqq l$. In view of **(21.11.5.3)**, the relation $C \subset L$ is a cons uence of the following lemma:

**(21.14.6.1)** *In a real Hilbert space* E, *let* $(\beta_j)_{1 \leq j \leq n}$ *be a finite free family such that* $(\beta_j | \beta_k) \leqq 0$ *whenever* $j \neq k$. *Then, if* $\lambda = \sum_{j=1}^{n} c_j \beta_j$ *is such that* $(\lambda | \beta_j) \geqq 0$ *for* $1 \leqq j \leqq n$, *we must have* $c_j \geqq 0$ *for* $1 \leqq j \leqq n$.

The result is obvious if $n = 1$, and we proceed by induction on $n$. It is not possible that $c_j < 0$ for *all* $j$, because it would then follow that $(\lambda | c_j \beta_j) \leqq 0$ for all $j$, and therefore

$$(\lambda | \lambda) = \sum_{j=1}^{n} (\lambda | c_j \beta_j) \leqq 0,$$

so that $\lambda = 0$, contradicting the hypothesis that $c_j \neq 0$ for all $j$. Suppose therefore, without loss of generality, that $c_n \geqq 0$. Then, for $1 \leqq j \leqq n-1$, we have

$$c_1(\beta_1 \mid \beta_j) + \cdots + c_{n-1}(\beta_{n-1} \mid \beta_j) \geqq -c_n(\beta_n \mid \beta_j) \geqq 0,$$

and by applying the inductive hypothesis to $c_1\beta_1 + \cdots + c_{n-1}\beta_{n-1}$, we deduce that $c_j \geqq 0$ for $1 \leqq j \leqq n$.

If $\mu > 0$, it follows from the definition **(21.14.4)** that we may write $\mu = \gamma + \sum_{j=1}^{l} t_j \beta_j$, where $\gamma \in i\mathfrak{c}^*$ and $t_j \geqq 0$ for $1 \leqq j \leqq l$, and at least one of the $t_j$ is $> 0$. If then $\lambda \in \mathrm{C}$, we have $(\lambda \mid \mu) = \sum_{j=1}^{l} t_j(\lambda \mid \beta_j) > 0$, because $(\lambda \mid \beta_j) > 0$ for $1 \leqq j \leqq l$.

(ii) If $w \cdot \lambda < \lambda$ for all $w \neq 1$ in W, then in particular **(21.10.6)** $s_\alpha \cdot \lambda = \lambda - \lambda(\mathbf{h}_\alpha)\alpha < \lambda$ for all positive roots $\alpha$, which is possible only if $\lambda(\mathbf{h}_j) > 0$ for $1 \leqq j \leqq l$, in other words if $\lambda \in \mathrm{C}$. To prove the converse, put $s_j = s_{\beta_j}$ for $1 \leqq j \leqq l$; then W is generated by the reflections $s_j$ **(21.11.8)**, and we shall argue by induction on the smallest number $p$ such that $w$ can be written in the form $w = s_{j_1} s_{j_2} \cdots s_{j_p}$. The result is clear if $p = 1$; suppose therefore that it is true for all products of at most $p-1$ reflections $s_j$, and put $w = w' s_{j_p}$, where $w' = s_{j_1} s_{j_2} \cdots s_{j_{p-1}}$. Then we have $w \cdot \lambda = w' \cdot \lambda - \lambda(\mathbf{h}_{j_p}) w' \cdot \beta_{j_p}$. We distinguish two cases, according as the root $w' \cdot \beta_{j_p}$ is positive or negative. In the first case, the hypothesis $\lambda(\mathbf{h}_{j_p}) > 0$ implies that $w \cdot \lambda < \lambda$. Consider therefore the second case, and let $r$ be the least integer such that for all $k \geqq r$ the root $\alpha_k = s_{j_k} s_{j_{k+1}} \cdots s_{j_{p-1}} \cdot \beta_{j_p}$ is positive. This number $r$ always exists (if we agree to put $r = p$ and $\alpha_p = \beta_{j_p}$ when $\alpha_k < 0$ for $1 \leqq k \leqq p-1$), and we have $r > 1$ because $w' \cdot \beta_{j_p} < 0$. By definition, we have $\alpha_r > 0$ and $\alpha_{r-1} = s_{j_{r-1}} \cdot \alpha_r < 0$, and by virtue of **(21.11.6)**, this is possible only if $\alpha_r = \beta_{j_{r-1}}$. Now put $w_1 = s_{j_1} \cdots s_{j_{r-2}}$, $w_2 = s_{j_r} \cdots s_{j_{p-1}}$, so that $w' = w_1 s_{j_{r-1}} w_2$, and $w_2 \cdot \beta_{j_p} = \beta_{j_{r-1}}$. Since $w_2 s_\alpha w_2^{-1} = s_{w_2 \cdot \alpha}$ for all roots $\alpha$, we have $w_2 s_{j_p} = s_{j_{r-1}} w_2$ and therefore

$$w = w' s_{j_p} = w_1 s_{j_{r-1}}^2 w_2 = w_1 w_2;$$

in other words, $w$ can be written as a product of $p-2$ reflections $s_j$, and hence $w \cdot \lambda < \lambda$ by virtue of the inductive hypothesis. For the relations $w \cdot \lambda \leqq \lambda$ and $\lambda \in \bar{\mathrm{C}}$, the proof is the same.

**(21.14.6.2)** Let $\lambda \in i\mathfrak{t}^*$ be such that $\lambda(\mathbf{h}_j)$ is an *integer* $\geqq 0$ for $1 \leqq j \leqq l$ (or, equivalently, such that $\lambda(\mathbf{h}_\alpha)$ is an integer $\geqq 0$ for all positive roots $\alpha$ (relative to **B**), since the $\mathbf{h}_j$ form a basis of the root system $\mathbf{S}^\vee$ formed by the $\mathbf{h}_\alpha$ (21.11.5.5)). Then for each $w \in \mathrm{W}$ we have

(21.14.6.3) $$w \cdot \lambda = \lambda - \sum_{j=1}^{l} n_j \beta_j$$

where the $n_j$ are *integers* $\geqq 0$. We may proceed by induction as in the proof of (21.14.6(ii)), since the result is obvious when $w = s_j$. With the same notation, the case in which $w' \cdot \beta_{j_p}$ is a negative root can be eliminated, because $w$ is then a product of $p - 2$ reflections $s_j$; and if $w' \cdot \beta_{j_p}$ is a positive root, we may write $w' \cdot \beta_{j_p} = \sum_{j=1}^{l} n'_j \beta_j$, where the $n'_j$ are integers $\geqq 0$, and $w' \cdot \lambda = \lambda - \sum_{j=1}^{l} n''_j \beta_j$ where the $n''_j$ are integers $\geqq 0$. From these two equations we obtain (21.14.6.3), with $n_j = n'_j \lambda(h_{j_p}) + n''_j$.

(21.14.6.4) It follows from (21.11.5.3) that if there are two roots $\beta_j, \beta_k \in \mathbf{B}$ such that $(\beta_j | \beta_k) \neq 0$, then they cannot belong to $\bar{\mathrm{C}}$. In all the examples considered in (21.12), with the exception of $\mathbf{SU}(2)$, *none* of the basis roots therefore belongs to $\bar{\mathrm{C}}$.

(21.14.7) For each root $\alpha \in \mathbf{S}$, let $\mathrm{H}_\alpha$ be the hyperplane in $it^*$ defined by the equation $\lambda(\mathbf{h}_\alpha) = 0$. A linear form $\lambda \in it^*$ is said to be *singular* if it belongs to at least one of the $\mathrm{H}_\alpha$, and *regular* if it does not. Clearly the Weyl group transforms regular forms into regular forms, and singular forms into singular forms.

(21.14.7.1) *For each regular linear form $\lambda \in it^*$, there exists one and only one element $w$ of the Weyl group $\mathrm{W}$ such that $w \cdot \lambda \in \mathrm{C}$. For each linear form $\lambda \in it^*$ there exists one and only one $w \cdot \lambda$ in the $\mathrm{W}$-orbit of $\lambda$ that belongs to $\bar{\mathrm{C}}$.*

Suppose first that $\lambda$ is regular. We may write $\lambda = \gamma + \mu$, where $\gamma \in ic^*$ and $\mu \in it'^*$, and since $\gamma(\mathbf{h}_\alpha) = 0$ for all $\alpha \in \mathbf{S}$ we have $\mu(\mathbf{h}_\alpha) \neq 0$ for all $\alpha \in \mathbf{S}$. It follows then from (21.11.5.2), applied to the dual root system $\mathbf{S}^\vee$, that $\mu$ defines a basis $\mathbf{B}_\mu^\vee$ of $\mathbf{S}^\vee$. By virtue of (21.11.8), there exists $w \in \mathrm{W}$ such that $w(\mathbf{B}_\mu^\vee) = \mathbf{B}^\vee$; and since $w(\mathbf{B}_\mu^\vee) = \mathbf{B}_{w \cdot \mu}^\vee$ and $w \cdot \gamma = \gamma$, this implies that $w \cdot \lambda \in \mathrm{C}$, by definition. The uniqueness of $w$ follows from the same argument, in conjunction with the fact that W acts simply transitively on the set of bases of $\mathbf{S}^\vee$ (21.11.10.1).

Now let $\lambda$ be any element of $it^*$, and let $\lambda_0$ be a regular linear form. Since the number of hyperplanes $H_\alpha$ is finite, the linear form $\lambda + t(\lambda_0 - \lambda)$ is singular for only finitely many values of $t \in \mathbf{R}$, and we may therefore assume that it is regular for $0 < t \leqq 1$. Let $w \in W$ be such that $w \cdot \lambda_0 \in C$; since $w \cdot (\lambda + t(\lambda_0 - \lambda))$ is regular for $0 < t \leqq 1$, all these linear forms belong to C, and therefore $w \cdot \lambda$ must belong to the closure $\bar{C}$.

If $\lambda \in \bar{C}$ and if there were an element $w \in W$ such that $w \cdot \lambda \in \bar{C}$ and $w \cdot \lambda \neq \lambda$, we should have $w \cdot \lambda \leqq \lambda$ by **(21.14.6)**, hence $w \cdot \lambda < \lambda$. But since $\lambda = w^{-1} \cdot (w \cdot \lambda)$, the same argument shows that $w \cdot \lambda > \lambda$, which is absurd.

**(21.14.8)** (i) *The half-sum $\delta$ of the positive roots of* **S** **(21.11.7)** *is such that $\delta(\mathbf{h}_j) = 1$ for $1 \leqq j \leqq l$, and hence belongs to $C \cap P(\mathfrak{g})$.*

(ii) *Every element of $P(\mathfrak{g}) \cap C$ is of the form $\delta + p$, where $p \in P(\mathfrak{g}) \cap \bar{C}$.*

(iii) *For each $p \in P \cap \bar{C}$, the set of linear forms $q \in P \cap \bar{C}$ such that $q \leqq p$ is finite.*

(i) We have seen in **(21.11.7)** that $s_j \cdot \delta = \delta - \delta(\mathbf{h}_j) \cdot \beta_j = \delta - \beta_j$, hence $\delta(\mathbf{h}_j) = 1$ for $1 \leqq j \leqq l$.

(ii) If $\lambda \in P(\mathfrak{g}) \cap C$, we have $\lambda(\mathbf{h}_j) > 0$ for $1 \leqq j \leqq l$ and moreover $\lambda(\mathbf{h}_j)$ is an integer, hence $\lambda(\mathbf{h}_j) \geqq 1$ for $1 \leqq j \leqq l$. Consequently $p = \lambda - \delta$ is such that $p(\mathbf{h}_j) \geqq 0$ for all $j$, hence $p \in P(\mathfrak{g}) \cap \bar{C}$. The converse is obvious.

(iii) Since $p - q \geqq 0$ and $p$, $q$ are in $\bar{C}$, we have $(p \mid p - q) \geqq 0$ and $(q \mid p - q) \geqq 0$ **(21.14.6)**, so that $(q \mid q) \leqq (p \mid q) \leqq (p \mid p)$. But since P is a discrete subspace of $it^*$, its intersection with the closed ball with center 0 and radius $(p \mid p)^{1/2}$ is finite **(3.16.3)**, whence the result.

**(21.14.8.1)** If the compact connected group G is *semisimple*, the set $P(\mathfrak{g})$ is also *discrete*, because $\mathfrak{c} = \{0\}$. The proof of (iii) above then applies without any modification to show that, for each $p \in P(\mathfrak{g}) \cap \bar{C}$, the set of $q \in P(\mathfrak{g}) \cap \bar{C}$ such that $q \leqq p$ is *finite*.

**(21.14.9)** The elements of the Weyl group, considered as endomorphisms of $it^*$, belong to the orthogonal group relative to the scalar product $(\lambda \mid \mu)$, hence have determinant equal to $\pm 1$. An element $\Phi$ of the free **Z**-module $\mathbf{Z}[P]$ (or of $\mathbf{Z}[cP]$, where $c$ is a nonzero real number) is said to be *anti-invariant* under W if $w \cdot \Phi = \det(w)\Phi$ for all $w \in W$. For each $p \in P$, the element

**(21.14.9.1)** $$J(e^p) = \sum_{w \in W} \det(w) e^{w \cdot p}$$

of $\mathbf{Z}[\mathrm{P}]$ is anti-invariant, because for each $w' \in \mathrm{W}$ we have

$$\begin{aligned} w' \cdot \mathrm{J}(e^p) &= \sum_{w \in \mathrm{W}} \det(w)(w' \cdot e^{w \cdot p}) \\ &= \sum_{w \in \mathrm{W}} \det(w) e^{(w'w) \cdot p} \\ &= \det(w') \sum_{w \in \mathrm{W}} \det(w'w) e^{(w'w) \cdot p} \\ &= \det(w') \mathrm{J}(e^p). \end{aligned}$$

**(21.14.10)** (i) *If the weight* $p \in \mathrm{P}$ *is a singular linear form* **(21.14.6.3)**, *we have* $\mathrm{J}(e^p) = 0$.

(ii) *As* $p$ *runs through* $\mathrm{P} \cap \mathrm{C}$, *the elements* $\mathrm{J}(e^p)$ *form a basis of the* $\mathbf{Z}$*-module* $\mathbf{Z}[\mathrm{P}]^{a\mathrm{W}}$ *of anti-invariant elements of* $\mathbf{Z}[\mathrm{P}]$.

(i) Suppose that $p(\mathbf{h}_\alpha) = 0$ for some root $\alpha \in \mathbf{S}$; then we have $s_\alpha \cdot p = p$, where $s_\alpha$ is the corresponding reflection. If $\mathrm{W}'$ is a set of representatives of the left cosets of the subgroup $\{1, s_\alpha\}$ in $\mathrm{W}$, we have

$$\begin{aligned} \mathrm{J}(e^p) &= \sum_{w' \in \mathrm{W}'} (\det(w') e^{w' \cdot p} + \det(w's_\alpha) e^{(w's_\alpha) \cdot p}) \\ &= 0 \end{aligned}$$

because $\det(w's_\alpha) = -\det(w')$ and $(w's_\alpha) \cdot p = w' \cdot p$.

(ii) To say that an element $\sum_{p \in \mathrm{P}} z_p e^p$ of $\mathbf{Z}[\mathrm{P}]$ (where $z_p \in \mathbf{Z}$ for all $p \in \mathrm{P}$) is anti-invariant means that $z_{w \cdot p} = \det(w) z_p$ for all $w \in \mathrm{W}$, and consequently the $\mathrm{J}(e^p)$ generate the $\mathbf{Z}$-module $\mathbf{Z}[\mathrm{P}]^{a\mathrm{W}}$. It follows from **(21.14.7)** that the group $\mathrm{W}$ acts *freely* on the set $\mathrm{P}_{\mathrm{reg}}$ of weights that are *regular* linear forms, so that $\mathrm{J}(e^p) \neq 0$ for all $p \in \mathrm{P}_{\mathrm{reg}}$; furthermore, each $\mathrm{W}$-orbit in $\mathrm{P}_{\mathrm{reg}}$ intersects $\mathrm{C}$ in exactly one point **(21.14.7)**, hence the $\mathrm{J}(e^p)$ with $p \in \mathrm{P} \cap \mathrm{C} \doteq \mathrm{P}_{\mathrm{reg}} \cap \mathrm{C}$ are linearly independent over $\mathbf{Z}$. In view of (i), this proves (ii).

The results of **(21.14.10)** apply unchanged to $\mathbf{Z}[c\mathrm{P}]$ if $c > 0$.

**(21.14.11)** Given an element $\Phi = \sum_{p \in \mathrm{P}} z_p e^p$ of $\mathbf{Z}[\mathrm{P}]$, we shall say that $z_p e^p$ is the *leading term* of $\Phi$ if $z_p \neq 0$ and if $p' < p$ for all other $p' \in \mathrm{P}$ such that $z_{p'} \neq 0$. It is clear that if $z_p e^p$ is the leading term of $\Phi$, and if $\Phi' = \sum_{p \in \mathrm{P}} z'_p e^p$ is another element of $\mathbf{Z}[\mathrm{P}]$, with leading term $z'_q e^q$, then $z_p z'_q e^{p+q}$ is the leading term of $\Phi\Phi'$. This definition and this remark apply without change to $\mathbf{Z}[c\mathrm{P}]$, $c > 0$.

It follows from **(21.14.7)** that each orbit $\Pi \in \mathrm{P}/\mathrm{W}$ intersects $\bar{\mathrm{C}}$ in exactly one point $p$. For $p \in P \cap \bar{C}$ we therefore denote the sum $S(\Pi)$ by $S(p)$. Since $w \cdot p \leqq p$ for all $w \in W$ (21.14.6), it follows that $e^p$ is the *leading term* of $S(p)$. Every element $\Psi$ of $\mathbf{Z}[P]^W$ that has leading term $z_p e^p$ may therefore be written uniquely in the form $\Psi = z_p S(p) + \sum_{q \in P \cap \bar{C},\, q < p} z_q S(q)$, where $z_p$ and the $z_q$ are integers.

**(21.14.12)** Since the roots $\alpha \in \mathbf{S}$ belong to P (21.14.2), the element

**(21.14.12.1)**
$$\Delta = \prod_{\alpha \in \mathbf{S}^+} (e^{\alpha/2} - e^{-\alpha/2})$$

(where $\mathbf{S}^+$ is the set of positive roots, relative to the basis $\mathbf{B}$) belongs to $\mathbf{Z}[\frac{1}{2}P]$, but not necessarily to $\mathbf{Z}[P]$ (cf. (21.16.10)). We have

**(21.14.12.2)**
$$\Delta = e^{\delta} \prod_{\alpha \in \mathbf{S}^+} (1 - e^{-\alpha}) = e^{-\delta} \prod_{\alpha \in \mathbf{S}^+} (e^{\alpha} - 1);$$

by virtue of (21.14.8), this shows that $\Delta$ belongs to $\mathbf{Z}[P(\mathfrak{g})]$ in any case, and that $e^{-\delta}\Delta$ belongs to $\mathbf{Z}[P]$; moreover the first of the formulas (21.14.12.2) shows immediately that $e^{\delta}$ is the *leading term* of $\Delta$ (21.14.11). By virtue of the formula (21.14.12.1), $\Delta$ is *anti-invariant*. Indeed, it is enough to show that $s_j \cdot \Delta = -\Delta$ for $1 \leqq j \leqq l$, because W is generated by the reflections $s_j$, in the notation of (21.14.7); but by virtue of (21.11.6), $s_j$ changes the sign of the factor $e^{\beta_j/2} - e^{-\beta_j/2}$ and permutes the other factors of $\Delta$, whence the result.

**(21.14.13)** (i) *We have* $\Delta = J(e^{\delta})$ *in* $\mathbf{Z}[\frac{1}{2}P]$.

(ii) *For each weight* $p \in P \cap \bar{C}$, *the element* $J(e^{p+\delta})/J(e^{\delta})$ *is an invariant element of* $\mathbf{Z}[P]$, *with leading term equal to* $e^p$.

(iii) *For each* $p \in P \cap \bar{C}$, *let* $\Psi_p$ *be an element of* $\mathbf{Z}[P]^W$ *with leading term equal to* $e^p$; *then the* $\Psi_p$ *form a* $\mathbf{Z}$*-basis of the* $\mathbf{Z}$*-algebra* $\mathbf{Z}[P]^W$. (In particular, this is so for the elements $J(e^{p+\delta})/J(e^{\delta})$.)

(i) Since $\Delta$ is anti-invariant and belongs to $\mathbf{Z}[\frac{1}{2}P]$, it is a linear combination with integral coefficients of the $J(e^q)$ with $q \in \frac{1}{2}P \cap C$ (21.14.10); but since $e^{-\delta}\Delta \in \mathbf{Z}[P]$, we must have $q - \delta \in P$ for each of the $J(e^q)$ appearing in $\Delta$ with a nonzero coefficient; hence $\Delta = \sum_{p \in P \cap \bar{C}} z_p J(e^{p+\delta})$, with $z_p \in \mathbf{Z}$. The coefficient of $e^{p+\delta}$ in $\Delta$ is therefore $z_p$. Now, for $p \in P \cap \bar{C}$ and $p \neq 0$, $p + \delta$ is not comparable with $\delta$ with respect to the ordering if $p \in i\mathfrak{c}^*$, and is $\geqq \delta$ otherwise. Since all the terms of $\Delta$ other than $e^{\delta}$ are of the form $z'_q e^q$ with $q < \delta$, we must have $z_p = 0$ for $p \neq 0$, and $z_0 = 1$.

(ii) Since $w \cdot \delta - \delta \in \mathrm{P}$ for all $w \in \mathrm{W}$ (21.11.7), it follows that $e^{-\delta}\mathrm{J}(e^{p+\delta})$ belongs to $\mathbf{Z}[\mathrm{P}]$. Hence, by virtue of (21.14.12.2) and (21.14.1), it is enough to show that $e^{-\delta}\mathrm{J}(e^{p+\delta})$ is divisible by each of the elements $1 - e^{-\alpha}$ with $\alpha \in \mathbf{S}^+$ (bearing in mind that no two distinct elements of $\mathbf{S}^+$ are proportional). If W′ is a set of representatives of the right cosets of the subgroup $\{1, s_\alpha\}$ in W, we have

$$e^{-\delta}\mathrm{J}(e^q) = \sum_{w' \in \mathrm{W}'} \det(w')(e^{w' \cdot q} - e^{s_\alpha \cdot (w' \cdot q)})e^{-\delta}$$

where $q = p + \delta$. But since $q \in \mathrm{P}(\mathfrak{g})$, we have $w' \cdot q \in \mathrm{P}(\mathfrak{g})$ and therefore $s_\alpha \cdot (w' \cdot q) = w' \cdot q - m(w')\alpha$, where $m(w')$ is an *integer*. We reduce therefore to showing that $1 - e^{-n\alpha}$ is divisible by $1 - e^{-\alpha}$ in $\mathbf{Z}[\mathrm{P}]$ for all integers $n \in \mathbf{Z}$. This is clear if $n \geqq 0$, and if $n < 0$ we have only to remark that $1 - e^{-n\alpha} = e^{-n\alpha}(e^{n\alpha} - 1)$, and that $e^{-\alpha}$ is invertible in $\mathbf{Z}[\mathrm{P}]$.

Now put $\mathrm{J}(e^{p+\delta})/\mathrm{J}(e^\delta) = \sum_{r \in \mathrm{P}} z_r e^r$, and let $u$ be a *maximal* element of the finite set of $r \in \mathrm{P}$ such that $z_r \neq 0$. We shall show that $\delta + u$ is maximal among the elements $v \in \frac{1}{2}\mathrm{P}$ such that the coefficient of $e^v$ in $\mathrm{J}(e^{p+\delta})$ is $\neq 0$. Indeed, if $z'_t e^t$ is a term of $\mathrm{J}(e^\delta)$ other than $e^\delta$, then $t < \delta$; if we had $t + r > \delta + u$, it would follow that $r > u + \delta - t > u$, contrary to the hypothesis that $u$ is maximal. Since $p + \delta \in \mathrm{C}$, we have $p + \delta > w \cdot (p + \delta)$ for all $w \neq 1$ in W (21.14.6), and consequently $e^{p+\delta}$ is the leading term of $\mathrm{J}(e^{p+\delta})$; hence $u = p$ and $z_p = 1$, and $e^p$ is the leading term of $\mathrm{J}(e^{p+\delta})/\mathrm{J}(e^\delta)$.

(iii) The hypothesis implies that for each $p \in \mathrm{P} \cap \bar{\mathrm{C}}$ we may write

(21.14.13.1) $$\Psi_p - \mathrm{S}(p) = \sum_{q \in \mathrm{P} \cap \bar{\mathrm{C}},\, q < p} z_{pq}\, \mathrm{S}(q).$$

We shall first show that the $\Psi_p$ are linearly independent. If not, there would exist a finite nonempty subset I of $\mathrm{P} \cap \bar{\mathrm{C}}$, and for each $p \in \mathrm{I}$ an integer $c_p \neq 0$ such that $\sum_{p \in \mathrm{I}} c_p \Psi_p = 0$, and therefore, by (21.14.13.1),

(21.14.13.2) $$\sum_{p \in \mathrm{I}} c_p \Bigl(\mathrm{S}(p) + \sum_{q \in \mathrm{P} \cap \bar{\mathrm{C}},\, q < p} z_{pq}\, \mathrm{S}(q)\Bigr) = 0.$$

There exists in the finite set I a *maximal* element $r$. For each $p \in \mathrm{I}$ distinct from $r$ (resp. each $q \in \mathrm{P} \cap \bar{\mathrm{C}}$ such that $q < r$ or $q < p$ for $p \in \mathrm{I}$ and $p \neq r$) there cannot appear in $\mathrm{S}(p)$ (resp. $\mathrm{S}(q)$) a term in $e^r$ with a nonzero coefficient, because this would imply that $r < p$ (resp. $r \leqq q < r$ or $r \leqq q < p$) by virtue of (21.14.11), contradicting the definition of $r$. Hence we cannot have $c_r \neq 0$, which proves our assertion.

Next we shall prove that each $S(p)$, for $p \in P \cap \bar{C}$, is a linear combination of the $\Psi_q$ with $q \in P \cap \bar{C}$. Suppose that this were not the case, and let $p_0 \in P \cap \bar{C}$ be an element for which the assertion is not true. Since the set of weights $p \leqq p_0$ in $P \cap \bar{C}$ is *finite* (21.14.8), we may assume that $p_0$ is *minimal*, in other words, that for each $p \in P \cap \bar{C}$ such that $p < p_0$, $S(p)$ is a linear combination of the $\Psi_q$. But then the relation (21.14.13.1), with $p = p_0$, would show that the difference $\Psi_{p_0} - S(p_0)$ was a linear combination of the $\Psi_q$ with $q \in P \cap \bar{C}$, and we should arrive at a contradiction.

## 15. WEYL'S FORMULAS

(21.15.1) *Let* $G$ *be a compact connected Lie group,* $T$ *a maximal torus of* $G$. *Then the* $C^\infty$ *mapping* $(s, t) \mapsto sts^{-1}$ *of* $G \times T$ *into* $G$ *is a submersion* (16.7.1) *at all points* $(s, t)$ *such that* $t$ *is regular* (21.7.13).

For each $s_0 \in G$, we have $sts^{-1} = s_0((s_0^{-1}s)t(s_0^{-1}s)^{-1})s_0^{-1}$; therefore, since $x \mapsto s_0 x s_0^{-1}$ is an automorphism of $G$, it is enough to prove the proposition at the point $(e, t)$ of $G \times T$. Let $\mathfrak{t}$ be the Lie algebra of $T$ and let $\mathfrak{m}$ be the subspace of the Lie algebra $\mathfrak{g}$ of $G$ that is the direct sum of the $(\mathfrak{g}_\alpha \oplus \mathfrak{g}_{-\alpha}) \cap \mathfrak{g}$ (in the notation of (21.8.1)) and supplementary to $\mathfrak{t}$. Since the exponential mapping is a diffeomorphism of a neighborhood of 0 in $\mathfrak{g}$ onto a neighborhood of $e$ in $G$, it will be enough to show that the $C^\infty$ mapping $\varphi: (\mathbf{u}, t) \mapsto \exp(\mathbf{u})t(\exp(\mathbf{u}))^{-1}$ is a submersion of $\mathfrak{m} \times T$ into $G$ at the point $(0, t_0)$ when $t_0 \in T$ is regular. Since the dimensions of $\mathfrak{m} \times T$ and $G$ are equal, it comes to the same thing to show that the tangent linear mapping $T_{(0, t_0)}(\varphi)$ is *injective* (A.4.11). The tangent vectors in $T_{(0, t_0)}(\mathfrak{m} \times T)$ are of the form $(\mathbf{v}, t_0 \cdot \mathbf{w})$ with $\mathbf{v} \in \mathfrak{m}$ and $\mathbf{w} \in \mathfrak{t}$. Let us apply (16.6.6) to the functions $\varphi(0, \cdot): t \mapsto t$ and $\varphi(\cdot, t_0): \mathbf{u} \mapsto \exp(\mathbf{u})t_0(\exp(\mathbf{u}))^{-1}$: the second of these is the composition of the left translation $z \mapsto t_0 z$, the mapping $(x, y) \mapsto xy$ of $G \times G$ into $G$ and the mapping $\mathbf{u} \mapsto (t_0^{-1} \exp(\mathbf{u})t_0, \exp(\mathbf{u})^{-1})$ of $\mathfrak{m}$ into $G \times G$. Using the formulas of (16.9.9), we obtain

(21.15.1.1) $$T_{(0, t_0)}(\varphi) \cdot (\mathbf{v}, t_0 \cdot \mathbf{w}) = t_0 \cdot (\mathrm{Ad}(t_0^{-1}) \cdot \mathbf{v} - \mathbf{v} + \mathbf{w}).$$

If this tangent vector is zero, then $\mathbf{w} = \mathbf{v} - \mathrm{Ad}(t_0^{-1}) \cdot \mathbf{v}$ belongs to both $\mathfrak{t}$ and $\mathfrak{m}$, because the choice of $\mathfrak{m}$ ensures that it is stable under $\mathrm{Ad}(t)$ for all $t \in T$; consequently $\mathbf{w} = 0$ and $\mathbf{v} = \mathrm{Ad}(t_0^{-1}) \cdot \mathbf{v}$. But the second of these relations implies that $\mathbf{v} = 0$, by reason of the hypothesis that $t_0$ is *regular* (21.8.4), and the proof is complete.

**(21.15.2)** For each $t' \in T$, we have $(st')t(st')^{-1} = sts^{-1}$ for all $s \in G$ and $t \in T$, because T is commutative. If $\pi$: $G \to G/T$ is the canonical projection, it follows therefore that the mapping $(s, t) \mapsto sts^{-1}$ factorizes as

$$G \times T \xrightarrow{\pi \times 1_T} (G/T) \times T \xrightarrow{f} G \tag{21.15.2.1}$$

where $f$ is of class $C^\infty$ (**(16.10.4)** and **(16.10.5)**). Let $T_{reg}$ be the set of regular points of T; it is a dense open subset of T, whose complement N is the union of the tori $U_\alpha$ of dimension $\dim(T) - 1$, as $\alpha$ runs through the set $\mathbf{S}^+$ of positive roots of G relative to T (with respect to an arbitrary basis $\mathbf{B}$ of the root system $\mathbf{S}$) **(21.8.4)**. Hence $(G/T) \times N$ is *negligible* in $(G/T) \times T$ **(16.22.2)**, and it follows from Sard's theorem that the compact set $f((G/T) \times N)$ is *negligible* in G **(16.23.2)**. Next, the restriction of $f$ to $(G/T) \times T_{reg}$ is a *submersion* of this open set onto an open subset V in G, by virtue of **(21.15.1)** and **(16.7.5)**. Finally, the mapping $f$ is surjective by **(21.7.4)**, hence G is the union of V and $f((G/T) \times N)$, from which it follows that *the complement of* V *is negligible.*

Let W be the Weyl group of G relative to T. We shall show that $(G/T) \times T_{reg}$ is *a covering of the open set* V, *with* Card(W) *sheets*, the projection being the restriction of $f$. For this purpose, we shall show that W acts differentiably and *freely* on $(G/T) \times T_{reg}$, so that the orbits are precisely the intersections of this submanifold with the inverse images $f^{-1}(x)$ for $x \in V$. In the first place, since W permutes the roots **(21.8.6)**, it leaves $T_{reg}$ stable and acts differentiably on this manifold by virtue of **(16.10.4)** and the definition of W **(21.7.16)**. Next, the normalizer $\mathcal{N}(T)$ of T in G acts on G/T on the right, by the rule $\pi(s) \cdot x = \pi(sx)$ for $s \in G$ and $x \in \mathcal{N}(T)$, because we have $sTx = sxT$ since $x$ normalizes T. Furthermore, if $x' = xt$, where $t \in T$, then $\pi(sx') = \pi(sx)$, and therefore for each coset $w \in W = \mathcal{N}(T)/T$ we may define $\pi(s) \cdot w$ as the common value of $\pi(sx)$ for all $x \in w$. It is clear that this action of W on G/T is differentiable **(16.10.4)** and *free*, because the relation $\pi(sx) = \pi(s)$ implies that $x \in T$. We now define a left action of W on $(G/T) \times T_{reg}$ by the rule $w \cdot (\pi(s), t) = (\pi(s) \cdot w^{-1}, w \cdot t)$ for $w \in W$, $s \in G$, and $t \in T_{reg}$. Clearly this is a free action of W. Moreover, if $sts^{-1} = s't's'^{-1}$, where $s, s' \in G$ and $t, t' \in T$, there exists $w \in W$ such that $t' = w \cdot t$ **(21.7.17)**, and it follows that $\pi(s') = \pi(s) \cdot w^{-1}$. The restriction of $f$ to $(G/T) \times T_{reg}$ therefore factorizes as

$$(G/T) \times T_{reg} \xrightarrow{\pi'} ((G/T) \times T_{reg})/W \xrightarrow{f_0} V,$$

and since $f$ maps each open set in $(G/T) \times T_{reg}$ to an open set in V, the mapping $f_0$ is a homeomorphism of the orbit space $((G/T) \times T_{reg})/W$ onto V. But since the restriction of $f$ to $(G/T) \times T_{reg}$ is a submersion, we deduce from **(16.10.3)** that the orbit *manifold* exists, and that $f_0$ is a diffeomorphism. The result now follows from **(16.14.1)**.

**(21.15.3)** With the notation of **(21.15.1)**, let $l = \dim(\mathrm{T})$ (the rank of G), and $2n = \dim(\mathrm{G/T}) = \dim(\mathfrak{m})$. Let $\upsilon_{\mathrm{G}}$ and $\upsilon_{\mathrm{T}}$ denote the translation-invariant volume forms on G and T, respectively, corresponding **(16.24.1)** to the *normalized* Haar measures $m_{\mathrm{G}}$ and $m_{\mathrm{T}}$ on G and T, respectively. We shall show that there is a canonically determined volume form $\upsilon_{\mathrm{G/T}}$ on G/T, invariant under the action of G. For this purpose, we observe that the tangent space† $\mathrm{T}_{\pi(e)}(\mathrm{G/T})$ may be canonically identified with $\mathfrak{g}/\mathfrak{t}$, and hence also with the supplement $\mathfrak{m}$ of $\mathfrak{t}$ in $\mathfrak{g}$. The image of a $2n$-vector $\mathbf{z} \in \bigwedge^{2n} \mathrm{T}_{\pi(e)}(\mathrm{G/T})$ under the diffeomorphism $x \mapsto s \cdot x$ is $\bigwedge^{2n} \mathrm{T}_s(\pi) \cdot (s \cdot \mathbf{z}) \in \bigwedge^{2n} \mathrm{T}_{s \cdot \pi(e)}(\mathrm{G/T})$; it depends only on the point $s \cdot \pi(e) \in \mathrm{G/T}$, not on $s \in \mathrm{G}$. For the relation $s \cdot \pi(e) = s' \cdot \pi(e)$ is equivalent to $s' = st$ for some $t \in \mathrm{T}$, and we therefore have to see that $\bigwedge^{2n} \mathrm{T}_t(\pi) \cdot (t \cdot \mathbf{z}) = \mathbf{z}$. Now, for each vector $\mathbf{u} \in \mathfrak{m}$, we have $\mathrm{T}_{t^{-1}}(\pi) \cdot (\mathbf{u} \cdot t^{-1}) = \mathbf{u}$, with the identification made above, and therefore $\mathrm{T}_t(\pi) \cdot (t \cdot \mathbf{u}) = \mathrm{Ad}(t) \cdot \mathbf{u} \in \mathfrak{m}$ (since $\mathfrak{m}$ is stable under $\mathrm{Ad}(t)$); but $t \mapsto \mathrm{Ad}(t) | \mathfrak{m}$ is a homomorphism of T into the orthogonal group of the restriction to $\mathfrak{m}$ of an Ad(G)-invariant scalar product on $\mathfrak{g}$; since T is *connected*, the determinant of $\mathrm{Ad}(t) | \mathfrak{m}$ is necessarily equal to 1, and therefore we have

$$\bigwedge^{2n} \mathrm{T}_t(\pi) \cdot (t \cdot \mathbf{z}) = \bigwedge^{2n} \mathrm{Ad}(t) \cdot \mathbf{z} = \mathbf{z}.$$

Let now $\mathbf{e}^*$ be a $2n$-covector on $\mathfrak{m}$ such that $\mathbf{e}^* \wedge \upsilon_{\mathrm{T}}(e) = \upsilon_{\mathrm{G}}(e)$ (we are identifying $\mathbf{e}^*$ and $\upsilon_{\mathrm{T}}(e)$ with their canonical images in $\bigwedge^{2n+l} \mathfrak{g}^*$ under $\bigwedge^{2n} {}^t\mathrm{pr}_1$ and $\bigwedge^{l} {}^t\mathrm{pr}_2$); for each $s \in \mathrm{G}$ the $2n$-covector $s \cdot \mathbf{e}^* \in \bigwedge^{2n} \mathrm{T}_{s \cdot \pi(e)}(\mathrm{G/T})$ depends on $x = s \cdot \pi(e)$ and not on $s$, by the remarks above and **(19.1.9.1)**. We may therefore define $\upsilon_{\mathrm{G/T}}(x) = s \cdot \mathbf{e}^*$. We denote by $m_{\mathrm{G/T}}$ the positive measure on G/T corresponding to $\upsilon_{\mathrm{G/T}}$ **(16.24.1)**, which is therefore G-*invariant*.

By abuse of notation, we denote by $\upsilon_{\mathrm{G/T}} \wedge \upsilon_{\mathrm{T}}$ the volume form on $(\mathrm{G/T}) \times \mathrm{T}$ that is equal to $\bigwedge^{2n} {}^t\mathrm{pr}_1(\upsilon_{\mathrm{G/T}}) \wedge \bigwedge^{l} {}^t\mathrm{pr}_2(\upsilon_{\mathrm{T}})$, to which corresponds the product measure $m_{\mathrm{G/T}} \otimes m_{\mathrm{T}}$. For the $C^\infty$ mapping $f$ defined in **(21.15.2.1)**, ${}^tf(\upsilon_{\mathrm{G}})$ is a $(2n + l)$-form on $(\mathrm{G/T}) \times \mathrm{T}$, and we may therefore write

**(21.15.3.1)** $$ {}^tf(\upsilon_{\mathrm{G}}) = \Theta \cdot (\upsilon_{\mathrm{G/T}} \wedge \upsilon_{\mathrm{T}}).$$

† The use of the letter T (with indices) to denote tangent spaces and tangent linear mappings should not be confused with the use of the same letter (without indices) to denote a maximal torus in G.

We propose to calculate the numerical function $\Theta$ on $(G/T) \times T$. For this purpose, we first remark that since $f(\pi(s), t) = sf(\pi(e), t)s^{-1}$ by definition (i.e., $f \circ \gamma(s) = \text{Int}(s) \circ f$, where $\gamma(s)$ denotes the diffeomorphism

$$(x, t) \mapsto (s \cdot x, t)$$

of $(G/T) \times T$ onto itself), by transport of structure we have also ${}^t f(s \cdot \upsilon_G \cdot s^{-1}) = s \cdot {}^t f(\upsilon_G)$, and $s \cdot \upsilon_G \cdot s^{-1} = \upsilon_G$ since G is unimodular. Also $s \cdot \upsilon_{G/T} = \upsilon_{G/T}$, and therefore we see that for all $s \in G$ we have

**(21.15.3.2)** $$\Theta(\pi(s), t) = \Theta(\pi(e), t).$$

If as before we identify $\mathfrak{m}$ and $T_{\pi(e)}(G/T)$, the calculation in **(21.15.1)** shows that

$$T_{(\pi(e), t)}(f) \cdot (\mathbf{v}, t \cdot \mathbf{w}) = t \cdot (\text{Ad}(t^{-1}) \cdot \mathbf{v} - \mathbf{v} + \mathbf{w})$$

for $\mathbf{v} \in \mathfrak{m}$ and $\mathbf{w} \in \mathfrak{t}$, since $T_0(\exp_G)$ is the identity mapping. The definition of ${}^t f(\upsilon_G)$ **(16.20.9.3)** and the choice of $\mathbf{e}^*$ then show that

**(21.15.3.3)** $$\Theta(\pi(e), t) = \det((\text{Ad}(t^{-1}) | \mathfrak{m}) - 1_{\mathfrak{m}}).$$

Now take in $\mathfrak{m}$ the basis consisting of the vectors $\mathbf{y}_\alpha$ and $\mathbf{z}_\alpha$ defined in **(21.8.3)**, for $\alpha \in \mathbf{S}^+$. Relative to this basis, $\text{Ad}(t^{-1}) | \mathfrak{m}$ is defined by the formulas **(21.8.3.3)**, with $\exp(u)$ replaced by $t^{-1}$. Since

$$\begin{vmatrix} \cos\theta - 1 & \sin\theta \\ -\sin\theta & \cos\theta - 1 \end{vmatrix} = 2(1 - \cos\theta) = 4\sin^2 \tfrac{1}{2}\theta = |e^{i\theta/2} - e^{-i\theta/2}|^2,$$

it follows from the formula **(21.14.12.1)** that we have

**(21.15.3.4)** $$\Theta(\pi(e), \exp(\mathbf{u})) = |\Delta(-\mathbf{u})|^2 = |\Delta(\mathbf{u})|^2.$$

This shows in the first place that the restriction of $f$ to $(G/T) \times T_{\text{reg}}$ is a local diffeomorphism onto V that preserves the orientation, when $(G/T) \times T$ and G are oriented by the forms $\upsilon_{G/T} \wedge \upsilon_T$ and $\upsilon_G$, respectively. Secondly, we deduce *Weyl's integration formula*:

**(21.15.4)** *Considering the function* $|\Delta|^2 = \prod_{\alpha \in \mathbf{S}^+} |e^\alpha - 1|^2$ *as a function on* T **(21.13.6)**, *for each continuous complex-valued function $g$ on* G *we have*

**(21.15.4.1)**

$$\int_G g(s)\, dm_G(s) = (\text{Card}(W))^{-1} \int_{G/T} \int_T g(f(x, t)) |\Delta(t)|^2\, dm_{G/T}(x)\, dm_T(t),$$

*and if g is a central function* (21.2.2)

(21.15.4.2) $$\int_G g(s)\,dm_G(s) = (\mathrm{Card(W)})^{-1}\int_T g(t)\,|\Delta(t)|^2\,dm_T(t).$$

Apply the formula of successive integrations (16.24.8.1) with $(G/T) \times T_{reg}$ in place of X, V in place of Y, the restriction to V of the $(2n + l)$-form $gv_G$ in place of $\zeta$, and the inverse image $(g \circ f) \cdot {}^t f(v_G)$ of this restriction in place of $v$. Each of the fibers $f^{-1}(y)$ is a finite set of cardinality Card(W) (21.15.2), and at each point $(x, t)$ of this fiber, $v/\zeta(y)$ is the number 1 by virtue of (16.21.9.2); hence we have

$$\int_V g(s)\,dm_G(s) = (\mathrm{Card(W)})^{-1}\int_{G/T}\int_{T_{reg}} g(f(x, t))\,|\Delta(t)|^2\,dm_{G/T}(x)\,dm_T(t).$$

But since the complement of V in G and the complement of $(G/T) \times T_{reg}$ in $(G/T) \times T$ are negligible (21.15.2), we may replace V by G and $T_{reg}$ by T in this formula, which gives (21.15.4.1). In particular, putting $g = 1$ and remembering that $m_G$ is normalized, we obtain

$$1 = (\mathrm{Card(W)})^{-1} m_{G/T}(G/T)\int_{T_{reg}} |\Delta(t)|^2\,dm_T(t).$$

Since $\Delta = J(e^{\delta}) = e^{\delta} \sum_{w \in W} \det(w) e^{w \cdot \delta - \delta}$ on t, and since the $w \cdot \delta - \delta$ are pairwise distinct weights in $P = 2\pi i \Gamma_T^*$, it follows from the orthogonality relations for characters (21.3.2.4) applied to T that

$$\int_T |\Delta(t)|^2\,dm_T(t) = \mathrm{Card(W)}$$

and consequently that $m_{G/T}(G/T) = 1$ (cf. (22.3.7.4)). If now $g$ is a central function, we have $g(f(x, t)) = g(t)$ for all $x \in G/T$, and the formula (21.15.4.2) follows from (21.15.4.1) applied to $g$, together with the above evaluation of $m_{G/T}(G/T)$ and the Lebesgue–Fubini theorem.

We can now describe completely the characters of a compact connected Lie group G in terms of the weight lattice $P = P(G)$ and the half-sum $\delta$ of the positive roots:

(21.15.5) (Weyl's theorem) *The mapping that sends each character $\chi_\rho$ of* G *to its restriction $\chi_\rho | T$ is a bijection onto the set of elements $J(e^{p+\delta})/J(e^{\delta})$ of $\mathbf{Z}[P]^W$, where p runs through the set $P(G) \cap \bar{C}$, and each of the weights $p \in P(G) \cap \bar{C}$ has multiplicity* 1 *in the character $\chi_\rho$ to which it corresponds. The canonical mapping $\mathbf{Z}^{(R(G))} \to \mathbf{Z}[P]^W$ is a ring isomorphism.*

For each weight $p \in \mathrm{P} \cap \bar{\mathrm{C}}$, let $g_p$ denote the unique continuous central function on G whose restriction to T is $\mathrm{J}(e^{p+\delta})/\mathrm{J}(e^{\delta})$ (21.13.5). Then the $g_p$ form an *orthonormal system* in $\mathrm{L}^2_{\mathbf{C}}(\mathrm{G}, m_{\mathrm{G}})$. For since $\Delta = \mathrm{J}(e^{\delta})$, it follows from Weyl's integration formula (21.15.4.2) that

(21.15.5.1) $$\int_{\mathrm{G}} g_p \bar{g}_q \, dm_{\mathrm{G}} = (\mathrm{Card}(\mathrm{W}))^{-1} \int_{\mathrm{T}} (e^{-\delta}\mathrm{J}(e^{p+\delta}))(e^{\delta}\overline{\mathrm{J}(e^{q+\delta})}) \, dm_{\mathrm{T}}.$$

As $w$ runs through W, the weights $w \cdot (p + \delta) - \delta$ are all distinct (21.14.5), and if $p \neq q$ and $q \in \mathrm{P} \cap \bar{\mathrm{C}}$, all the weights $w \cdot (q + \delta) - \delta$ are distinct from the weights $w \cdot (p + \delta) - \delta$ (21.14.7). By virtue of the orthogonality relations for the characters of T (21.3.2.4), the right-hand side of (21.15.5.1) therefore vanishes if $p \neq q$. Furthermore, when $q = p$, since $e^{-\delta}\mathrm{J}(e^{p+\delta})$ is a linear combination of Card(W) characters of T with coefficients $\pm 1$, the right-hand side of (21.15.5.1) is equal to 1.

This being so, it follows from (21.14.13(iii)) and (21.13.6) that we may write

$$\chi_{\boldsymbol{\rho}} | \mathrm{T} = \sum_{p \in \mathrm{P} \cap \bar{\mathrm{C}}} n(p)\mathrm{J}(e^{p+\delta})/\mathrm{J}(e^{\delta}),$$

where the $n(p)$ are *integers* $\geqq 0$; by (21.13.5), this implies that $\chi_{\boldsymbol{\rho}} = \sum_{p \in \mathrm{P} \cap \bar{\mathrm{C}}} n(p) g_p$. Since $\int_{\mathrm{G}} |\chi_{\boldsymbol{\rho}}|^2 \, dm_{\mathrm{G}} = 1$ (21.3.2.4), it follows therefore from the orthonormality of the $g_p$ that $\sum_{p \in \mathrm{P} \cap \bar{\mathrm{C}}} (n(p))^2 = 1$, and hence we have $n(p) = 0$ except for *one* weight $p_0 \in \mathrm{P} \cap \bar{\mathrm{C}}$, for which $n(p_0) = 1$. Consequently $\chi_{\boldsymbol{\rho}} = g_{p_0}$, and the weight $p_0$ occurs with multiplicity 1 in the character $\chi_{\boldsymbol{\rho}}$, because $e^{p_0}$ occurs with coefficient 1 in $\mathrm{J}(e^{p_0+\delta})/\mathrm{J}(e^{\delta})$ (21.14.13(ii)).

In view of (21.14.13(iii)), the proof will be complete if we show that each $g_p$ is a character of G. If it were not so, there would exist a weight $p_0 \in \mathrm{P} \cap \bar{\mathrm{C}}$ such that $g_{p_0}$ were orthogonal to *all* the characters of G, by the previous part of the proof. Since $g_{p_0} \neq 0$, this would contradict the fact that the characters of G form a Hilbert basis of the center of $\mathrm{L}^2_{\mathbf{C}}(\mathrm{G}, m_{\mathrm{G}})$ (21.3.2).

A class of irreducible representations $\boldsymbol{\rho} \in \mathrm{R}(\mathrm{G})$ is therefore determined by the *highest weight* $p \in \mathrm{P}(\mathrm{G})$ contained in $\chi_{\boldsymbol{\rho}}$. This weight $p$ is called the *dominant weight* of the class $\boldsymbol{\rho}$ (or of any representation in this class), relative to the chosen basis **B** of **S**. Every weight $p \in \mathrm{P} \cap \bar{\mathrm{C}}$ is therefore the dominant weight of a unique class $\boldsymbol{\rho} \in \mathrm{R}(\mathrm{G})$. Moreover, we have

(21.15.5.2) *If $p$ is the dominant weight of $\boldsymbol{\rho}$, the other weights contained in $\chi_{\boldsymbol{\rho}}$ are all of the form $p - \sum_j n_j \beta_j$, where the $n_j$ are integers $\geqq 0$.*

If $\chi_{\mathbf{p}} | \mathrm{T} = \sum_{q \in \mathrm{P}} n(q) e^{q}$, we have

(21.15.5.3) $$e^{-\delta}\mathrm{J}(e^{\delta}) \cdot \sum_{q \in \mathrm{P}} n(q) e^{q-p} = e^{-(p+\delta)}\mathrm{J}(e^{p+\delta}).$$

Since $e^{-\delta}\mathrm{J}(e^{\delta})$ is a polynomial in the $e^{-\beta_j}$ **(21.14.12.2)**, it is enough to show that $e^{-(p+\delta)}\mathrm{J}(e^{p+\delta})$ is also a polynomial in the $e^{-\beta_j}$, for it will then follow from **(21.15.5.3)** that if $n(q) \neq 0$ we must have $q - p = \sum_j z_j \beta_j$ with coefficients $z_j \in \mathbf{Z}$, and since we know that $q \leqq p$, the $z_j$ must all be $\leqq 0$. From the definition of $\mathrm{J}(e^{p+\delta})$, it is enough to verify that for all $r \in \mathrm{P} \cap \bar{\mathrm{C}}$ and all $w \in \mathrm{W}$, the weight $r - w \cdot r$ is a linear combination of the $\beta_j$ with coefficients that are *integers* $\geqq 0$. But this result is precisely **(21.14.6.2)**, since the numbers $r(\mathbf{h}_j)$ are *integers* $\geqq 0$.

**(21.15.5.4)** For $p \in i\mathfrak{c}^*$ we have $w \cdot p = p$ for all $w \in \mathrm{W}$, hence $\mathrm{J}(e^{p+\delta})/\mathrm{J}(e^{\delta}) = e^{p}$. These are the only weights $p \in \mathrm{P(G)}$ such that the character $e^{p}$ of T is the restriction of a character of G. For if $p$ is such that $w \cdot p = p$ for all $w \in \mathrm{W}$, then $p$ must be orthogonal to all elements $w \cdot \mathbf{u} - \mathbf{u} \in i\mathfrak{t}$, for all $\mathbf{u} \in i\mathfrak{t}$ and all $w \in \mathrm{W}$; but these elements span $i\mathfrak{t}'$ because $s_\alpha \cdot \mathbf{h}_\alpha = -\mathbf{h}_\alpha$, and therefore we must have $p \in i\mathfrak{c}^*$.

**(21.15.5.5)** Let $\mathbf{p}', \mathbf{p}''$ be two classes of irreducible representations in R(G), with dominant weights $p', p''$, respectively. Since $e^{p'}$ (resp. $e^{p''}$) is the leading term in $\chi_{\mathbf{p}'} | \mathrm{T}$ (resp. $\chi_{\mathbf{p}''} | \mathrm{T}$), it follows that $e^{p'+p''}$ is the leading term in the restriction to T of the product $\chi_{\mathbf{p}'}\chi_{\mathbf{p}''}$. If $\mathbf{p}$ is the class of irreducible representations with dominant weight $p' + p''$, it follows therefore that $\mathbf{p}$ is contained in $\mathbf{p}'\mathbf{p}''$ *with multiplicity* 1, and that every other class $\mathbf{p}_1 \in \mathrm{R(G)}$ contained in $\mathbf{p}'\mathbf{p}''$ corresponds to a dominant weight $< p' + p''$.

**(21.15.6)** *The dimension of the representations in the class* $\mathbf{p} \in \mathrm{R(G)}$ *with dominant weight* $p \in \mathrm{P} \cap \bar{\mathrm{C}}$ *is given by the formula* (Weyl's dimension formula)

**(21.15.6.1)**

$$n_{\mathbf{p}} = \prod_{\alpha \in \mathbf{S}^+} (p + \delta | \alpha) \Big/ \prod_{\alpha \in \mathbf{S}^+} (\delta | \alpha) = \prod_{\alpha \in \mathbf{S}^+} \langle p + \delta, \mathbf{h}_\alpha \rangle \Big/ \prod_{\alpha \in \mathbf{S}^+} \langle \delta, \mathbf{h}_\alpha \rangle$$

(where $(\lambda | \mu)$ is a W-invariant scalar product on $i\mathfrak{t}^*$).

The problem here is to calculate the value of the character $\chi_{\mathbf{p}}$ at the identity element of G **(21.3.2.8)**, i.e., to calculate the value of $\mathrm{J}(e^{p+\delta})/\mathrm{J}(e^{\delta})$ at the identity element of T (**21.15.5**), or equivalently at the point $0 \in \mathfrak{t}$, by considering the elements of **Z**[P] as functions defined on $\mathfrak{t}$. We shall calculate the value of $J(e^{p+\delta})/J(e^{\delta})$ at a point of the form $\xi i\mathbf{h}_\delta$, where $\mathbf{h}_\delta$ is the element of $i\mathfrak{t}$ defined by the relation $\lambda(\mathbf{h}_\delta) = (\lambda \mid \delta)$ for all $\lambda \in i\mathfrak{t}^*$, and $\xi$ is a nonzero real number; and then we shall take the limit of this as $\xi \to 0$. Now, for each linear form $q \in \frac{1}{2}P$, the value of $J(e^q)$ at the point $\xi i\mathbf{h}_\delta$ is by definition $\sum_{w \in W} \det(w) e^{i\xi(w \cdot \delta \mid q)}$, i.e., it is the value of $\Delta = J(e^\delta)$ at the point $\xi i\mathbf{h}_q \in \mathfrak{t}$, where $\mathbf{h}_q$ is the element of $i\mathfrak{t}$ defined by the relation $\lambda(\mathbf{h}_q) = (\lambda \mid q)$ for all $\lambda \in i\mathfrak{t}^*$. From the formula (**21.14.12.1**) for $\Delta$, we therefore have

(**21.15.6.2**) $$\Delta(\xi i\mathbf{h}_q) = \prod_{\alpha \in \mathbf{S}^+} (e^{i\xi(q \mid \alpha)/2} - e^{-i\xi(q \mid \alpha)/2}),$$

a function of $\xi$ whose principal part, as $\xi \to 0$, is $\prod_{\alpha \in \mathbf{S}^+} (i\xi(q \mid \alpha))$. The formula (**21.15.6.1**) now follows immediately.

## PROBLEMS

1. Let G be a compact connected group, and let P be the weight lattice of G (relative to a maximal torus T), $\mathbf{S} \subset P$ the root system of G relative to T, and W the Weyl group of G. Let $V$ be an irreducible representation of G, and let $p$ be a weight of $V$.
(a) Let $\alpha \in \mathbf{S}$, and let I be the set of integers $t \in \mathbf{Z}$ such that $p + t\alpha$ is a weight of $V$; let $n(p + t\alpha)$ be the multiplicity of this weight in $V$. Let $b$ (resp. $-a$) be the largest (resp. smallest) element of I. Prove that $I = [-a, b]$ and that $a - b = p(\mathbf{h}_\alpha)$. (Consider the restriction of $V$ to the subgroup $K_\alpha$ of G (**21.8.5**) and use (**21.9.3**).)
(b) Show that for each integer $u \in [0, a + b]$ we have $s_\alpha \cdot (p + (b - u)\alpha) = p - (a - u)\alpha$, and hence that the weights $p + (b - u)\alpha$, $p - (a - u)\alpha$ have the same multiplicity in $V$.
(c) Show that the function $t \mapsto n(p + t\alpha)$ is increasing in the interval $[-a, \frac{1}{2}(b - a)]$ and decreasing in the interval $[\frac{1}{2}(b - a), b]$. (Use (**21.9.3**).)
(d) A subset X of P is said to be **S**-*saturated* if, for each $p \in X$ and each root $\alpha \in \mathbf{S}$, we have $p - t\alpha \in X$ for all integers $t$ lying between 0 and $p(\mathbf{h}_\alpha)$ inclusive.
Show that every **S**-saturated subset of P is stable under the Weyl group. For each integer $d \geqq 1$, the set of weights $p$ of $V$ with multiplicity $n(p) \geqq d$ is **S**-saturated. (Use (c).)

2. With the notation of Problem 1, let E be the complex vector space of the representation $V$. Then E is a *simple* $U(\mathfrak{g}_{(\mathbf{C})})$-module (**21.9.1**), hence is generated by any $v \neq 0$ in V.
(a) For each weight $p$ of $V$, let $E_p$ be the set of vectors $v \in E$ such that $\mathbf{h} \cdot v = p(\mathbf{h})v$ for all $\mathbf{h} \in \mathfrak{h}$; then E is the direct sum of the $E_p$ as $p$ runs through the set of weights of $V$. Show that for each root $\alpha \in \mathbf{S}$, either $p + \alpha$ is not a weight of $V$, in which case $\mathbf{x}_\alpha \cdot E_p = \{0\}$, or else $\mathbf{x}_\alpha \cdot E_p \subset E_{p+\alpha}$.
(b) Let **B** be a basis of **S**, and let $\alpha_1, \dots, \alpha_n$ be the positive roots in **S** relative to this basis. Let $p$ be a weight of $V$ such that none of the weights $p + \alpha_j$ $(1 \leqq j \leqq n)$ is a weight of $V$. Then every weight of $V$ is of the form $q = p - m_1\alpha_1 - \cdots - m_n\alpha_n$, where the $m_j$ are integers $\geqq 0$, and $p$ is the dominant weight of $V$. (If $v \neq 0$ is a vector in $E_p$, remark that $v$ generates E as $U(\mathfrak{g}_{(\mathbf{C})})$-module, and use the basis (**21.16.3.4**) of $U(\mathfrak{g}_{(\mathbf{C})})$.)

3. The notation is the same as in Problems 1 and 2. Let Y be a subset of P; an element $p \in \mathrm{Y}$ is said to be **S**-*extremal in* Y if, for each root $\alpha \in \mathbf{S}$, we have either $p + \alpha \notin \mathrm{Y}$ or $p - \alpha \notin \mathrm{Y}$.
(a) Let $p$ be the dominant weight of $V$, and let X be the set of weights of $V$. Show that the **S**-extremal elements of X are the transforms $w \cdot p$ of $p$ under the Weyl group. (Let $q$ be an **S**-extremal element of X; without loss of generality, we may assume that $q \in \mathrm{P} \cap \bar{\mathrm{C}}$. If $\alpha$ is a positive root, show that in the notation of Problem 1 we must have $b = 0$, and then use Problem 2.)
(b) Show that X is the smallest **S**-saturated subset of P that contains $p$. (If X′ is this set, we have $\mathrm{X}' \subset \mathrm{X}$ by virtue of Problem 1(d). Assume that $\mathrm{X} \neq \mathrm{X}'$, and choose in $\mathrm{X} - \mathrm{X}'$ a maximal element $q$ (relative to the ordering defined by **B**). Then there exists a positive root $\alpha$ such that $q + \alpha \in \mathrm{X}$; deduce that, in the notation of Problem 1, we have $q + b\alpha \in \mathrm{X}'$, and hence (by using the definition of a saturated subset and Problem 1) that $q \in \mathrm{X}'$, which is a contradiction.)

4. With the same notation, show that for each weight $q$ of $V$ other than the dominant weight $p$, we have $(q\,|\,q) \leqq (p\,|\,p)$ and $(q + \delta\,|\,q + \delta) < (p + \delta\,|\,p + \delta)$. (Reduce to the case where $q \in \mathrm{P} \cap \bar{\mathrm{C}}$, and use **(21.14.6)**.)

5. With the same notation, show that there exists a weight in X that is the smallest element of X relative to the ordering defined by **B**, and that this smallest weight has multiplicity 1. (Observe that there exists an element $w_0$ in the Weyl group that transforms **B** into $-$**B**.)

6. With the same notation, show that for each finite nonempty **S**-saturated subset X of P, there exists a linear representation $V$ of G such that X is the union of the sets of weights of the irreducible components of $V$. (For each $p \in \mathrm{X}$ consider the weight $w \cdot p$ that lies in $\mathrm{P} \cap \bar{\mathrm{C}}$ **(21.14.7)** and the irreducible representation with dominant weight $w \cdot p$.)

7. (a) With the same notation, let $p \in \mathrm{P}$ and let X be the smallest **S**-saturated subset of P containing $p$. Prove that the following conditions are equivalent:

($\alpha$) $\mathrm{X} = \mathrm{W} \cdot p$.
($\beta$) $(q\,|\,q) = (p\,|\,p)$ for all $q \in \mathrm{X}$.
($\gamma$) For all roots $\alpha \in \mathbf{S}$ and all integers $t$ between 0 and $p(\mathbf{h}_\alpha)$ inclusive, we have $(p - t\alpha\,|\,p - t\alpha) \geqq (p\,|\,p)$.
($\delta$) For all roots $\alpha \in \mathbf{S}$, $p(\mathbf{h}_\alpha)$ is equal to 0, 1, or $-1$.

(To show that ($\gamma$) implies ($\delta$), observe that $(p - p(\mathbf{h}_\alpha)\alpha\,|\,p - p(\mathbf{h}_\alpha)\alpha) = (p\,|\,p)$ and that the Euclidean ball is strictly convex. To show that ($\delta$) implies ($\alpha$), observe that for each $w \in \mathrm{W}$, $(w \cdot p)(\mathbf{h}_\alpha)$ is also equal to 0, 1, or $-1$, and deduce that for each integer $t$ lying between 0 and $(w \cdot p)(\mathbf{h}_\alpha)$ inclusive, $w \cdot p - t\alpha$ is equal to either $w \cdot p$ or $(s_\alpha w) \cdot p$.)

Deduce that every nonempty **S**-saturated subset Y of P contains an element $p$ satisfying these conditions.
(b) Let $U$ be an irreducible representation of G, let E be the representation space of $U$, and let $p$ be the dominant weight of $U$. Show that the following conditions are equivalent:

($\alpha$) $p$ satisfies the equivalent conditions of (a) above.
($\beta$) All the weights of $U$ are of the form $w \cdot p$ for some $w \in \mathrm{W}$.
($\gamma$) For each root $\alpha \in \mathbf{S}$ and each $v \in \mathrm{E}$, we have $(\mathbf{x}_\alpha)^2 \cdot v = 0$.

(To show that ($\alpha$) and ($\beta$) are equivalent, use Problem 3(b). To show that ($\alpha$) implies ($\gamma$), use Problem 2(a). To show that if there exists $\alpha \in \mathbf{S}$ such that $p(\mathbf{h}_\alpha) \geqq 2$ we cannot have

$(\mathbf{x}_\alpha^2) \cdot v = 0$ for all $v \in \mathrm{E}$, consider the restriction of $U$ to the subgroup $\mathrm{K}_\alpha$ of G (21.8.5) and use (21.9.3).

**8.** (a) On the model of the algebra of formal power series (A.21), show that it is possible to define a **Z**-algebra $\mathbf{Z}[[\mathrm{P}^-]]$ whose elements are "formal sums" $\sum_{p \leq \lambda} c_p e^p$ (where $\lambda \in it^*$), the multiplication being defined by

$$\left(\sum_{p \leq \lambda} c_p e^p\right)\left(\sum_{p \leq \mu} c'_p e^p\right) = \sum_{p \leq \lambda+\mu} \left(\sum_{q+r=p} c_q c'_r\right) e^p.$$

With the notation of Problem 2(b), for each weight $p \in \mathrm{P}$ let $\nu(p)$ denote the number of systems of integers $(m_1, \ldots, m_n) \in \mathbf{N}^n$ such that $p = m_1 \alpha_1 + \cdots + m_n \alpha_n$ (so that $\nu(p)$ is the "number of partitions of $p$ into positive roots"). To say that $\nu(p) > 0$ means that $p$ is a linear combination of positive roots with coefficients that are integers $\geqq 0$. Then we have

$$1\Big/\prod_{\alpha \in \mathbf{S}^+} (1 - e^{-\alpha}) = \sum_{p \in \mathrm{P}} \nu(p) e^{-p}$$

in the ring $\mathbf{Z}[[\mathrm{P}^-]]$.

(b) For each weight $p \in \mathrm{P} \cap \bar{\mathrm{C}}$, show that the multiplicity $n_p(q)$ of a weight $q$ in the irreducible representation with dominant weight $p$ is given by the formula

$$(1) \qquad n_p(q) = \sum_{w \in \mathrm{W}} \det(w) \cdot \nu(w \cdot (p + \delta) - (q + \delta)).$$

(Use (a) above and Weyl's theorem.)

(c) Deduce from (b) that for each weight $q \neq p$ of the representation with dominant weight $p$ we have

$$n_p(q) = - \sum_{w \in \mathrm{W},\, w \neq 1} \det(w) \cdot n_p(q + \delta - w \cdot \delta).$$

(Use the formula (1) with $p = 0$.)

(d) Let $\boldsymbol{\rho}_p$ be the class of irreducible representations of G with dominant weight $p \in \mathrm{P} \cap \bar{\mathrm{C}}$. Show that the formula (21.4.7.1) can be written explicitly as

$$\boldsymbol{\rho}_p \boldsymbol{\rho}_q = \sum_{r \in \mathrm{P} \cap \bar{\mathrm{C}}} c(p, q, r) \boldsymbol{\rho}_r$$

where

$$(2) \qquad c(p, q, r) = \sum_{w \in \mathrm{W},\, w' \in \mathrm{W}} \det(ww') \cdot \nu(w \cdot (p + \delta) + w' \cdot (q + \delta) - (r + 2\delta)).$$

(Observe that by virtue of (21.4.6.2) we have

$$(3) \qquad \sum_{r \in \mathrm{P} \cap \bar{\mathrm{C}}} c(p, q, r) \mathrm{J}(e^{r+\delta}) = \left(\sum_{s \in \mathrm{P}} n_p(s) e^s\right) \mathrm{J}(e^{q+\delta})$$

and note that for $r \in \mathrm{P} \cap \bar{\mathrm{C}}$, $w \cdot (r + \delta)$ does not belong to $\mathrm{P} \cap \bar{\mathrm{C}}$; consequently $c(p, q, r)$ is equal to the coefficient of $e^{r+\delta}$ in the right-hand side of (3).)

9. With the notation of Section **21.15**, take as scalar product $(\lambda|\mu)$ on $it^*$ the canonical scalar product (Section **21.11**, Problem 11). Show that if the representations of the class $\boldsymbol{\rho} \in R(G)$ have dominant weight $p \in P \cap \bar{C}$ and dimension $n_{\boldsymbol{\rho}}$ (given by (**21.15.6.1**)), then

$$\sum_q (q|\delta)^2 = \tfrac{1}{24} n_{\boldsymbol{\rho}}((p+\delta|p+\delta) - (\delta|\delta))$$

where the sum on the left-hand side is over all the weights $q$ of $\boldsymbol{\rho}$, counted according to their multiplicities. (Expand the right-hand side of (**21.15.6.2**) and $\chi_{\boldsymbol{\rho}}(\xi i\mathbf{h}_q)$ as far as terms in $\xi^3$.)

10. Let G be an *almost simple* compact connected Lie group.
(a) With the notation of Section **21.14**, let **B** be a basis of **S**. Show that for the ordering defined by **B** there exists a *highest root* $\mu = n_1\beta_1 + \cdots + n_l\beta_l$, such that for each other root $\alpha = p_1\beta_1 + \cdots + p_l\beta_l$ in **S**, we have $p_j \leqq n_j$ for $1 \leqq j \leqq l$. (Observe that the adjoint represehtation of G on $\mathfrak{g}_{(\mathbf{C})}$ is irreducible, and consider its dominant weight.) The root $\mu$ lies in C, the Weyl chamber relative to **B**, and we have $(\alpha|\alpha) \leqq (\mu|\mu)$ for all roots $\alpha \in \mathbf{S}$. Also, for each root $\alpha > 0$ other than $\mu$, the Cartan integer $n(\alpha, \mu)$ is equal to 0 or 1 (use Section **21.11**, Problem 1).
(b) Let $\mathbf{h}_0$ be the half-sum of the $\mathbf{h}_\alpha \in it$ for all positive roots $\alpha$. For each root $\alpha = p_1\beta_1 + \cdots + p_l\beta_l$, we have $\alpha(\mathbf{h}_0) = p_1 + p_2 + \cdots + p_l = \frac{1}{2}\sum_{\beta \in \mathbf{S}^+} n(\alpha, \beta)$. (Use (**21.11.5.5**) and (**21.11.7**).)
(c) If $\mu = n_1\beta_1 + \cdots + n_l\beta_l$ is the highest root (relative to **B**), show that

$$n_1 + \cdots + n_l = h - 1,$$

where $h$ is the Coxeter number. (Use Section **21.11**, Problem 16(e), and observe that $n(\alpha, \mu)^2 = n(\alpha, \mu)$ for all roots $\alpha > 0$.)

11. Let G be an *almost simple* compact connected Lie group, and retain the notation of Problem 10. For each $\alpha \in \mathbf{S}$ and $k \in \mathbf{Z}$, let $\mathfrak{u}_{\alpha,k}$ denote the affine hyperplane $\alpha^{-1}(2\pi k)$ in $it$. Also let $\{\mathbf{p}_1, \ldots, \mathbf{p}_l\}$ be the basis of $it$ dual to the basis $\mathbf{B} = \{\beta_1, \ldots, \beta_l\}$ of $it^*$, so that $\beta_j(\mathbf{p}_k) = \delta_{jk}$ (Kronecker delta). Then the element $\mathbf{h}_0 \in it$ (Problem 10) is equal to $\mathbf{p}_1 + \cdots + \mathbf{p}_l$ (cf. (**21.16.5.2**).)

The Weyl group W, considered as a group of automorphisms of the vector space $it$, is generated by the reflections $s_\alpha$: $\mathbf{u} \mapsto \mathbf{u} - \alpha(\mathbf{u})\mathbf{h}_\alpha$ (**21.8.7**), where $\alpha \in \mathbf{S}$.
(a) The group $W_a$ generated by the orthogonal reflections in the affine hyperplanes $\mathfrak{u}_{\alpha,k}$ is called the *affine Weyl group*. Show that $W_a$ is the semidirect product of W by the group $P_0$ generated by the translations $\mathbf{u} \mapsto \mathbf{u} + 2\pi\mathbf{h}_\alpha$, $\alpha \in \mathbf{S}$ (or, equivalently, by the translations $\mathbf{u} \mapsto \mathbf{u} + 2\pi\mathbf{h}_j$, $1 \leqq j \leqq l$). (Observe that the translation $\mathbf{u} \mapsto \mathbf{u} + 2\pi\mathbf{h}_\alpha$ is the product of the reflections in the hyperplanes $\mathfrak{u}_{\alpha,0}$ and $\mathfrak{u}_{\alpha,1}$, and that W leaves invariant the root system $\mathbf{S}^\vee$ formed by the $\mathbf{h}_\alpha$.) The group $W_a$ leaves globally invariant the union of the hyperplanes $\mathfrak{u}_{\alpha,k}$.
(b) The set $C^*$ of vectors $\mathbf{u} \in it$ such that $\beta_j(\mathbf{u}) > 0$ for $1 \leqq j \leqq l$ is called the *Weyl chamber* of $\mathbf{S}^\vee$ (relative to the basis **B**), and the set $A^*$ of vectors $\mathbf{u} \in C^*$ such that $\mu(\mathbf{u}) < 2\pi$, where $\mu$ is the highest root of **S** relative to **B** (Problem 10) is called the *principal alcove* of $C^*$. Show that $A^*$ intersects none of the hyperplanes $\mathfrak{u}_{\alpha,k}$. The set $A^*$ is the interior of the *simplex* constructed on the vectors $2\pi\mathbf{p}_j/n_j$ $(1 \leqq j \leqq l)$ (**14.3.10**).
(c) Show that if an element of $W_a$ fixes a vector $\mathbf{z}$ *not lying in any of the hyperplanes* $\mathfrak{u}_{\alpha,k}$, then it is the identity element of $W_a$. (The element of $W_a$ in question is the product of an element $w \in W$ by a translation $\mathbf{u} \mapsto \mathbf{u} + \mathbf{h}$ of $P_0$, and must satisfy $w \cdot \mathbf{z} = \mathbf{z} + \mathbf{h}$; deduce that $\exp(iw \cdot \mathbf{z}) = \exp(i\mathbf{z})$ and use (**21.7.14**).)

(d) Deduce from (c) that the only element of $W_a$ that leaves $A^*$ globally invariant is the identity. (Remark that the subgroup $W_0$ of $W_a$ leaving $A^*$ globally invariant is finite, and consider the barycenter of the transforms under $W_0$ of a vector in $A^*$.)
(e) The connected components of the complement in $it$ of the union $iD(G)$ (Section **21.10**, Problem 2) of the hyperplanes $\mathfrak{u}_{\alpha, k}$ are called the *alcoves* of S. Show that each alcove is of the form $v(A^*)$ for a *unique* element $v \in W_a$. (The uniqueness follows from (d) above. To prove the existence of $v$, take a point $\mathbf{x}$ in an alcove and a point $\mathbf{a} \in A^*$; consider the point $\mathbf{y}$ of the $W_a$-orbit of $\mathbf{x}$ that is nearest to $\mathbf{a}$, and show that $\mathbf{y} \in A^*$. To do this, show that $\mathbf{y}$ and $\mathbf{a}$ lie on the same side of each of the "walls" of $A^*$, i.e., the $l + 1$ hyperplanes $\mathfrak{u}_{\beta_j, 0}$ and $\mathfrak{u}_{\mu, 1}$.)
(f) Show that each point of G is conjugate to a point of the form $\exp_G(i\mathbf{u})$, where $\mathbf{u} \in \overline{A^*}$. (Observe that $\exp(2\pi i \mathbf{h}_\alpha) = e$ for all roots $\alpha$, and use (e) above and (**21.8.7**).)
(g) Show that, for $1 \leqq j \leqq l$, the points of the form $2\pi t \mathbf{p}_j$ with $0 \leqq t \leqq 1$ that belong to $iD(G)$ are the points $2\pi \mathbf{p}_j/m$, where $1 \leqq m \leqq n_j$. (Use Section **21.11**, Problem 3(c).)

**12.** We retain the hypotheses and notation of Problem 11, and the notation of (**21.10.6**). Suppose that the vectors $\mathbf{x}_\alpha$, together with a basis of $it$, from a *Weyl basis* of $\mathfrak{g}_{(\mathbf{C})}$ (**21.10.6**). Consider the vector $\mathbf{h}_0 \in it$, and take an element in $\mathfrak{g}$ of the form

$$\mathbf{z} = \sum_{j=1}^{l} (z_j \mathbf{x}_{\beta_j} - \bar{z}_j \mathbf{x}_{-\beta_j})$$

with $z_j \in \mathbf{C}$ $(1 \leqq j \leqq l)$.
(a) Show that if we put

$$\mathbf{z}' = \sum_{j=1}^{l} i(z_j \mathbf{x}_{\beta_j} + \bar{z}_j \mathbf{x}_{-\beta_j}),$$

which belongs to $\mathfrak{g}$, then we have

$$[i\mathbf{h}_0, \mathbf{z}] = \mathbf{z}', \qquad [i\mathbf{h}_0, \mathbf{z}'] = -\mathbf{z},$$

and

$$[\mathbf{z}, \mathbf{z}'] = 2i \sum_{j=1}^{l} z_j \bar{z}_j \mathbf{h}_{\beta_j}.$$

(Use the fact that $\beta_j - \beta_k$ is not a root.) Show that we may choose the coefficients $z_j \neq 0$ so that $[\mathbf{z}, \mathbf{z}'] = 2i\mathbf{h}_0$. The vectors $i\mathbf{h}_0$, $\mathbf{z}$, and $\mathbf{z}'$ then generate a three-dimensional Lie subalgebra $\mathfrak{k}_0$ of $\mathfrak{g}$, which is the Lie algebra of an almost simple compact subgroup $K_0$ of G.
(b) Show that $K_0$ is a *principal nice* subgroup of G, of rank 1. (If G′ is a connected closed subgroup of G that contains $K_0$ and T, and if $\mathfrak{g}'$ is its Lie algebra, observe that $\mathfrak{g}'_{(\mathbf{C})}$ must be the direct sum of $\mathfrak{h} = \mathfrak{t}_{(\mathbf{C})}$ and a certain number of subspaces $\mathfrak{g}_\alpha$, and that the sum of these $\mathfrak{g}_\alpha$ must contain $\mathbf{z}$ and $\mathbf{z}'$; note also that the $\mathfrak{g}_{\beta_j}$ and $\mathfrak{g}_{-\beta_j}$ must occur among these $\mathfrak{g}_\alpha$, because the $z_j$ are $\neq 0$.)
(c) Show that a nice subgroup of G *of rank* 1 that contains a regular element of G is necessarily conjugate to the group $K_0$ (use Problems 8 and 12 of Section **21.11**). Such a subgroup is called *miniprincipal*.
(d) Let $G_1$ be a connected closed subgroup of G. Show that for $G_1$ to be a principal nice subgroup of G, it is necessary and sufficient that $G_1$ contain a miniprincipal subgroup $K_0$ of G. (To show that the condition is necessary, consider a miniprincipal subgroup $K'_0$ of $G_1$, and use Section **21.11**, Problem 8 to show that a principal diagonal of $K'_0$ is a principal

diagonal of G. To show that the condition is sufficient, observe that if $G_2$ is a connected closed proper subgroup of $G_1$ containing $K_0$, then the rank of $G_2$ cannot be equal to that of $G_1$, by virtue of Section **21.11**, Problem 7(d), and conclude that $K_0$ is miniprincipal in $G_1$, by virtue of the fact that an element of $G_1$ that is regular in G is also regular in $G_1$.)

(e) Let $\varphi$ be an automorphism of G. Show that there exists an element $z \in G$ such that $\mathrm{Int}(z) \circ \varphi$ fixes all the elements of a given *miniprincipal* subgroup $K_0$ of G. Deduce that there exists a finite subgroup F of Aut(G) such that Aut(G) is the semidirect product of its identity component Int(G) and F.

**13.** Let G be an *almost simple* compact connected Lie group. With the notation of Problem 11, if $\sigma$ is a *Coxeter element* in the Weyl group $W_{\mathbf{S}}$ (Section **21.11**, Problem 14), there exists an element $s \in \mathcal{N}(T)$ such that the restriction of $\mathrm{Ad}(s) \otimes 1_{\mathbf{C}}$ to $\mathfrak{t}_{(\mathbf{C})}$ is equal to that of the contragradient ${}^t\sigma^{-1}$ of $\sigma$. For each root $\alpha \in \mathbf{S}$, $\mathrm{Ad}(s) \otimes 1_{\mathbf{C}}$ transforms $\mathfrak{g}_\alpha$ into $\mathfrak{g}_{\sigma(\alpha)}$. Deduce that the eigenvalues of this transformation on the stable subspace $\sum_{\alpha \in \mathbf{S}} \mathfrak{g}_\alpha$ are $h$th roots of unity, each occurring with multiplicity $l$ (Section **21.11**, Problem 16(d)); the eigenvalues of $\mathrm{Ad}(s) \otimes 1_{\mathbf{C}}$ on $\mathfrak{t}_{(\mathbf{C})}$ are the complex numbers $\exp(2\pi i m_j/h)$ (Section **21.11**, Problem 14(c)). The element $s$ is called a *Coxeter element* of G (relative to T). Show that any two such elements are conjugate in G. (Use the fact that 1 is not an eigenvalue of $\sigma$.)

**14.** Let G be an *almost simple* compact connected Lie group, and retain the notation of Problem 11. Each Coxeter element of G (Problem 13) is regular in G and of finite order, equal to $h$. Show that each element $s \in G$ that is regular and of order $h$ is conjugate to a Coxeter element, and that no regular element $\neq e$ in G has finite order $\leqq h - 1$. (We may assume that $s = \exp(i\mathbf{u})$, where $\mathbf{u}$ lies in the principal alcove A* (Problem 11) in $it$. Use the fact that $s^h = e$ to show that $\beta_j(\mathbf{u}) = 2\pi p_j/h$ with $p_j$ a positive integer, and $\mu(\mathbf{u}) = 2\pi(\sum_{j=1}^{l} p_j n_j)/h$; then express that $\mu(\mathbf{u}) < 2\pi$ and use Problem 10(c) to deduce that $p_j = 1$ for $1 \leqq j \leqq l$ and hence that $\mathbf{u} = (2\pi/h)\mathbf{h}_0$ (Problem 10(b)).

**15.** With the same hypotheses and notation as in Problem 13, let $s \in \mathcal{N}(T)$ be the Coxeter element considered there. Since $s$ is regular in G, it is contained in a unique maximal torus T′ of G. Let $\mathfrak{t}'$ be the Lie algebra of T′, let $\mathbf{S}'$ be the corresponding root system, and let $\{\beta'_1, \ldots, \beta'_l\}$ be a basis of $\mathbf{S}'$. If $\mathbf{h}'_0 \in i\mathfrak{t}'$ is such that $\beta'_j(\mathbf{h}'_0) = 1$ for $1 \leqq j \leqq l$, we may assume that $s = \exp(i\mathbf{u}')$ with $\mathbf{u}' = (2\pi/h)\mathbf{h}'_0$ (Problem 14). For each integer $j \neq 0$, let $\mathfrak{a}'_j$ denote the direct sum of the $\mathfrak{g}_{\alpha'}$ for all $\alpha' \in \mathbf{S}'$ such that $\alpha' = p_1\beta'_1 + \cdots + p_l\beta'_l$ with $p_1 + \cdots + p_l = j$, so that we have $\mathfrak{a}'_j = \{0\}$ if $|j| \geqq h$. Also put $\mathfrak{a}'_0 = \mathfrak{t}'_{(\mathbf{C})}$, so that $\mathfrak{g}_{(\mathbf{C})}$ is the direct sum of the $\mathfrak{a}'_j$. For all $\mathbf{x} \in \mathfrak{a}'_j$, we have $[\mathbf{h}'_0, \mathbf{x}] = j\mathbf{x}$.

(a) Show that

$$\dim \mathfrak{a}'_j + \dim \mathfrak{a}'_{j-h} = \dim \mathfrak{a}'_j + \dim \mathfrak{a}'_{h-j} = l + \varphi(j)$$

for $0 \leqq j \leqq h - 1$, where $\varphi(j)$ is the multiplicity of $\exp(2\pi ij/h)$ as an eigenvalue of $\sigma$ (and therefore $\varphi(j) = 0$ if $j$ is not one of the $m_k$ (Section **21.11**, Problem 14)). (Use Problem 13.)

(b) Let $\mathfrak{b}_j = \mathfrak{t}_{(\mathbf{C})} \cap (\mathfrak{a}'_j + \mathfrak{a}'_{j-h})$, so that $\mathfrak{b}_j$ is the eigenspace of $\sigma$ for the eigenvalue $\exp(2\pi ij/h)$, and $\dim \mathfrak{b}_j = \varphi(j)$. We have $\mathfrak{b}_0 = \{0\}$, and $\mathfrak{b}_1$ contains a regular element $\mathbf{x}$ (Section **21.11**, Problems 14(b) and 16(c)). Let $\mathfrak{g}'_+$ (resp. $\mathfrak{g}'_-$) denote the direct sum of the $\mathfrak{a}'_j$ with $j > 0$ (resp. $j < 0$), so that $\mathfrak{g}_{(\mathbf{C})}$ is the direct sum of $\mathfrak{g}'_+$, $\mathfrak{g}'_-$, and $\mathfrak{a}'_0$. Let $\pi_+$ and $\pi_-$ denote the projections of $\mathfrak{g}_{(\mathbf{C})}$ onto $\mathfrak{g}'_+$ and $\mathfrak{g}'_-$ defined by this direct sum decomposition. Then by definition we have $\mathbf{x} = \pi_+(\mathbf{x}) + \pi_-(\mathbf{x})$, $\pi_+(\mathbf{x}) = \sum_{j>0} r_j \mathbf{x}'_{\beta'_j}$ and $\pi_-(\mathbf{x}) = r\mathbf{x}'_{-\mu'}$, where $\{\mathbf{x}'_{\alpha'}\}$ is a basis of $\mathfrak{g}_{\alpha'}$, and $\mu'$ is the highest root of the system $\mathbf{S}'$, and $r_j$, $r$ are complex numbers. Show that the $r_j$ are all $\neq 0$. (Note that the centralizer of $\mathbf{x}$ in $\mathfrak{g}_{(\mathbf{C})}$ is $\mathfrak{t}_{(\mathbf{C})}$, and that $[\mathbf{x}'_{\beta'_j}, \mathbf{x}'_{-\beta'_k}] = 0$ for $k \neq 0$, and $[\mathbf{x}'_{-\mu'}, \mathbf{x}'_{-\beta'_k}] = 0$.) Putting $\mathbf{e} = \pi_+(\mathbf{x})$, deduce that there exists an element $\mathbf{f} \in \mathfrak{g}'_-$ such that $[\mathbf{e}, \mathbf{f}] = \mathbf{h}'_0$, $[\mathbf{h}'_0, \mathbf{e}] = \mathbf{e}$, and $[\mathbf{h}'_0, \mathbf{f}] = -\mathbf{f}$.

(c) The vector subspace $\mathfrak{s}$ of $\mathfrak{g}_{(\mathbf{C})}$ spanned by $\mathbf{e}$, $\mathbf{f}$, and $\mathbf{h}'_0$ is a Lie subalgebra isomorphic to $\mathfrak{sl}(2, \mathbf{C})$. Show that $\mathfrak{g}_{(\mathbf{C})}$, considered as $\mathrm{U}(\mathfrak{s})$-module corresponding to the adjoint representation restricted to $\mathfrak{s}$, is the direct sum of $l$ simple $\mathrm{U}(\mathfrak{s})$-modules that are odd-dimensional as complex vector spaces. (Observe that all the eigenvalues of $\mathrm{ad}(\mathbf{h}'_0)$ are integers, and use **(21.9.3)**.)

(d) Show that the mapping $\pi_+$, restricted to $\mathfrak{t}_{(\mathbf{C})}$, is injective. (Observe that $\mathfrak{b}_0 = \{0\}$ and that $\mathrm{ad}(\mathbf{z})$ is nilpotent for each $\mathbf{z} \in \mathfrak{g}'_-$, whereas $\mathrm{ad}(\mathbf{u})$ is not nilpotent for $\mathbf{u} \neq 0$ in $\mathfrak{t}_{(\mathbf{C})}$.) Show that $\pi_+(\mathfrak{t}_{(\mathbf{C})})$ is contained in the kernel of $\mathrm{ad}(\mathbf{e})$. (We have $[\mathbf{x}, \mathbf{u}] = 0$ for $\mathbf{u} \in \mathfrak{t}_{(\mathbf{C})}$; deduce that $[\pi_+(\mathbf{x}), \pi_+(\mathbf{u})] = 0$, by writing $\mathbf{x} = \pi_+(\mathbf{x}) + \pi_-(\mathbf{x})$ and $\mathbf{u} = \pi_+(\mathbf{u}) + \pi_-(\mathbf{u})$.) Deduce that $\pi_+(\mathfrak{t}_{(\mathbf{C})})$ is equal to the kernel of $\mathrm{ad}(\mathbf{e})$ (use (c) above).

(e) Deduce from (d) that the image of $\mathfrak{g}'_+ + \mathfrak{a}'_0$ under $\mathrm{ad}(\mathbf{e})$ is $\mathfrak{g}'_+$, and hence that the image of $\mathfrak{a}'_j$ under $\mathrm{ad}(\mathbf{e})$ is $\mathfrak{a}'_{j+1}$. Hence, by using (b) above, obtain *Kostant's formula*

$$\varphi(j) = \dim \mathfrak{a}'_j - \dim \mathfrak{a}'_{j+1}$$

for $0 \leqq j \leqq h - 1$.

(f) Show that the $l$ simple $\mathrm{U}(\mathfrak{s})$-modules into which $\mathfrak{g}_{(\mathbf{C})}$ splits have dimensions $2m_j + 1$ $(1 \leqq j \leqq l)$, where the $m_j$ are the integers defined in Section **21.11**, Problem 14(c). (Note that the kernel of $\mathrm{ad}(\mathbf{e})$ is the direct sum of the $\pi_+(\mathfrak{t}_{(\mathbf{C})}) \cap \mathfrak{a}'_j$, and that in $\mathfrak{a}'_j$ the restriction of $\mathrm{ad}(\mathbf{h}'_0)$ is multiplication by $j$; then use **(21.9.4)**.)

**16.** Let G be an *almost simple* compact connected Lie group, and let T be a maximal torus of G. We retain the notation of Problem 11.

(a) The identity component of the centralizer $\mathscr{Z}(s)$ in G of an arbitrary element $s \in \mathrm{G}$ is conjugate to that of the centralizer of an element of the form $\exp(i\mathbf{u})$, where $\mathbf{u}$ lies in the closure of the principal alcove $\mathrm{A}^*$.

(b) Let $\mathfrak{F}$ be the set of connected closed proper subgroups of G. The identity components of the centralizers $\mathscr{Z}(s)$ that are *maximal* elements of $\mathfrak{F}$ are the conjugates of certain of the identity components $(\mathscr{Z}(\exp(2\pi i t\mathbf{p}_j)))_0$, where $1 \leqq j \leqq l$ and $0 < t < 1/n_j$. (Use Section **21.11**, Problem 4.)

(c) If $n_j = 1$, the identity component $\mathrm{G}' = (\mathscr{Z}(\exp(2\pi i t\mathbf{p}_j)))_0$ is the same for all $t$ such that $0 < t < 1$, and is maximal in $\mathfrak{F}$. (Observe that if $\mathrm{G}'' \supset \mathrm{G}'$ is a connected Lie subgroup, distinct from G or G′, then $\mathbf{S}(\mathrm{G}'')$ must be equal to $\mathbf{S}(\mathrm{G}')$, by using Problem 10(a).) The center of G′ is 1-dimensional.

(d) If $n_j > 1$ and is not a prime number, the group $\mathrm{G}' = (\mathscr{Z}(\exp(2\pi i t\mathbf{p}_j)))_0$ is not a maximal element of $\mathfrak{F}$, for any value of $t$ such that $0 < t \leqq 1/n_j$. (It is sufficient to consider the case $t = 1/n_j$; the roots in $\mathbf{S}_+(\mathrm{G}')$ are the roots in $\mathbf{S}_+(\mathrm{G})$ that are of the form $\sum_{k=1}^{l} m_k \beta_k$, where $m_j = n_j$ or $m_j = 0$. If $n_j = ab$ where $a$, $b$ are integers $> 1$, and if $\mathrm{G}'' = (\mathscr{Z}(\exp(2\pi i\mathbf{p}_j/b)))_0$, show that $\mathbf{S}(\mathrm{G}'') \supset \mathbf{S}(\mathrm{G}')$ and that $\mathbf{S}(\mathrm{G}'')$ contains at least one root that does not belong to $\mathbf{S}(\mathrm{G}')$, by using Problem 11(g).)

(e) If $n_j > 1$ and is prime, then G′ is maximal in $\mathfrak{F}$. (Argue by contradiction, by supposing that $\mathrm{G}' \subset \mathrm{G}'' = (\mathscr{Z}(\exp(2\pi i\mathbf{v})))_0$ with G″ distinct from G′ and G. We should then have $\mathbf{v} = \sum_{k=1}^{l} a_k \mathbf{p}_k$, where the $a_k$ are integers except for $a_j = q/n_j$, where $q$ is not a multiple of $n_j$. If $r$ is an integer such that $qr \equiv 1 \pmod{n_j}$, consider the centralizer of $\exp(2\pi i r\mathbf{v})$ and thus obtain a contradiction.) The group G′ is semisimple.

(f) Describe the maximal elements of $\mathfrak{F}$ that have rank equal to the rank of G, when G is an almost simple classical group.

## 16. CENTER, FUNDAMENTAL GROUP, AND IRREDUCIBLE REPRESENTATIONS OF SEMISIMPLE COMPACT CONNECTED GROUPS

Throughout this section, G denotes a compact connected *semisimple* Lie group, so that its center Z is *finite* and is contained in all the maximal tori of G **(21.7.11)**. *All other notation is the same as in Sections* **21.13–21.15**.

**(21.16.1)** *The lattice* P(G/Z) *of weights of* G/Z *relative to* T/Z *is the lattice* Q(𝔤) *generated by the roots* $\alpha \in$ **S** (and is therefore a free **Z**-module with basis any basis **B** of **S**), *and* Z *is isomorphic to the quotient* P(G)/P(G/Z) = P(G)/Q(𝔤).

Since the center of G/Z consists only of the identity element **(21.6.9)**, the adjoint representation of G/Z is *faithful*; hence **(21.13.2)** every irreducible representation of G/Z is contained in the tensor product of a certain number of linear representations equal to $\mathrm{Ad} \otimes 1_{\mathbf{C}}$ and a certain number of linear representations equal to its conjugate. By definition, the restriction to T of the character of an irreducible representation contained in $\mathrm{Ad} \otimes 1_{\mathbf{C}}$ is a linear combination of characters of T of the form $e^{\alpha}$, where $\alpha$ is a root **(21.8.1)**. These remarks, coupled with **(21.15.5)**, show that P(G/Z) = Q(𝔤). Next, the groups T and T/Z have the same Lie algebra t, and the exponential mapping $\exp_{T/Z}$ is the composition of the canonical homomorphism $T \to T/Z$ with $\exp_T$; consequently the lattice $\Gamma_{T/Z}$ **(21.7.5)** is the inverse image $\exp_T^{-1}(Z)$, and Z is isomorphic to $\Gamma_{T/Z}/\Gamma_T$. The elementary theory of free **Z**-modules **(A.26.5)** then shows that Z is also isomorphic to $\Gamma_T^*/\Gamma_{T/Z}^*$, i.e. to P(G)/P(G/Z).

**(21.16.2)** The result just proved may be applied to the universal covering group $\tilde{G}$ of G; if D is the center of $\tilde{G}$, then $\tilde{G}/D$ is isomorphic to G/Z, and G is isomorphic to $\tilde{G}/Z_G$, where $Z_G$ is a subgroup of the center D of $\tilde{G}$ **(16.30.4)**; also Z is isomorphic to $D/Z_G$ **(20.22.5.1)**. It follows that *the fundamental group* $\pi_1(G)$ *is isomorphic to* $P(\tilde{G})/P(G)$, hence to $\Gamma_T/\Gamma_{T_1}$, where $T_1$ is the maximal torus of $\tilde{G}$ that is the inverse image of T.

Since $\tilde{G}$ has the same Lie algebra as G, it follows that $P(\tilde{G})$ is contained in the lattice P(𝔤) **(21.14.3)**. But from the fact that $\tilde{G}$ is semisimple and simply connected, *it follows that* $P(\tilde{G})$ = P(𝔤). In other words:

**(21.16.3)** *Let* G *be a simply connected compact semisimple group and* 𝔤 *its Lie algebra. Let* P(𝔤) *be the lattice of linear forms* $\lambda \in i\mathfrak{t}^*$ *such that* $\lambda(\mathbf{h}_\alpha)$ *is an integer for all* $\alpha \in$ **S** (i.e., the lattice *dual* to the lattice in *i*t having as basis the $\mathbf{h}_j$ **(21.14.2)** for $1 \leq j \leq l$). *Then* P(G) = P(𝔤).

Let $p \neq 0$ be a linear form belonging to $P(\mathfrak{g})$. It is enough to construct a linear representation $V$ of G on a finite-dimensional complex vector space E, such that there exists a vector $x \neq 0$ in E with the property that $V(\exp(\mathbf{u})) \cdot x = e^{p(\mathbf{u})}x$ for all $\mathbf{u} \in \mathfrak{t}$ (21.13.6). Since G is *simply connected*, it comes to the same thing (19.7.6) to define a **C**-homomorphism of Lie algebras $\rho \colon \mathfrak{g}_{(\mathbf{C})} \to \mathfrak{gl}(E)$ such that, for all $\mathbf{u} \in \mathfrak{h} = \mathfrak{t} \oplus i\mathfrak{t}$, we have

(21.16.3.1)
$$\rho(\mathbf{u}) \cdot x = p(\mathbf{u})x.$$

If U denotes the *enveloping algebra* $U(\mathfrak{g}_{(\mathbf{C})})$ (19.6.3), it again comes to the same thing (21.9.1) to construct a *left* U-*module* E of *finite* dimension over **C**, and an element $x \neq 0$ in E such that

(21.16.3.2)
$$(\mathbf{u} - p(\mathbf{u}) \cdot 1) \cdot x = 0$$

for all $\mathbf{u} \in \mathfrak{h}$.

Since P(G) and $P(\mathfrak{g})$ are both invariant under the action of the Weyl group, we may, by replacing $p$ by $w \cdot p$ for a suitably chosen $w \in W$, assume that $p \in \bar{C}$, or in other words that

(21.16.3.3)
$$m_j = p(\mathbf{h}_j) \geqq 0 \qquad (1 \leqq j \leqq l).$$

The construction of E is in several steps.

(A) Arrange the positive roots (relative to **B**) in a sequence $(\alpha_k)_{1 \leq k \leq n}$. With the notation of (21.10.3), the algebra U has a basis over **C** consisting of the elements

(21.16.3.4)
$$\mathbf{x}_{-\alpha_1}^{a_1} \mathbf{x}_{-\alpha_2}^{a_2} \cdots \mathbf{x}_{-\alpha_n}^{a_n} \mathbf{h}_1^{c_1} \cdots \mathbf{h}_l^{c_l} \mathbf{x}_{\alpha_1}^{b_1} \mathbf{x}_{\alpha_2}^{b_2} \cdots \mathbf{x}_{\alpha_n}^{b_n}$$

where the $a_k$, $b_k$, and $c_j$ $(1 \leqq k \leqq n,\ 1 \leqq j \leqq l)$ are arbitrary integers $\geqq 0$ (19.6.2). Let $U_-$ denote the vector subspace of U generated by the basis elements (21.16.3.4) for which $b_k = 0$ for $1 \leqq k \leqq n$ and $c_j = 0$ for $1 \leqq j \leqq l$, and let $U_0$ denote the vector subspace generated by the basis elements (21.16.3.4) for which $a_k = 0$ for $1 \leqq k \leqq n$. Then the vector space U may be identified with $U_- \otimes_{\mathbf{C}} U_0$ (A.20.2). We remark that the elements $\mathbf{x}_\alpha$ for $\alpha \in \mathbf{S}^+$ and the $\mathbf{h}_j$ for $1 \leqq j \leqq l$ form a basis of a *Lie subalgebra* $\mathfrak{b}$ of the complex Lie algebra $\mathfrak{g}_{(\mathbf{C})}$, because if $\alpha$ and $\beta$ are any two positive roots, we have $[\mathbf{x}_\alpha, \mathbf{x}_\beta] = 0$ if $\alpha + \beta$ is not a root, and $[\mathbf{x}_\alpha, \mathbf{x}_\beta] \in \mathfrak{g}_{\alpha+\beta} \subset \mathfrak{b}$ if $\alpha + \beta$ is a root (21.10.5). The vector subspace $U_0$ of U may therefore be identified with the *enveloping algebra* $U(\mathfrak{b})$ (19.6.2).

Let $\mathfrak{n}$ denote the *left ideal* of U generated by the elements

(21.16.3.5) $$\mathbf{x}_{\alpha_k} \quad (1 \leqq k \leqq n), \qquad \mathbf{u} - p(\mathbf{u}) \cdot 1 \quad (\mathbf{u} \in \mathfrak{h}).$$

Let M be the quotient U-module $\mathrm{U}/\mathfrak{n}$. We shall first show that

(21.16.3.6) *The* U-*module* M *is nonzero, and if* $v$ *is the canonical image in* M *of the identity element* 1 *of* U, *then* M *has a* **C**-*basis consisting of the elements*

(21.16.3.7) $$\mathbf{x}^{a_1}_{-\alpha_1} \cdots \mathbf{x}^{a_n}_{-\alpha_n} \cdot v$$

*where the exponents* $a_k$ $(1 \leqq k \leqq n)$ *are arbitrary integers* $\geqq 0$.

The elements (21.16.3.5) belong to $\mathrm{U}_0$; let $\mathfrak{n}_0$ be the left ideal *of* $\mathrm{U}_0$ that they generate. Clearly we have

(21.16.3.8) $$\mathfrak{n} = \mathrm{U}_- \otimes_{\mathbf{C}} \mathfrak{n}_0, \qquad \mathrm{U}/\mathfrak{n} = \mathrm{U}_- \otimes_{\mathbf{C}} (\mathrm{U}_0/\mathfrak{n}_0).$$

To prove (21.16.3.6), it is enough to show that $\mathrm{U}_0/\mathfrak{n}_0$ is 1-dimensional over **C**. Now, we may define a **C**-homomorphism of Lie algebras $\theta \colon \mathfrak{b} \to \mathbf{C}$ by setting $\theta(\mathbf{u}) = p(\mathbf{u})$ for $\mathbf{u} \in \mathfrak{h}$, and $\theta(\mathbf{x}_\alpha) = 0$ for $\alpha \in \mathbf{S}^+$; we have only to verify that $\theta([\mathbf{u}, \mathbf{x}_\alpha]) = 0$ for $\mathbf{u} \in \mathfrak{h}$ and $\alpha \in \mathbf{S}^+$, and this follows from the fact that $[\mathbf{u}, \mathbf{x}_\alpha]$ is a scalar multiple of $\mathbf{x}_\alpha$. Since $p \neq 0$, the homomorphism $\theta$ extends to a *surjective* homomorphism $\theta \colon \mathrm{U}(\mathfrak{b}) \to \mathbf{C}$ (19.6.4), the kernel of which evidently contains $\mathfrak{n}_0$; but since the $\mathbf{h}_j$ commute with each other, every element $\mathbf{h}_1^{c_1} \cdots \mathbf{h}_l^{c_l} \mathbf{x}_{\alpha_1}^{b_1} \cdots \mathbf{x}_{\alpha_n}^{b_n}$ is congruent modulo $\mathfrak{n}_0$ to $p(\mathbf{h}_1)^{c_1} \cdots p(\mathbf{h}_l)^{c_l} \mathbf{x}_{\alpha_1}^{b_1} \cdots \mathbf{x}_{\alpha_n}^{b_n}$ (and hence to 0 unless all the integers $b_k$ are zero), and therefore the kernel of $\theta$ is *equal to* $\mathfrak{n}_0$, which proves our assertion.

(B) For each $\mathbf{u} \in \mathfrak{h}$ we have $\mathbf{u}\mathbf{x}_{-\alpha_k} - \mathbf{x}_{-\alpha_k}\mathbf{u} = -\alpha_k(\mathbf{u})\mathbf{x}_{-\alpha_k}$ in the algebra U, from which it follows immediately by induction on $a_1 + \cdots + a_n$ that

(21.16.3.9)
$$\mathbf{u} \cdot (\mathbf{x}^{a_1}_{-\alpha_1} \cdots \mathbf{x}^{a_n}_{-\alpha_n} \cdot v) = (p(\mathbf{u}) - a_1\alpha_1(\mathbf{u}) - \cdots - a_n\alpha_n(\mathbf{u}))(\mathbf{x}^{a_1}_{-\alpha_1} \cdots \mathbf{x}^{a_n}_{-\alpha_n} \cdot v)$$

in the module M; and since we have (21.11.5)

(21.16.3.10) $$\alpha_k = \sum_{j=1}^{l} d_{kj}\beta_j$$

where the $d_{kj}$ are *integers* $\geqq 0$, we may write

(21.16.3.11)

$$\mathbf{u} \cdot (\mathbf{x}_{-\alpha_1}^{a_1} \cdots \mathbf{x}_{-\alpha_n}^{a_n} \cdot v) = (p(\mathbf{u}) - n_1 \beta_1(\mathbf{u}) - \cdots - n_l \beta_l(\mathbf{u}))(\mathbf{x}_{-\alpha_1}^{a_1} \cdots \mathbf{x}_{-\alpha_n}^{a_n} \cdot v)$$

where

(21.16.3.12)
$$n_j = \sum_{k=1}^{n} d_{kj} a_k .$$

For each multi-index $\mathbf{n} = (n_j) \in \mathbf{N}^l$, put

(21.16.3.13)
$$p_{\mathbf{n}} = p - n_1 \beta_1 - \cdots - n_l \beta_l ,$$

so that $p_{\mathbf{0}} = p$ and $p_{\mathbf{n}} < p$ (for the ordering defined in (21.14.6)) for all $\mathbf{n} \neq \mathbf{0}$. Next, for each $q \in \mathrm{P}(\mathfrak{g})$, let $\mathrm{M}(q)$ be the vector subspace of M consisting of the vectors $z$ satisfying

(21.16.3.14)
$$\mathbf{u} \cdot z = q(\mathbf{u}) z$$

for all $\mathbf{u} \in \mathfrak{h}$. Then:

(21.16.3.15) *The vector space* M *is the direct sum of the* $\mathrm{M}(q)$ *for all* $q \in \mathrm{P}(\mathfrak{g})$ *such that* $q \leqq p$*; the subspaces* $\mathrm{M}(q)$ *are finite-dimensional, and* $\mathrm{M}(p)$ *is 1-dimensional. For each root* $\alpha \in \mathbf{S}$, *we have* $\mathbf{x}_\alpha \cdot \mathrm{M}(q) \subset \mathrm{M}(q + \alpha)$.

The first assertion is a consequence of what has already been established; by virtue of (21.16.3.6), M is the direct sum of the $\mathrm{M}(p_{\mathbf{n}})$ for $\mathbf{n} \in \mathbf{N}^l$, and $\mathrm{M}(q) = \{0\}$ if $q$ is not one of the forms $p_{\mathbf{n}}$ (A.24.4). For each given $\mathbf{n} = (n_j) \in \mathbf{N}^l$, there is only a finite number of systems of integers $a_k \geqq 0$ ($1 \leqq k \leqq n$) satisfying the equations (21.16.3.12), hence the subspaces $\mathrm{M}(p_{\mathbf{n}})$ are finite-dimensional; also it is clear that the only element of the basis (21.16.3.7) that belongs to $\mathrm{M}(p)$ is $v$, and hence $\mathrm{M}(p)$ is one-dimensional. Since $[\mathbf{u}, \mathbf{x}_\alpha] = \alpha(\mathbf{u})\mathbf{x}_\alpha$, (21.16.3.14) implies that

(21.16.3.16)
$$\mathbf{u} \cdot (\mathbf{x}_\alpha \cdot z) = \mathbf{x}_\alpha \cdot (\mathbf{u} \cdot z) + [\mathbf{u}, \mathbf{x}_\alpha] \cdot z = (q(\mathbf{u}) + \alpha(\mathbf{u}))(\mathbf{x}_\alpha \cdot z),$$

which completes the proof.

(C) The integers $m_j$ being those defined in (21.16.3.3), we shall now show that:

(21.16.3.17) *The* U*-submodule* N *of* M *generated by the* $l$ *elements*

(21.16.3.18) $$\mathbf{x}_{-\beta_j}^{m_j+1} \cdot v \qquad (1 \leqq j \leqq l)$$

*does not contain* $v$*; moreover, for each* $z \in$ M *and each index* $j \in [1, l]$, *there exists an integer* $s \geqq 0$ *such that* $\mathbf{x}_{-\beta_j}^s \cdot z \in$ N.

We first remark that in the algebra U, every element $\mathbf{x}_\alpha$ (resp. $\mathbf{x}_{-\alpha}$) for $\alpha \in \mathbf{S}^+$ is a linear combination of products $\mathbf{x}_{\gamma_1}\mathbf{x}_{\gamma_2}\cdots\mathbf{x}_{\gamma_r}$ (resp. $\mathbf{x}_{-\gamma_1}\mathbf{x}_{-\gamma_2}\cdots\mathbf{x}_{-\gamma_r}$), where $(\gamma_1, \ldots, \gamma_r)$ is a sequence of roots all belonging to the *basis* $\mathbf{B} = \{\beta_1, \ldots, \beta_l\}$. Indeed, this is obvious if $\alpha \in \mathbf{B}$; if not, $\alpha$ is of the form $\sum_{j=1}^{l} d_j \beta_j$ with each $d_j$ an integer $\geqq 0$ (21.11.5), and we can proceed by induction on $\sum_{j=1}^{l} d_j$. By hypothesis, we have $\alpha = \lambda + \mu$, where $\lambda = \sum_{j=1}^{l} d'_j \beta_j$ and $\mu = \sum_{j=1}^{l} d''_j \beta_j$ are two roots in $\mathbf{S}^+$ with $\sum_{j=1}^{l} d'_j < \sum_{j=1}^{l} d_j$ and $\sum_{j=1}^{l} d''_j < \sum_{j=1}^{l} d_j$ (21.11.5); hence $\mathbf{x}_\alpha$ is a scalar multiple of $[\mathbf{x}_\lambda, \mathbf{x}_\mu] = \mathbf{x}_\lambda \mathbf{x}_\mu - \mathbf{x}_\mu \mathbf{x}_\lambda$ (21.10.5), and our assertion follows from the inductive hypothesis.

To prove the first assertion of (21.16.3.17), it will be enough to show that N is contained in the sum of the M$(q)$ with $q < p$. In view of what has already been established, and the form of the elements of the basis (21.16.3.4) of U, this will result from the following properties:

(1) For $\mathbf{u} \in \mathfrak{h}$ and $\alpha \in \mathbf{S}^+$ we have

(21.16.3.19) $$\mathbf{u} \cdot (\mathbf{x}_\alpha^r \cdot v) = (p(\mathbf{u}) + r\alpha(\mathbf{u}))(\mathbf{x}_\alpha^r \cdot v)$$

by induction on $r$, starting from (21.16.3.16).

(2) For $1 \leqq k \leqq l$, we have $\mathbf{x}_{\beta_k} \cdot (\mathbf{x}_{-\beta_j}^{m_j+1} \cdot v) = 0$. Indeed, if $k \neq j$, $\beta_k - \beta_j$ is not a root (21.11.5), hence $[\mathbf{x}_{\beta_k}, \mathbf{x}_{-\beta_j}] = 0$ (21.10.5): in other words, $\mathbf{x}_{\beta_k}$ and $\mathbf{x}_{-\beta_j}$ commute in U, and the assertion follows from the fact that $\mathbf{x}_{\beta_k} \cdot v = 0$ for all $k$, by the definition of M. If $k = j$, we observe that (in the notation of (21.10.3)) the algebra U$(\mathfrak{s}_{\beta_j})$ is a subalgebra of U, and that if we apply the formulas (21.9.3.6) to the U$(\mathfrak{s}_{\beta_j})$-module generated by the element $v \in$ M, we have

(21.16.3.20) $$\mathbf{x}_{\beta_j} \cdot (\mathbf{x}_{-\beta_j}^r \cdot v) = r(p(\mathbf{h}_j) - r + 1)(\mathbf{x}_{-\beta_j}^{r-1} \cdot v)$$

for all $r \geqq 1$, and therefore when $r = m_j + 1 = p(\mathbf{h}_j) + 1$ we obtain $\mathbf{x}_{\beta_j} \cdot (\mathbf{x}_{-\beta_j}^{m_j+1} \cdot v) = 0$.

(3) For $1 \leqq k \leqq l$ we have, by definition,

$$\mathbf{x}_{-\beta_k} \cdot (\mathbf{x}_{-\beta_j}^{m_j+1} \cdot v) \in \mathrm{M}(q)$$

where $q = p - (m_j + 1)\beta_j - \beta_k$, by virtue of **(21.16.3.15)**.

To prove the second assertion of **(21.16.3.17)**, we need only consider the case where

$$z = \mathbf{x}_{-\gamma_1} \mathbf{x}_{-\gamma_2} \cdots \mathbf{x}_{-\gamma_r} \cdot v$$

in which the $\gamma_k$ belong to **B**; when $r = 0$, the assertion follows from the definition of N, and we proceed by induction on $r$. The only positive integral values of $k$ for which $k\beta_j + \gamma_1$ can be a root are 0, 1, 2, 3 at most **(21.11.11)**. In view of **(21.10.5)** we may therefore write

$$\mathbf{x}_{-\beta_j}^{s+3} \mathbf{x}_{-\gamma_1} = t_s \mathbf{x}_{-\gamma_1} \mathbf{x}_{-\beta_j}^{s+3} + t_s' \mathbf{x}_{-\gamma_1-\beta_j} \mathbf{x}_{-\beta_j}^{s+2} + t_s'' \mathbf{x}_{-\gamma_1-2\beta_j} \mathbf{x}_{-\beta_j}^{s+1} + t_s''' \mathbf{x}_{-\gamma_1-3\beta_j} \mathbf{x}_{-\beta_j}^{s}$$

for all integers $s > 0$, with scalar coefficients $t_s, t_s', t_s'', t_s'''$ and $\mathbf{x}_\lambda$ replaced by 0 if $\lambda$ is not a root. The result now follows from the inductive hypothesis.

(D) Now consider the U-module $\mathrm{E} = \mathrm{M}/\mathrm{N}$. By virtue of **(21.16.3.17)**, the image $\bar{v}$ of $v$ in E is not zero, and satisfies the relation $\mathbf{u} \cdot \bar{v} = p(\mathbf{u})\bar{v}$ for all $\mathbf{u} \in \mathfrak{h}$. It is therefore enough to show that E is *finite-dimensional* over **C**. Let $\mathrm{E}_q$ denote the canonical image of $\mathrm{M}(q)$ in E, for all $q \in \mathrm{P}(\mathfrak{g})$ such that $q \leqq p$; then $\mathrm{E}_q$ is *finite*-dimensional, and for each $y \in \mathrm{E}_q$ and $\mathbf{u} \in \mathfrak{h}$ we have $\mathbf{u} \cdot y = q(\mathbf{u})y$; furthermore, E is the sum of the subspaces $\mathrm{E}_q$ (hence in fact the *direct sum*, cf. **(A.24.4)**). Hence we have to prove that $\mathrm{E}_q = \{0\}$ for all but a *finite* number of values of $q \in \mathrm{P}(\mathfrak{g})$. This is a consequence of the following proposition:

**(21.16.3.21)** *If* $\mathrm{E}_q \neq \{0\}$, *then also* $\mathrm{E}_{w \cdot q} \neq \{0\}$ *for all elements w of the Weyl group* W.

Assume this result for a moment. For each $q \in \mathrm{P}(\mathfrak{g})$ such that $\mathrm{E}_q \neq \{0\}$, there exists $w \in \mathrm{W}$ such that $w \cdot q \in \bar{\mathrm{C}}$ **(21.14.5.1)**. Since $\mathrm{E}_{w \cdot q} \neq \{0\}$, we must have $w \cdot q \leqq p$ (in fact, $w \cdot q$ must be one of the $p_\mathbf{n}$, cf. **(A.24.4)**). But the set of $p_\mathbf{n}$ such that $p_\mathbf{n} \in \bar{\mathrm{C}}$ is *finite*, because they satisfy $p_\mathbf{n} \leqq p$ **(21.14.8.1)**. Since W is a finite group, it follows that the set of $q \in \mathrm{P}(\mathfrak{g})$ such that $\mathrm{E}_q \neq \{0\}$ is finite, as required.

It remains therefore to prove **(21.16.3.21)**. Let $y \neq 0$ be an element of $\mathrm{E}_q$. It is enough to show that $\mathrm{E}_{w \cdot q} \neq \{0\}$ for $w = s_j$ $(1 \leqq j \leqq l)$, because W is generated by the $s_j$ **(21.11.8)**. Consider the $\mathrm{U}(\mathfrak{s}_{\beta_j})$-submodule F of E generated by $y$; we claim that it is *finite*-dimensional over $\mathbf{C}$. Indeed, it is clear that the vector space F is spanned by the elements $\mathbf{x}_{-\beta_j}^a \mathbf{x}_{\beta_j}^b \cdot y$, where $a$, $b$ are integers $\geqq 0$. But $\mathbf{x}_{\beta_j}^b \cdot y \in E_{q+b\beta_j}$ by virtue of (21.16.3.15), and since by definition and the fact that $\beta_j(\mathbf{h}_{\beta_j}) = 2$ we cannot have $q + b\beta_j \leqq p$ for more than a *finite* number of integers $b \geqq 0$, it follows that $\mathbf{x}_{\beta_j}^b \cdot y = 0$ for all but finitely many values of $b$. It then follows from (21.16.3.17) that for each of these values of $b$ we have $\mathbf{x}_{-\beta_j}^a \mathbf{x}_{\beta_j}^b \cdot y = 0$ for all sufficiently large $a$, and therefore F is indeed finite-dimensional. The $U(\mathfrak{s}_{\beta_j})$-module F is thus a direct sum of submodules isomorphic to the modules $L_m$ (21.9.3) (if we identify $\mathfrak{s}_{\beta_j}$ with $\mathfrak{su}(2)$). By hypothesis, there is an element $y' \neq 0$ in one of these submodules such that $\mathbf{h}_j \cdot y' = q(\mathbf{h}_j)y'$, and therefore it follows from the first of the formulas (21.9.3.1) that there is also an element $y'' \neq 0$ in this submodule that belongs to an $E_{q'}$ with $q' = q + k\beta_j$ for some integer $k \in \mathbf{Z}$ and $q'(\mathbf{h}_j) = -q(\mathbf{h}_j)$. This last relation may also be written as $(q + q' \mid \beta_j) = 0$; now $\frac{1}{2}(q + q')$ cannot be orthogonal to $\beta_j$, for $q'$ of the form $q + k\beta_j$, unless $q$ and $q'$ are images of each other under the orthogonal reflection $\sigma_{\beta_j}$ (21.11.2.1); hence we have $q' = s_j \cdot q$. This completes the proof of (21.16.3.21) and hence also of (21.16.3).

With the same notation, the results of (21.16.2) and (21.16.3) can be stated as follows:

(21.16.4) *Let* G *be a simply connected semisimple compact Lie group,* Z *its center,* $\mathfrak{g}$ *its Lie algebra,* $P(G) = P(\mathfrak{g})$ *the lattice of weights of* G *relative to a maximal torus* T *of* G, *and* $Q(\mathfrak{g})$ *the sublattice of* $P(\mathfrak{g})$ *generated by the roots of* $\mathfrak{g}$ *(relative to* T).

(i) *The lattice* $P(\mathfrak{g})$ *is the dual in* $i\mathfrak{t}^*$ *of the lattice in* $i\mathfrak{t}$ *generated by the* $\mathbf{h}_\alpha$, *which is the lattice* $(2\pi i)^{-1}\Gamma_T = (2\pi i)^{-1} \exp_T^{-1}(e)$ (21.7.5). *The lattice* $Q(\mathfrak{g})$ *is the dual of the lattice* $(2\pi i)^{-1} \exp_T^{-1}(Z)$.

(ii) *There is a canonical one-to-one correspondence between the quotients* $G_0 = G/D$ *of* G *by a subgroup* D *of the center* Z, *and the lattices* $\Gamma^*$ *in* $i\mathfrak{t}^*$ *such that* $Q(\mathfrak{g}) \subset \Gamma^* \subset P(\mathfrak{g})$. *If* $T_0 = T/D$, *a maximal torus of* $G_0$, *then* $\Gamma^*$ *is the dual of the lattice* $(2\pi i)^{-1}\Gamma_{T_0}$, *the quotient* $P(\mathfrak{g})/\Gamma^*$ *is isomorphic to the fundamental group* $\pi_1(G_0)$, *and* $\Gamma^*/Q(\mathfrak{g})$ *is isomorphic to the center* $Z_0 = Z/D$ *of* $G_0$.

(21.16.5) Since the $\mathbf{h}_j$ $(1 \leqq j \leqq l)$ form a basis of the lattice $(2\pi i)^{-1}\Gamma_T$ generated by the $\mathbf{h}_\alpha$ (21.11.5.5), the lattice $P(\mathfrak{g})$ admits as $\mathbf{Z}$-basis the basis of the vector space $i\mathfrak{t}^*$ *dual* to the basis $(\mathbf{h}_j)$, i.e., the basis consisting of the linear forms $\varpi_j$ $(1 \leqq j \leqq l)$ such that

(21.16.5.1) $$\varpi_j(\mathbf{h}_k) = \delta_{jk} \qquad (1 \leqq j, k \leqq l).$$

The $\varpi_j$ $(1 \leqq j \leqq l)$ are called the *fundamental weights* of $\mathfrak{g}$ (or of the corresponding *simply connected* group G) relative to the basis **B** of the root system **S**. From the definition of $P(\mathfrak{g})$, all the elements of $P(\mathfrak{g}) \cap \bar{C}$ are linear combinations $\sum_{j=1}^{l} n_j \varpi_j$ in which the $n_j$ are *integers* $\geqq 0$. By virtue of (21.11.7), the half-sum of the positive roots is given by the formula

(21.16.5.2) $$\delta = \varpi_1 + \varpi_2 + \cdots + \varpi_l .$$

This is the *smallest* element of $P(\mathfrak{g}) \cap C$ (21.14.8).

It should be noted, however, that the $\varpi_j$ are not necessarily minimal elements of $(P(\mathfrak{g}) - \{0\}) \cap \bar{C}$.

If $\boldsymbol{\rho}_j \in R(G)$ is the class of irreducible representations of G with dominant weight $\varpi_j$ $(1 \leqq j \leqq l)$, the classes $\boldsymbol{\rho}_j$ (or the representations belonging to these classes) are called *fundamental.*

(21.16.6) *Let G be a simply connected semisimple compact Lie group of rank l. Then the* **Z**-*algebra homomorphism*

(21.16.6.1) $$\mathbf{Z}[X_1, X_2, \ldots, X_l] \to \mathbf{Z}^{(R(G))}$$

*which maps identity element to identity element, and each indeterminate* $X_j$ *to the fundamental class* $\boldsymbol{\rho}_j$ $(1 \leqq j \leqq l)$, *is bijective.*

We know from (21.15.5) that there is a canonical isomorphism $\mathbf{Z}^{(R(G))} \to \mathbf{Z}[P(\mathfrak{g})]^W$, because G is simply connected (21.16.3), and that the elements $v_p = J(e^{p+\delta})/J(e^\delta)$ form a basis of $\mathbf{Z}[P(\mathfrak{g})]^W$ as $p$ runs through $P(\mathfrak{g}) \cap \bar{C}$ (21.14.13). Composing this isomorphism with (21.16.6.1), we obtain a homomorphism of $\mathbf{Z}[X_1, \ldots, X_l]$ into $\mathbf{Z}[P(\mathfrak{g})]^W$ that, for each multi-index $\mathbf{n} = (n_1, \ldots, n_l) \in \mathbf{N}^l$, maps $X_\mathbf{n} = X_1^{n_1} \cdots X_l^{n_l}$ to the element $u_\mathbf{n} = v_{\varpi_1}^{n_1} \cdots v_{\varpi_l}^{n_l}$. Since in the expression of $v_p$ as a linear combination of the $e^q$ with $q \in P(\mathfrak{g})$, the term $e^p$ is the *leading term* (21.14.13), it is clear that $u_\mathbf{n}$ has a leading term equal to $e^{\varpi(\mathbf{n})}$, where $\varpi(\mathbf{n}) = n_1\varpi_1 + \cdots + n_l \varpi_{Hl}$. As $\mathbf{n}$ runs through $\mathbf{N}^l$, $\varpi(\mathbf{n})$ runs through the set $P(\mathfrak{g}) \cap \bar{C}$, by virtue of (21.16.5.1) and the definition of $\bar{C}$. It follows therefore from (21.14.13) that the $u_\mathbf{n}$ form a basis of $\mathbf{Z}[P(\mathfrak{g})]^W$, and the proof is complete.

*Remarks*

(21.16.7) (i) It should be obser·ed that the restrictions $\chi_{\rho_j}|T = v_{\varpi_j} = J(e^{\varpi_j + \delta})/J(e^\delta)$ of the characters of the fundamental representations are not necessarily equal to the elements $S(\varpi_j)$ (21.16.10).

(ii) Once the linear representations of a simply-connected semisimple compact Lie group G are known, the linear representations of any quotient G/D of G by a discrete subgroup of its center (i.e., of a Lie group locally isomorphic to G) can be deduced: they are the representations obtained on passing to the quotient from a representation $U: \mathrm{G} \to \mathbf{GL}(\mathrm{E})$ such that the image of D under $U$ is the identity element of $\mathbf{GL}(\mathrm{E})$.

*Examples: The Fundamental Representations of the Classical Groups*

(21.16.8) I. *Representations of* $\mathbf{SU}(n)$. With the notation of (21.12.1), the elements $\mathbf{h}_j$ of the basis $\mathbf{B}^\vee$ of $\mathbf{S}^\vee$ are the matrices

$$E_{jj} - E_{j+1,\, j+1} \qquad (1 \leqq j \leqq n-1);$$

the $i\mathbf{h}_j$ $(1 \leqq j \leqq n-1)$ form a basis of the Lie algebra $\mathfrak{t}'$ of the maximal torus $\mathrm{T}' = \mathrm{T} \cap \mathbf{SU}(n)$, the intersection of $\mathbf{SU}(n)$ with the maximal torus T of $\mathbf{U}(n)$ defined in (21.12.1); the lattice $\Gamma_{\mathrm{T}'}$, the kernel of the exponential mapping $\exp_{\mathrm{T}'}$, has as a $\mathbf{Z}$-basis the elements $2\pi i\mathbf{h}_j$ $(1 \leqq j \leqq n-1)$; the lattice $2\pi i\Gamma_{\mathrm{T}'}^* = \mathrm{P}(\mathbf{SU}(n))$ therefore has as a basis the fundamental weights $\varpi_j$ $(1 \leqq j \leqq n-1)$, and we regain the fact that $\mathrm{P}(\mathbf{SU}(n)) = \mathrm{P}(\mathfrak{su}(n))$ (21.16.3).

The restrictions $\varepsilon_j'$ to $i\mathfrak{t}'$ of the $n$ linear forms $\varepsilon_j$ $(1 \leqq j \leqq n)$ generate the dual $i\mathfrak{t}'^*$ of the real vector space $i\mathfrak{t}'$, and satisfy the relation

$$\varepsilon_1' + \varepsilon_2' + \cdots + \varepsilon_n' = 0;$$

the $\varepsilon_j'$ for $1 \leqq j \leqq n-1$ form a basis of $i\mathfrak{t}'^*$ over $\mathbf{R}$. A simple calculation then shows that the linear form $\varpi_j$ on $i\mathfrak{t}'$ $(1 \leqq j \leqq n-1)$ is given by the formula

(21.16.8.1) $$\varpi_j = \varepsilon_1' + \varepsilon_2' + \cdots + \varepsilon_j' \qquad (1 \leqq j \leqq n-1).$$

The vector space $i\mathfrak{t}$ is the direct sum of $i\mathfrak{t}'$ and $i\mathfrak{c} = \mathbf{R}\sum_{j=1}^{n} E_{jj}$; hence the dual $i\mathfrak{t}'^*$ may be canonically identified with the annihilator of $i\mathfrak{c}$ in $i\mathfrak{t}^*$, and with this identification it is easily verified that

(21.16.8.2) $$\varepsilon_j' = \varepsilon_j - \frac{1}{n}\sum_{h=1}^{n} \varepsilon_h \qquad (1 \leqq j \leqq n).$$

We shall now determine *explicitly* the irreducible representations of $\mathbf{SU}(n)$ with dominant weights $\varpi_j$ $(1 \leqq j \leqq n-1)$. Since the functions $\varepsilon'_j$ take values belonging to $2\pi i\mathbf{Z}$ at the points of the lattice $\Gamma_{T'}$ (generated by the $2\pi i\mathbf{h}_j$), we may identify $e^{\varepsilon'_j}$ with a function on T′ (21.13.6). Let $U_1$ be the canonical injection $s \mapsto s$ of $\mathbf{SU}(n)$ into $\mathbf{GL}(n, \mathbf{C})$; it is clearly a linear representation of $\mathbf{SU}(n)$, and its restriction to T′ is the Hilbert sum of $n$ one-dimensional representations on spaces $\mathbf{C}\mathbf{a}_j$, where $(\mathbf{a}_j)_{1 \leq j \leq n}$ is the canonical basis of $\mathbf{C}^n$. The representation on $\mathbf{C}\mathbf{a}_j$ maps $s \in$ T′ to the homothety with ratio $e^{\varepsilon'_j(s)}$, hence we have $\operatorname{Tr}(U_1(s)) = \sum_{j=1}^{n} e^{\varepsilon'_j(s)}$ for $s \in$ T′; as a function on T′ this is just the sum $S(\varpi_1)$, because the Weyl group is equal to $\mathfrak{S}_n$. It follows (21.13.7) that $U_1$ is *irreducible*; and from the expressions for the roots forming the basis (21.12.1.3), that $\varpi_1$ is indeed the dominant weight of $U_1$ relative to this basis.

For $2 \leqq j \leqq n-1$, we now define $U_j(s) = \bigwedge^j U_1(s)$, so that $U_j$ is a linear representation of $\mathbf{SU}(n)$ on the space $\bigwedge^j \mathbf{C}^n$ of dimension $\binom{n}{j}$. The canonical basis of this space consists of the $j$-vectors

$$\mathbf{a}_{\mathrm{H}} = \mathbf{a}_{k_1} \wedge \mathbf{a}_{k_2} \wedge \cdots \wedge \mathbf{a}_{k_j}$$

where H is the set of elements of the strictly increasing sequence of integers $k_1 < k_2 < \cdots < k_j$ in the interval $[1, n]$, and H runs through all $j$-element subsets of $[1, n]$. It is clear that the restriction of $U_j$ to T′ is the Hilbert sum of $\binom{n}{j}$ one-dimensional representations on spaces $\mathbf{C}\mathbf{a}_{\mathrm{H}}$, and that the representation on $\mathbf{C}\mathbf{a}_{\mathrm{H}}$ maps $s \in$ T′ to the homothety with ratio $e^{\varepsilon'_{\mathrm{H}}(s)}$, where $\varepsilon'_{\mathrm{H}} = \varepsilon'_{k_1} + \varepsilon'_{k_2} + \cdots + \varepsilon'_{k_j}$; from this it follows that $\operatorname{Tr}(U_j(s)) = S(\varpi_j)$. We deduce as above that $U_j$ is irreducible and that $\varpi_j$ is its dominant weight.

It is easily verified (for example with the help of (21.8.4.2)) that the center of $\mathbf{SU}(n)$ is the subgroup of T′ formed by the scalar matrices $\omega I_n$, where $\omega$ runs through the set of $n$th roots of unity. We may also calculate directly $P(\mathfrak{su}(n))/Q(\mathfrak{su}(n))$ by using the preceding results (21.16.4).

(21.16.9) II. *Representations of* $\mathbf{U}(n, \mathbf{H})$. In view of the description of the maximal torus T considered in (21.12.2), the elements $2\pi i\mathbf{h}_j$ $(1 \leqq j \leqq n)$ form a basis of the lattice $\Gamma_{\mathrm{T}}$; as in (21.16.8) we recover the fact that $P(\mathbf{U}(n, \mathbf{H})) = P(\mathfrak{u}(n, \mathbf{H}))$ (21.16.3), and verify that

(21.16.9.1) $$\varpi_j = \varepsilon_1 + \varepsilon_2 + \cdots + \varepsilon_j \qquad (1 \leqq j \leqq n).$$

Since $\mathbf{U}(n, \mathbf{H})$ is simply connected, the canonical injection

$$\mathfrak{u}(n, \mathbf{H}) \to \mathfrak{sp}(2n, \mathbf{C})$$

described in (21.12.2) corresponds to a linear representation $U_1$ of $\mathbf{U}(n, \mathbf{H})$ on $\mathbf{C}^{2n}$. If $(\mathbf{a}_j)_{1 \leq j \leq 2n}$ is the canonical basis of $\mathbf{C}^{2n}$, the restriction of $U_1$ to T is the Hilbert sum of $2n$ one-dimensional representations on the spaces $\mathbf{C}\mathbf{a}_j$ $(1 \leqq j \leqq n)$; for $1 \leqq j \leqq n$, the representation on $\mathbf{C}\mathbf{a}_j$ maps $s \in \mathrm{T}$ to the homothety with ratio $e^{\varepsilon_j(s)}$, and the representation on $\mathbf{C}\mathbf{a}_{n+j}$ maps $s \in \mathrm{T}$ to the homothety with ratio $e^{-\varepsilon_j(s)}$. From the description of the Weyl group (21.12.2), it follows that the function $s \mapsto \mathrm{Tr}(U_1(s))$ on T is equal to the sum $\mathrm{S}(\varpi_1)$; hence $U_1$ is irreducible (21.13.7) with dominant weight $\varpi_1$. Likewise, if we define $U_j(s) = \bigwedge^j U_1(s)$ for $2 \leqq j \leqq n$, then $U_j$ is a linear representation of $\mathbf{U}(n, \mathbf{H})$ on the space $\bigwedge^j (\mathbf{C}^{2n})$ of dimension $\binom{2n}{j}$, whose canonical basis consists of the $j$-vectors $\mathbf{a}_{\mathrm{H}}$. The restriction of $U_j$ to T is the Hilbert sum of $\binom{2n}{j}$ one-dimensional representations of T on the spaces $\mathbf{C}\mathbf{a}_{\mathrm{H}}$: the representation on $\mathbf{C}\mathbf{a}_{\mathrm{H}}$ maps each $s \in \mathrm{T}$ to the homothety with ratio $e^{\varepsilon_{\mathrm{H}}(s)}$, where $\varepsilon_{\mathrm{H}}$ is defined as follows: if the elements of H are

$$k_1 < k_2 < \cdots < k_j,$$

then $\varepsilon_{\mathrm{H}} = \varepsilon'_{k_1} + \varepsilon'_{k_2} + \cdots + \varepsilon'_{k_j}$, where $\varepsilon'_{k_h} = \varepsilon_{k_h}$ if $1 \leqq k_h \leqq n$, and $\varepsilon'_{k_h} = -\varepsilon_{k_h - n}$ if $n + 1 \leqq k_h \leqq 2n$.

However, when $j \geqq 2$, the representation $U_j$ so defined *is not irreducible*. For example, since $\mathbf{U}(n, \mathbf{H})$ may be identified with a subgroup of the symplectic group $\mathbf{Sp}(2n, \mathbf{C})$, it follows from (A.16.4) that $U_2$ leaves *invariant* the bivector $\mathbf{a}_1 \wedge \mathbf{a}_{n+1} + \cdots + \mathbf{a}_n \wedge \mathbf{a}_{2n}$ in $\bigwedge^2 (\mathbf{C}^{2n})$. Nevertheless, we shall show that the representation $U_j$ decomposes into a Hilbert sum of irreducible representations, of which *exactly one* has $\varpi_j$ as dominant weight (cf. Problem 3). Indeed, the dominant weights of the irreducible representations into which $U_j$ splits must be certain of the $\varepsilon_{\mathrm{H}}$; if we observe that $\varepsilon_r > \varepsilon_{r'}$ in $it$* if $r < r'$, and that $\varepsilon_r > -\varepsilon_{r'}$ for all $r, r'$ (21.12.2.8), we see that the set of weights $\varepsilon_{\mathrm{H}}$ contains a *greatest element*, corresponding to the subset $\mathrm{H}_0 = \{1, 2, \ldots, j\}$, and that $\varepsilon_{\mathrm{H}_0} = \varpi_j$. Hence, among the irreducible components of the representation $U_j$, there is a unique $V_j$ whose dominant weight is $\varpi_j$. The space $\mathrm{E}_j$ of this representation is the subspace of $\bigwedge^j (\mathbf{C}^{2n})$ generated by the transforms $S \cdot \mathbf{a}_{\mathrm{H}_0}$ of $\mathbf{a}_{\mathrm{H}_0}$ by all symplectic matrices $S \in \mathbf{Sp}(2n, \mathbf{C})$. By construction, $\mathbf{a}_{\mathrm{H}_0}$ is a decomposable $j$-vector corresponding to a *totally isotropic* $j$-dimensional subspace of $\mathbf{C}^{2n}$; since the symplectic group acts transitively on the set of totally isotropic subspaces of dimension $j \leqq n$, it follows that $E_j$ is the subspace of $\bigwedge^j (\mathbf{C}^{2n})$ spanned by *all* the decomposable $j$-vectors (called *totally isotropic* $j$-vectors), which correspond to the totally isotropic subspaces of dimension $j$.

(21.16.10) III. *Representations of* $\mathbf{SO}(m)$. With the notation of (21.12.3) and (21.12.4), in both cases the lattice $\Gamma_T$ has as a basis the $2\pi i H_j$ for $1 \leqq j \leqq n$. For $\mathbf{SO}(2n)$ we have

(21.16.10.1)
$$\begin{aligned} H_j &= \mathbf{h}_j + \mathbf{h}_{j+1} + \cdots + \mathbf{h}_{n-2} + \tfrac{1}{2}(\mathbf{h}_{n-1} + \mathbf{h}_n) \\ &\qquad (1 \leqq j \leqq n-2), \\ H_{n-1} &= \tfrac{1}{2}(\mathbf{h}_{n-1} + \mathbf{h}_n), \\ H_n &= \tfrac{1}{2}(\mathbf{h}_n - \mathbf{h}_{n-1}), \end{aligned}$$

and for $\mathbf{SO}(2n+1)$

(21.16.10.2)
$$\begin{aligned} H_j &= \mathbf{h}_j + \mathbf{h}_{j+1} + \cdots + \mathbf{h}_{n-1} + \tfrac{1}{2}\mathbf{h}_n \qquad (1 \leqq j \leqq n-1), \\ H_n &= \tfrac{1}{2}\mathbf{h}_n . \end{aligned}$$

In both cases, we see therefore that $\mathbf{SO}(m)$, for $m \geqq 3$, *is not simply connected* (21.16.4). We denote by $\mathbf{Spin}(m)$ the Lie group that is the universal covering of $\mathbf{SO}(m)$. If $T_1$ is the inverse image of the torus T in $\mathbf{Spin}(m)$, the formulas above show that the lattice $(2\pi i)^{-1}\Gamma_T$ is generated by $(2\pi i)^{-1}\Gamma_{T_1}$ and the element $\frac{1}{2}(\mathbf{h}_n - \mathbf{h}_{n-1})$ in the case of $\mathbf{SO}(2n)$, and by $(2\pi i)^{-1}\Gamma_{T_1}$ and $\frac{1}{2}\mathbf{h}_n$ in the case of $\mathbf{SO}(2n+1)$; in both cases, it follows that the fundamental group $\pi_1(\mathbf{SO}(m))$ is a *group with two elements* (cf. (16.30.6)).

The fundamental weights are given by the following formulas, for $\mathbf{Spin}(2n)$:

(21.16.10.3)
$$\begin{aligned} \varpi_j &= \varepsilon_1 + \cdots + \varepsilon_j \qquad (1 \leqq j \leqq n-2), \\ \varpi_{n-1} &= \tfrac{1}{2}(\varepsilon_1 + \cdots + \varepsilon_{n-2} + \varepsilon_{n-1} - \varepsilon_n), \\ \varpi_n &= \tfrac{1}{2}(\varepsilon_1 + \cdots + \varepsilon_{n-2} + \varepsilon_{n-1} + \varepsilon_n), \end{aligned}$$

and for $\mathbf{Spin}(2n+1)$:

(21.16.10.4)
$$\begin{aligned} \varpi_j &= \varepsilon_1 + \cdots + \varepsilon_j \qquad (1 \leqq j \leqq n-1), \\ \varpi_n &= \tfrac{1}{2}(\varepsilon_1 + \cdots + \varepsilon_{n-1} + \varepsilon_n). \end{aligned}$$

Consider the canonical injection $\mathbf{SO}(m) \to \mathbf{SO}(m, \mathbf{C})$, which defines a linear representation $V_1$ of $\mathbf{SO}(m)$ on $\mathbf{C}^m$. For $j \leqq m$ we obtain as in (21.16.8) a linear representation $V_j = \bigwedge^j V_1$; by composing these with the canonical homomorphism $\mathbf{Spin}(m) \to \mathbf{SO}(m)$, we obtain linear representations $U_j$ $(1 \leqq j \leqq m)$ of $\mathbf{Spin}(m)$. We shall now study these representations directly, and show that *for* $m = 2n$ *and* $j \leqq n - 1$, *or* $m = 2n + 1$ *and* $j \leqq n$, the representation $V_j$ (and hence also $U_j$) is irreducible. Let $(\mathbf{a}_k)_{1 \leq k \leq m}$ be the canonical basis of $\mathbf{R}^m$, identified with the canonical basis of $\mathbf{C}^m$; then the canonical basis of $\bigwedge^j (\mathbf{C}^m)$ consists of the $j$-vectors

$$\mathbf{a}_{\mathrm{H}} = \mathbf{a}_{k_1} \wedge \mathbf{a}_{k_2} \wedge \cdots \wedge \mathbf{a}_{k_j},$$

where H is the set of elements $k_1 < k_2 < \cdots < k_j$ in the interval $[1, m]$, and H runs through the set of all $j$-element subsets of $[1, m]$. We shall show (under the above restrictions on $j$) that the subspace $\mathrm{F}(\mathbf{z})$ of $\bigwedge^j (\mathbf{C}^m)$ stable under $V_j$, generated by an arbitrary $j$-vector $\mathbf{z} \neq 0$, is the whole space $\bigwedge^j (\mathbf{C}^m)$. Put $\mathbf{z} = \sum_{\mathrm{H}} c_{\mathrm{H}} \mathbf{a}_{\mathrm{H}}$, where $c_{\mathrm{H}} \in \mathbf{C}$; we shall argue by induction on the number $r$ of coefficients $c_{\mathrm{H}}$ that are $\neq 0$. The assertion is obvious when $r = 1$: indeed, for each permutation $\pi \in \mathfrak{S}_m$, the automorphism of $\mathbf{C}^m$ that transforms $\mathbf{a}_k$ into $\pm \mathbf{a}_{\pi(k)}$ for $1 \leqq k \leqq m$ belongs to the image of $\mathbf{SO}(m)$ under $V_1$, provided that the product of the minus signs is equal to the signature of $\pi$. Since $\mathrm{F}(\mathbf{z})$ contains the element $\mathbf{a}_{\mathrm{H}}$ of the canonical basis of $\bigwedge^j (\mathbf{C}^m)$, it therefore contains also all elements $\mathbf{a}_{\pi(\mathrm{H})}$, and hence is the whole of $\bigwedge^j (\mathbf{C}^m)$. Suppose now that the assertion has been proved for some value of $r \geqq 1$, and for all values $< r$, and suppose that the number of nonzero coefficients $c_{\mathrm{H}}$ in $\mathbf{z}$ is $r + 1$. Then there exist two distinct $j$-element subsets H, L of $[1, m]$ such that $c_{\mathrm{H}} c_{\mathrm{L}} \neq 0$. Let $p$ be an element of $\mathrm{L} \cap \complement \mathrm{H}$. Next, *since* $2j < m$, there exists $q \in [1, m]$ that does not belong to $\mathrm{H} \cup \mathrm{L}$. The automorphism $T$ of $\mathbf{C}^m$ that leaves $\mathbf{a}_k$ fixed for $k$ not equal to $p$ or $q$, and transforms $\mathbf{a}_p$ into $-\mathbf{a}_p$ and $\mathbf{a}_q$ into $-\mathbf{a}_q$, is in the image of $\mathbf{SO}(m)$ under $V_1$ and transforms $\mathbf{a}_{\mathrm{H}}$ into itself, $\mathbf{a}_{\mathrm{L}}$ into $-\mathbf{a}_{\mathrm{L}}$, and each other $\mathbf{a}_{\mathrm{M}}$ into $\pm \mathbf{a}_{\mathrm{M}}$. It follows immediately that in the $j$-vector $\mathbf{z} + T \cdot \mathbf{z}$, which belongs to $\mathrm{F}(\mathbf{z})$, the number of coefficients $\neq 0$ is $\geqq 1$ and $\leqq r$; we may therefore apply the inductive hypothesis to complete the proof.

Put $\mathbf{b}_{2r-1} = \mathbf{a}_{2r-1} - i\mathbf{a}_{2r}$, $\mathbf{b}_{2r} = \mathbf{a}_{2r-1} + i\mathbf{a}_{2r}$ for $2r \leqq m$. When $m = 2n$, the $\mathbf{b}_k$ for $1 \leqq k \leqq 2n$ form a basis of $\mathbf{C}^{2n}$; the restriction of $V_1$ to T is the Hilbert sum of $2n$ one-dimensional representations on subspaces $\mathbf{C}\mathbf{b}_k$ $(1 \leqq k \leqq 2n)$, and the representation on $\mathbf{C}\mathbf{b}_{2r-1}$ (resp. $\mathbf{C}\mathbf{b}_{2r}$) is the homothety with ratio $e^{\varepsilon_r}$ (resp. $e^{-\varepsilon_r}$). For $j \leqq 2n$, a basis of $\bigwedge^j (\mathbf{C}^{2n})$ is formed by the $j$-vectors $\mathbf{b}_H = \mathbf{b}_{k_1} \wedge \mathbf{b}_{k_2} \wedge \cdots \wedge \mathbf{b}_{k_j}$, where H is the set of elements $k_1 < k_2 < \cdots < k_j$ in the interval $[1, 2n]$, and H runs through all $j$-element subsets of $[1, 2n]$. Then the restriction of $V_j(s)$ to T transforms $\mathbf{b}_H$ into $e^{\varepsilon_H(s)}\mathbf{b}_H$, where $\varepsilon_H = \varepsilon''_{k_1} + \varepsilon''_{k_2} + \cdots + \varepsilon''_{k_j}$, the $\varepsilon''_k$ being defined by $\varepsilon''_{2r-1} = \varepsilon_r$ and $\varepsilon''_{2r} = -\varepsilon_r$. Since, for the ordering on $it^*$ (21.14.5), we have $\varepsilon_r > \varepsilon_{r'}$ if $r < r'$, and $\varepsilon_r > -\varepsilon_{r'}$ for all $r, r' \in [1, n]$ by virtue of (21.12.3.5), it follows that, *for* $j \leqq n - 2$, the representation $V_j$ has dominant weight $\varpi_j$.

When $m = 2n + 1$, a basis of $\bigwedge^j (\mathbf{C}^{2n+1})$ is formed by the $\mathbf{b}_H$ defined above and the $\mathbf{b}_{H'} \wedge \mathbf{a}_{2n+1}$, where H' is a subset of $j - 1$ elements of $[1, 2n]$; it follows as above that $V_j$ has dominant weight $\varpi_j$ *for* $j \leqq n - 1$.

For the irreducible representations of **Spin**$(2n)$ with dominant weights $\varpi_{n-1}$ and $\varpi_n$, and the irreducible representation of **Spin**$(2n + 1)$ with dominant weight $\varpi_n$, see Problem 7.

## PROBLEMS

1. With the notation of (21.16.8), show that the complex conjugate of the irreducible representation $U_j$ is equivalent to $U_{n-j}$ $(1 \leqq j \leqq n - 1)$. If $n = 2m$ is even, the representation $U_m^{(\mathbf{R})}$ (Section 21.1, Problem 9) is defined when $m$ is even, and the representation $U_m^{(\mathbf{H})}$ is defined when $m$ is odd.

2. With the notation of (21.16.9), show that each of the representations $U_j$ (or $V_j$) is equivalent to its complex conjugate, $U_j^{(\mathbf{R})}$ is defined for even $j$, and $U_j^{(\mathbf{H})}$ for odd $j$.

3. (a) Let B be a nondegenerate alternating bilinear form on $\mathbf{C}^{2n}$; let $(\mathbf{e}_j)_{1 \leq j \leq 2n}$ be a symplectic basis of $\mathbf{C}^{2n}$, and let $(\mathbf{e}_j^*)_{1 \leq j \leq 2n}$ be the dual basis, so that

$$\mathrm{B} = \mathbf{e}_1^* \wedge \mathbf{e}_2^* + \cdots + \mathbf{e}_{2n-1}^* \wedge \mathbf{e}_{2n}^*$$

in the vector space $\bigwedge^2 (\mathbf{C}^{2n})^*$. Put

$$\mathrm{B}^* = \mathbf{e}_1 \wedge \mathbf{e}_2 + \cdots + \mathbf{e}_{2n-1} \wedge \mathbf{e}_{2n}$$

in the vector space $\bigwedge^2 (\mathbf{C}^{2n})$; the bivector B* is independent of the symplectic basis $(\mathbf{e}_j)$ chosen.

For each subset H of $[1, 2n]$, we have

$$(\mathbf{e}_{2j-1} \wedge \mathbf{e}_{2j}) \wedge \mathbf{e}_H = 0 \qquad \text{if } \{2j-1, 2j\} \cap \mathrm{H} \neq \varnothing,$$

$$(\mathbf{e}_{2j-1} \wedge \mathbf{e}_{2j}) \wedge \mathbf{e}_H = \mathbf{e}_H \wedge (\mathbf{e}_{2j-1} \wedge \mathbf{e}_{2j}) = \mathbf{e}_{H \cup \{2j-1, 2j\}}$$

if $\{2j-1, 2j\} \cap \mathrm{H} = \varnothing$. If we consider $\mathbf{C}^{2n}$ as the dual of $(\mathbf{C}^{2n})^*$, we have likewise

$$(\mathbf{e}_{2j-1}^* \wedge \mathbf{e}_{2j}^*) \lrcorner \mathbf{e}_{\mathrm{H}} = 0 \qquad \text{if } \{2j-1, 2j\} \cap \complement\mathrm{H} \neq \varnothing,$$

$$(\mathbf{e}_{2j-1}^* \wedge \mathbf{e}_{2j}^*) \lrcorner \mathbf{e}_{\mathrm{H}} = \mathbf{e}_{\mathrm{H}-\{2j-1, 2j\}} \qquad \text{if } \{2j-1, 2j\} \subset \mathrm{H}.$$

(b) In the endomorphism ring of the vector space $\bigwedge(\mathbf{C}^{2n})$, let $Y^+$ denote the mapping $z \mapsto \mathrm{B} \lrcorner z$, and $Y^-$ the mapping $z \mapsto \mathrm{B}^* \wedge z$; also put $Z = [Y^+, Y^-]$. For each subset H of $[1, 2n]$, let $c_{\mathrm{H}}^+$ (resp. $c_{\mathrm{H}}^-$) denote the number of subsets $\{2j-1, 2j\}$ contained in $\complement\mathrm{H}$ (resp. H). Show that

$$Z \cdot \mathbf{e}_{\mathrm{H}} = (c_{\mathrm{H}}^+ - c_{\mathrm{H}}^-)\mathbf{e}_{\mathrm{H}}$$

(use (a) above). Deduce that

$$[Z, Y^+] = 2Y^+, \qquad [Z, Y^-] = -2Y^-$$

and hence that the Lie subalgebra of $\mathfrak{gl}(\bigwedge(\mathbf{C}^{2n}))$ spanned by $Y^+$, $Y^-$, and $Z$ is isomorphic to $\mathfrak{sl}(2, \mathbf{C})$.

(c) With the notation of (**21.16.9**), show that for each $p \in [1, n]$ the restrictions of $Y^+$ and $Y^-$ to $\bigwedge^p(\mathbf{C}^{2n})$ commute with all the automorphisms $U_p(s)$. Use (**21.9.3**) to deduce that the subspace $\mathrm{E}_p$ of $\bigwedge^p(\mathbf{C}^{2n})$ spanned by the totally isotropic decomposable $p$-vectors consists of the $p$-vectors $z$ such that $Y^+ \cdot z = 0$ and $Z \cdot z = (n-p)z$. Deduce that if $p < n$, the mapping $z \mapsto (Y^-)^{n-p} \cdot z = (\mathrm{B}^*)^{n-p} \wedge z$ (where $(\mathrm{B}^*)^h$ denotes the $2h$-vector that is the product of $h$ factors equal to $\mathrm{B}^*$ in the exterior algebra $\bigwedge(\mathbf{C}^{2n})$) is injective on $\bigwedge^p(\mathbf{C}^{2n})$.

(d) Hence show that, for $p \leqq n$, $\bigwedge^p(\mathbf{C}^{2n})$ is the direct sum of the subspaces

$$\mathrm{E}_p, \quad (\mathrm{B}^*) \wedge \mathrm{E}_{p-2}, \quad (\mathrm{B}^*)^2 \wedge \mathrm{E}_{p-4}, \ldots$$

each stable under the representation $U_p$, and that the restriction of $U_p$ to $(\mathrm{B}^*)^h \wedge \mathrm{E}_{p-2h}$ is irreducible and similar to $V_{p-2h}$ (Lepage's decomposition). The dimension of $\mathrm{E}_p$ is

$$\binom{2n}{p} - \binom{2n}{p-2}.$$

**4.** There exists a **C**-algebra $\mathrm{C}_m$ (the *Clifford algebra*) of dimension $2^m$, having a basis consisting of the identity and all products $a_{i_1} a_{i_2} \cdots a_{i_p}$ for $1 \leqq i_1 < i_2 < \cdots < i_p \leqq m$, where the $a_j$ $(1 \leqq j \leqq m)$ are $m$ elements such that $a_j^2 = 1$ and $a_j a_k = -a_k a_j$ whenever $j \neq k$ (cf. Section **16.15**, Problem 2). The algebra $\mathrm{C}_m$ is the direct sum of the vector subspace $\mathrm{C}_m^+$ spanned by the products $a_{i_1} a_{i_2} \cdots a_{i_p}$ with $p$ even, and the subspace $\mathrm{C}_m^-$ spanned by the analogous products with $p$ odd; also $\mathrm{C}_m^+$ is a subalgebra of $\mathrm{C}_m$.

(a) If $m$ is even, the center of $\mathrm{C}_m$ is $\mathbf{C} \cdot 1$, and the center of $\mathrm{C}_m^+$ is spanned by 1 and $a_1 a_2 \cdots a_m$. If $m$ is odd, the center of $\mathrm{C}_m$ is spanned by 1 and $a_1 a_2 \cdots a_m$, and the center of $\mathrm{C}_m^+$ is $\mathbf{C} \cdot 1$.

(b) Let E be the **C**-vector subspace of $\mathrm{C}_m$ spanned by $a_1, \ldots, a_m$ and let $\Phi$ be the symmetric bilinear form on E such that $\Phi(a_j, a_k) = \delta_{jk}$ (Kronecker delta). For each $x \in \mathrm{E}$, we have $x^2 = \Phi(x, x) \cdot 1$ and $xy + yx = 2\Phi(x, y) \cdot 1$ in the algebra $\mathrm{C}_m$. Show that if A is a **C**-algebra and $f$ a **C**-linear mapping of E into A such that $f(x)^2 = \Phi(x, x) \cdot 1$ for all $x \in \mathrm{E}$, then $f$ has a unique extension to a homomorphism of $\mathrm{C}_m$ into A.

(c) Show that there exists an isomorphism $\beta$ of $\mathrm{C}_m$ onto the algebra opposite to $\mathrm{C}_m$ (i.e., $\beta$ is an *antiautomorphism* of $\mathrm{C}_m$) such that $\beta(x) = x$ for all $x \in \mathrm{E}$.

(d) Let G be the group of invertible elements $s \in C_m$ such that $sEs^{-1} = E$, and let $G^+ = G \cap C_m^+$. For each $s \in G$, let $\varphi(s)$ denote the linear mapping $x \mapsto sxs^{-1}$ of E into itself. Show that $E \cap G$ is the set of vectors in E that are nonisotropic for $\Phi$, and that for each $x \in E \cap G$, $-\varphi(x)$ is the reflection in the hyperplane in E orthogonal to $x$ (relative to $\Phi$). Deduce that $\varphi$ is a homomorphism of G into the orthogonal group $\mathbf{O}(\Phi)$, whose kernel is the set of invertible elements of the center Z of $C_m$. We have $\varphi(G) = \mathbf{O}(\Phi)$ if $m$ is even, $\varphi(G) = \mathbf{SO}(\Phi)$ if $m$ is odd, and $\varphi(G^+) = \mathbf{SO}(\Phi)$ in either case.
(e) For each $s \in G^+$, show that $N(s) = \beta(s)s$ is a scalar, and that $s \mapsto N(s)$ is a homomorphism of $G^+$ into $\mathbf{C}^*$.

5. With the notation of Problem 4, suppose that $m = 2n$ is even, and put $m_j = a_{2j-1} - ia_{2j}$, $p_j = a_{2j-1} + ia_{2j}$ for $1 \leqq j \leqq n$. The $m_j$ (resp. the $p_j$) form a basis of a totally isotropic subspace M (resp. P) (relative to $\Phi$), and E is the direct sum of M and P. We may identify M with the dual of P by identifying each $m \in M$ with the linear form $p \mapsto \Phi(m, p)$ on P. The subalgebra of $C_{2n}$ generated by M has as a basis the elements $m_H = m_{i_1} m_{i_2} \cdots m_{i_k}$ for each subset H of $I = \{1, 2, \ldots, n\}$, the $i_k$ being the elements of H arranged in ascending order; this subalgebra may be identified with the exterior algebra $S = \bigwedge M$ on the vector space M.

Show that there exists a unique homomorphism $\rho$ of $C_{2n}$ into the algebra End(S) of endomorphisms of the vector space S, such that for each $m \in M$ the image $\rho(m)$ is the linear mapping $z \mapsto mz$ of S into itself, and such that for each $p \in P$, $\rho(p)$ is the interior product $i(p)$ (A.15.3), M being identified with the dual of P, and S with $\bigwedge M$. (Use Problem 4(b).) Put $p_I = p_1 p_2 \cdots p_n$, and for each pair of subsets H, K of I, put $z_{H,K} = m_H p_I m_{I-K}$. Show that, for each subset L of I, we have $\rho(z_{H,K})m_L = 0$ if $K \neq L$, and $\rho(z_{H,K})m_K = \pm m_H$. Deduce that $\rho$ is an *isomorphism* of $C_{2n}$ onto End(S), which is isomorphic to the matrix algebra $\mathbf{M}_{2^n}(\mathbf{C})$.

The vector space S is the direct sum of $S^+ = S \cap C_{2n}^+$ and $S^- = S \cap C_{2n}^-$, having as respective bases the set of $m_H$ for subsets H with an even number and an odd number of elements. The subalgebra $\rho(C_{2n}^+)$ of End(S) leaves invariant the subspaces $S^+$ and $S^-$, and is isomorphic to $\operatorname{End}(S^+) \times \operatorname{End}(S^-)$.

6. With the notation of Problem 4, suppose that $m = 2n + 1$ is odd; the algebra $C_{2n}$ may be canonically identified with the subalgebra of $C_{2n+1}$ generated by the $a_j$ with $j \leqq 2n$. Show that the mapping $y \mapsto iya_{2n+1}$ of the vector space $F \subset E$ spanned by the $a_j$ with $j \leqq 2n$, into the algebra $C_{2n+1}^+$, extends to an isomorphism $\theta$ of $C_{2n}$ onto $C_{2n+1}^+$ (use Problem 4(b)). Deduce that $C_{2n+1}$ is isomorphic to the product of two algebras isomorphic to $\mathbf{M}_{2^n}(\mathbf{C})$.

7. With the notation of Problem 4, let $E_0$ be the real vector space spanned by $a_1, \ldots, a_m$. Then $E_0 \cap G$ is the set of vectors $\neq 0$ in $E_0$. Let $G_0$ be the subgroup of $G^+$ generated by the products of an even number of vectors $x \in E_0$ such that $N(x) = \Phi(x, x) = 1$.
(a) Show that $G_0$ is connected. (If $x, y$ are two distinct vectors in $E_0$ such that $\Phi(x, x) = \Phi(y, y) = 1$, consider the plane in $E_0$ spanned by $x$ and $y$, and a vector $x'$ in that plane orthogonal to $x$ and such that $\Phi(x', x') = 1$, and the vectors $z = x \cos t + x' \sin t$ for $t \in \mathbf{R}$.) Deduce that $G_0$ is isomorphic to $\mathbf{Spin}(m)$, by observing that $\varphi(G_0)$ is isomorphic to $\mathbf{SO}(m)$ and that $\varphi$ makes $G_0$ a double covering of $\mathbf{SO}(m)$.
(b) Deduce from Problems 5 and 6 that for $m = 2n$ the representations $s \mapsto \rho(s) | S^+$ and $s \to \rho(s) | S^-$ are irreducible representations of $\mathbf{Spin}(2n)$ (identified with $G_0$) of dimension $2^{n-1}$; for $m = 2n + 1$, the representation $s \mapsto \rho(\theta^{-1}(s))$ is an irreducible representation of $\mathbf{Spin}(2n + 1)$ of dimension $2^n$.

(c) If $m = 2n$, put $r_j(\theta_j) = a_{2j-1} \cos \theta_j - a_{2j} \sin \theta_j$ for $1 \leqq j \leqq n$. Then the elements $a_1 r_1(\theta_1) a_2 r_2(\theta_2) \cdots a_n r_n(\theta_n)$ form a maximal torus $T_0$ of **Spin**$(2n)$, whose image $T = \varphi(T_0)$ is the torus described in **(21.12.3)**, when $\varphi(G^+)$ is identified with $\mathbf{SO}(2n)$. For $m = 2n + 1$, the same torus $T_0$ (when $C_{2n}$ is canonically identified with a subalgebra of $C_{2n+1}$) is a maximal torus of **Spin**$(2n+1)$, whose image $T = \varphi(T_0)$ is the torus described in **(21.12.4)**. In both cases, the vectors $m_H \in S$ are eigenvectors for the restriction of $\rho$ (or of $\rho \circ \theta^{-1}$) to $T_0$. In particular, for the vector $m_1 m_2 \cdots m_n$, the corresponding weight is $\frac{1}{2}(\varepsilon_1 + \cdots + \varepsilon_n)$: in other words, for $s = \exp_{T_0} \mathbf{u}$, where $\mathbf{u} \in \mathfrak{so}(2n)$, the corresponding eigenvalue is $\frac{1}{2}(\varepsilon_1(\mathbf{u}) + \cdots + \varepsilon_n(\mathbf{u}))$. Likewise, for $m_1 m_2 \cdots m_{n-1}$, the corresponding weight is $\frac{1}{2}(\varepsilon_1 + \cdots + \varepsilon_{n-1} - \varepsilon_n)$. Deduce that when $m = 2n + 1$ the dominant weight of the irreducible representation $s \mapsto \rho(\theta^{-1}(s))$ is $\varpi_n$, given by **(21.16.10.4)**; when $m = 2n$, if $n$ is even the dominant weight of $s \mapsto \rho(s) \mid S^+$ is $\varpi_n$ and the dominant weight of $s \mapsto \rho(s) \mid S^-$ is $\varpi_{n-1}$; but when $m = 2n$ with $n$ odd, the dominant weight of $s \mapsto \rho(s) \mid S^+$ is $\varpi_{n-1}$ and the dominant weight of $s \mapsto \rho(s) \mid S^-$ is $\varpi_n$ (where $\varpi_{n-1}$ and $\varpi_n$ are given by **(21.16.10.3)**).

**8.** Let $(\mathbf{a}_j)_{1 \leq j \leq 2n}$ be the canonical basis of $\mathbf{C}^{2n}$, and let $\Phi$ be the symmetric bilinear form on $\mathbf{C}^{2n}$ such that $\Phi(\mathbf{a}_j, \mathbf{a}_k) = \delta_{jk}$, so that $\mathbf{O}(\Phi) = \mathbf{O}(2n, \mathbf{C})$.
(a) Consider the basis $(\mathbf{a}_j)_{1 \leq j \leq 2n}$ also as an orthonormal basis of $\mathbf{R}^{2n}$, relative to the restriction of $\Phi$ to $\mathbf{R}^{2n}$. Define a mapping T of $(\mathbf{R}^{2n})^n$ into $\bigwedge^n (\mathbf{R}^{2n})$ as follows: if $\mathbf{x}_1, \ldots, \mathbf{x}_n$ are linearly dependent in $\mathbf{R}^{2n}$, then $T(\mathbf{x}_1, \ldots, \mathbf{x}_n) = 0$; if they are linearly independent, then we may write $\mathbf{x}_1 \wedge \mathbf{x}_2 \wedge \cdots \wedge \mathbf{x}_n = \lambda \mathbf{y}_1 \wedge \mathbf{y}_2 \wedge \cdots \wedge \mathbf{y}_n$, where the vectors $\mathbf{y}_j$ $(1 \leqq j \leqq n)$ form an orthonormal basis of the subspace of dimension $n$ in $\mathbf{R}^{2n}$ spanned by the $\mathbf{x}_j$, and $\lambda \in \mathbf{R}$; then there exists an element $u \in \mathbf{SO}(2n)$ such that $u(\mathbf{a}_{2j-1}) = \mathbf{y}_j$ for $1 \leqq j \leqq n$, and we define $T(\mathbf{x}_1, \ldots, \mathbf{x}_n) = \lambda u(\mathbf{a}_2) \wedge u(\mathbf{a}_4) \wedge \cdots \wedge u(\mathbf{a}_{2n})$. Show that this value depends neither on the choice of the $\mathbf{y}_j$ nor on the choice of $u$, and that T is an alternating $n$-linear mapping, which therefore factorizes uniquely into $(\mathbf{R}^{2n})^n \to \bigwedge^n (\mathbf{R}^{2n}) \xrightarrow{\tau} \bigwedge^n (\mathbf{R}^{2n})$, where $\tau$ is a linear bijection. This bijection extends uniquely to a bijection of $\bigwedge^n (\mathbf{C}^{2n})$ onto itself, also denoted by $\tau$. We have $\tau^2 = (-1)^n \cdot 1$. For each $u \in \mathbf{SO}(2n, \mathbf{C})$, we have $\tau \circ \bigwedge^n (u) = \bigwedge^n (u) \circ \tau$; but if $u \in \mathbf{O}(2n, \mathbf{C})$ has determinant equal to $-1$, then $\tau \circ \bigwedge^n (u) = -\bigwedge^n (u) \circ \tau$. Deduce that $\bigwedge^n (\mathbf{C}^{2n})$ is the direct sum of two subspaces $F^+$, $F^-$ of the same dimension, such that the restriction of $\tau$ to $F^+$ (resp. $F^-$) is the homothety with ratio 1 (resp. $-1$) if $n$ is even, the homothety of ratio $i$ (resp. $-i$) if $n$ is odd.
(b) Put

$$\mathbf{m}_j = \mathbf{a}_{2j-1} - i\mathbf{a}_{2j}, \qquad \mathbf{p}_j = \mathbf{a}_{2j-1} + i\mathbf{a}_{2j}, \qquad m_{\text{I}} = \mathbf{m}_1 \wedge \mathbf{m}_2 \wedge \cdots \wedge \mathbf{m}_n,$$

$$m'_{\text{I}} = \mathbf{m}_1 \wedge \mathbf{m}_2 \wedge \cdots \wedge \mathbf{m}_{n-1} \wedge \mathbf{p}_n.$$

Show that $\tau(m_{\text{I}}) = i^n m_{\text{I}}$, $\tau(m'_{\text{I}}) = -i^n m'_{\text{I}}$.

If we define a *totally isotropic* $n$-vector in $\bigwedge^n (\mathbf{C}^{2n})$ (relative to $\Phi$) to be a decomposable $n$-vector $z$ corresponding to a totally isotropic subspace $V_z$ of $\mathbf{C}^{2n}$, deduce from these results that every totally isotropic $n$-vector belongs either to $F^+$ or to $F^-$. Let $N^+$ (resp. $N^-$) denote the set of those which belong to $F^+$ (resp. $F^-$) (cf. Section **16.14**, Problem 18). If $z$ and $z'$ belong both to $N^+$ or both to $N^-$, show that $V_z \cap V_{z'}$ has even codimension in $V_z$ (and in $V_{z'}$); if on the other hand one of $z$, $z'$ belongs to $N^+$ and the other to $N^-$, then $V_z \cap V_{z'}$ has odd codimension in $V_z$ (and in $V_{z'}$).
(c) Show that the $n$-vectors belonging to $N^+$ (resp. $N^-$) span the $\mathbf{C}$-vector space $F^+$

(resp. $F^-$), and deduce that the representation $V_n$ of $\mathbf{SO}(2n)$ on $\bigwedge^n (\mathbf{C}^{2n})$ (in the notation of **(21.16.10)**) splits into two inequivalent irreducible representations on the subspaces $F^+$ and $F^-$, respectively. The dominant weights of these two representations are $2\varpi_{n-1}$ and $2\varpi_n$. (To show that $N^+$ spans $F^+$ and $N^-$ spans $F^-$, prove that $N^+ \cup N^-$ spans the whole space $\bigwedge^n (\mathbf{C}^{2n})$, by using the irreducibility of the representation $V_{n-1}$ of $\mathbf{SO}(2n-1)$.)

**9.** (a) For the group $\mathbf{SO}(2n+1)$, the weight lattice P is generated by $\varpi_1, \dots, \varpi_{n-1}$ and $2\varpi_n$, and another basis of P is $\varepsilon_1, \dots, \varepsilon_n$. The elements of $\mathbf{Z}[P]^W$ **(21.13.6)** are of the form

$$G(e^{\varepsilon_1}, e^{-\varepsilon_1}, \dots, e^{\varepsilon_n}, e^{-\varepsilon_n}),$$

where $G(T_1, T_2, \dots, T_{2n})$ is a symmetric polynomial with integer coefficients. In particular, let $\sigma_j$ $(0 \leq j \leq 2n)$ be the elementary symmetric functions of $T_1, \dots, T_{2n}$, i.e., the coefficients of the polynomial $(X + T_1)(X + T_2) \cdots (X + T_{2n})$ in X. Then the character of the representation $V_j$ $(1 \leq j \leq n)$ is

$$\sigma_j(e^{\varepsilon_1}, e^{-\varepsilon_1}, \dots, e^{\varepsilon_n}, e^{-\varepsilon_n}) + \sigma_{j-1}(e^{\varepsilon_1}, e^{-\varepsilon_1}, \dots, e^{\varepsilon_n}, e^{-\varepsilon_n}).$$

Deduce that if $\boldsymbol{\rho}_j$ is the class of the representation $V_j$, the ring $\mathbf{Z}^{(R(G))}$ for the group $G = \mathbf{SO}(2n+1)$ is isomorphic to $\mathbf{Z}[\boldsymbol{\rho}_1, \dots, \boldsymbol{\rho}_n]$, the $\boldsymbol{\rho}_i$ being algebraically independent over $\mathbf{Z}$.

(b) For the group $\mathbf{SO}(2n)$, the weight lattice P is generated by $\varpi_1, \dots, \varpi_{n-2}, 2\varpi_{n-1}$, and $2\varpi_n$, and has as a basis $\varepsilon_1, \varepsilon_2, \dots, \varepsilon_n$. Let $\mathscr{H}$ be the vector space of polynomials $G(T_1, T_2, \dots, T_{2n})$ with rational coefficients that are invariant (i) under the product of transpositions that interchange $T_{2i-1}$ and $T_{2j-1}$, and $T_{2i}$ and $T_{2j}$, where $i \neq j$; (ii) under the product of an even number of transpositions that interchange $T_{2i-1}$ and $T_{2i}$. The space $\mathscr{H}$ is the direct sum of the space $\mathscr{H}^+$ of symmetric polynomials in $T_1, \dots, T_{2n}$ and the space $\mathscr{H}^-$ of polynomials in $\mathscr{H}$ that change sign under interchange of $T_{2n-1}$ and $T_{2n}$ (observe that $\mathscr{H}$ remains globally invariant when this interchange is made on every polynomial in $\mathscr{H}$). Every polynomial in $\mathscr{H}^-$ is of the form $(T_1 - T_2)(T_3 - T_4) \cdots (T_{2n-1} - T_{2n})F$, where $F \in \mathscr{H}^+$. Show that $\mathbf{Z}[P]^W$ is the set of elements $G(e^{\varepsilon_1}, e^{-\varepsilon_1}, \dots, e^{\varepsilon_n}, e^{-\varepsilon_n})$ where G runs through the set of polynomials $G(T_1, \dots, T_{2n})$ with integer coefficients that belong to $\mathscr{H}$. Deduce that the ring $\mathbf{Z}^{(R(G))}$ for the group $G = \mathbf{SO}(2n)$ is a free module over the ring $\mathbf{Z}[\boldsymbol{\rho}_1, \dots, \boldsymbol{\rho}_n]$ (where $\boldsymbol{\rho}_j$ is the class of the representation $V_j$, the $\boldsymbol{\rho}_j$ being algebraically independent); a basis of this module is formed by 1 and the class $\boldsymbol{\rho}_n^+$ of the restriction of $V_n$ to $F^+$ (in the notation of Problem 8). This implies the existence of a relation $(\boldsymbol{\rho}_n^+)^2 = \boldsymbol{\alpha} + \boldsymbol{\beta}\boldsymbol{\rho}_n^+$, where $\boldsymbol{\alpha}$ and $\boldsymbol{\beta}$ lie in the ring $\mathbf{Z}[\boldsymbol{\rho}_1, \dots, \boldsymbol{\rho}_n]$.

**10.** Let G be a *simply connected* almost simple compact group. We retain the notation of Section **21.15**, Problem 11.

(a) Consider the composite mapping

$$g\colon (G/T) \times \mathfrak{t} \xrightarrow{1 \times \exp_T} (G/T) \times T \xrightarrow{f} G$$

where $f$ is the mapping defined in **(21.15.2.1)**. Show that the affine Weyl group $W_a$ acts differentiably and freely on $(G/T) \times (\mathfrak{t} - D(G))$, and that $g$ makes this space into a covering of the open subset V of G that is the image of $(G/T) \times T_{\text{reg}}$ under $g$. Use Section **21.15**, Problem 11(e) to show that $(G/T) \times (\mathfrak{t} - D(G))$ is the disjoint union of the open sets $(G/T) \times iv(A^*)$, where $v \in W_a$, and that the restriction of $g$ to $(G/T) \times iv(A^*)$ is a diffeomorphism onto V for each $v \in W_a$. (Note that the lattice $(2\pi i)^{-1}\Gamma_T$ is generated by the $\mathbf{h}_\alpha$.)

(b) Show that none of the vertices of the simplex $A^*$, other than 0, can belong to the

lattice $i\Gamma_T$. (Suppose if possible that there exists a vertex $\mathbf{a}_j = 2\pi\mathbf{p}_j/n_j$ of A* in $i\Gamma_T$. Then $A^* - \mathbf{a}_j = w(A^*)$ for some $w \in W$; show that if $\mathbf{u} \in A^*$ is sufficiently close to 0, we have $\mathbf{u} \neq \mathbf{a}_j + w(\mathbf{u})$, and obtain a contradiction by observing that there exist $\bar{s}_1$ and $\bar{s}_2$ in G/T such that $g(\bar{s}_1, i\mathbf{u}) = g(\bar{s}_2, i(\mathbf{a}_j + w(\mathbf{u})))$. Deduce that each orbit of $W_a$ in $i\mathfrak{t}$ meets the closure of A* in exactly one point.
(c) If Z is the center of G, show that Card(Z) − 1 is the number of integers $n_j$ that are equal to 1 in the expression $\mu = \sum_{j=1}^{l} n_j \beta_j$, where $\mu$ is the highest root (Section **21.15**, Problem 10). (Observe that the vectors $\mathbf{p}_j$ form a basis of the lattice $(2i\pi)^{-1} \exp_T^{-1}(Z)$.)

**11.** (a) The hypotheses on G and the notation are the same as in Problem 10. Show that for each automorphism $v$ of G, the group F of fixed points of $v$ is *connected*. (Use Section **21.11**, Problem 19 to reduce to showing that each $x \in F$ that is regular in G is contained in the identity component of F. Having chosen a maximal torus T in G, we may write $x = \exp_T(i\mathbf{u})$, where $\mathbf{u}$ belongs to the principal alcove A*; we then have $v_*(\mathbf{u}) - \mathbf{u} = \mathbf{z}$, where $i\mathbf{z} \in \exp_T^{-1}(Z)$. Use Problem 10 to show that $\mathbf{z} = 0$, and deduce that the one-parameter subgroup consisting of the $\exp_T(i\xi\mathbf{u})$ with $\xi \in \mathbf{R}$ is contained in F.)
(b) Give an example of an involutory automorphism of the group **SO**(3) whose set of fixed points is not connected.

**12.** With the notation of the proof of (**21.16.3**), let $\mathfrak{n}_+$ and $\mathfrak{n}_-$ be the Lie subalgebras of $\mathfrak{g}_{(\mathbf{C})}$ spanned respectively by the elements $\mathbf{x}_{\alpha_k}$ $(1 \leq k \leq n)$ and $\mathbf{x}_{-\alpha_k}$ $(1 \leq k \leq n)$. Let $e_{a,b,c}$ denote the element (**21.16.3.4**), where $a = (a_1, \ldots, a_n)$, $b = (b_1, \ldots, b_n)$, $c = (c_1, \ldots, c_l)$.
(a) Let $U^0$ be the commutator of $\mathfrak{h}$ in U, or equivalently the commutator of the subalgebra $U(\mathfrak{h})$ in $U = U(\mathfrak{g}_{(\mathbf{C})})$. Show that $U^0$ has a basis consisting of the $e_{a,b,c}$ such that $\sum_k a_k \alpha_k = \sum_k b_k \alpha_k$.
(b) Show that $\mathfrak{L} = (\mathfrak{n}_- U) \cap U^0 = U^0 \cap (U\mathfrak{n}_+)$ is a two-sided ideal in $U^0$, and that $U^0 = U(\mathfrak{h}) \oplus \mathfrak{L}$.

**13.** With the same notation as in Problem 12, for each integer $r > 0$ let $U^{(r)}$ be the vector subspace of U spanned by the $e_{a,b,c}$ such that

$$a_1 + \cdots + a_n + b_1 + \cdots + b_n + c_1 + \cdots + c_l \leq r.$$

For each $s \in G$ the automorphism Ad($s$) of $\mathfrak{g}$ has a unique extension to an automorphism, also written Ad($s$), of the algebra U, which leaves invariant each $U^{(r)}$, and $s \mapsto \mathrm{Ad}(s) \mid U^{(r)}$ is a continuous linear representation of G on $U^{(r)}$. The derived homomorphism is $\mathbf{u} \mapsto \mathrm{ad}(\mathbf{u})$, where ad($\mathbf{u}$) denotes the mapping $z \mapsto \mathbf{u}z - z\mathbf{u}$ of $U^{(r)}$ into itself (cf. Section **19.11**, Problem 1).
(a) Let [U, U] denote the subspace of U spanned by the elements $[x, x'] = xx' - x'x$ for all $x$, $x'$ in U. Likewise let $[\mathfrak{g}, U]$ denote the subspace of [U, U] spanned by all $[\mathbf{u}, x]$ with $\mathbf{u} \in \mathfrak{g}$ and $x \in U$. Show that $[U, U] = [\mathfrak{g}, U]$.
(b) If Z is the center of the algebra U, show that $U = Z \oplus [U, U]$. (Using (a) and the complete reducibility of the linear representation $s \mapsto \mathrm{Ad}(s) \mid U^{(r)}$ of G, show that

$$U^{(r)} = (Z \cap U^{(r)}) \oplus ([U, U] \cap U^{(r)}).$$

If the component of $x \in U$ in Z is denoted by $x^\natural$, show that $(xy)^\natural = (yx)^\natural$ and that $(zx)^\natural = zx^\natural$ if $z \in Z$.

(c) Each element $x = \sum_c \xi_c \mathbf{h}^c$ (where $\mathbf{h}^c = \mathbf{h}_1^{c_1} \cdots \mathbf{h}_l^{c_l}$) in $\mathrm{U}(\mathfrak{h})$ may be canonically identified with the polynomial function

$$\lambda \mapsto \mathrm{H}_x(\lambda) = \sum_c \xi_c \langle \lambda, \mathbf{h}_1 \rangle^{c_1} \cdots \langle \lambda, \mathbf{h}_l \rangle^{c_l}$$

on the dual $\mathfrak{h}^*$ of $\mathfrak{h}$, with values in $\mathbf{C}$. For each irreducible representation $V_p$ of G on a vector space $\mathrm{E}_p$, with dominant weight $p \in \mathrm{P}(\mathfrak{g})$, the homomorphism $(V_p)_* \otimes 1$: $\mathfrak{g}_{(\mathbf{C})} \to \mathfrak{gl}(\mathrm{E}_p)$ extends to a homomorphism $R_p$: $\mathrm{U}(\mathfrak{g}_{(\mathbf{C})}) \to \mathrm{End}(\mathrm{E}_p)$ of $\mathbf{C}$-algebras. For each $x \in \mathrm{U}$, put $\Xi_p(x) = (\dim \mathrm{E}_p)^{-1} \operatorname{Tr}(R_p(x))$. Show that $\Xi_p(x) = \Xi_p(x^\natural)$. If $\Phi$ is the canonical homomorphism of $\mathrm{U}^0$ onto $\mathrm{U}(\mathfrak{h})$ with kernel $\mathfrak{L}$ (Problem 12(b)), show that $\Xi_p(x) = \mathrm{H}_{\Phi(x^\natural)}(p)$ for all $x \in \mathrm{U}$. (Observe that if $y \in \mathrm{E}_p$ is such that $R_p(\mathbf{u}) \cdot y = p(\mathbf{u})y$ for all $\mathbf{u} \in \mathfrak{h}$, then we have $R_p(z) \cdot y = \mathrm{H}_z(p)y$ for all $z \in \mathrm{Z}$, and $R_p(x) \cdot y = 0$ for all $x \in \mathfrak{L}$.)

**14.** (a) With the same notation, put $\mathrm{D}(\lambda) = \prod_{\alpha \in \mathbf{S}^+} \langle \lambda, \mathbf{h}_\alpha \rangle$ for all $\lambda \in \mathfrak{h}^*$. Show that for each $\mathbf{u} \in \mathfrak{h}$ and each $\lambda \in \mathfrak{h}^*$, the series $\sum_{n=0}^{\infty} (1/n!)\mathrm{H}_{\mathbf{u}^n}(\lambda)$ is absolutely convergent in $\mathbf{C}$, and that

$$\mathrm{D}(\lambda + \delta)\left(\sum_{n=0}^{\infty} \frac{1}{n!} \mathrm{H}_{\mathbf{u}^n}(\lambda)\right)\left(\sum_{w \in \mathrm{W}} \det(w) e^{\langle w \cdot (\lambda+\delta), \mathbf{u} \rangle}\right) = \mathrm{D}(\delta) \sum_{w \in \mathrm{W}} \det(w) e^{\langle w \cdot \delta, \mathbf{u} \rangle}.$$

(Use Weyl's formulas to show that the formula is true for all $\lambda = p \in \mathrm{P}(\mathfrak{g})$, by using **(21.13.6.1)** and the power-series expansion of $e^{\langle p, \mathbf{u} \rangle}$.)
(b) Deduce from this formula that, for each integer $n \geqq 0$ and each $\mathbf{u} \in \mathfrak{h}$, the rational function on $\mathfrak{h}^*$

$$\text{(1)} \qquad \lambda \mapsto \frac{1}{\mathrm{D}(\lambda)} \sum_{w \in \mathrm{W}} \det(w) \langle w \cdot \lambda, \mathbf{u} \rangle^n$$

is in fact a polynomial function, and is a linear combination of the polynomial functions $\lambda \mapsto \mathrm{H}_{\mathbf{u}^q}(\lambda - \delta)$ for $0 \leqq q \leqq n$. Deduce from the same formula that each of the polynomial functions $\lambda \mapsto \mathrm{H}_{\mathbf{u}^n}(\lambda - \delta)$ is invariant under the action of W on $\mathfrak{h}^*$. Consequently $\lambda \mapsto \mathrm{H}_x(\lambda - \delta)$ is invariant under the action of W on $\mathfrak{h}^*$, for each $x \in \mathrm{U}(\mathfrak{h})$.
(c) Show that $\mathrm{U}(\mathfrak{h}) + [\mathrm{U}, \mathrm{U}] = \mathrm{U}$. (Consider as in Problem 13 the representation $s \mapsto \mathrm{Ad}(s) | \mathrm{U}^{(r)}$ and show, by considering the derived homomorphism, that the image of $\mathrm{U}(\mathfrak{h}) \cap \mathrm{U}^{(r)}$ under $\mathrm{Ad}(s)$ is contained in $[\mathrm{U}, \mathrm{U}] \cap \mathrm{U}^{(r)}$. Next, using the conjugacy of the maximal commutative subalgebras of $\mathfrak{g}$, show that for each $\mathbf{v} \in \mathfrak{g}$ there exists $s \in \mathrm{G}$ such that $\mathrm{Ad}(s) \cdot \mathbf{v}^m \in \mathrm{U}(\mathfrak{h})$ for all $m$, and deduce that $\mathrm{U}(\mathfrak{h}) + [\mathrm{U}, \mathrm{U}]$ contains the vector subspace V of U spanned by the $\mathbf{v}^m$ for $\mathbf{v} \in \mathfrak{g}$ and $m \geqq 0$. Finally, prove that this vector subspace is the whole of U, by showing by induction on $r$ that it contains all products $\mathbf{v}_1 \mathbf{v}_2 \cdots \mathbf{v}_r$ of $r$ elements $\mathbf{v}_j \in \mathfrak{g}$. For this purpose, observe that for each permutation $\pi \in \mathfrak{S}_r$ the difference $\mathbf{v}_{\pi(1)} \mathbf{v}_{\pi(2)} \cdots \mathbf{v}_{\pi(r)} - \mathbf{v}_1 \mathbf{v}_2 \cdots \mathbf{v}_r$ belongs to $\mathrm{U}^{(r-1)}$, and that $(\xi_1 \mathbf{v}_1 + \cdots + \xi_r \mathbf{v}_r)^r \in \mathrm{V} \cap \mathrm{U}^{(r)}$ for all systems of scalars $\xi_1, \ldots, \xi_r$.)
(d) Deduce from (c) that if we put $\mathrm{H}_x^0(\lambda) = \mathrm{H}_x(\lambda - \delta)$ for all $x \in \mathrm{U}(\mathfrak{h})$, the mapping $z \mapsto \mathrm{H}_{\Phi(z)}^0$ is a *surjective* homomorphism of the center Z of $\mathrm{U}(\mathfrak{g})$ onto the algebra $\mathrm{I}(\mathfrak{h}, \mathrm{W})$ of polynomial functions on $\mathfrak{h}^*$ that are *invariant* under the action of W. (Use the fact that each function in $\mathrm{I}(\mathfrak{h}, \mathbf{W})$ is a linear combination of polynomial functions of the form (1).)

**15.** (a) With the same notation, there exists a canonical homomorphism

$$\psi \colon \mathbf{T}(\mathfrak{g}_{(\mathbf{C})}) \to \mathrm{U}(\mathfrak{g}_{(\mathbf{C})}) = \mathrm{U},$$

of the tensor algebra of $\mathfrak{g}_{(C)}$ onto the algebra U, which is the identity on $\mathfrak{g}_{(C)}$ and maps $\mathbf{x} \otimes \mathbf{y} - \mathbf{y} \otimes \mathbf{x}$ to $[\mathbf{x}, \mathbf{y}]$ for all $\mathbf{x}, \mathbf{y} \in \mathfrak{g}_{(C)}$. Show that the restriction of $\psi$ to the space $\mathbf{S}_r(\mathfrak{g}_{(C)})$ of symmetric tensors of order $r$ on $\mathfrak{g}_{(C)}$ is an isomorphism of this vector space onto a supplement of $U^{(r-1)}$ in $U^{(r)}$, and deduce that the restriction of $\psi$ to the symmetric algebra $\mathbf{S}(\mathfrak{g}_{(C)})$ (A.17.5) is an isomorphism of *vector spaces* (but not of algebras) of $\mathbf{S}(\mathfrak{g}_{(C)})$ onto U. For each $s \in G$, the automorphism Ad($s$) of $\mathfrak{g}$ extends to an algebra automorphism, also denoted by Ad($s$), of $\mathbf{S}(\mathfrak{g}_{(C)})$. For each $z \in \mathbf{S}(\mathfrak{g}_{(C)})$, we have $\psi(\mathrm{Ad}(s) \cdot z) = \mathrm{Ad}(s) \cdot \psi(z)$. Let $I(\mathfrak{g}_{(C)}, G)$ denote the set of elements of $\mathbf{S}(\mathfrak{g}_{(C)})$ that are invariant under Ad($s$) for all $s \in G$. Show that the restriction of $\psi$ to $I(\mathfrak{g}_{(C)}, G)$ is a vector space isomorphism of $I(\mathfrak{g}_{(C)}, G)$ onto the *center* Z of U, but is not necessarily an isomorphism of algebras (take $G = \mathbf{SU}(2)$).

(b) For each linear form $\mathbf{u}^*$ on the vector space $\mathfrak{g}_{(C)}$, let $\beta(\mathbf{u}^*)$ be the element of $\mathfrak{g}_{(C)}$ such that $B_{\mathfrak{g}_{(C)}}(\beta(\mathbf{u}^*), \mathbf{v}) = \langle \mathbf{u}^*, \mathbf{v} \rangle$ for all $\mathbf{v} \in \mathfrak{g}_{(C)}$, so that $\beta$ is an isomorphism of the vector space $\mathfrak{g}^*_{(C)}$ (the dual of $\mathfrak{g}_{(C)}$) onto $\mathfrak{g}_{(C)}$. Show that the annihilator in $\mathfrak{g}^*_{(C)}$ of the subspace $\mathfrak{n}_+ + \mathfrak{n}_-$ of $\mathfrak{g}_{(C)}$ is mapped onto $\mathfrak{h}$ by $\beta$, and may therefore be identified with the dual $\mathfrak{h}^*$ of $\mathfrak{h}$. The isomorphism $\beta$ extends uniquely to an isomorphism (also denoted by $\beta$) of the symmetric algebra $\mathbf{S}(\mathfrak{g}^*_{(C)})$ onto $\mathbf{S}(\mathfrak{g}_{(C)})$, which transforms $\mathbf{S}(\mathfrak{h}^*)$ into $\mathbf{S}(\mathfrak{h}) = U(\mathfrak{h})$. If $\mathfrak{J}$ is the ideal of $\mathbf{S}(\mathfrak{g}_{(C)})$ generated by $\mathfrak{n}_+ + \mathfrak{n}_-$, we have $\mathbf{S}(\mathfrak{g}_{(C)}) = U(\mathfrak{h}) \oplus \mathfrak{J}$. If $j$ is the homomorphism $\mathbf{S}(\mathfrak{g}_{(C)}) \to U(\mathfrak{h})$ defined by this decomposition, then $\beta^{-1}(\mathfrak{J})$ is the kernel of the restriction homomorphism $i$: $\mathbf{S}(\mathfrak{g}^*_{(C)}) \to \mathbf{S}(\mathfrak{h}^*)$ obtained by considering $\mathbf{S}(\mathfrak{g}^*_{(C)})$ (resp. $\mathbf{S}(\mathfrak{h}^*)$) as the algebra of polynomial functions on $\mathfrak{g}_{(C)}$ (resp. $\mathfrak{h}$), and we have $i = \beta^{-1} \circ j \circ \beta$. For each $s \in G$, Ad($s$) acts on $\mathfrak{g}^*_{(C)}$ and extends to an algebra automorphism of $\mathbf{S}(\mathfrak{g}^*_{(C)})$, again denoted by Ad($s$). If $I(\mathfrak{g}^*_{(C)}, G)$ is the subalgebra of $\mathbf{S}(\mathfrak{g}^*_{(C)})$ consisting of the elements that are invariant under Ad($s$) for all $s \in G$, then the image of $I(\mathfrak{g}^*_{(C)}, G)$ under $\beta$ is $I(\mathfrak{g}_{(C)}, G)$. Likewise, if $I(\mathfrak{h}^*, W)$ is the subalgebra of $\mathbf{S}(\mathfrak{h}^*)$ consisting of the elements that are invariant under the action of W on $\mathfrak{h}^*$, then the image of $I(\mathfrak{h}^*, W)$ under $\beta$ is $I(\mathfrak{h}, W) \subset U(\mathfrak{h})$.

(c) For each linear representation $V$ of G on a complex vector space, the polynomial functions $\mathbf{u} \mapsto \mathrm{Tr}((V_*(\mathbf{u}))^n)$ on $\mathfrak{g}_{(C)}$ belong to $I(\mathfrak{g}^*_{(C)}, G)$, and their restrictions to $\mathfrak{h}$ belong to $I(\mathfrak{h}^*, W)$. Show conversely that every polynomial function in $I(\mathfrak{h}^*, W)$ is a linear combination of these restrictions. (Use the fact that the weights $p \in P(\mathfrak{g})$ span the vector space $\mathfrak{h}^*$, and the isomorphism $\mathbf{Z}^{(R(G))} \to \mathbf{Z}[P]^W$ of (21.15.5).)

(d) Show that the restriction to $I(\mathfrak{g}^*_{(C)}, G)$ of the homomorphism $i$ in (b) above is an isomorphism of this algebra onto $I(\mathfrak{h}^*, W)$. (To show that $i$ is injective, note that if $i(f) = 0$ for $f \in I(\mathfrak{g}^*_{(C)}, G)$, then $f = 0$ on $\mathfrak{h}$ and also on $\mathrm{Ad}(s) \cdot \mathfrak{h}$ for each $s \in G$, and use the conjugacy theorem. To show that $i$ is surjective, note that if L is the set of all linear combinations of the polynomial functions $\mathbf{u} \mapsto \mathrm{Tr}((V_*(\mathbf{u})^n))$ for all linear representations $V$ of G, then $i(L) = I(\mathfrak{h}^*, W)$ by virtue of (c) above.) Deduce that the homomorphism

$$j: I(\mathfrak{g}_{(C)}, G) \to I(\mathfrak{h}, W)$$

defined in (b) above is bijective.

(e) Show that the composite isomorphism

$$Z \xrightarrow{\psi^{-1}} I(\mathfrak{g}_{(C)}, G) \xrightarrow{j} I(\mathfrak{h}, W)$$

is the same as the isomorphism $z \mapsto H^0_{\Phi(z)}$ defined in Problem 14(d). (Show that for each $r$ these two homomorphisms define the same mapping of $(Z \cap U^{(r)})/(Z \cap U^{(r-1)})$ onto $(I(\mathfrak{h}, W) \cap \mathbf{S}_r(\mathfrak{h}))/(I(\mathfrak{h}, W) \cap \mathbf{S}_{r-1}(\mathfrak{h}))$, and that this mapping is bijective; for this purpose, use the basis $(e_{a,b,c})$ of U defined in (21.16.3.4).)

16. Let E be a complex vector space of dimension $n$. If we identify the symmetric algebra $\mathbf{S}(E^*)$ with the algebra of polynomial functions on E, the group $\mathbf{GL}(E)$ acts on $\mathbf{S}(E^*)$ by

the rule $(s \cdot \mathrm{P})(x) = \mathrm{P}(s^{-1} \cdot x)$ for all $\mathrm{P} \in \mathbf{S}(\mathrm{E}^*)$, $s \in \mathbf{GL}(\mathrm{E})$, and $x \in \mathrm{E}$. If G is a subgroup of $\mathbf{GL}(\mathrm{E})$, denote by $\mathrm{I}(\mathrm{E}, \mathrm{G})$ the subalgebra of $\mathbf{S}(\mathrm{E}^*)$ consisting of the G-invariant polynomial functions P, i.e., those for which $s \cdot \mathrm{P} = \mathrm{P}$ for all $s \in \mathrm{G}$. If G is finite, with order N, then for each $\mathrm{P} \in \mathbf{S}(\mathrm{E}^*)$ the polynomial function $\mu(\mathrm{P}) = \mathrm{N}^{-1} \sum_{s \in \mathrm{G}} s \cdot \mathrm{P}$ is G-invariant. We have $\mu(\mathrm{P}) = \mathrm{P}$ if $\mathrm{P} \in \mathrm{I}(\mathrm{E}, \mathrm{G})$.

(a) Let $\mathrm{P} = \mathrm{P}_0 + \mathrm{P}_1 + \cdots + \mathrm{P}_h$ be a polynomial function on E that is invariant under a subgroup G of $\mathbf{GL}(\mathrm{E})$, where $\mathrm{P}_j$ is the homogeneous component of degree $j$ in P. Show that each $\mathrm{P}_j$ is G-invariant (consider $\mathrm{P}(tx)$ for $t \in \mathbf{C}$).

(b) Suppose that G is finite. A rational function R on E is said to be G-invariant if $\mathrm{R}(s^{-1} \cdot x) = \mathrm{R}(x)$ whenever both sides are defined. Show that R is then of the form P/Q, where P and Q are G-invariant polynomial functions.

(c) The ring $\mathrm{I}(\mathrm{E}, \mathrm{G}) = \mathbf{C} \oplus \mathrm{J}_1 \oplus \cdots \oplus \mathrm{J}_h \oplus \cdots$ is graded, $\mathrm{J}_h$ being the vector space spanned by the $\mathrm{P} \in \mathrm{I}(\mathrm{E}, \mathrm{G})$, which are homogeneous of degree $h$ (by virtue of (a) above); we define $\mathrm{J}_0$ to be $\mathbf{C}$. Let $\mathfrak{J}_+ = \mathrm{J}_1 \oplus \cdots \oplus \mathrm{J}_h \oplus \cdots$, which is an ideal of $\mathrm{I}(\mathrm{E}, \mathrm{G})$, and let $\mathfrak{N}$ be the graded ideal of $\mathbf{S}(\mathrm{E}^*)$ generated by $\mathfrak{J}_+$.

Suppose from now on that $\mathrm{G} \subset \mathbf{GL}(\mathrm{E})$ is a *finite* group, generated by orthogonal *reflections* $r_1, \ldots, r_m$ in hyperplanes in E (relative to a scalar product on E). In order that a homogeneous polynomial function P of degree $> 0$ should belong to $\mathfrak{N}$, it is necessary and sufficient that $r_j \cdot \mathrm{P} - \mathrm{P} \in \mathfrak{N}$ for $1 \leqq j \leqq m$. (Observe that this condition implies that $s \cdot \mathrm{P} - \mathrm{P} \in \mathfrak{N}$ for all $s \in \mathrm{G}$, and that $\mu(\mathrm{P}) \in \mathfrak{N}$.)

(d) Let $\mathrm{U}_1, \mathrm{U}_2, \ldots, \mathrm{U}_p$ be elements of $\mathfrak{J}_+$ such that $\mathrm{U}_1$ does not belong to the ideal in $\mathrm{I}(\mathrm{E}, \mathrm{G})$ generated by $\mathrm{U}_2, \ldots, \mathrm{U}_p$. Let $\mathrm{P}_1, \mathrm{P}_2, \ldots, \mathrm{P}_p$ be homogeneous elements of $\mathbf{S}(\mathrm{E}^*)$ such that $\mathrm{P}_1\mathrm{U}_1 + \mathrm{P}_2\mathrm{U}_2 + \cdots + \mathrm{P}_p\mathrm{U}_p = 0$. Show that $\mathrm{P}_1 \in \mathfrak{N}$. (Proceed by induction on the degree of $\mathrm{P}_1$. If $\mathrm{P}_1$ is a constant, observe that $\sum_j \mu(\mathrm{P}_j)\mathrm{U}_j = 0$, and deduce that $\mathrm{P}_1 = 0$. In general, show that $r_j \cdot \mathrm{P}_1 - \mathrm{P}_1 \in \mathfrak{N}$ for $1 \leqq j \leqq m$, by observing that there is a linear form $\mathrm{L}_j \neq 0$ such that $r_j \cdot \mathrm{P} - \mathrm{P}$ is divisible by $\mathrm{L}_j$, for all $\mathrm{P} \in \mathbf{S}(\mathrm{E}^*)$.)

(e) Let $(\mathrm{I}_1, \mathrm{I}_2, \ldots, \mathrm{I}_q)$ be a *minimal* system of generators of the ideal $\mathfrak{N}$, consisting of homogeneous invariant polynomial functions. Let $d_k > 0$ be the degree of $\mathrm{I}_k$. Show that $\mathrm{I}_1, \ldots, \mathrm{I}_q$ are algebraically independent over the field $\mathbf{C}$. (Suppose not, and let $\mathrm{H}(\mathrm{Y}_1, \ldots, \mathrm{Y}_q) \in \mathbf{C}[\mathrm{Y}_1, \ldots, \mathrm{Y}_q]$ be a nonzero polynomial of *smallest degree* such that $\mathrm{H}(\mathrm{I}_1, \ldots, \mathrm{I}_q) = 0$; we may also assume that all the monomials $\mathrm{Y}_1^{\nu_1} \cdots \mathrm{Y}_q^{\nu_q}$ appearing in H are such that $\nu_1 d_1 + \cdots + \nu_q d_q$ has the same value. Then the invariants $\mathrm{H}_k = (\partial\mathrm{H}/\partial\mathrm{Y}_k)(\mathrm{I}_1, \ldots, \mathrm{I}_q)$ are not all zero. If $\mathfrak{U}$ is the ideal that they generate in $\mathrm{I}(\mathrm{E}, \mathrm{G})$, we may assume that $\mathrm{H}_1, \ldots, \mathrm{H}_s$ is a minimal set of generators of $\mathfrak{U}$, so that we have equations $\mathrm{H}_{s+j} = \sum_{k=1}^{s} \mathrm{V}_{jk}\mathrm{H}_k$, where the $\mathrm{V}_{jk}$ are homogeneous $(1 \leqq j \leqq q - s)$ and belong to $\mathrm{I}(\mathrm{E}, \mathrm{G})$. Identifying the functions in $\mathbf{S}(\mathrm{E}^*)$ with polynomials in the coordinates $x_h$ $(1 \leqq h \leqq n)$ of a point $x \in \mathrm{E}$, show that we have

$$\frac{\partial \mathrm{I}_k}{\partial x_h} + \sum_{j=1}^{q-s} \mathrm{V}_{jk} \frac{\partial \mathrm{I}_{s+j}}{\partial x_h} \in \mathfrak{N}$$

for $1 \leqq h \leqq n$ and $1 \leqq k \leqq s$ (use (d) above). Use Euler's identity to deduce that for $1 \leqq k \leqq s$ we have

$$d_k \mathrm{I}_k + \sum_{j=1}^{q-s} d_{s+j} \mathrm{V}_{jk} \mathrm{I}_{s+k} = \sum_{l=1}^{q} \mathrm{W}_{jl} \mathrm{I}_l$$

where the $\mathrm{W}_{jl}$ are homogeneous of degree $> 0$, and all the polynomials $\mathrm{I}_k$, $\mathrm{V}_{jk}\mathrm{I}_{s+k}$, and $\mathrm{W}_{jl}\mathrm{I}_l$ are *homogeneous of the same degree*; this implies in particular that $\mathrm{W}_{jk} = 0$ and leads to a contradiction.)

(f) Show that 1 and the $I_k$ $(1 \leqq k \leqq q)$ generate the algebra I(E, G). (For each element $P \in J_h$ show, by induction on $h$, that P is a polynomial in $I_1, \ldots, I_q$, by expressing that $P \in \mathfrak{N}$ and $P = \mu(P)$.) By virtue of (e) above, we have $q \leqq n$. Prove that $q \geqq n$ by noting that if L is the field of fractions of $\mathbf{S}(E^*)$, on which G acts, then the field $K \subset L$ of G-invariant rational functions is the field of fractions of I(E, G) (use (b)), and has the same transcendence degree $n$ over **C** as does L. The algebra I(E, G) is therefore generated by $n$ algebraically independent homogeneous elements (*Chevalley's theorem*).

**17.** Let V be a finite-dimensional complex vector space, $s$ an endomorphism of V, and $s_m$ the canonical extension of $s$ to the $m$th symmetric power $\mathbf{S}_m(V)$. Show that we have

$$\sum_{m=0}^{\infty} \mathrm{Tr}(s_m)T^m = (\det(1 - sT))^{-1}$$

in the formal power series ring **C**[[T]]. (Choose a basis of V with respect to which the matrix of $s$ is triangular.)

**18.** With the notation of Problem 16, suppose that G is a finite group of order N.

(a) Show that the endomorphism $f = N^{-1} \sum_{s \in G} s$ is a projection of V onto the subspace F of V consisting of G-invariant elements. Hence we have $\dim(F) = \mathrm{Tr}(f)$.

(b) Show that

$$\sum_{m=0}^{\infty} (\dim(J_m))T^m = N^{-1} \sum_{s \in G} (\det(1 - sT))^{-1}$$

in the formal power series ring **C**[[T]]. (Apply (a) to each of the spaces $\mathbf{S}_m(V^*)$, and use Problem 17.)

(c) Suppose from now on that G is generated by orthogonal reflections, and let R be the set of orthogonal reflections belonging to G. We have $s \in R$ if and only if $\det(1 - sT)$ is divisible by $(1 - T)^{n-1}$ but not by $(1 - T)^n$. Use Problem 16(f) to show that

$$\sum_{m=0}^{\infty} (\dim(J_m))T^m = \prod_{k=1}^{n} (1 - T^{d_k})^{-1}.$$

(d) If $r$ is the number of elements in R, show that

$$\prod_{k=1}^{n} d_k = N, \qquad \sum_{k=1}^{n} (d_k - 1) = r.$$

(Equate the constant terms and the coefficients of T on either side of the identity

$$(1 - T)^n \prod_{k=1}^{n} (1 - T^{d_k})^{-1} = N^{-1}(1 - T)^n \sum_{s \in G} (\det(1 - sT))^{-1}.)$$

(e) Let $\lambda_j(x) = 0$ $(1 \leqq j \leqq r)$ be the equations of the hyperplanes of fixed points of the reflections belonging to G. Show that the Jacobian $\partial(I_1, \ldots, I_n)/\partial(x_1, \ldots, x_n)$ is a polynomial proportional to $\prod_{j=1}^{r} \lambda_j(x)$. (Observe that both polynomials have the same degree, by (d) above, and that the mapping $(x_1, \ldots, x_n) \mapsto (I_1(x_1, \ldots, x_n), \ldots, I_n(x_1, \ldots, x_n))$ of $\mathbf{C}^n$ into itself is not invertible at any point of any of the hyperplanes $\lambda_j(x) = 0$ $(1 \leqq j \leqq r)$.)

**19.** With the notation of Problem 15, suppose that G is *almost simple*, and let $I_1, \ldots, I_l$ be algebraically independent homogeneous polynomial functions on $\mathfrak{h}$, of degrees $d_1, \ldots, d_l$, respectively, which generate the algebra $\mathrm{I}(\mathfrak{h}, \mathrm{W})$ (Problem 16(f)). Take a basis of $\mathfrak{h}$ consisting of eigenvectors $\mathbf{e}_j$ ($1 \leqq j \leqq l$) of a Coxeter element $\sigma$ (which is diagonalizable), so that $\sigma \cdot \mathbf{e}_j = \exp(2\pi i m_j/h)\mathbf{e}_j$ in the notation of Section **21.11**, Problem 14. Use Problem 18(e) and Section **21.11**, Problem 16(c) to show that, for the chosen basis of $\mathfrak{h}$, there exists a permutation $\pi \in \mathfrak{S}_l$ such that $(\partial \mathrm{I}_j/\partial x_{\pi(j)})(1, 0, \ldots, 0) \neq 0$, and hence that in $\mathrm{I}_j$ the monomial $x_1^{d_j - 1} x_{\pi(j)}$ appears with nonzero coefficient. By expressing that $\mathrm{I}_j$ is invariant under $\sigma$, deduce that we have $d_j - 1 + m_{\pi(j)} \equiv 0 \pmod{h}$. Using the relation $m_j + m_{l+1-j} = h$, deduce that by renumbering the $\mathrm{I}_j$ we may assume that we have $d_j - 1 \equiv m_j \pmod{h}$; and since $d_j - 1 \geqq 0$ and $m_j < h$, we have $d_j - 1 = m_j + \mu_j h$ with $\mu_j$ an integer $\geqq 0$. Finally, use the relation $m_1 + m_2 + \cdots + m_l = \frac{1}{2}lh$, the fact that the number of roots is $lh$ (Section **21.11**, Problem 16(b)), and Problem 18(d) to show that $d_j = m_j + 1$ for $1 \leqq j \leqq l$.

**20.** (a) When $\mathrm{G} = \mathbf{SU}(n)$, show that for the basis of $\mathfrak{u}(n)$ defined in (**21.12.1**) the polynomial functions $\mathrm{I}_1, \ldots, \mathrm{I}_{n-1}$ generating the algebra $\mathrm{I}(\mathfrak{h}, \mathrm{W})$ may be taken to be the restrictions to $\mathfrak{h} \subset \mathfrak{t}_{(\mathbf{C})}$ of the elementary symmetric functions

$$s_j(x_1, \ldots, x_n) = \sum_{\pi \in \mathfrak{S}_n} x_{\pi(1)} x_{\pi(2)} \cdots x_{\pi(j)}$$

for $2 \leqq j \leqq n$, so that the $m_j$ are the numbers $1, 2, \ldots, n-1$.
(b) When $\mathrm{G} = \mathbf{U}(n, \mathbf{H})$ or $\mathrm{G} = \mathbf{SO}(2n+1)$, the rings $\mathrm{I}(\mathfrak{h}, \mathrm{W})$ corresponding to these two groups are isomorphic. For the bases of $\mathfrak{t}$ given in (**21.12.2**) and (**21.12.4**), the polynomial functions $\mathrm{I}_1, \ldots, \mathrm{I}_n$ that generate $\mathrm{I}(\mathfrak{h}, \mathrm{W})$ may be taken to be the elementary symmetric functions

$$s_j(x_1^2, \ldots, x_n^2) = \sum_{\pi \in \mathfrak{S}_n} x_{\pi(1)}^2 x_{\pi(2)}^2 \cdots x_{\pi(j)}^2$$

for $1 \leqq j \leqq n$, so that the $m_j$ are the numbers $1, 3, 5, \ldots, 2n-1$.
(c) When $\mathrm{G} = \mathbf{SO}(2n)$, for the basis of $\mathfrak{t}$ defined in (**21.12.3**), the polynomial functions $\mathrm{I}_1, \ldots, \mathrm{I}_n$ that generate $\mathrm{I}(\mathfrak{h}, \mathrm{W})$ may be taken to be the elementary symmetric functions $s_j(x_1^2, \ldots, x_n^2)$ for $1 \leqq j \leqq n-1$, and the function $x_1 x_2 \cdots x_n$. The $m_j$ are the numbers 1, 3, ..., $2n-3$, and $n-1$.

**21.** Let G be a denumerable group of displacements in a real vector space E endowed with a scalar product, and let $\mu$ be a positive G-invariant measure on E (for example, Lebesgue measure if E is identified with $\mathbf{R}^n$, with the usual scalar product).
(a) Let U, U′ be two open subsets of E, of finite measure. Suppose that the sets $s \cdot \mathrm{U}$ (resp. $s \cdot \mathrm{U}'$) are pairwise disjoint for all $s \in \mathrm{G}$, and that the complement of their union is $\mu$-negligible. Show that $\mu(\mathrm{U}) = \mu(\mathrm{U}')$. (If V (resp. V′) is the union of the sets $s \cdot \mathrm{U}$ (resp. $s \cdot \mathrm{U}'$) for $s \in \mathrm{G}$, then $\mathrm{V} \cap \mathrm{V}'$ is G-stable and has a negligible complement, and $\mathrm{U} \cap \mathrm{V}'$ and $\mathrm{U}' \cap \mathrm{V}$ are two G-tessellations of $\mathrm{V} \cap \mathrm{V}'$ (Section **14.1**, Problem 6).)
(b) Let $\mathrm{G}_0$ be a subgroup of G, and suppose that there exists an open subset $\mathrm{U}_0$ of E, of finite measure, such that the $t \cdot \mathrm{U}_0$ for $t \in \mathrm{G}_0$ are pairwise disjoint and the complement of their union is negligible. Then the index $(\mathrm{G} : \mathrm{G}_0)$ is finite and equal to $\mu(\mathrm{U}_0)/\mu(\mathrm{U})$ (Section **14.1**, Problem 6(d)).

**22.** The hypotheses and notation are as in Section **21.15**, Problem 11. Apply the results of Problem 21 to the group $\mathrm{W}_a$ and the open set $\mathrm{A}^*$, to the group $\mathrm{P}_0$ and the open parallelotope constructed on the vectors $2\pi\mathbf{h}_j$ ($1 \leqq j \leqq l$), and finally to the subgroup $\mathrm{Q}_0$ of $\mathrm{P}_0$ generated by the translations $\mathbf{u} \mapsto \mathbf{u} + 2\pi\mathbf{p}_j$ and the parallelotope constructed on the vectors $2\pi\mathbf{p}_j$ ($1 \leqq j \leqq l$). Deduce that the order of the Weyl group W is $l!\, n_1 n_2 \cdots n_l f$, where $f - 1$ is the number of indices $j$ such that $n_j = 1$.

## 17. COMPLEXIFICATIONS OF COMPACT CONNECTED SEMISIMPLE GROUPS

(21.17.1) Let $\tilde{K}$ be a *simply connected* compact semisimple Lie group, $K = Ad(\tilde{K})$ its adjoint group, the quotient $\tilde{K}/C$ of $\tilde{K}$ by its (finite) center, and $\mathfrak{k}$ the Lie algebra of $\tilde{K}$ and K. The group K is the *identity component* of the closed subgroup $Aut(\mathfrak{k})$ of $\mathbf{GL}(\mathfrak{k})$ (21.6.9). The complexification $\mathfrak{g} = \mathfrak{k}_{(\mathbf{C})} = \mathfrak{k} \oplus i\mathfrak{k}$ of $\mathfrak{k}$ is a *complex semisimple Lie algebra* (21.6.1). We shall denote by $c$ the semilinear bijection of $\mathfrak{g}$ onto itself defined by $c(\mathbf{y} + i\mathbf{z}) = \mathbf{y} - i\mathbf{z}$ for $\mathbf{y}, \mathbf{z} \in \mathfrak{k}$, so that $c^2 = 1_{\mathfrak{g}}$. It is immediately verified that

$$c([\mathbf{u}, \mathbf{v}]) = [c(\mathbf{u}), c(\mathbf{v})] \tag{21.17.1.1}$$

for all $\mathbf{u}, \mathbf{v} \in \mathfrak{g}$. The Lie subalgebra $\mathfrak{k}$ of $\mathfrak{g}_{|\mathbf{R}}$ is the set of all $\mathbf{u} \in \mathfrak{g}$ such that $c(\mathbf{u}) = \mathbf{u}$.

We propose to describe (up to isomorphism) the complex connected semisimple Lie groups having $\mathfrak{g}$ as Lie algebra. If $\tilde{G}$ is the *simply connected* complex group with $\mathfrak{g}$ as Lie algebra (19.11.9), then the adjoint group $G = Ad(\tilde{G})$, the quotient of $\tilde{G}$ by its (discrete) center, may be identified with the identity component of the closed subgroup $Aut(\mathfrak{g})$ of $\mathbf{GL}(\mathfrak{g})$ (21.6.8), and its center consists only of $e$ (20.22.5.1). We shall first study the group $Aut(\mathfrak{g})$, whose Lie algebra is the image $ad(\mathfrak{g})$ of $\mathfrak{g}$ under the homomorphism $\mathbf{u} \mapsto ad(\mathbf{u})$, a Lie subalgebra of $\mathfrak{gl}(\mathfrak{g}) = End_{\mathbf{C}}(\mathfrak{g})$, isomorphic to $\mathfrak{g}$ (21.6.3). Since every automorphism $u$ of $\mathfrak{k}$ extends uniquely to an automorphism $u \otimes 1_{\mathbf{C}}$ of $\mathfrak{g}$, the group $Aut(\mathfrak{k})$ may be identified with the subgroup of $Aut(\mathfrak{g})$ consisting of the automorphisms that leave $\mathfrak{k}$ globally invariant.

(21.17.2) The Killing form $B_{\mathfrak{k}}$ of $\mathfrak{k}$ is the restriction to $\mathfrak{k} \times \mathfrak{k}$ of the Killing form $B_{\mathfrak{g}}$ of $\mathfrak{g}$ (21.6.1), and $B_{\mathfrak{k}}$ is *negative definite*. It follows that the mapping

$$(\mathbf{u}, \mathbf{v}) \mapsto -B_{\mathfrak{g}}(\mathbf{u}, c(\mathbf{v})) \tag{21.17.2.1}$$

is a *scalar product*, which gives $\mathfrak{g}$ the structure of a finite-dimensional Hilbert space. For if $\mathbf{u} = \mathbf{y} + i\mathbf{z}$ and $\mathbf{u}' = \mathbf{y}' + i\mathbf{z}'$ with $\mathbf{y}, \mathbf{z}, \mathbf{y}', \mathbf{z}' \in \mathfrak{k}$, we have

$$\begin{aligned} B_{\mathfrak{g}}(\mathbf{u}, c(\mathbf{u}')) &= B_{\mathfrak{k}}(\mathbf{y}, \mathbf{y}') + B_{\mathfrak{k}}(\mathbf{z}, \mathbf{z}') + i(B_{\mathfrak{k}}(\mathbf{y}', \mathbf{z}) - B_{\mathfrak{k}}(\mathbf{y}, \mathbf{z}')) \\ &= B_{\mathfrak{g}}(\mathbf{u}', c(\mathbf{u})) \end{aligned}$$

and $B_{\mathfrak{g}}(\mathbf{u}, c(\mathbf{u})) = B_{\mathfrak{k}}(\mathbf{y}, \mathbf{y}) + B_{\mathfrak{k}}(\mathbf{z}, \mathbf{z})$, which vanishes only if $\mathbf{y} = \mathbf{z} = 0$, i.e., $\mathbf{u} = 0$. For each endomorphism $V$ of the complex vector space $\mathfrak{g}$, let $V^*$ denote the endomorphism *adjoint* to $V$, relative to this Hilbert space structure on $\mathfrak{g}$ (11.5.1). Relative to any orthonormal basis of $\mathfrak{g}$, the matrix of $V^*$ is the conjugate transpose of the matrix of $V$.

(21.17.3) (i) *For each automorphism* $U \in \mathrm{Aut}(\mathfrak{g})$, *we have*

(21.17.3.1)
$$U^* = c \circ U^{-1} \circ c.$$

(ii) *For each* $\mathbf{u} \in \mathfrak{g}$ *we have*

(21.17.3.2)
$$(\mathrm{ad}(\mathbf{u}))^* = -\mathrm{ad}(c(\mathbf{u})).$$

(i) For $\mathbf{x}, \mathbf{y} \in \mathfrak{g}$ we have

$$\mathrm{B}_{\mathfrak{g}}(U \cdot \mathbf{x}, c(\mathbf{y})) = \mathrm{B}_{\mathfrak{g}}(\mathbf{x}, U^{-1} \cdot c(\mathbf{y})) = \mathrm{B}_{\mathfrak{g}}(\mathbf{x}, c((c \circ U^{-1} \circ c) \cdot \mathbf{y}))$$

since $\mathrm{B}_{\mathfrak{g}}$ is invariant under $U$ (21.5.6.2). This proves (21.17.3.1).

(ii) By virtue of (21.5.6.1), we have

$$\mathrm{B}_{\mathfrak{g}}(\mathrm{ad}(\mathbf{u}) \cdot \mathbf{x}, c(\mathbf{y})) + \mathrm{B}_{\mathfrak{g}}(\mathbf{x}, c((c \circ \mathrm{ad}(\mathbf{u}) \circ c) \cdot \mathbf{y})) = 0.$$

But it follows from (21.17.1.1) and the fact that $c^2 = 1$, that $c \circ \mathrm{ad}(\mathbf{u}) \circ c = \mathrm{ad}(c(\mathbf{u}))$. This proves (21.17.3.2)

(21.17.4) *An automorphism* $U \in \mathrm{Aut}(\mathfrak{g})$ *is unitary* (relative to the Hilbert space structure of $\mathfrak{g}$) *if and only if* $U \in \mathrm{Aut}(\mathfrak{k})$.

By virtue of (21.17.3.1), to say that $U$ is unitary signifies that $U^{-1}$ commutes with $c$, or again that $U$ commutes with $c$; this implies that $U$ leaves invariant the subspace $\mathfrak{k}$ of fixed points of $c$. The converse is obvious.

(21.17.5) We shall now characterize those automorphisms of $\mathfrak{g}$ that are *positive* and *self-adjoint* relative to the Hilbert space structure on $\mathfrak{g}$. For that purpose we shall first examine, from the viewpoint of the theory of Lie groups, the decomposition of an endomorphism of a complex vector space as the product of a self-adjoint operator and a unitary operator (cf. Section 11.5, Problem 15). Let E be a complex Hilbert space of finite dimension $n$, and let $\mathfrak{a}(\mathrm{E}) \subset \mathfrak{gl}(\mathrm{E}) = \mathrm{End}_{\mathbf{C}}(\mathrm{E})$ be the set of self-adjoint endomorphisms of E; it is a real vector space, which may be identified (Section 11.5) with the space $\mathscr{H}(\mathrm{E})$ of Hermitian forms on $\mathrm{E} \times \mathrm{E}$, under the mapping that replaces each $H \in \mathfrak{a}(\mathrm{E})$ by the Hermitian form $(x, y) \mapsto (H \cdot x | y)$. Under this mapping, the set $\mathfrak{a}_+(\mathrm{E})$ of *positive self-adjoint endomorphisms*, characterized by the relation $(H \cdot x | x) > 0$ for all $x \neq 0$ (or, equivalently, by the condition that their spectra should contain only numbers $> 0$ (11.5.7)), is identified with the subset $\mathscr{H}_{n,0}(\mathrm{E})$ of Hermitian forms of signature $(n, 0)$ on $\mathrm{E} \times \mathrm{E}$, which is an *open* subset of the vector space $\mathscr{H}(\mathrm{E})$ (16.11.3). Also, let $\mathbf{U}(\mathrm{E})$ denote the unitary group (isomorphic to $\mathbf{U}(n, \mathbf{C})$) of the form $\Phi(x, y) = (x \mid y)$, the scalar product on E. Then:

**(21.17.6)** (i) *The mapping $H \mapsto \exp(H)$ is a diffeomorphism of $\mathfrak{a}(\mathrm{E})$ onto the submanifold $\mathfrak{a}_+(\mathrm{E})$ of $\mathbf{GL}(\mathrm{E}) \subset \mathrm{End}(\mathrm{E})$.*

(ii) *The mapping $(H, U) \mapsto \exp(H) \cdot U$ is a diffeomorphism of*

$$\mathfrak{a}(\mathrm{E}) \times \mathbf{U}(\mathrm{E})$$

*onto the Lie group* $\mathbf{GL}(\mathrm{E})$.

(Here exp is the exponential mapping $H \mapsto \sum_{n=0}^{\infty} \frac{1}{n!} H^n$ of the group $\mathbf{GL}(\mathrm{E})$ (19.8.7.2).)

(i) The fact that $H \mapsto \exp(H)$ is a bijection of $\mathfrak{a}(\mathrm{E})$ onto $\mathfrak{a}_+(\mathrm{E})$ is a particular case of (15.11.11), applied to the function $x \mapsto e^x$, which is a homeomorphism of $\mathbf{R}$ onto $\mathbf{R}_+^* = ]0, +\infty[$. To show that $H \mapsto \exp(H)$ is a diffeomorphism of $\mathfrak{a}(\mathrm{E})$ onto $\mathfrak{a}_+(\mathrm{E})$, it is enough to prove that the tangent linear mapping $\mathrm{T}_H(\exp)$ is bijective for all $H \in \mathfrak{a}(\mathrm{E})$ (16.8.8(iv)); by virtue of (19.16.6), this reduces to showing that no nonzero eigenvalue of the endomorphism $\mathrm{ad}(H)$ of $\mathfrak{gl}(\mathrm{E})$ is of the form $2\pi i k$ with $k \in \mathbf{Z}$. Now, relative to a suitably chosen orthonormal basis of E, the matrix of $H$ is a diagonal matrix $(\lambda_1, \lambda_2, \ldots, \lambda_n)$, with $\lambda_j$ *real* (11.5.7), and therefore by (19.4.2.2) we have

$$\mathrm{ad}(H) \cdot E_{jk} = (\lambda_j - \lambda_k) E_{jk} \tag{21.17.6.1}$$

for all the matrix units $E_{jk}$ $(1 \leq j, k \leq n)$. This shows that the eigenvalues of $\mathrm{ad}(H)$ are the real numbers $\lambda_j - \lambda_k$, and completes the proof of (i).

(ii) The relation $X = \exp(H) \cdot U$, where $H \in \mathfrak{a}(\mathrm{E})$ and $U \in \mathbf{U}(\mathrm{E})$, implies that $X^* = U^* \cdot \exp(H) = U^{-1} \cdot \exp(H)$, and therefore

$$XX^* = \exp(2H).$$

Now, for each automorphism $X \in \mathbf{GL}(\mathrm{E})$, $XX^*$ is a positive self-adjoint automorphism of E (11.5.3). Hence, by virtue of (i) above, there exists a unique $H \in \mathfrak{a}(\mathrm{E})$ satisfying the equation $\exp(2H) = XX^*$, which we write as $H = \frac{1}{2}\log(XX^*)$. If we put $U = (\exp(H))^{-1} \cdot X$, it is immediately verified that we have $UU^* = I$, that is to say, $U \in \mathbf{U}(\mathrm{E})$. Since $H \mapsto \exp(H)$ is a diffeomorphism of $\mathfrak{a}(\mathrm{E})$ onto $\mathfrak{a}_+(\mathrm{E})$, and $A \mapsto \log(A)$ is the inverse diffeomorphism, (ii) is established.

We now return to the determination of the positive self-adjoint automorphisms of $\mathfrak{g}$. By virtue of **(21.17.6)**, such an automorphism is uniquely expressible as $\exp(H)$, where $H \in \mathfrak{a}(\mathfrak{g})$.

**(21.17.7)** *For a self-adjoint endomorphism $H$ of the vector space* $\mathfrak{g}$ (relative to the Hilbert space structure defined in **(21.17.2)**) *to be such that* $\exp(H)$ *is an automorphism of the Lie algebra* $\mathfrak{g}$, *it is necessary and sufficient that* $H = \operatorname{ad}(i\mathbf{u})$ *with* $\mathbf{u} \in \mathfrak{k}$.

To say that $\exp(H) \in \operatorname{Aut}(\mathfrak{g})$ signifies that $[\exp(H) \cdot \mathbf{u}, \exp(H) \cdot \mathbf{v}] = \exp(H) \cdot [\mathbf{u}, \mathbf{v}]$ for all $\mathbf{u}, \mathbf{v} \in \mathfrak{g}$, or equivalently

$$\exp(H) \circ \operatorname{ad}(\mathbf{u}) \circ \exp(H)^{-1} = \operatorname{ad}(\exp(H) \cdot \mathbf{u})$$

in $\operatorname{End}(\mathfrak{g})$. If we put $\mathfrak{m} = \operatorname{ad}(\mathfrak{g})$, this therefore implies **(19.11.2.5)** that

**(21.17.7.1)** $$\operatorname{Ad}(\exp(H)) \cdot \mathfrak{m} \subset \mathfrak{m}$$

in $\mathfrak{gl}(\mathfrak{g}) = \operatorname{End}(\mathfrak{g})$, which can also be written **(19.11.2.2)** as

**(21.17.7.2)** $$\exp(\operatorname{ad}(H)) \cdot \mathfrak{m} \subset \mathfrak{m},$$

the exponential here being that of the group $\mathbf{GL}(\operatorname{End}(\mathfrak{g}))$. Relative to a suitably chosen orthonormal basis of $\mathfrak{g}$, $\operatorname{ad}(H)$ acts on $\operatorname{End}(\mathfrak{g})$ according to the formulas **(21.17.6.1)**; hence, relative to the basis $(E_{jk})$, its matrix is the diagonal matrix formed by the $\lambda_j - \lambda_k$, and the matrix of $\exp(\operatorname{ad}(H))$ is therefore the diagonal matrix formed by the $e^{\lambda_j - \lambda_k}$. From this it follows that the subspaces of the vector space $\operatorname{End}(\mathfrak{g})$ that are stable under $\exp(\operatorname{ad}(H))$ are the same as those which are stable under $\operatorname{ad}(H)$ **(A.24.3)**, and hence

**(21.17.7.3)** $$\operatorname{ad}(H) \cdot \mathfrak{m} \subset \mathfrak{m}.$$

This signifies also that $X \mapsto [H, X]$ is a *derivation* of the Lie algebra $\mathfrak{m} = \operatorname{ad}(\mathfrak{g})$; but $\operatorname{ad}(\mathfrak{g})$ is isomorphic to $\mathfrak{g}$, hence semisimple, and therefore every derivation of $\operatorname{ad}(\mathfrak{g})$ is inner **(21.6.7)**. In other words, there exists a unique $\mathbf{u}_0 \in \mathfrak{g}$ such that, putting $H_0 = \operatorname{ad}(\mathbf{u}_0)$, we have $[H - H_0, X] = 0$ for all $X \in \operatorname{ad}(\mathfrak{g})$. Since $\operatorname{ad}(\mathfrak{g})$ is stable under the mapping $X \mapsto X^*$ **(21.17.3.2)**, we have also $[H - H_0^*, X] = 0$ for all $X \in \operatorname{ad}(\mathfrak{g})$, because $H$ is self-adjoint. From this we conclude that $H_0^* = H_0$ and therefore **(21.17.3.2)** $c(\mathbf{u}_0) = -\mathbf{u}_0$, that is to say, $\mathbf{u}_0 \in i\mathfrak{k}$. Since $H_0 \in \operatorname{ad}(\mathfrak{g})$, we have $[H, H_0] = 0$, so that $H$ and $H_0$ commute, and consequently

$$\exp(H) = \exp(H - H_0)\exp(H_0);$$

and clearly

$$\exp(H_0) = \exp(\operatorname{ad}(\mathbf{u}_0)) = \operatorname{Ad}(\exp(\mathbf{u}_0)) \in \operatorname{Aut}(\mathfrak{g}),$$

so that the hypothesis $\exp(H) \in \operatorname{Aut}(\mathfrak{g})$ implies that $\exp(H - H_0) \in \operatorname{Aut}(\mathfrak{g})$.

Let $\alpha_1, \ldots, \alpha_m$ be the *distinct* eigenvalues of the selfadjoint endomorphism $Z = H - H_0$ of the Hilbert space $\mathfrak{g}$, and let $\mathfrak{g}_1, \ldots, \mathfrak{g}_m$ be the corresponding eigenspaces, so that $\mathfrak{g}$ is the Hilbert sum of the $\mathfrak{g}_j$ $(1 \leqq j \leqq m)$ (11.5.7). Since, for each $\mathbf{u} \in \mathfrak{g}$, $\mathrm{ad}(\mathbf{u})$ *commutes* with $Z$ in $\mathrm{End}(\mathfrak{g})$, we must have $\mathrm{ad}(\mathbf{u}) \cdot \mathfrak{g}_j \subset \mathfrak{g}_j$ for $1 \leqq j \leqq m$; in other words, the $\mathfrak{g}_j$ are *ideals* of the algebra $\mathfrak{g}$. Moreover, for each $\mathbf{x} \in \mathfrak{g}_j$, we have $\exp(Z) \cdot \mathbf{x} = e^{\alpha_j}\mathbf{x}$; but since $\exp(Z) \in \mathrm{Aut}(\mathfrak{g})$, we have $\exp(Z) \cdot [\mathbf{x}, \mathbf{y}] = [\exp(Z) \cdot \mathbf{x}, \exp(Z) \cdot \mathbf{y}]$ for $\mathbf{x}$ and $\mathbf{y}$ in the same $\mathfrak{g}_j$, and therefore $e^{\alpha_j}[\mathbf{x}, \mathbf{y}] = e^{2\alpha_j}[\mathbf{x}, \mathbf{y}]$. This is possible only if either $\alpha_j = 0$ or else $[\mathbf{x}, \mathbf{y}] = 0$ for all $\mathbf{x}, \mathbf{y} \in \mathfrak{g}_j$. The second alternative is ruled out by virtue of (21.6.2(i)), hence we have $\alpha_j = 0$ for all $j$, which means that $Z = 0$; in other words, $H = H_0 = \mathrm{ad}(i\mathbf{u})$ with $\mathbf{u} \in \mathfrak{k}$. The converse follows immediately from (21.17.3.2) and (21.17.6(i)).

(21.17.8) *The mapping* $(\mathbf{u}, U) \mapsto \exp(\mathrm{ad}(i\mathbf{u})) \cdot U$ *is a diffeomorphism of* $\mathfrak{k} \times \mathrm{Aut}(\mathfrak{k})$ *onto* $\mathrm{Aut}(\mathfrak{g})$.

Since $\mathrm{Aut}(\mathfrak{k})$ consists of unitary endomorphisms of the Hilbert space $\mathfrak{g}$ (21.17.4), and since $\mathbf{u} \mapsto \exp(\mathrm{ad}(i\mathbf{u}))$ is a diffeomorphism of $\mathfrak{k}$ onto a submanifold of the vector space of Hermitian endomorphisms of the Hilbert space $\mathfrak{g}$ (21.17.6(i)), it follows from (21.17.6(ii)) that is enough to show that the image of the mapping $(\mathbf{u}, U) \mapsto \exp(\mathrm{ad}(i\mathbf{u})) \cdot U$ of $\mathfrak{k} \times \mathrm{Aut}(\mathfrak{k})$ into $\mathbf{GL}(\mathfrak{g})$ is exactly equal to $\mathrm{Aut}(\mathfrak{g})$. Now, if $X \in \mathrm{Aut}(\mathfrak{g})$, then also $X^* \in \mathrm{Aut}(\mathfrak{g})$ by virtue of (21.17.3.1) and (21.17.1.1); hence $XX^* \in \mathrm{Aut}(\mathfrak{g})$. We have seen in (21.17.6(i)) that there exists a unique self-adjoint endomorphism $H$ of the Hilbert space $\mathfrak{g}$ such that $\exp(2H) = XX^*$; it follows from (21.17.7) that $H = \mathrm{ad}(i\mathbf{u})$ with $\mathbf{u} \in \mathfrak{k}$, and the calculation made in the course of the proof of (21.17.6(ii)) then shows that $U = (\exp(H))^{-1}X$ is unitary; but since $U \in \mathrm{Aut}(\mathfrak{g})$, it follows from (21.17.4) that $U \in \mathrm{Aut}(\mathfrak{k})$ and therefore $X = \exp(\mathrm{ad}(i\mathbf{u})) \cdot U$. The converse inclusion is obvious from the identification of $\mathrm{Aut}(\mathfrak{k})$ with a subgroup of $\mathrm{Aut}(\mathfrak{g})$.

(21.17.9) *The mapping* $(\mathbf{u}, U) \mapsto \exp(\mathrm{ad}(i\mathbf{u})) \cdot U$ *is a diffeomorphism of* $\mathfrak{k} \times \mathrm{Ad}(\tilde{\mathrm{K}})$ *onto* $\mathrm{Ad}(\tilde{\mathrm{G}})$.

This follows from the fact that $\mathfrak{k}$ is connected and therefore $\mathfrak{k} \times \mathrm{Ad}(\tilde{\mathrm{K}})$ is the identity component of $\mathfrak{k} \times \mathrm{Aut}(\mathfrak{k})$.

(21.17.10) *Let* $\pi \colon \tilde{\mathrm{G}} \to \mathrm{G} = \mathrm{Ad}(\tilde{\mathrm{G}})$ *denote the canonical projection* (so that $\pi(s) = \mathrm{Ad}(s)$).

(i) *The inverse image* $\pi^{-1}(\mathrm{K})$ (*where* $\mathrm{K} = \mathrm{Ad}(\tilde{\mathrm{K}})$) *may be identified with the simply connected compact group* $\tilde{\mathrm{K}}$, *and with the Lie subgroup of* $\tilde{\mathrm{G}}_{|\mathbf{R}}$ *corresponding to the Lie subalgebra* $\mathfrak{k}$ *of* $\mathfrak{g}_{|\mathbf{R}}$ (19.7.4); *in particular, the center* C *of* $\tilde{\mathrm{K}}$ *may be identified with the center of* $\tilde{\mathrm{G}}$.

(ii) *The mapping* $i\mathbf{u} \mapsto \exp_{\tilde{G}}(i\mathbf{u})$ *is a diffeomorphism of the vector subspace* $i\mathfrak{k}$ *of* $\mathfrak{g}_{|\mathbf{R}}$ *onto a submanifold* $\tilde{P}$ *of* $\tilde{G}$, *such that* $\tilde{P} \cap \tilde{K} = \{\tilde{e}\}$ (*the identity element of* $\tilde{G}$).

(iii) *The mapping* $(y, z) \mapsto yz$ *of* $P \times K$ *into* $G$ *is a diffeomorphism of* $\tilde{P} \times \tilde{K}$ *onto* $\tilde{G}$.

Let P be the image of $i\mathfrak{k}$ in G under the mapping $i\mathbf{u} \mapsto \exp_G(i\mathbf{u})$, which is the same as the mapping $i\mathbf{u} \mapsto \exp(\operatorname{ad}(i\mathbf{u}))$ by definition of $G = \operatorname{Ad}(\tilde{G})$; P is therefore a submanifold of G diffeomorphic to $i\mathfrak{k}$. If $\tilde{P}$ is the connected component of $\tilde{e}$ in $\pi^{-1}(P)$, then $\tilde{P}$ is a covering of P ((16.12.9) and (16.28.6)); but since P, being homeomorphic to a vector space, is simply connected, the restriction of $\pi$ to $\tilde{P}$ is a *diffeomorphism* of $\tilde{P}$ onto P (16.28.6), and the intersection $\tilde{P} \cap \pi^{-1}(e)$ of $\tilde{P}$ and the center of $\tilde{G}$ consists only of the identity element. Furthermore, for each $\mathbf{u} \in \mathfrak{k}$ we have $\pi(\exp_{\tilde{G}}(i\mathbf{u})) = \exp_G(i\mathbf{u})$; since the one-parameter subgroup of $\tilde{G}$ that is the image of $\mathbf{R}$ under the mapping $t \mapsto \exp_{\tilde{G}}(it\mathbf{u})$ is connected, we have $\exp_{\tilde{G}}(i\mathbf{u}) \in \tilde{P}$, and consequently $\tilde{P}$ is the image of $i\mathfrak{k}$ under the restriction to $i\mathfrak{k}$ of the mapping $\exp_{\tilde{G}}$, which is a diffeomorphism.

Consider now the Lie subgroup $K' = \pi^{-1}(K)$, which is a covering of K and contains the center $\pi^{-1}(e)$ of $\tilde{G}$. We shall show that every $x \in \tilde{G}$ can be written uniquely in the form $yz$ with $y \in \tilde{P}$ and $z \in K'$. We have $\pi(x) = y_0 z_0$ with $y_0 \in P$ and $z_0 \in K$, and this decomposition is unique (21.17.9); we may write $y_0 = \pi(y)$ and $z_0 = \pi(z')$, with $y \in \tilde{P}$ and $z' \in K'$; hence $x = yz'w$, where $w \in \pi^{-1}(e)$; but since $\pi^{-1}(e) \subset K'$, it follows that $z = z'w \in K'$ and we have $x = yz$ as required. As to the uniqueness of this factorization, if $x = y_1 z_1$ with $y_1 \in \tilde{P}$ and $z_1 \in K'$, then $\pi(y)\pi(z) = \pi(y_1)\pi(z_1)$, whence $\pi(y) = \pi(y_1)$ (21.17.9), which as above implies that $y_1 = y$ and therefore $z_1 = z$.

Next we shall show that the bijection $(y, z) \mapsto yz$ of $\tilde{P} \times K'$ onto $\tilde{G}$ is a diffeomorphism. If $(a, b) \in \tilde{P} \times K'$ and $c = ab$, there exist open neighborhoods U, V, W of $a$ in $\tilde{P}$, $b$ in $K'$, and $c$ in $\tilde{G}$, respectively, such that the restrictions of $\pi$ to U, V, W are diffeomorphisms onto the open sets $\pi(U)$, $\pi(V)$, $\pi(W)$ in P, K, and G, respectively. Since we may assume that U and V are so small that the mapping $(y_0, z_0) \mapsto y_0 z_0$ of $\pi(U) \times \pi(V)$ into $\pi(W)$ is a diffeomorphism onto an open subset of $\pi(W)$ (21.17.9), the result now follows immediately.

We see therefore that $\tilde{P} \times K'$ is diffeomorphic to $\tilde{G}$. This implies that $K'$ is *simply connected* (16.27.10), hence isomorphic to $\tilde{K}$. If we identify $\tilde{K}$ with $K'$, the center of $\tilde{K}$ contains $\pi^{-1}(e)$, and since $K = \tilde{K}/\pi^{-1}(e)$ has center $\{e\}$, it follows that $\pi^{-1}(e)$ is in fact the center C of $\tilde{K}$ (20.22.5.1).

(21.17.11) It is now easy to deduce from (21.17.10) the determination of *all* the complex connected Lie groups that have $\mathfrak{g}$ as Lie algebra. Indeed, such a group is isomorphic to a quotient $G_1 = \tilde{G}/D$ of $\tilde{G}$ by a subgroup D of its (finite) center C (**16.30.4**); the center $C_1$ of $G_1$ is C/D. If $\pi_1 : G_1 \to G = G_1/C_1$ is the canonical projection, then $\pi_1^{-1}(K)$ may be identified with the compact group $K_1 = \tilde{K}/D$ with center $C_1$, and K with $K_1/C_1$. We may therefore repeat without any changes the argument of (**21.17.10**); if $P_1$ is the connected component of the identity element $e_1$ of $G_1$ in $\pi^{-1}(P)$, the restriction of $\pi_1$ to $P_1$ is a diffeomorphism of $P_1$ onto P, and $i\mathbf{u} \mapsto \exp_{G_1}(i\mathbf{u})$ is a diffeomorphism of $i\mathfrak{k}$ onto $P_1$. We have $P_1 \cap K_1 = \{e_1\}$, and the mapping $(y, z) \mapsto yz$ is a diffeomorphism of $P_1 \times K_1$ onto $G_1$.

There is therefore a canonical one-to-one correspondence between the *compact connected semisimple* Lie groups with Lie algebra $\mathfrak{k}$, and the *complex connected semisimple* Lie groups with Lie algebra $\mathfrak{k}_{(\mathbf{C})} = \mathfrak{g}$.

(**21.17.12**) (i) The exponential mapping of $\tilde{G}$ maps $\mathfrak{k}$ *onto* $\tilde{K}$ (**21.7.4**) and $i\mathfrak{k}$ *onto* $\tilde{P}$; nevertheless, it is *not* necessarily a surjection of $\mathfrak{g} = \mathfrak{k} \oplus i\mathfrak{k}$ onto $\tilde{G}$ (Section **19.8**, Problem 2).

(ii) With the notation of (**21.17.11**), the subgroup $K_1$ is *maximal* among the *compact* subgroups of $G_1$. For if an element $yz$, with $y \in P_1$ and $z \in K_1$, belongs to a compact subgroup $K_1' \supset K_1$, then $y \in K_1'$; but if $y = \exp(i\mathbf{u})$ with $u \in \mathfrak{k}$, the subgroup of $G_1$ generated by $y$ is the image under the exponential mapping of the subgroup $\mathbf{Z}i\mathbf{u}$ of $i\mathfrak{k}$; this subgroup is closed and not compact in $i\mathfrak{k}$ if $\mathbf{u} \neq 0$, and therefore the subgroup generated by $y$ would also be closed and noncompact in $K_1'$, which is impossible. Hence we must have $\mathbf{u} = 0$ and therefore $K_1' = K_1$.

## PROBLEMS

1. Let $G_0$ be a connected (real) Lie group, $\mathfrak{g}_0$ its Lie algebra, $\mathfrak{g} = \mathfrak{g}_0 \otimes_{\mathbf{R}} \mathbf{C}$ the complexification of $\mathfrak{g}_0$, and G the *simply connected* complex Lie group with Lie algebra $\mathfrak{g}$ (**21.23.4**). If $\tilde{G}_0$ is the simply connected universal covering Lie group of $G_0$, then the canonical injection $\mathfrak{g}_0 \to \mathfrak{g}_{|\mathbf{R}}$ is the derived homomorphism of a unique homomorphism $h: \tilde{G}_0 \to G_{|\mathbf{R}}$. For each Lie group homomorphism $u: G_0 \to H_{|\mathbf{R}}$, where H is a complex Lie group, there exists a unique homomorphism $u^*: G \to H$ of complex Lie groups such that $u^* \circ h = u \circ p$, where $p: \tilde{G}_0 \to G_0$ is the canonical homomorphism. Let $G^+$ be the quotient of G by the intersection N of the kernels of the homomorphisms $u^*$ corresponding to all homomorphisms $u: G_0 \to H_{|\mathbf{R}}$. Show that if D is the kernel of $p$, then $h(D) \subset N$. (Consider the composite homomorphism $G_0 \xrightarrow{\text{Ad}} \text{Aut}(\mathfrak{g}_0) \to \text{Aut}(\mathfrak{g})_{|\mathbf{R}}$.) Deduce that there exists a canonical homomorphism $\varphi: G_0 \to G^+_{|\mathbf{R}}$ such that every homomorphism $u: G_0 \to H_{|\mathbf{R}}$ (where H is a complex Lie group) factorizes uniquely as $G_0 \xrightarrow{\varphi} G^+_{|\mathbf{R}} \xrightarrow{u^+} H_{|\mathbf{R}}$, where $u^+: G^+ \to H$ is a homomorphism of complex Lie groups.

Show that if there exists an *injective* homomorphism $u: G_0 \to H_{|\mathbf{R}}$, where H is a complex Lie group, then the homomorphism $\varphi: G_0 \to G^+_{|\mathbf{R}}$ is injective and $\mathfrak{g}$ is the Lie algebra of $G^+$; the group $G^+$ is said to be the *complexification* of $G_0$. If we identify $G_0$ with a subgroup of $G^+_{|\mathbf{R}}$, there exists no complex Lie subgroup of $G^+$ containing $G_0$, other than $G^+$ itself.

2. Let K be a compact connected Lie group of dimension $n$, which we may assume to be a subgroup of $\mathbf{O}(N, \mathbf{R})$ (21.13.1); K is then the set of real matrices whose components are the zeros of some family of polynomials in $\mathbf{R}[T_{11}, T_{12}, \ldots, T_{NN}]$ (Section 21.13, Problem 2). Let $\mathfrak{a}$ denote the ideal of $\mathbf{R}[T_{11}, \ldots, T_{NN}]$ formed by the polynomials that vanish at all points of K.
(a) Let G be the set of complex matrices in $\mathbf{GL}(N, \mathbf{C})$ for which all the polynomials in $\mathfrak{a}$ vanish; G is also the set of complex matrices for which the polynomials in the ideal $\mathfrak{a}^+ = \mathfrak{a} + i\mathfrak{a}$ in $\mathbf{C}[T_{11}, \ldots, T_{NN}]$ vanish. Show that G is a closed subgroup of $\mathbf{GL}(N, \mathbf{C})$. (First prove that if $s \in K$ and $t \in G$, then $st \in G$.) We have $K = G \cap \mathbf{GL}(N, \mathbf{R}) = G \cap \mathbf{O}(N, \mathbf{R}) = G \cap \mathbf{U}(N, \mathbf{C})$. (Observe that $\mathbf{O}(N, \mathbf{C}) \cap \mathbf{U}(N, \mathbf{C}) = \mathbf{O}(N, \mathbf{R})$.)
(b) If $X$ is a matrix belonging to G, then ${}^t\bar{X}$ also belongs to G. If we write $X = HU$, where $U$ is unitary and $H$ is hermitian and positive definite (21.17.6), then the matrices $H$, $U$ also belong to G. (Note that $H^2 = XX^* \in G$ and therefore $H^{2k} \in G$ for all integers $k \in \mathbf{Z}$. If we write $H = A \cdot \exp(D) \cdot A^{-1}$ where $D = \operatorname{diag}(a_1, \ldots, a_N)$, the $a_j$ being *real*, then for each polynomial $P \in \mathfrak{a}^+$ and each $z \in \mathbf{C}$, $P(A \cdot \exp(zD) \cdot A^{-1})$ is a linear combination of exponentials $e^{c_k z}$ with $c_k \in \mathbf{R}$. By observing that this function of $z$ vanishes for all $z \in 2\mathbf{Z}$, show that it vanishes identically, and hence in particular is zero for $z = 1$.) If $S = ADA^{-1}$, so that $H = \exp(S)$, then $\exp(zS) \in G$ for all $z \in \mathbf{C}$.
(c) Let $S$ be a hermitian matrix. Show that $\exp(S) \in G$ if and only if $iS \in \mathfrak{k}$, the Lie algebra of K. (Observe that if $\exp(itS)$ is a zero of all the polynomials in $\mathfrak{a}$, where $t \in \mathbf{R}$, then the same is true of $\exp(zS)$ for $z \in \mathbf{C}$.) Deduce from (b) above and from (21.17.6) that G is diffeomorphic to $K \times \mathbf{R}^n$ and that its Lie algebra is $\mathfrak{k} \oplus i\mathfrak{k}$. The group G may therefore be identified with the *complexification* of the compact group K; its Lie algebra $\mathfrak{g}$ is the direct sum of its center $\mathfrak{c}$ and its derived algebra $\mathfrak{D}(\mathfrak{g})$, which is semisimple, and the universal covering $\tilde{G}$ of G is therefore isomorphic to the product of $\mathbf{C}^m$ (for some positive integer $m$) and a complex semisimple Lie group, which is the complexification of a compact semisimple Lie group.

## 18. REAL FORMS OF THE COMPLEXIFICATIONS OF COMPACT CONNECTED SEMISIMPLE GROUPS AND SYMMETRIC SPACES

(21.18.1) We have already observed in two contexts ((21.8.2) and (21.17.1)) that if $\mathfrak{a}$ is a real Lie algebra and $\mathfrak{b} = \mathfrak{a}_{(\mathbf{C})}$ is its complexification, then the bijection $c: \mathbf{y} + i\mathbf{z} \mapsto \mathbf{y} - i\mathbf{z}$ of $\mathfrak{b}$ onto itself (where $\mathbf{y}, \mathbf{z} \in \mathfrak{a}$) is a *semilinear* involution that satisfies the relation $c([\mathbf{u}, \mathbf{v}]) = [c(\mathbf{u}), c(\mathbf{v})]$ for all $\mathbf{u}, \mathbf{v} \in \mathfrak{b}$ (i.e., it is an automorphism of the *real* Lie algebra $\mathfrak{b}_{|\mathbf{R}}$). For the sake of brevity, a bijection of a complex Lie algebra $\mathfrak{b}$ onto itself that has these properties will be called a *conjugation*. Conversely, a conjugation $c$ in a complex Lie algebra $\mathfrak{b}$ determines uniquely a real Lie subalgebra $\mathfrak{a}$ of $\mathfrak{b}_{|\mathbf{R}}$ such that $\mathfrak{b}$ is isomorphic to the complexification of $\mathfrak{a}$. For since $c$ is $\mathbf{R}$-linear and $c^2 = 1_{\mathfrak{b}}$, the vector space $\mathfrak{b}_{|\mathbf{R}}$ is the direct sum $\mathfrak{a} \oplus \mathfrak{a}'$ of two real vector subspaces $\mathfrak{a}$, $\mathfrak{a}'$, such that $c(\mathbf{u}) = \mathbf{u}$ for $\mathbf{u} \in \mathfrak{a}$, and $c(\mathbf{u}) = -\mathbf{u}$ for $\mathbf{u} \in \mathfrak{a}'$. Since also $c(i\mathbf{u}) = -ic(\mathbf{u})$ for all $\mathbf{u} \in \mathfrak{b}$, we have $i\mathfrak{a} \subset \mathfrak{a}'$ and $i\mathfrak{a}' \subset \mathfrak{a}$, from which it follows that $\mathfrak{a}' = i\mathfrak{a}$. Finally, since $c$ is an automorphism of $\mathfrak{b}_{|\mathbf{R}}$, the subspace $\mathfrak{a}$ is a Lie subalgebra of $\mathfrak{b}_{|\mathbf{R}}$, and it is immediately seen that $\mathfrak{b}$ is the complexification of $\mathfrak{a}$. There is therefore a canonical one-to-one correspondence between conjugations of $\mathfrak{b}$ and *real forms* of $\mathfrak{b}$. Further, if $\varphi$ is an automorphism of the complex Lie algebra $\mathfrak{b}$, and $c$ is a conjugation of $\mathfrak{b}$, it is clear that $\varphi \circ c \circ \varphi^{-1} = c_1$ is also a conjugation of $\mathfrak{b}$, and that if $\mathfrak{a}$ and $\mathfrak{a}_1$ are the real forms of $\mathfrak{b}$ corresponding to $c$ and $c_1$, respectively, then $\mathfrak{a}_1 = \varphi(\mathfrak{a})$.

(21.18.2) Changing the notation of (21.17), let $\tilde{G}_u$ be a simply connected compact semisimple Lie group, $\mathfrak{g}_u$ its Lie algebra, $\mathfrak{g} = (\mathfrak{g}_u)_{(\mathbf{C})}$ the complexification of $\mathfrak{g}_u$, and $c_u$ the conjugation of $\mathfrak{g}$ corresponding to $\mathfrak{g}_u$. We propose to determine, up to isomorphism, all the *real forms* of the complex semisimple algebra $\mathfrak{g}$, and we shall show that this is equivalent to the following problem relative to the algebra $\mathfrak{g}_u$: to determine the *involutory automorphisms* of this Lie algebra.

This will result from the following proposition:

(21.18.3) *With the notation of* (21.18.2), *let $c$ be a conjugation of $\mathfrak{g}$. Then there exists an automorphism $\varphi$ of $\mathfrak{g}$ such that $c_u$ commutes with $\varphi \circ c \circ \varphi^{-1}$.*

We have seen (21.17.2.1) that $(\mathbf{x}|\mathbf{y}) = -B_{\mathfrak{g}}(\mathbf{x}, c_u(\mathbf{y}))$ is a *scalar product* that makes $\mathfrak{g}$ a finite-dimensional Hilbert space. The mapping $H = cc_u$ is an *automorphism* of the complex Lie algebra $\mathfrak{g}$; it is also a *self-adjoint* endomorphism of the Hilbert space $\mathfrak{g}$, because we have

$$B_{\mathfrak{g}}(H \cdot \mathbf{x}, c_u \cdot \mathbf{y}) = B_{\mathfrak{g}}(\mathbf{x}, H^{-1}c_u \cdot \mathbf{y}) = B_{\mathfrak{g}}(\mathbf{x}, c_u H \cdot \mathbf{y})$$

since $H$ leaves invariant the Killing form of $\mathfrak{g}$, and $c$, $c_u$ are involutions. Hence there exists an orthonormal basis $(\mathbf{e}_j)_{1 \le j \le n}$ of $\mathfrak{g}$ with respect to which the matrix of $H$ is diagonal and invertible. Consequently the matrix of $H^2 = A$ with respect to this basis is of the form $\operatorname{diag}(\lambda_1, \lambda_2, \ldots, \lambda_n)$, where the $\lambda_j$ are real and $> 0$. For each real number $t > 0$, let $A^t$ be the automorphism of the vector space $\mathfrak{g}$ defined by the matrix $\operatorname{diag}(\lambda_1^t, \lambda_2^t, \ldots, \lambda_n^t)$ (cf. (15.11.11)); these automorphisms commute with $H$, and moreover they are automorphisms of the complex *Lie algebra* $\mathfrak{g}$. For if the multiplication table of $\mathfrak{g}$, relative to the basis $(\mathbf{e}_j)$, is

$$[\mathbf{e}_j, \mathbf{e}_k] = \sum_l a_{jkl}\, \mathbf{e}_l$$

then the fact that $A$ is an automorphism of $\mathfrak{g}$ is expressed by the relations

$$\lambda_j \lambda_k a_{jkl} = a_{jkl} \lambda_l \qquad (1 \leqq j, k, l \leqq n),$$

which evidently imply, for all $t > 0$, that

$$\lambda_j^t \lambda_k^t a_{jkl} = a_{jkl} \lambda_l^t,$$

thereby proving our assertion. Now consider the conjugation $c' = A^t c_u A^{-t}$ of $\mathfrak{g}$, and note that by definition we have $c_u H c_u^{-1} = c_u c = H^{-1}$, so that $c_u A c_u^{-1} = A^{-1}$. But if we put $L = \operatorname{diag}(\log \lambda_1, \log \lambda_2, \ldots, \log \lambda_n)$, then $A = e^L$, so that $A^t = e^{tL}$ and therefore $c_u A^t c_u^{-1} = A^{-t}$. Consequently

$$cc' = cA^t c_u A^{-t} = cc_u A^{-2t} = HA^{-2t},$$

$$c'c = (cc')^{-1} = A^{2t} H^{-1} = H^{-1} A^{2t},$$

and therefore when $t = \frac{1}{4}$ we have $cc' = c'c = H^{-1}A^{1/2}$, because $HA^{-1} = H^{-1}$. Hence $\varphi = A^{-1/4}$ satisfies the conditions of the proposition.

**(21.18.4)** In the determination of all conjugations of $\mathfrak{g}$, we may therefore limit our search to conjugations $c_0$ that *commute with* $c_u$, and therefore leave $\mathfrak{g}_u$ and $i\mathfrak{g}_u$ globally invariant. The restriction of $c_0$ to $\mathfrak{g}_u$ is then an *involutory automorphism* of this real Lie algebra. Consequently, $\mathfrak{g}_u$ is the direct sum of a *real Lie subalgebra* $\mathfrak{k}_0$, consisting of the $\mathbf{x} \in \mathfrak{g}_u$ such that $c_0(\mathbf{x}) = \mathbf{x}$, and a *real vector subspace*, denoted by $i\mathfrak{p}_0$, consisting of the $\mathbf{x} \in \mathfrak{g}_u$ such that $c_0(\mathbf{x}) = -\mathbf{x}$. It follows that $i\mathfrak{g}_u$ is the direct sum of $i\mathfrak{k}_0$ and $\mathfrak{p}_0$, and because $c_0$ is a conjugation of $\mathfrak{g}$ we have $c_0(\mathbf{x}) = \mathbf{x}$ for $\mathbf{x} \in \mathfrak{p}_0$ and $c_0(\mathbf{x}) = -\mathbf{x}$ for $\mathbf{x} \in i\mathfrak{k}_0$. The real form $\mathfrak{g}_0$ of $\mathfrak{g}$ corresponding to $c_0$ is therefore

**(21.18.4.1)**
$$\mathfrak{g}_0 = \mathfrak{k}_0 \oplus \mathfrak{p}_0 .$$

Since the Killing form $\mathrm{B}_{\mathfrak{g}_u}$ is the restriction of $\mathrm{B}_{\mathfrak{g}}$ to $\mathfrak{g}_u \times \mathfrak{g}_u$, it follows from the definition of the scalar product $(\mathbf{x} \mid \mathbf{y})$ on $\mathfrak{g}$ that $(\mathbf{x} \mid \mathbf{y}) = -\mathrm{B}_{\mathfrak{g}_u}(\mathbf{x}, \mathbf{y})$ for $\mathbf{x}, \mathbf{y} \in \mathfrak{g}_u$. Since the restriction of $c_0$ to $\mathfrak{g}_u$ is an automorphism of this Lie algebra, it leaves invariant its Killing form **(21.5.6)**; for $\mathbf{x} \in \mathfrak{k}_0$ and $\mathbf{y} \in \mathfrak{p}_0$, we have therefore $(\mathbf{x} \mid \mathbf{y}) = (c_0(\mathbf{x}) \mid c_0(\mathbf{y})) = -(\mathbf{x} \mid \mathbf{y})$, whence $(\mathbf{x} \mid \mathbf{y}) = 0$. It follows that $\mathrm{B}_{\mathfrak{g}}(\mathbf{x}, \mathbf{y}) = 0$ and hence also $\mathrm{B}_{\mathfrak{g}}(\mathbf{x}, i\mathbf{y}) = 0$. Since the Killing form $\mathrm{B}_{\mathfrak{g}_0}$ is the restriction of $\mathrm{B}_{\mathfrak{g}}$, it follows that in the decomposition **(21.18.4.1)**, $\mathfrak{k}_0$ and $\mathfrak{p}_0$ are *orthogonal subspaces relative to the Killing form of* $\mathfrak{g}_0$ (hence are *nonisotropic*). Further, the restriction of $\mathrm{B}_{\mathfrak{g}_0}$ to $\mathfrak{k}_0 \times \mathfrak{k}_0$ is *negative definite*, because it is also the restriction of $\mathrm{B}_{\mathfrak{g}_u}$ **(21.6.9)**; by contrast, its restriction to $\mathfrak{p}_0 \times \mathfrak{p}_0$ is positive definite, because for $\mathbf{x} \in i\mathfrak{p}_0$ we have $\mathrm{B}_{\mathfrak{g}_0}(i\mathbf{x}, i\mathbf{x}) = \mathrm{B}_{\mathfrak{g}}(i\mathbf{x}, i\mathbf{x}) = -\mathrm{B}_{\mathfrak{g}}(\mathbf{x}, \mathbf{x}) = -\mathrm{B}_{\mathfrak{g}_u}(\mathbf{x}, \mathbf{x})$. Finally, we have

**(21.18.4.2)**
$$[\mathfrak{k}_0, \mathfrak{p}_0] \subset \mathfrak{p}_0, \qquad [\mathfrak{p}_0, \mathfrak{p}_0] \subset \mathfrak{k}_0 .$$

For if $\mathbf{x} \in \mathfrak{k}_0$ and $\mathbf{y} \in i\mathfrak{p}_0$, then $c_0([\mathbf{x}, \mathbf{y}]) = [c_0(\mathbf{x}), c_0(\mathbf{y})] = -[\mathbf{x}, \mathbf{y}]$, and since $[\mathbf{x}, \mathbf{y}] \in \mathfrak{g}_u$ we have $[\mathbf{x}, \mathbf{y}] \in i\mathfrak{p}_0$; this shows that $[\mathfrak{k}_0, i\mathfrak{p}_0] \subset i\mathfrak{p}_0$ and therefore also that $[\mathfrak{k}_0, \mathfrak{p}_0] \subset \mathfrak{p}_0$. Likewise, if $\mathbf{x}, \mathbf{y} \in i\mathfrak{p}_0$, then $c_0([\mathbf{x}, \mathbf{y}]) = [\mathbf{x}, \mathbf{y}]$ and hence $[\mathbf{x}, \mathbf{y}] \in \mathfrak{k}_0$, because $[\mathbf{x}, \mathbf{y}] \in \mathfrak{g}_u$; this proves the relation $[i\mathfrak{p}_0, i\mathfrak{p}_0] \subset \mathfrak{k}_0$, whence $[\mathfrak{p}_0, \mathfrak{p}_0] \subset \mathfrak{k}_0$.

**(21.18.5)** Let $\tilde{\mathrm{G}}$ be the simply connected complex (semisimple) Lie group of which $\mathfrak{g}$ is the Lie algebra **(19.11.9)**, and let $\tilde{\mathrm{P}}$ be the closed submanifold of $\tilde{\mathrm{G}}$ that is the image of $i\mathfrak{g}_u$ under the mapping $i\mathbf{u} \mapsto \exp_{\tilde{\mathrm{G}}}(i\mathbf{u})$. From **(21.17.10)**, the mapping $(y, z) \mapsto yz$ of $\tilde{\mathrm{P}} \times \tilde{\mathrm{G}}_u$ into $\tilde{\mathrm{G}}$ is a *diffeomorphism*.

To the automorphism $c_0$ of the real Lie algebra $\mathfrak{g}_{|\mathbf{R}}$ there corresponds a unique *involutory automorphism* $\sigma$ of $\tilde{\mathrm{G}}_{|\mathbf{R}}$ such that the derived automorphism $\sigma_* = c_0$ **(19.7.6)**; $\sigma$ therefore leaves $\tilde{\mathrm{G}}_u$ and $\tilde{\mathrm{P}}$ stable, because $c_0$ leaves $\mathfrak{g}_u$ and $i\mathfrak{g}_u$ stable. Let $\mathrm{G}_0$ be the *Lie subgroup* of $\tilde{\mathrm{G}}_{|\mathbf{R}}$ consisting of the points fixed by $\sigma$ **(19.10.1)**; its Lie algebra is $\mathfrak{g}_0$ **(20.4.3)**, hence is *semisimple*, and it evidently contains the *compact subgroup* $\mathrm{K}_0 = \mathrm{G}_0 \cap \tilde{\mathrm{G}}_u$ consisting of the fixed points of the restriction of $\sigma$ to $\tilde{\mathrm{G}}_u$, because $\tilde{\mathrm{G}}_u$ is stable under $\sigma$. Likewise, $\mathrm{G}_0$ contains the image $\mathrm{P}_0$ under the exponential mapping $\mathbf{u} \mapsto \exp_{\tilde{\mathrm{G}}}(\mathbf{u}) = \exp_{\mathrm{G}_0}(\mathbf{u})$ of the vector subspace $\mathfrak{p}_0$ of $i\mathfrak{g}_u$, and since $\exp(c_0(\mathbf{u})) = \sigma(\exp(\mathbf{u}))$, P is the set of points of $\tilde{\mathrm{P}}$ fixed by $\sigma$. Furthermore:

**(21.18.5.1)** $\mathrm{P}_0$ *is a closed submanifold of* $\mathrm{G}_0$; *the mapping* $\mathbf{u} \mapsto \exp_{\mathrm{G}_0}(\mathbf{u})$ *is a diffeomorphism of* $\mathfrak{p}_0$ *onto* $\mathrm{P}_0$, *and the mapping* $(y, z) \mapsto yz$ *of* $\mathrm{P}_0 \times \mathrm{K}_0$ *into* $\mathrm{G}_0$ *is a diffeomorphism of* $\mathrm{P}_0 \times \mathrm{K}_0$ *onto* $\mathrm{G}_0$.

The first two assertions are obvious, since $\mathfrak{p}_0$ is a vector subspace (and hence a closed submanifold) of $i\mathfrak{g}_u$. Again, it is clear that the restriction to $\mathrm{P}_0 \times \mathrm{K}_0$ of the diffeomorphism $(y, z) \mapsto yz$ of $\tilde{\mathrm{P}} \times \tilde{\mathrm{G}}_u$ onto $\tilde{\mathrm{G}}$ is a diffeomorphism onto its image in $\tilde{\mathrm{G}}$, and it remains to show that this image is the whole of $\mathrm{G}_0$. Each element $x \in \mathrm{G}_0$ is uniquely expressible in the form $yz$ with $y \in \tilde{\mathrm{P}}$ and $z \in \tilde{\mathrm{G}}_u$; since $\sigma(x) = x$, we have $\sigma(y)\sigma(z) = yz$, and since $\sigma(y) \in \tilde{\mathrm{P}}$ and $\sigma(z) \in \tilde{\mathrm{G}}_u$, we must have $y = \sigma(y)$ and $z = \sigma(z)$, whence $y \in \mathrm{P}_0$ and $z \in \mathrm{K}_0$.

**(21.18.5.2)** *If* C *is the center of* $\tilde{\mathrm{G}}$ *(identified with the center of* $\tilde{\mathrm{G}}_u$ **(21.17.10)**), *the center of* $\mathrm{G}_0$ *is* $\mathrm{C} \cap \mathrm{G}_0$.

For if $s \in \mathrm{G}_0$, the restriction of $\mathrm{Ad}(s)$ to $\mathfrak{g}_0$ is the identity if and only if the restriction of $\mathrm{Ad}(s)$ to $\mathfrak{g}$ is the identity, because $\mathfrak{g}$ is the complexification of $\mathfrak{g}_0$; the result therefore follows from **(19.11.6)**.

**(21.18.6)** It can be proved that the compact group $K_0$ is *connected* (Section 21.16, Problem 11); we shall assume this result in the rest of this section.† On the other hand, $K_0$ *is not necessarily semisimple or simply connected.* The group $G_0$ is therefore connected, and the same reasoning as in **(21.17.10)**, with G replaced by $G_0$, and $\tilde{G}$ by $\tilde{G}_0$, the universal covering group of $G_0$, shows that:

**(21.18.7)** *The inverse image* $\pi^{-1}(K_0)$ *of* $K_0$ *under the canonical projection* $\pi$ *of* $\tilde{G}_0$ *onto* $G_0$ *is isomorphic to the simply connected group* $\tilde{K}_0$, *the universal covering of the compact group* $K_0$. *The mapping* $\mathbf{u} \mapsto \exp_{\tilde{G}_0}(\mathbf{u})$ *is a diffeomorphism of* $\mathfrak{p}_0$ *onto a closed submanifold* $\tilde{P}_0$ *of* $\tilde{G}_0$, *such that* $\tilde{P}_0 \cap \tilde{K}_0$ *consists only of the identity element of* $\tilde{G}_0$. *The mapping* $(y, z) \mapsto yz$ *is a diffeomorphism of* $\tilde{P}_0 \times \tilde{K}_0$ *onto* $\tilde{G}_0$.

The center Z of $\tilde{G}_0$ is $\pi^{-1}(C_0)$, a discrete subgroup contained in the center of $\tilde{K}_0$, but distinct from the latter if $K_0$ is not semisimple (in which case $\tilde{K}_0$ is not compact **(21.6.9)**).

**(21.18.8)** Finally, the reasoning of **(21.17.11)** gives the determination (up to isomorphism) of *all* the connected real Lie groups that have $\mathfrak{g}_0$ as their Lie algebra: such a group is isomorphic to a quotient $G_1 = \tilde{G}_0/D$, where D is a (discrete) subgroup of the center Z of $\tilde{G}_0$, and the center $C_1$ of $G_1$ is Z/D. If $\pi_1 : G_1 \to \mathrm{Ad}(G_0) = G_1/C_1$ is the canonical projection, $\pi_1^{-1}(\mathrm{Ad}(K_0))$ may be identified with the group $K_1 = \tilde{K}_0/D$, the connected Lie subgroup of $G_1$ with Lie algebra $\mathfrak{k}_0$; it contains $C_1$ (which is *not* in general the center of $K_1$), and is *compact* if and only if $C_1$ is *finite*. If $P_1$ is the connected component of the identity element $e_1 \in G_1$ in $\pi_1^{-1}(\mathrm{Ad}(P_0))$, the restriction of $\pi_1$ to $P_1$ is a diffeomorphism of $P_1$ onto $\mathrm{Ad}(P_0)$, and $\mathbf{u} \mapsto \exp_{G_1}(\mathbf{u})$ is a diffeomorphism of $\mathfrak{p}_0$ onto $P_1$; we have $P_1 \cap K_1 = \{e_1\}$, and the mapping $(y, z) \mapsto yz$ is a diffeomorphism of $P_1 \times K_1$ onto $G_1$.

The decomposition **(21.18.4.1)** is called the *Cartan decomposition* of the semisimple real Lie algebra $\mathfrak{g}_0$. The corresponding decomposition as a product $P_1 \times K_1$, for a connected Lie group $G_1$ having $\mathfrak{g}_0$ as its Lie algebra, is called a *Cartan decomposition* of $G_1$. Since $\mathrm{Ad}(K_0) = \mathrm{Ad}(K_1)$ is compact, $K_1$ is in any case isomorphic to the product of a compact group and a vector group $\mathbf{R}^m$ **(21.6.9)**, hence $G_1$ is diffeomorphic to the product of a compact group and a vector group $\mathbf{R}^N$; and the same argument as in **(21.17.12)** proves that the compact subgroup in this product decomposition is *maximal* in $G_1$.

† We shall not make use of this result anywhere except in this section.

*Examples*

**(21.18.9)** Consider a *Weyl basis* of $\mathfrak{g}$ **(21.10.6)**, consisting of a basis of a maximal commutative subalgebra $\mathfrak{t}$ of $\mathfrak{g}_u$, together with elements $\mathbf{x}_\alpha$ ($\alpha \in \mathbf{S}$) satisfying **(21.10.6.4)**. Since the numbers $N_{\alpha,\beta}$ are *real*, it is clear that the *real* vector subspace $\mathfrak{g}_0$ of $\mathfrak{g}$ spanned by this Weyl basis is a *real Lie algebra* having $\mathfrak{g}$ as its complexification; this real Lie algebra is called a *normal* real form of $\mathfrak{g}$. One sees immediately that in the corresponding Cartan decomposition $\mathfrak{g}_0 = \mathfrak{k}_0 + \mathfrak{p}_0$, the elements $\mathbf{x}_\alpha - \mathbf{x}_{-\alpha}$ form a basis of $\mathfrak{k}_0$, the subspace $\mathfrak{p}_0$ contains $\mathfrak{t}$ and is spanned by $\mathfrak{t}$ and the elements $\mathbf{x}_\alpha + \mathbf{x}_{-\alpha}$.

**(21.18.10)** Consider the complex Lie group $H = \tilde{G} \times \tilde{G}$, whose Lie algebra is $\mathfrak{g} \oplus \mathfrak{g}$, the complexification of the Lie algebra $\mathfrak{g}_u \oplus \mathfrak{g}_u$ of $G_u \times G_u$. Let $c_0$ be the conjugation of $\mathfrak{g} \oplus \mathfrak{g}$ defined by

$$c_0(\mathbf{x} + i\mathbf{y}, \mathbf{x}' + i\mathbf{y}') = (\mathbf{x}' - i\mathbf{y}', \mathbf{x} - i\mathbf{y})$$

for $\mathbf{x}, \mathbf{y}, \mathbf{x}', \mathbf{y}' \in \mathfrak{g}_u$. It is clear that the set of $(\mathbf{v}, \mathbf{w}) \in \mathfrak{g} \oplus \mathfrak{g}$ fixed by $c_0$ is the set of elements $(\mathbf{z}, c_u(\mathbf{z}))$ for $\mathbf{z} \in \mathfrak{g}$, and hence is *isomorphic to* $\mathfrak{g}_{|\mathbf{R}}$. In this way the Lie algebra $\mathfrak{g}_{|\mathbf{R}}$ appears as a real form of $\mathfrak{g} \oplus \mathfrak{g}$; the corresponding Cartan decomposition $\mathfrak{k}_0 \oplus \mathfrak{p}_0$ is such that $\mathfrak{k}_0 = \mathfrak{g}_u$ and $\mathfrak{p}_0 = i\mathfrak{g}_u$.

**(21.18.11)** Let us take $\tilde{G}_u$ to be the almost simple compact group $\mathbf{SU}(n)$ **(21.12.1)**, which is simply connected **(16.30.6)**. We have seen in **(21.12.1)** that the complexification $\mathfrak{g}$ of $\mathfrak{g}_u = \mathfrak{su}(n)$ may be identified with $\mathfrak{sl}(n, \mathbf{C})$. We shall show that the corresponding group $\mathbf{SL}(n, \mathbf{C})$ is *simply connected*. By virtue of **(21.17.6)**, $\mathbf{SL}(n, \mathbf{C})$ is diffeomorphic to the submanifold of $\mathfrak{a}(\mathbf{C}^n) \times \mathbf{U}(n)$ consisting of pairs of matrices $(H, U)$ such that $\det(\exp(H)) \cdot \det(U) = 1$, or equivalently $e^{\mathrm{Tr}(H)} \cdot \det(U) = 1$; since $\mathrm{Tr}(H)$ is a real number, and the only unitary matrices with a positive real determinant are those with determinant 1, it follows that $\mathbf{SL}(n, \mathbf{C})$ is diffeomorphic to $V \times \mathbf{SU}(n)$, where $V$ is the hyperplane in $\mathfrak{a}(\mathbf{C}^n)$ defined by the equation $\mathrm{Tr}(H) = 0$. This proves our assertion **(16.27.10)**; the group denoted by $\tilde{G}$ in **(21.18.5)** is here $\mathbf{SL}(n, \mathbf{C})$.

The conjugation $c_u$ corresponding to the real form $\mathfrak{g}_u$ is the involutory bijection $X \mapsto -{}^t\bar{X}$ of $\mathfrak{sl}(n, \mathbf{C})$ onto itself. Among the conjugations of $\mathfrak{g}$ that commute with $c_u$, there are the following three types:

(I) $c_0: X \mapsto \bar{X}$; $\mathfrak{g}_0$ is therefore the set of *real* matrices in $\mathfrak{sl}(n, \mathbf{C})$, hence is the Lie algebra $\mathfrak{sl}(n, \mathbf{R})$ (the *normal* real form of $\mathfrak{sl}(n, \mathbf{C})$ **(21.18.9)**); the subalgebra $\mathfrak{k}_0$ of $\mathfrak{g}_u$ is the set of real matrices in $\mathfrak{su}(n)$, so that $\mathfrak{g}_u = \mathfrak{so}(n)$ and therefore is semisimple if $n \geq 3$ (**(21.12.3)** and **(21.12.4)**); $\mathfrak{p}_0$ is the space of real symmetric $n \times n$ matrices with trace 0, and $P_0$ is the set of *positive definite* real symmetric matrices with determinant 1. The automorphism $\sigma$ of $\tilde{G}_{|\mathbf{R}}$ is again the mapping $X \mapsto \bar{X}$ on $\mathbf{SL}(n, \mathbf{C})$, and therefore we have $G_0 = \mathbf{SL}(n, \mathbf{R})$ and $K_0 = \mathbf{SO}(n)$. When $n = 2$, $K_0$ is isomorphic to $\mathbf{T}$, and $\tilde{K}_0$ to $\mathbf{R}$, so that $\tilde{G}_0$, the universal covering of $\mathbf{SL}(2, \mathbf{R})$, is diffeomorphic to $\mathbf{R}^3$; when $n \geqq 3$, the group $K_0$ is not simply connected, and $\tilde{K}_0$ is isomorphic to $\mathbf{Spin}(n)$ (21.16.10); hence $\tilde{G}_0$ has finite center, but it can be shown that $\tilde{G}_0$ is *not* isomorphic to any Lie subgroup of a linear group $\mathbf{GL}(N, \mathbf{R})$ (Problem 1).

(II) Suppose that $n = 2m$ is even, and consider the mapping $c_0$: $X \mapsto J\bar{X}J^{-1}$, where

$$J = \begin{pmatrix} 0 & I_m \\ -I_m & 0 \end{pmatrix}.$$

Since $\bar{J} = J$ and $J^{-1} = -J = {}^tJ$, it is immediately verified that $c_0$ is a conjugation that commutes with $c_u$. The corresponding automorphism $\sigma$ of $\tilde{G}_{|\mathbf{R}} = \mathbf{SL}(2m, \mathbf{C})$ is the same mapping $X \mapsto J\bar{X}J^{-1}$, and it is easily verified that the matrices fixed by $\sigma$ are the matrices in $\mathbf{SL}(2m, \mathbf{C})$ of the form (21.12.2.2), in other words, the matrices of the form $\begin{pmatrix} U & V \\ -\bar{V} & \bar{U} \end{pmatrix}$ of determinant 1, with $U$ and $V$ in $\mathbf{GL}(m, \mathbf{C})$. It follows therefore from (21.12.2) that $K_0$ is semisimple and simply connected, and is isomorphic to $\mathbf{U}(m, \mathbf{H})$; the group $G_0$ is therefore simply connected and may be identified with the intersection of $\mathbf{GL}(m, \mathbf{H})$ and $\mathbf{SL}(2m, \mathbf{C})$; its center consists of $\pm I$.

(III) Let $p$, $q$ be two integers such that $p \geqq q \geqq 1$ and $p + q = n$. Consider the $n \times n$ matrix

$$I_{p,q} = \begin{pmatrix} I_p & 0 \\ 0 & -I_q \end{pmatrix};$$

the mapping $c_0$: $X \mapsto -I_{p,q} \cdot {}^t\bar{X} \cdot I_{p,q}$ is a conjugation that commutes with $c_u$, by reason of the relations $\bar{I}_{p,q} = I_{p,q}$ and $I_{p,q}^{-1} = {}^tI_{p,q} = I_{p,q}$. The restriction of $c_0$ to $\mathfrak{g}_u$ is the automorphism $X \mapsto I_{p,q} X I_{p,q}$ of this real Lie algebra; the restriction to $\tilde{G}_u = \mathbf{SU}(n)$ of the corresponding automorphism $\sigma$ of $\tilde{G}_{|\mathbf{R}}$ is the same mapping $X \mapsto I_{p,q} X I_{p,q}$, and it follows that the group $K_0$ is the set of matrices $\begin{pmatrix} U & 0 \\ 0 & V \end{pmatrix}$, where $U \in \mathbf{U}(p)$, $V \in \mathbf{U}(q)$, and $\det(U)\det(V) = 1$. One sees immediately that such a matrix can be uniquely expressed as a product

$$\begin{pmatrix} U_1 & 0 \\ 0 & I_q \end{pmatrix} D \begin{pmatrix} I_p & 0 \\ 0 & V_1 \end{pmatrix}$$

where $U_1 \in \mathbf{SU}(p)$, $V_1 \in \mathbf{SU}(q)$, and $D$ is a diagonal matrix of the form

$$D = \operatorname{diag}(\delta, 1, \ldots, 1, \delta^{-1}, 1, \ldots, 1)$$

with $\delta^{-1}$ in the $(p+1)$th place, and $|\delta| = 1$. Consequently $K_0$ is *diffeomorphic* to $\mathbf{SU}(p) \times \mathbf{T} \times \mathbf{SU}(q)$, hence is not simply connected. Its Lie algebra $\mathfrak{k}_0$ consists of the matrices of the form $\begin{pmatrix} X & 0 \\ 0 & Y \end{pmatrix}$ with $X \in \mathfrak{u}(p)$, $Y \in \mathfrak{u}(q)$, and $\operatorname{Tr}(X) + \operatorname{Tr}(Y) = 0$; such a matrix can be written uniquely in the form

$$\begin{pmatrix} X_1 & 0 \\ 0 & 0 \end{pmatrix} + \begin{pmatrix} p^{-1}\alpha I_p & 0 \\ 0 & -q^{-1}\alpha I_q \end{pmatrix} + \begin{pmatrix} 0 & 0 \\ 0 & Y_1 \end{pmatrix}$$

where $\alpha \in i\mathbf{R}$, $X_1 \in \mathfrak{su}(p)$, and $Y_1 \in \mathfrak{su}(q)$; and it is immediately verified that this decomposition $\mathfrak{k}_0 = \mathfrak{su}(p) \oplus \mathbf{R} \oplus \mathfrak{su}(q)$ is a decomposition into *ideals*. The simply connected group $\tilde{K}_0$, the universal covering of $K_0$, is therefore isomorphic to $\mathbf{SU}(p) \times \mathbf{R} \times \mathbf{SU}(q)$. The group $G_0$ is the set of matrices $X \in \mathbf{SL}(n, \mathbf{C})$ such that ${}^t\bar{X} \cdot I_{p,q} \cdot X = I_{p,q}$, i.e., it is the subgroup $\mathbf{SU}(p, q)$ of matrices with determinant 1 in the unitary group $\mathbf{U}(p, q)$ of a sesquilinear Hermitian form of signature $(p, q)$ on $\mathbf{C}^n$ **(16.11.3)**; the foregoing remarks show that $G_0$ is not simply connected.

It can be shown (Problem 3; also [62], [85]) that every conjugation of $\mathfrak{sl}(n, \mathbf{C})$ that commutes with $c_u$ is of the form $\varphi \circ c_0 \circ \varphi^{-1}$, where $\varphi$ is an automorphism of $\mathfrak{sl}(n, \mathbf{C})$ and $c_0$ is one of the three types of conjugation just described.

**(21.18.12)** We retain the notation of **(21.18.8)**. If $\tau$ is the involutory automorphism of the simply connected Lie group $\tilde{G}_0$ that corresponds to the automorphism $c_u | \mathfrak{g}_0$ of $\mathfrak{g}_0$, then $\tau$ fixes each element of $\tilde{K}_0$ and transforms each element of $\tilde{P}_0$ into its inverse. Since the center Z of $\tilde{G}_0$ is contained in $\tilde{K}_0$, it follows that, on passing to the quotient in $G_1 = \tilde{G}_0/D$, $\tau$ gives rise to an involutory automorphism $\tau_1$ of $G_1$ which fixes the elements of $K_1 = \tilde{K}_0/D$ and transforms each element of $P_1$ into its inverse. We conclude that $K_1$ is exactly the subgroup of $G_1$ consisting of the fixed points of $\tau_1$, by virtue of the relation $G_1 = P_1 K_1$ and the fact that no element of $P_1$ has order 2, because of the existence of the diffeomorphism $\mathbf{u} \mapsto \exp_{G_1}(\mathbf{u})$ of $\mathfrak{p}_0$ onto $P_1$.

Suppose now that the algebra $\mathfrak{g}$ is *simple*; this implies that every real form of $\mathfrak{g}$, and in particular $\mathfrak{g}_0$, is simple, and consequently the only normal Lie subgroups of $G_1$ are the subgroups of the center $C_1$. In order that the group $K_1$ should contain no normal subgroup of $G_1$ other than $\{e\}$, we must therefore take $D = Z$, i.e., $G_1 = \operatorname{Ad}(G_0)$ and $K_1 = \operatorname{Ad}(K_0)$. The composite canonical mapping

**(21.18.12.1)** $$\mathfrak{p}_0 \to P_1 = \exp_{G_1}(\mathfrak{p}_0) \to G_1/K_1,$$

in which the left-hand arrow is the exponential mapping and the right-hand arrow is the restriction to $P_1$ of the canonical projection $G_1 \to G_1/K_1$, is a diffeomorphism. Identifying all the spaces $G_1/K_1$ with $Ad(G_0)/Ad(K_0)$, we see that, since the conditions of **(20.11.1)** are satisfied for the latter space, we may define a structure of a *Riemannian symmetric space* on the spaces $P_1$, or $G_1/K_1$, for which the Levi-Civita connection is entirely determined by the conjugation $c_0$. But in fact we can define *canonically* a $G_1$-invariant Riemannian *metric* on $G_1/K_1$: since the restriction to $\mathfrak{p}_0$ of the Killing form $B_{\mathfrak{g}_0}$ (or $B_{\mathfrak{g}}$) is positive definite and invariant under $Ad(t)$ for all $t \in K_1$, we may take this restriction as the value of the Riemannian metric tensor on $G_1/K_1$ at the point $x_0$ that is the image of the identity element **(20.11.1)**.

With this choice of metric, the *sectional curvature* $A(\mathbf{u}, \mathbf{v})$ is easily calculated, where $\mathbf{u}, \mathbf{v}$ are any two vectors in $\mathfrak{p}_0 = T_{x_0}(G_1/K_1)$: for by virtue of **(20.21.2.1)** and the invariance of $B_{\mathfrak{g}}$, we have

**(21.18.12.2)**
$$A(\mathbf{u}, \mathbf{v}) = -B_{\mathfrak{g}}([\mathbf{u}, \mathbf{v}], [\mathbf{u}, \mathbf{v}])/\|\mathbf{u} \wedge \mathbf{v}\|^2 = -\|[\mathbf{u}, \mathbf{v}]\|^2/\|\mathbf{u} \wedge \mathbf{v}\|^2.$$

Hence $G_1/K_1$ is a Riemannian manifold with sectional curvature *everywhere* $\leqq 0$.

**(21.18.13)** The existence of the involutory automorphism $\sigma$ of $\tilde{G}_u$ corresponding to the conjugation $c_0$ **(21.18.5)** gives rise to other Riemannian symmetric spaces. Supposing always that $\mathfrak{g}$ is *simple*, the largest normal subgroup of $\tilde{G}_u$ contained in $K_0$ is $C \cap G_0$ **(21.18.5.2)**. Let $G_2 = \tilde{G}_u/(C \cap G_0)$ and $K_2 = K_0/(C \cap G_0)$. On passing to the quotients, $\sigma$ defines an involutory automorphism $\sigma_2$ of $G_2$ that fixes the points of $K_2$; but here $K_2$ is only the *identity component* of the subgroup $K_2'$ of fixed points of $\sigma_2$, and may well be distinct from $K_2'$, as the example $G_2 = \mathbf{SO}(n+1)$, $K_2 = \mathbf{SO}(n)$ ($n$ even) shows **(20.11.4)**. For each subgroup $K_2''$ such that $K_2 \subset K_2'' \subset K_2'$, the symmetric pair $(G_2, K_2'')$ therefore fulfills the conditions of **(20.11.1)** and defines a *compact Riemannian symmetric space* $G_2/K_2''$. The tangent space to this manifold at the point $x_0$, the image of the identity element of $G_2$, may be identified with the subspace $i\mathfrak{p}_0$ of $\mathfrak{g}_u$. The restriction of $B_{\mathfrak{g}}$ to $i\mathfrak{p}_0$ is *negative definite* **(21.18.4)**; on the other hand, for each $t \in K_2'$, the space $i\mathfrak{p}_0$ is stable under $Ad(t)$, and $B_{\mathfrak{g}}$ is invariant under $Ad(t)$, so that we may again define *canonically* a $G_2$-invariant Riemannian metric on $G_2/K_2''$, by taking the restriction of $-B_{\mathfrak{g}}$ to $i\mathfrak{p}_0$ as the value of the metric tensor at the point $x_0$ (so that the spaces $G_2/K_2''$, for all the different possible choices of $K_2''$, are locally isometric). The same calculation as in **(21.18.2.2)** now gives the sectional curvature $A(\mathbf{u}, \mathbf{v})$, for $\mathbf{u}, \mathbf{v} \in i\mathfrak{p}_0$:

**(21.18.13.1)**
$$A(\mathbf{u}, \mathbf{v}) = \|[\mathbf{u}, \mathbf{v}]\|^2/\|\mathbf{u} \wedge \mathbf{v}\|^2$$

so that the Riemannian manifolds $G_2/K''_2$ have sectional curvature *everywhere* $\geqq 0$. It can be shown that $G_2/K_2$ is simply connected (Problem 5), so that it is a finite covering of each of the spaces $G_2/K''_2$.

The direct sum decomposition of the Lie algebra $\mathfrak{g}_u$,

**(21.18.13.2)** $$\mathfrak{g}_u = \mathfrak{k}_0 \oplus i\mathfrak{p}_0\,,$$

is again called the *Cartan decomposition* of $\mathfrak{g}_u$ corresponding to $c_0$. The image of $\mathfrak{k}_0$ under the exponential mapping $\exp_{G_2}$ is equal to $K_2$ **(21.7.4)**. The image $P_2$ of $i\mathfrak{p}_0$ under the mapping $\exp_{G_2}$, however, has properties that are rather different from those of the set $P_1$ studied in **(21.18.8)**:

**(21.18.13.3)** *For each* $s \in G_2$, *let* $s^* = \sigma_2(s^{-1})$. *Then the group* $G_2$ *acts differentiably on itself by the action* $(s, t) \mapsto sts^*$. *For this action,* $P_2$ *is the orbit of e, and* $K'_2$ *is the stabilizer of e, so that* $P_2$ *is a compact submanifold of* $G_2$, *canonically diffeomorphic to* $G_2/K'_2$; *also we have* $K_2P_2 = P_2K_2 = G_2$.

We know from **(20.7.10.4)** that the geodesic trajectories on the compact Riemannian manifold $G_2/K_2$ that pass through $x_0$ are the images under $\pi: G_2 \to G_2/K_2$ of the 1-parameter subgroups corresponding to the tangent vectors belonging to $i\mathfrak{p}_0$. Since $G_2/K_2$ is compact and therefore complete, the union of these geodesic trajectories is the whole of $G_2/K_2$ **(20.18.5)**; in other words, $\pi(P_2) = G_2/K_2$, or equivalently $G_2 = P_2K_2$. Since the relation $x \in P_2$ implies $x^{-1} \in P_2$, it follows that also $G_2 = K_2P_2$.

The mapping $x \mapsto x^*$ clearly has the following properties:

$$x^{**} = x, \qquad (xy)^* = y^*x^*, \qquad e^* = e;$$

the relation $xx^* = e$ is equivalent to $x \in K'_2$; and for each $x \in P_2$ we have $x^* = x$, because $c_0(\mathbf{u}) = -\mathbf{u}$ for $\mathbf{u} \in i\mathfrak{p}_0$. Observe now that $\exp(\mathbf{u}) = (\exp(\frac{1}{2}\mathbf{u}))^2$; from this it follows that each $x \in P_2$ may be written as $x = y^2$ with $y \in P_2$, or equivalently $x = yy^*$. Conversely, for each $s \in G_2$ we may write $s = xz$ with $x \in P_2$ and $z \in K_2$, so that $ss^* = xzz^{-1}x = x^2 \in P_2$. This shows that $P_2$ is the orbit of $e$ for the action $(s, t) \mapsto sts^*$ of $G_2$ on itself. Since $G_2$ is compact, $P_2$ is a compact submanifold of $G_2$ **(16.10.7)**; moreover, we have seen above that the stabilizer of $e$ is $K'_2$, and therefore the corresponding canonical mapping $G_2/K'_2 \to P_2$ is a diffeomorphism **(16.10.7)**.

It should be carefully noted that in general the restriction to $P_2$ of the canonical mapping $\pi: G_2 \to G_2/K_2$ is *not* a diffeomorphism (Problem 6 and Section **21.21**, Problem 2).

**(21.18.14)** To summarize, we have shown that to each *involutory automorphism* of the Lie algebra $\mathfrak{g}_u$ (when $\mathfrak{g}$ is simple) there correspond:

(1) A real form $\mathfrak{g}_0$ of $\mathfrak{g}$, and the almost simple real Lie groups having $\mathfrak{g}_0$ as Lie algebra.

(2) A noncompact Riemannian symmetric space $G_1/K_1$, diffeomorphic to $\mathbf{R}^n$ for some $n$.

(3) A finite family of compact Riemannian symmetric spaces $G_2/K_2''$.

It can be shown that, together with the Euclidean spaces (20.11.2) and the almost simple compact groups, the Riemannian symmetric spaces of types (2) and (3) enable us to describe all Riemannian symmetric spaces (Problem 13). On the other hand, we shall see in (21.20.7) that *every* complex semisimple Lie algebra is isomorphic to the complexification of the Lie algebra of a compact semisimple group. It follows therefore that the determination of the almost simple compact groups and their involutory automorphisms implies *ipso facto* the determination of the real or complex semisimple groups and Riemannian symmetric spaces.

(21.18.15) *Let the symbols $\tilde{G}_u$, $\tilde{G}$, and $\tilde{G}_0$ have the same meanings as before. Then the linear representations of these three groups on the same finite-dimensional complex vector space* E *are in canonical one-to-one correspondence with each other, and are completely reducible* (H. Weyl's "unitary trick").

This is now obvious, because the linear representations of $\tilde{G}_u$ on E correspond one-to-one to the **R**-homomorphisms of $\mathfrak{g}_u$ into $\mathfrak{gl}(E)_{|\mathbf{R}}$, which in turn are in canonical one-to-one correspondence with the **C**-homomorphisms of $\mathfrak{g} = \mathfrak{g}_u \otimes_{\mathbf{R}} \mathbf{C}$ into $\mathfrak{gl}(E)$, by virtue of the fact that $\mathfrak{gl}(E)$ is a *complex* Lie algebra (21.9.1); and the same argument applies when we replace $\mathfrak{g}_u$ by any real form $\mathfrak{g}_0$ of $\mathfrak{g}$.

It can be shown that for $\tilde{G}$ and $\tilde{G}_0$ (when $\mathfrak{g}_0$ is not the Lie algebra of a compact group), no *finite*-dimensional linear representation can be equivalent to a *unitary* representation (Section 21.6, Problem 5). On the other hand, these groups admit many irreducible unitary representations of *infinite* dimension (cf. Chapter XXII).

## PROBLEMS

1. With the notation of (21.18.6), let Q be the kernel of the canonical homomorphism $\tilde{G}_0 \to G_0$. For each quotient $G_1 = \tilde{G}_0/D$, where D is a subgroup of the center Z of $\tilde{G}_0$, and each linear representation $\rho_1 : G_1 \to \mathbf{GL}(E)$ of $G_1$ on a finite-dimensional complex vector space E, show that the kernel of $\rho_1$ contains $p_1(Q)$, where $p_1 : \tilde{G}_0 \to G_1$ is the canonical homomorphism. (If $\sigma : \tilde{G}_0 \to G_0 \to \tilde{G}$ is the canonical homomorphism (with kernel Q), show that there exists a linear representation $\rho : \tilde{G} \to \mathbf{GL}(E)$ such that $\rho \circ \sigma =$

$\rho_1 \circ p_1$.) The only groups $G_1$ that admit a faithful linear representation on a *finite*-dimensional space are those for which $Q \subset D$ (use Section **21.17**, Problem 2); their centers are therefore *finite*.

2. With the notation of (**21.18.8**), show that the compact group $K_1$ is its own normalizer in $G_1$. (Reduce to showing that if an element $\mathbf{u} \in \mathfrak{p}_0$ is such that $p_1 = \exp_{G_1}(\mathbf{u})$ normalizes $K_1$, then $\mathbf{u} = 0$. Using the unique decomposition of an element of $G_1$ as a product $yz$, where $y \in P_1$ and $z \in K_1$, and the relation $[\mathfrak{k}_0, \mathfrak{p}_0] \subset \mathfrak{p}_0$, show first that $[\mathbf{u}, \mathbf{x}] = 0$ for all $\mathbf{x} \in \mathfrak{k}_0$; then use the invariance of $B_\mathfrak{g}$ (**21.5.6.1**) to deduce that $[\mathbf{u}, \mathbf{v}] \in \mathfrak{p}_0$ for all $\mathbf{v} \in \mathfrak{p}_0$; this implies that $[\mathbf{u}, \mathbf{v}] = 0$ and hence that $\mathbf{u}$ is in the center of $\mathfrak{g}$.)

3. (a) With the notation of (**21.8.2**), suppose that the compact group $\tilde{G}_u$ is almost simple. If $f$ is an involutory automorphism of $\mathfrak{g}_u$, there exists a regular element of $\mathfrak{g}_u$ invariant under $f$, and hence a maximal commutative subalgebra $\mathfrak{t}$ of $\mathfrak{g}_u$ stable under $f$, and a basis $\mathbf{B}$ of the system of roots of $\mathfrak{g}_u$ relative to $\mathfrak{t}$ that is stable under $f$ (Section **21.11**, Problem 19).
(b) Suppose that the transpose ${}^t(f \otimes 1)$ leaves invariant each of the roots of $\mathbf{B}$ in $(\mathfrak{t}_{(\mathbf{C})})^*$: this is the only possibility when $\mathfrak{g}$ is of type $B_l$ or $C_l$ (consider the Cartan integers for the basis $\mathbf{B}$). Then we have $f = \mathrm{Ad}(\exp(\mathbf{u}))$ with $\mathbf{u} \in \mathfrak{t}$ (cf. Section **21.11**, Problem 12); we may replace $\exp(\mathbf{u})$ by $z \cdot \exp(\mathbf{u})$, where $z$ is in the center of $\tilde{G}_u$, without changing $f$, and if we replace $\mathbf{u}$ by $w \cdot \mathbf{u}$, where $w$ is in the Weyl group W, then $f$ is replaced by $\varphi \circ f \circ \varphi^{-1}$, where $\varphi$ is an automorphism of $\mathfrak{g}_u$; we may therefore suppose that $i\mathbf{u}$ is in the closure of the principal alcove $A^*$ corresponding to $\mathbf{B}$ (Section **21.15**, Problem 11). By using the fact that $f^2 = 1$, show that either $i\mathbf{u} = \pi \mathbf{p}_j$ for some index $j$ such that $n_j = 1$, or $i\mathbf{u} = \pi \mathbf{p}_j$ for some index $j$ such that $n_j = 2$, or $i\mathbf{u} = \pi(\mathbf{p}_j + \mathbf{p}_k)$ for two indices $j$, $k$ such that $n_j = n_k = 1$ (cf. Section **21.16**, Problem 10). Show that this last case may be reduced to the first (observe that $2\pi(\mathbf{p}_j - \mathbf{p}_k)$ is a vertex of an alcove $w(A^*)$ for some $w \in W$).
(c) If $\mathfrak{g}$ is of type $A_l$ or $D_l$, there exists an involutory automorphism $f_0$ of $\mathfrak{g}_u$ such that ${}^t(f_0 \otimes 1)(\mathbf{B}) = \mathbf{B}$, but such that ${}^t(f_0 \otimes 1)$ does not fix every element of $\mathbf{B}$. (For type $A_l$, consider the automorphism $X \mapsto -{}^tX$ of $\mathfrak{u}(n, \mathbf{C})$, and for $D_l$ the automorphism defined in (**20.11.4**).) Furthermore, except for type $D_4$, if $f$ is another involutory automorphism of $\mathfrak{g}_u$ with the same property, then we must have $f = \mathrm{Ad}(\exp(\mathbf{u})) \circ f_0$ for some $\mathbf{u} \in \mathfrak{t}$, and we may again suppose that $\mathbf{u}$ lies in the closure of $A^*$; use the fact that $f^2 = 1$ to show that $\mathrm{Ad}(\exp(\mathbf{u} + f_0(\mathbf{u})))$ is the identity mapping. By observing that the indices $j$ such that $f_0(\mathbf{p}_j) \neq \mathbf{p}_j$ are such that $n_j = 1$ in both cases $A_l$ and $D_l$, show that $i\mathbf{u} = \pi \mathbf{p}_j$ for some index $j$ such that $f_0(\mathbf{p}_j) = \mathbf{p}_j$ and $n_j = 1$ or 2.
(d) Deduce from (b) and (c) that for the classical groups of types $B_l$, $C_l$, and $D_l$,† the compact real forms (up to isomorphism) correspond to the conjugation $c_u$: $X \mapsto \bar{X}$ in $\mathfrak{so}(n, \mathbf{C})$ for types $B_l$ and $D_l$, and to the conjugation $c_u$: $X \mapsto J\bar{X}J^{-1}$ in $\mathfrak{sp}(2n, \mathbf{C})$, where $J$ is the matrix (**21.12.2.4**). The noncompact real forms (up to isomorphism) correspond to the following conjugations:

| | |
|---|---|
| $c_0$: $X \mapsto I_{p,q}\bar{X}I_{p,q}$ | in $\mathfrak{so}(n, \mathbf{C})$ $(p + q = n)$, |
| $c_0$: $X \mapsto J\bar{X}J^{-1}$ | in $\mathfrak{so}(2n, \mathbf{C})$, |
| $c_0$: $X \mapsto \bar{X}$ | in $\mathfrak{sp}(2n, \mathbf{C})$, |
| $c_0$: $X \mapsto -K_{p,q} \cdot {}^t\bar{X} \cdot K_{p,q}$ | in $\mathfrak{sp}(2n, \mathbf{C})$ $(p + q = n)$, |

† It is necessary here to assume that $l \neq 4$ in order to apply (c), but it can be shown that the result remains true for $D_4$.

where $K_{p,q}$ is the matrix

$$\begin{pmatrix} -I_p & 0 & 0 & 0 \\ 0 & I_q & 0 & 0 \\ 0 & 0 & -I_p & 0 \\ 0 & 0 & 0 & I_q \end{pmatrix}$$

The compact symmetric spaces corresponding to the conjugations involving the matrices $I_{p,q}$ or $K_{p,q}$ include in particular the Grassmannians **(16.11.9)**.

4. With the notation and hypotheses of Problem 3, show that if the conjugation $c_0$ has as its restriction to $\mathfrak{g}_u$ an automorphism of the type considered in Problem 3(b), then the group $K_2'$ (in the notation of **(21.18.13)**) is connected. If on the other hand this restriction is of the type considered in Problem 3(c), then $(K_2' : K_2) = 2$.

5. With the hypotheses and notation of **(21.18.13)**, show that $G_2/K_2$ is simply connected. (Use **(16.14.9)** and Section **16.30**, Problem 11(a).)

6. With the hypotheses and notation of **(21.18.13)**, show that the mapping $(s, y) \mapsto sys^*$ is a submersion of $G_2 \times K_2$ into $G_2$ at the point $(e, y_0)$, for each $y_0 \in K_2$ such that $-1$ is not an eigenvalue of $\mathrm{Ad}(y_0)$. The set $P_2'$ of points $t \in G_2$ such that $t^* = t$ contains the union of the orbits (for the action $(s, t) \mapsto sts^*$) of the points $y \in K_2$ such that $y^2 = e$. Show that $P_2$ is the connected component of the point $e$ in $P_2'$, and is open in $P_2'$. For each $s \in G_2$, the mapping $z \mapsto szs^*$ is an isometry of $P_2$ onto itself; deduce that the geodesics in $P_2$ are the curves $\xi \mapsto s \cdot \exp_{G_2}(\xi \mathbf{u}) \cdot s^*$ for $\mathbf{u} \in i\mathfrak{p}_0$ and $s \in G_2$.
(b) In the case where $G_2 = \mathbf{SO}(n+1)$ and $\sigma_2$ is the automorphism defined in **(20.11.4)**, show that $P_2 \cap K_2'$ is the set consisting of $e$ and a submanifold diffeomorphic to $S_{n-1}$ that does not contain $e$, and that $P_2 \cap K_2 = \{e\}$. Determine the other connected components of $P_2'$.

7. With the notation and hypotheses of **(21.18.13)**, the mapping $s \mapsto ss^*$ of $G_2$ onto $P_2$ factorizes as $G_2 \xrightarrow{\pi'} G_2/K_2' \xrightarrow{\mu} P_2$, where $\pi'$ is the canonical mapping and $\mu$ is a diffeomorphism.
(a) Show that the composition $\mu \circ (\pi' \mid P_2)$ is the mapping $y \mapsto y^2$ of $P_2$ onto itself. Let $\exp_{G_2} : \mathfrak{g}_u \to G_2$ be the exponential mapping of the Lie group $G_2$ and $\exp_{x_0}$ the exponential mapping corresponding to the canonical connection on $G_2/K_2'$; then we have $\pi'(\exp_{G_2}(\mathbf{u})) = \exp_{x_0}(\mathbf{u})$ for $\mathbf{u} \in i\mathfrak{p}_0$ **(20.7.10.4)** ($i\mathfrak{p}_0$ being canonically identified with the tangent space at $x_0$ to $G_2/K_2'$). Show that $\mu(\exp_{x_0}(\mathbf{u})) = \exp_{G_2}(2\mathbf{u})$ for $\mathbf{u} \in i\mathfrak{p}_0$.
(b) Let $\mathbf{u} \in i\mathfrak{p}_0$ and let $y = \exp_{G_2}(\mathbf{u}) \in P_2$. Show that for each vector $\mathbf{v} \in i\mathfrak{p}_0$ we have

$$T_{\mathbf{u}}(\exp_{x_0}) \cdot \tau_{\mathbf{u}}^{-1}(\mathbf{v}) = T(\pi') \cdot \left( y \cdot \left( \left( \sum_{k=0}^{\infty} \frac{\mathrm{ad}(\mathbf{u})^{2k}}{(2k+1)!} \right) \cdot \mathbf{v} \right) \right).$$

(Use **(19.16.5.1)** and the relation $[\mathfrak{p}_0, \mathfrak{p}_0] \subset \mathfrak{k}_0$.)

8. With the notation and hypotheses of **(21.18.12)**, let $s^* = \tau_1(s^{-1})$ for all $s \in G_1$. State and prove for $G_1$ and $K_1$ the analogues of **(21.18.13.3)** and Problems 6(a) and 7.

9. With the notation and hypotheses of **(21.18.12)**, show that for a submanifold S of $P_1$ to be totally geodesic **(20.13.7)**, it is necessary and sufficient that the vector subspace $\mathfrak{s} = T_{x_0}(S)$

of $\mathfrak{p}_0$ should be such that the relations $\mathbf{u} \in \mathfrak{s}$, $\mathbf{v} \in \mathfrak{s}$, $\mathbf{w} \in \mathfrak{s}$ imply $[\mathbf{u}, [\mathbf{v}, \mathbf{w}]] \in \mathfrak{s}$; such an $\mathfrak{s}$ is called a *Lie triple system*, and S is the image of $\mathfrak{s}$ under $\exp_{G_1}$. (Using the definitions of the second fundamental forms (20.12.4) and of the parallel transport of a vector (18.6.3), show first that if S is totally geodesic, the parallel transport (*relative to* $P_1$) of a tangent vector to S along a curve in S is the same as the parallel transport of this vector *relative to* S, and therefore consists of tangent vectors to S. Then use (20.7.10.4) and Problem 8 to show that for all $\mathbf{u}$, $\mathbf{v}$ in $\mathfrak{s}$ we must have $(\mathrm{ad}(\mathbf{u}))^2 \cdot \mathbf{v} \in \mathfrak{s}$, and deduce that $\mathfrak{s}$ is a Lie triple system. Conversely, show that if $\mathfrak{s} \subset \mathfrak{p}_0$ is a Lie triple system, then $\mathfrak{g}' = \mathfrak{s} + [\mathfrak{s}, \mathfrak{s}]$ is a Lie subalgebra of $\mathfrak{g}_0$, stable under the conjugation $c_u$; if G′ is the connected Lie group immersed in $G_1$ that corresponds to $\mathfrak{g}'$, and if $K' = G' \cap K_1$, then K′ is closed for the proper topology of G′; the image S′ of $\mathfrak{s}$ in $P_1$ under $\exp_{G_1}$ is a closed submanifold of $P_1$, and the canonical mapping $G'/K' \to S'$ is a diffeomorphism (for the proper topology of G′); consequently S′ is a geodesic submanifold at the point $x_0$, and G′ acts on S′ as a transitive group of isometries.)

Show that the unique geodesic trajectory in $P_1$ that passes through two distinct points of S is contained in S.

**10.** In (21.18.12), take $G_1 = \mathbf{SL}(n, \mathbf{R})$ and $\tau_1$ to be the automorphism $X \mapsto {}^tX^{-1}$; its derived automorphism, the restriction of $c_u$ to $\mathfrak{sl}(n, \mathbf{R})$, is the automorphism $X \mapsto -{}^tX$. We have then $K_1 = \mathbf{SO}(n)$, and $P_1$ is the set S of positive definite symmetric matrices of determinant 1, which can also be written as $e^{\mathfrak{s}}$, where $\mathfrak{s}$ $(= \mathfrak{p}_0)$ is the space of symmetric matrices of trace 0. The geodesics in the Riemannian symmetric space S are the mappings $t \mapsto A \cdot e^{tX} \cdot {}^tA$ of $\mathbf{R}$ into S, where $A \in \mathbf{SL}(n, \mathbf{R})$ and $X \in \mathfrak{s}$ (Problem 8). Through any two points of S there passes one and only one geodesic trajectory.

Let $Q(X, Y) = \mathrm{Tr}(X^{-1}Y + Y^{-1}X)$ for any two matrices $X, Y \in S$.

(a) Show that $Q(A \cdot X \cdot {}^tA,\ A \cdot Y \cdot {}^tA) = Q(X, Y)$ for all $A \in \mathbf{SL}(n, \mathbf{R})$ and that $Q(X, Y) > 0$ for all $X, Y \in S$. (Use the fact that $X$ can be written as $Z^2$, where $Z \in S$.)

(b) Show that $Q(I, X) = 2 \sum_{j=1}^{n} \mathrm{ch}(\lambda_j)$, where $e^{\lambda_1}, \ldots, e^{\lambda_n}$ are the eigenvalues of the symmetric matrix $X \in S$ (use (a) above). Deduce that for each $X_0 \in S$ the mapping $X \mapsto Q(X_0, X)$ of S into $\mathbf{R}$ is proper (17.3.7).

(c) Let $t \mapsto G(t)$ be a geodesic in S. Show that for each $X_0 \in S$ the function $t \mapsto Q(X_0, G(t))$ is strictly convex on $\mathbf{R}$. (Reduce to the case where $G(t) = e^{tY}$, where $Y \in \mathfrak{s}$ is a diagonal matrix.)

**11.** (a) With the notation of Problem 10, let P be a totally geodesic submanifold of S, and let M be a compact subgroup of $\mathbf{SL}(n, \mathbf{R})$ leaving P globally invariant (for the action $(U, X) \mapsto U \cdot X \cdot {}^tU$ of $\mathbf{SL}(n, \mathbf{R})$ on S). Show that there exists $X_0 \in P$ that is invariant under M. (By (20.11.3.1) there exists $Z_0 \in S$ invariant under M. By using Problem 10, show that as $X$ runs through P the function $X \mapsto Q(Z_0, X)$ attains its lower bound at a unique point $X_0$: if the lower bound were attained at two distinct points, consider the unique geodesic trajectory joining them. Note also that $Q(Z_0, X_0) = Q(Z_0, U \cdot X_0 \cdot {}^tU)$ for all $U \in M$.)

(b) With the notation of (21.18.8), show that if $G_1 = \mathrm{Ad}(G_0)$, then for each compact subgroup M of $G_1$ there exists an inner automorphism of $G_1$ that transforms M into a subgroup of $K_1$ (*E. Cartan's conjugacy theorem*). (Using Section 21.17, show that if we identify $\mathrm{Aut}(\mathfrak{g}_0)$ with a subgroup of $\mathbf{GL}(n, \mathbf{R})$ (where $n = \dim(\mathfrak{g}_0)$), so that $K_1$ is identified with a subgroup of $\mathbf{O}(n)$ and $P_1$ with a submanifold of S, then there exists $y \in P_1$ such that $z \cdot y \cdot {}^tz = y$ for all $z \in M$, by using (a) above and Problem 10; then note that if $y = x^2$

with $x \in P_1$, the relation above takes the form $\tau_1(x^{-1}zx) = x^{-1}zx$ for all $z \in M$, in the notation of (21.18.12).)

**12.** (a) Let G/K be a symmetric Riemannian space (20.11.3), where G is a connected real Lie group and K is a compact subgroup of G that contains no normal subgroup of G other than $\{e\}$. Let $\sigma$ be the involutory automorphism of G for which K is contained in the subgroup of fixed points and contains the identity component of this subgroup. If $\mathfrak{g}$, $\mathfrak{k}$ are the Lie algebras of G and K, respectively, then $\mathfrak{k}$ is the subspace of vectors in $\mathfrak{g}$ fixed by $s = \sigma_*$, and contains no nonzero ideal of $\mathfrak{g}$. There exists a scalar product $(\mathbf{x} | \mathbf{y})$ on $\mathfrak{g}$ such that $(\mathrm{ad}(\mathbf{z}) \cdot \mathbf{x} | \mathbf{y}) + (\mathbf{x} | \mathrm{ad}(\mathbf{z}) \cdot \mathbf{y}) = 0$ for all $\mathbf{z} \in \mathfrak{k}$. A pair $(\mathfrak{g}, s)$ consisting of a finite-dimensional real Lie algebra $\mathfrak{g}$ and an involutory automorphism $s$ of $\mathfrak{g}$ having the above properties is called a *symmetrized* Lie algebra, and $s$ is called the *symmetrization* of $\mathfrak{g}$.
(b) Let $(\mathfrak{g}, s)$ be a symmetrized Lie algebra, $\mathfrak{k}$ the subspace of vectors fixed by $s$, and $\mathfrak{p}$ the vector subspace of $\mathfrak{g}$ consisting of all $\mathbf{x} \in \mathfrak{g}$ such that $s(\mathbf{x}) = -\mathbf{x}$. Then $\mathfrak{g} = \mathfrak{k} \oplus \mathfrak{p}$; we have $[\mathfrak{k}, \mathfrak{k}] \subset \mathfrak{k}$, $[\mathfrak{k}, \mathfrak{p}] \subset \mathfrak{p}$, $[\mathfrak{p}, \mathfrak{p}] \subset \mathfrak{k}$, and $\mathfrak{k}$, $\mathfrak{p}$ are orthogonal to each other with respect to the Killing form $B_{\mathfrak{g}}$. Show that $\mathfrak{k}$ is the Lie algebra of a compact group and that there exists a scalar product $Q(\mathbf{x}, \mathbf{y})$ on $\mathfrak{p}$ such that

$$Q(\mathrm{ad}(\mathbf{z}) \cdot \mathbf{x} | \mathbf{y}) + Q(\mathbf{x} | \mathrm{ad}(\mathbf{z}) \cdot \mathbf{y}) = 0$$

for all $z \in \mathfrak{k}$ (cf. Section 21.6, Problem 2). Furthermore, the restriction of $B_{\mathfrak{g}}$ to $\mathfrak{k}$ is a negative definite symmetric bilinear form.
(c) With the hypotheses of (b), let $A$ be the endomorphism of the vector space $\mathfrak{p}$ such that $Q(A \cdot \mathbf{x}, \mathbf{y}) = B_{\mathfrak{g}}(\mathbf{x}, \mathbf{y})$ (Section 11.5, Problem 3), so that $A$ is self-adjoint relative to the scalar product Q. Let $E_0$ be the kernel of $A$ (which may be zero) and $E_i$ $(1 \leqq i \leqq r)$ the eigenspaces of $A$ corresponding to the distinct nonzero eigenvalues $c_i$ of $A$, so that $\mathfrak{p}$ is the direct sum of the $E_i$ $(0 \leqq i \leqq r)$, which are pairwise orthogonal with respect to Q; also $B_{\mathfrak{g}}(\mathbf{x}, \mathbf{y}) = c_i Q(\mathbf{x}, \mathbf{y})$ for $\mathbf{x}$ and $\mathbf{y}$ in $E_i$, and $E_0$ is the subspace of $\mathfrak{p}$ orthogonal to $\mathfrak{p}$ with respect to $B_{\mathfrak{g}}$.
(d) The endomorphism $A$ commutes with $\mathrm{ad}(\mathbf{z})$ for all $\mathbf{z} \in \mathfrak{k}$, and therefore $[\mathfrak{k}, E_i] \subset E_i$ for $0 \leqq i \leqq r$. If K is a compact connected Lie group with $\mathfrak{k}$ as Lie algebra, then the sum F of $E_1, \dots, E_r$ is the direct sum of subspaces $\mathfrak{p}_j$ $(1 \leqq j \leqq m)$ stable under $\mathrm{Ad}(t)$ for all $t \in K$, each of which is contained in some $E_i$, and such that each representation $t \mapsto \mathrm{Ad}(t) | \mathfrak{p}_j$ of K is irreducible. The $\mathfrak{p}_j$ are pairwise orthogonal with respect to both Q and $B_{\mathfrak{g}}$; if we put $\mathfrak{p}_0 = E_0$, show that $[\mathfrak{p}_j, \mathfrak{p}_h] = 0$ for $0 \leqq j, h \leqq m$ and $j \neq h$. (If $\mathbf{u} \in \mathfrak{p}_j$, $\mathbf{v} \in \mathfrak{p}_h$, then we have $\mathbf{w} = [\mathbf{u}, \mathbf{v}] \in \mathfrak{k}$; show that $B_{\mathfrak{g}}(\mathbf{w}, \mathbf{w}) = 0$ and use (b) above.)
(e) Put $\mathfrak{g}_j = \mathfrak{p}_j + [\mathfrak{p}_j, \mathfrak{p}_j]$ for $1 \leqq j \leqq m$. Show that the $\mathfrak{g}_j$ are ideals of $\mathfrak{g}$ such that $[\mathfrak{g}_j, \mathfrak{g}_h] = 0$ for $j \neq h$, and that $s(\mathfrak{g}_j) = \mathfrak{g}_j$. By considering the restrictions of $B_{\mathfrak{g}}$ to $\mathfrak{g}_j \times \mathfrak{g}_j$, show that the $\mathfrak{g}_j$ are semisimple Lie algebras; $\mathfrak{g}$ is the direct sum of the $\mathfrak{g}_j$ $(1 \leqq j \leqq m)$ and the centralizer $\mathfrak{g}_0$ of the direct sum of the $\mathfrak{g}_j$ $(1 \leqq j \leqq m)$ (Section 21.6, Problem 4); and we have $s(\mathfrak{g}_0) = \mathfrak{g}_0$, $\mathfrak{p}_0 \subset \mathfrak{g}_0$ and $[\mathfrak{p}_0, \mathfrak{p}_0] = 0$.

**13.** With the notation of Problem 12, suppose that the decomposition of $\mathfrak{p}$ as the direct sum of the $\mathfrak{p}_j$ consists of only one term, and hence that $\mathfrak{g}$ is equal to one of the algebras $\mathfrak{g}_j$.
(a) If $[\mathfrak{p}, \mathfrak{p}] = 0$, then $\mathfrak{g}$ is the semidirect product of $\mathfrak{k}$ and the ideal $\mathfrak{p}$ (19.14.7). Hence there exists a connected Lie group G having $\mathfrak{g}$ as Lie algebra, and a compact subgroup K of G, such that G is the semidirect product of K and a commutative normal subgroup P (so that P is isomorphic to $\mathbf{R}^p \times \mathbf{T}^q$); we may further suppose that K contains no normal subgroup of G other than $\{e\}$. The corresponding Riemannian manifold G/K is the manifold P having as a Riemannian covering $\mathbf{R}^{p+q}$ with its canonical metric; G acts on this

manifold as a transitive group of isometries, containing always the translations of the group P.
(b) If $[\mathfrak{p}, \mathfrak{p}] \neq 0$, then $\mathfrak{g}$ is semisimple; hence there exists a connected semisimple group G having $\mathfrak{g}$ as Lie algebra, and a compact connected subgroup K of G having $\mathfrak{k}$ as Lie algebra, and containing no normal subgroup of G other than $\{e\}$. We have $B_{\mathfrak{g}}(\mathbf{x}, \mathbf{y}) = cQ(\mathbf{x}, \mathbf{y})$ for $\mathbf{x}, \mathbf{y} \in \mathfrak{p}$, with $c \neq 0$. If $c < 0$, then G is compact semisimple; if $\mathfrak{g}$ is not simple, its simple ideals must be permuted by $s$. Show that the irreducibility of the representation $t \mapsto \mathrm{Ad}(t) \mid \mathfrak{p}$ of K implies that $\mathfrak{g}$ has in this case two isomorphic simple ideals $\mathfrak{g}_1, \mathfrak{g}_2$ such that $s(\mathfrak{g}_1) = \mathfrak{g}_2$, with $\mathfrak{k}$ isomorphic to $\mathfrak{g}_1$ and $\mathfrak{g}_2$. The Riemannian symmetric space G/K is then isomorphic to a compact semisimple group with center $\{e\}$, endowed with a left- and right-invariant metric.

If $c < 0$ and $\mathfrak{g}$ is simple, we are in the situation described in **(21.18.13)**.

If $c > 0$, then G is semisimple and noncompact. Show that $\mathfrak{g}$ is necessarily simple, by showing that otherwise $\mathfrak{g}$ would be isomorphic to $\mathfrak{k} \times \mathfrak{k}$. In the complexified Lie algebra $\mathfrak{g}_{(\mathbf{C})}$, $\mathfrak{k} + i\mathfrak{p} = \mathfrak{g}_u$ is the Lie algebra of a compact group, and we are in the situation described in **(21.18.12)**.
(c) Deduce from (a) and (b) and Problem 12 that *every* symmetrized Lie algebra arises from a simply connected Riemannian symmetric space by the procedure of Problem 12(a).

**14.** (a) Let $X$ be a $C^\infty$ vector field on a differential manifold M, and let $F_X$ be the flow of the field **(18.2.1)**. Let $x_0$ be a point of M at which $X(x_0) = 0$. For each $C^\infty$ vector field $Y$ on M, the vector $(\theta_X \cdot Y)(x_0)$ depends only on $Y(x_0)$ (cf. **(17.14.11)**). For each $\mathbf{u} \in T_{x_0}(M)$, let $\tilde{\theta}_{X, x_0} \cdot \mathbf{u}$ denote the value of $(\theta_X \cdot Y)(x_0)$ for each vector field $Y$ such that $Y(x_0) = \mathbf{u}$. If we put $g_t(x) = F_X(x, t)$, we have $g_t(x_0) = x_0$ for all $t \in \mathbf{R}$; for sufficiently small values of $t$, $g_t$ is a diffeomorphism of an open neighborhood $U_0$ of $x_0$ in M onto another open neighborhood $U_t$ of $x_0$, and if $s, t \in \mathbf{R}$ are sufficiently small, then we have $g_{s+t} = g_s \circ g_t = g_t \circ g_s$. Hence if we put $V(t) = T_{x_0}(g_t) \in \mathbf{GL}(T_{x_0}(M))$, we have $V(s+t) = V(s)V(t)$ for all sufficiently small $s$ and $t$. Show that for sufficiently small $t$ we have $V(t) = \exp(t\tilde{\theta}_{X, x_0})$, the exponential being that of the group $\mathbf{GL}(T_{x_0}(M))$.
(b) Suppose that M is endowed with a principal connection $\boldsymbol{P}$ on R(M). If $X$ and $Y$ are infinitesimal automorphisms of the restrictions of $\boldsymbol{P}$ to two neighborhoods U, V of $x_0 \in M$ (Section **20.6**, Problem 6), then $X$ and $Y$ are said to be equivalent if they coincide on a neighborhood of $x_0$ contained in $U \cap V$. The equivalence classes (or *germs*) of infinitesimal automorphisms of restrictions of $\boldsymbol{P}$ to neighborhoods of $x_0$ form a Lie algebra $\mathfrak{g}_{x_0}$ of dimension $\leqq n(n+1)$, where $n = \dim_{x_0}(M)$. The classes of the $X$ such that $X(x_0) = 0$ form a Lie subalgebra $\mathfrak{k}_{x_0}$ of $\mathfrak{g}_{x_0}$. For each class $\xi \in \mathfrak{k}_{x_0}$, the mapping $\tilde{\theta}_{X, x_0} \in \mathrm{End}(T_{x_0}(M))$ is independent of the choice of $X \in \xi$, and the mapping $\xi \mapsto \tilde{\theta}_{X, x_0}$ is an injective homomorphism of $\mathfrak{k}_{x_0}$ into the Lie algebra $\mathfrak{gl}(T_{x_0}(M)) = \mathrm{End}(T_{x_0}(M))$.

**15.** Let M be a connected differential manifold endowed with a linear connection that is *invariant under parallelism* (Section **20.6**, Problem 18). Let U be an open neighborhood of $x_0 \in M$, determined as in Section **20.6**, Problem 15.
(a) For each vector $\mathbf{u} \in T_{x_0}(M)$ and each $t \in \mathbf{R}$ such that $\exp(t\mathbf{u}) \in U$, a *transvection* of the vector $t\mathbf{u}$ is by definition an isomorphism $\tau_{t\mathbf{u}}$ of a sufficiently small neighborhood of $x_0$ onto a sufficiently small neighborhood of $\exp(t\mathbf{u})$, such that $T_{x_0}(\tau_{t\mathbf{u}})$ is the parallel transport of $T_{x_0}(M)$ onto $T_{\exp(t\mathbf{u})}(M)$ along the geodesic $v$ for which $v(0) = x_0$ and $v'(0) = \mathbf{u}$ **(18.6.3)** (cf. Section **20.6**, Problem 18). We have $\tau_{(s+t)\mathbf{u}} = \tau_{s\mathbf{u}} \circ \tau_{t\mathbf{u}}$ if $s$ and $t$ are sufficiently small. For each $y$ in a sufficiently small neighborhood of $x_0$, let $X_{\mathbf{u}}(y)$ be the derivative at $t = 0$ of the mapping $t \mapsto \tau_{t\mathbf{u}}(y)$, so that $X_{\mathbf{u}}(x_0) = \mathbf{u}$; then $X_{\mathbf{u}}$ is an infinitesimal automorphism of the connection restricted to the neighborhood of $x_0$ under consideration. This

field $X_{\mathbf{u}}$ is called an *infinitesimal transvection* relative to $x_0$. Show that the mapping $\mathbf{u} \mapsto \xi_{\mathbf{u}}$ that sends each $\mathbf{u} \in T_{x_0}(M)$ to the germ $\xi_{\mathbf{u}}$ of the infinitesimal transvection $X_{\mathbf{u}}$ (Problem 14(b)) is injective, and therefore identifies $T_{x_0}(M)$ with a vector subspace $\mathfrak{p}_{x_0}$ of the Lie algebra $\mathfrak{g}_{x_0}$. An isomorphism (for the induced connections) of a neighborhood of $x_0$ onto a neighborhood of $x \in M$ that transforms $x_0$ into $x$ also transforms $\mathfrak{p}_{x_0}$ into $\mathfrak{p}_x$, by transport of structure.

For simplicity of notation, we shall henceforth write $\mathfrak{g}$, $\mathfrak{k}$, $\mathfrak{p}$ in place of $\mathfrak{g}_{x_0}$, $\mathfrak{k}_{x_0}$, $\mathfrak{p}_{x_0}$.

(b) Show that for all sufficiently small $t$ and all $C^\infty$ vector fields $Y$ on M, we have $(\theta_{X_{\mathbf{u}}} \cdot Y)(\exp(t\mathbf{u})) = (\nabla_{X_{\mathbf{u}}} \cdot Y)(\exp(t\mathbf{u}))$ (cf. Section **18.6**, Problem 6).

(c) Let $Z$ be an infinitesimal automorphism defined in a neighborhood of $x_0$ and such that $Z(x_0) = 0$, so that its germ $\zeta$ belongs to $\mathfrak{k}$; if we put $g_t(x) = F_Z(x, t)$, then $g_t$ leaves $\mathfrak{p}$ globally invariant, by transport of structure, and transforms a germ $\xi_{\mathbf{u}} \in \mathfrak{p}$ into $\xi_{V(t) \cdot \mathbf{u}}$ (in the notation of Problem 14(a)). Consequently, we have $[\mathfrak{k}, \mathfrak{p}] \subset \mathfrak{p}$, and for each infinitesimal transvection $X_{\mathbf{u}}$ relative to $x_0$ we have $[Z, X_{\mathbf{u}}](x_0) = \bar{\theta}_{X_{\mathbf{u}}, x_0} \cdot \mathbf{u}$.

(d) Show that $\mathfrak{g} = \mathfrak{k} \oplus \mathfrak{p}$. (For an infinitesimal automorphism $Z$ defined in a neighborhood of $x_0$, consider the infinitesimal transvection $X_{\mathbf{u}}$ for $\mathbf{u} = Z(x_0)$.)

(e) Identify $T_{x_0}(M)$ with $\mathfrak{p}$ under the bijection $\mathbf{u} \mapsto X_{\mathbf{u}}$; the bracket $[\mathbf{u}, \mathbf{v}]$ of two vectors $\mathbf{u}, \mathbf{v} \in T_{x_0}(M)$ is then defined by the requirement that $X_{[\mathbf{u}, \mathbf{v}]}$ should be equivalent (Problem 14(b)) to $[X_{\mathbf{u}}, X_{\mathbf{v}}]$. For all $\mathbf{u} \in \mathfrak{g}$, let $\mathbf{u}_{\mathfrak{k}}$ and $\mathbf{u}_{\mathfrak{p}}$ denote the components of $\mathbf{u}$ in $\mathfrak{k}$ and $\mathfrak{p}$, respectively. Show that, for $\mathbf{u}, \mathbf{v}, \mathbf{w}$ in $\mathfrak{p}$ $(= T_{x_0}(M))$, we have

$$\boldsymbol{t} \cdot (\mathbf{u} \wedge \mathbf{v}) = [\mathbf{u}, \mathbf{v}],$$

$$(\boldsymbol{r} \cdot (\mathbf{u} \wedge \mathbf{v})) \cdot \mathbf{w} = -[[\mathbf{u}, \mathbf{v}]_{\mathfrak{k}}, \mathbf{w}],$$

where $\boldsymbol{t}$, $\boldsymbol{r}$ are the torsion and curvature morphisms of M (Section **17.20**). (Use (b) and (c) above to calculate $\boldsymbol{t} \cdot (X_{\mathbf{u}} \wedge X_{\mathbf{v}})$ and $(\boldsymbol{r} \cdot (X_{\mathbf{u}} \wedge X_{\mathbf{v}})) \cdot X_{\mathbf{w}}$ by the formulas **(17.20.1.1)** and **(17.20.6.1)**.)

(f) Let M′ be another connected differential manifold endowed with a linear connection invariant under parallelism, $x'_0$ a point of M′, and $\mathfrak{g}'$, $\mathfrak{k}'$, $\mathfrak{p}'$ the Lie algebras and the vector space corresponding to $\mathfrak{g}$, $\mathfrak{k}$, $\mathfrak{p}$. Suppose that there exists an isomorphism of $\mathfrak{g}$ onto $\mathfrak{g}'$ that maps $\mathfrak{k}$ onto $\mathfrak{k}'$ and $\mathfrak{p}$ onto $\mathfrak{p}'$. Then there exists an isomorphism $f$ of a neighborhood of $x_0$ onto a neighborhood of $x'_0$ (for the connections of M and M′) such that $T_{x_0}(f) = F$ is the restriction to $\mathfrak{p}$ (identified with $T_{x_0}(M)$) of the given isomorphism of $\mathfrak{g}$ onto $\mathfrak{g}'$. (Use (e) above, together with Section **20.6**, Problem 17.) When this is so, for every star-shaped neighborhood U of $\mathbf{0}_{x_0}$ in $T_{x_0}(M)$, on which the exponential mapping is a diffeomorphism, and such that F(U) has the same property in M′, there exists an isomorphism $\bar{f}$ of exp(U) onto exp(F(U)) that extends the restriction of $f$ to a sufficiently small neighborhood of $x_0$. (Use the fact that in the linear differential equations of Section **20.6**, Problem 15, the coefficients $T^i_{lm}(t\mathbf{u})$ and $R^i_{jkl}(t\mathbf{u})$ are constants.)

**16.** (a) Let M be a connected differential manifold endowed with a linear connection **C**. Show that for **C** to be locally symmetric (Section **20.11**, Problem 7) it is necessary and sufficient that **C** be torsion-free and that the parallel transport along a geodesic arc joining two points $x$, $y$ be the tangent linear mapping of an isomorphism (for **C**) of a neighborhood of $x$ onto a neighborhood of $y$. If $s_x$ denotes the symmetry with center $x$ (Section **20.11**, Problem 7), then $s_{x_0}$ defines by transport of structure an involutory automorphism $\sigma$ of the Lie algebra $\mathfrak{g}$ (in the notation of Problem 15) such that $\sigma(\mathbf{u}) = \mathbf{u}$ for $\mathbf{u} \in \mathfrak{k}$ and $\sigma(\mathbf{u}) = -\mathbf{u}$ for $\mathbf{u} \in \mathfrak{p}$, which implies the condition $[\mathfrak{p}, \mathfrak{p}] \subset \mathfrak{k}$. Show that $\mathfrak{k}$ contains no nonzero ideal of $\mathfrak{g}$ (use Problem 14(b)).

(b) Let $a \in M$ be a point in the neighborhood of $x_0$ on which $s_{x_0}$ is defined, and let

$b = s_{x_0}(a)$. Show that $s_{x_0} \circ s_a = s_b \circ s_{x_0}$ in a sufficiently small neighborhood of $a$, and that in a sufficiently small neighborhood of $a$ this mapping coincides with the transvection corresponding to the geodesic arc passing through $x_0$ with endpoints $a$ and $b$ (Problem 15(a)); show that the tangent linear mappings $T_a(s_{x_0} \circ s_a)$ and $T_a(s_b \circ s_{x_0})$ coincide with the parallel transport from $a$ to $b$ along this geodesic arc.

(c) Suppose in addition that M is a pseudo-Riemannian manifold and that **C** is the corresponding Levi-Civita connection. Show that for each $x_0 \in M$ the symmetry $s_{x_0}$ is then an *isometry* of a neighborhood of $x_0$ onto itself. (Use (b) above, by noticing that $s_{x_0} = (s_{x_0} \circ s_a) \circ s_a$ and that a parallel transport along a geodesic arc joining $a$ and $b$ is an isometry of $T_a(M)$ onto $T_b(M)$.)

(d) With the hypotheses of (c), let $\mathfrak{g}_0 \subset \mathfrak{g}$ be the Lie algebra of the germs at $x_0$ of infinitesimal isometries (Section **20.9**, Problem 7). We have $\mathfrak{p} \subset \mathfrak{g}_0$, and if $\mathfrak{k}_0 = \mathfrak{g}_0 \cap \mathfrak{k}$, then $\mathfrak{g}_0 = \mathfrak{k}_0 \oplus \mathfrak{p}$. Furthermore, if $\Phi$ is the nondegenerate symmetric bilinear form on $\mathfrak{p} \times \mathfrak{p}$ (identified with $T_{x_0}(M) \times T_{x_0}(M)$) that is the value at $x_0$ of the metric tensor on M, then we have $\Phi([\mathbf{w}, \mathbf{u}], \mathbf{v}) + \Phi(\mathbf{u}, [\mathbf{w}, \mathbf{v}]) = 0$ for $\mathbf{u}, \mathbf{v} \in \mathfrak{p}$ and $\mathbf{w} \in \mathfrak{k}_0$.

Give an example where $\mathfrak{g}_0 \neq \mathfrak{g}$. (Cf. Section **20.9**, Problem 5.)

(e) Let M′ be another pseudo-Riemannian manifold, locally symmetric with respect to its Levi-Civita connection, and for a point $x'_0 \in M'$ let $\mathfrak{g}'_0$, $\mathfrak{k}'_0$, and $\mathfrak{p}'$ be the Lie algebras and the vector space corresponding to $\mathfrak{g}_0$, $\mathfrak{k}_0$, and $\mathfrak{p}$. For there to exist an isometry of a neighborhood of $x_0$ onto a neighborhood of $x'_0$, transforming $x_0$ into $x'_0$, it is necessary and sufficient that there exist an isomorphism of $\mathfrak{g}_0$ onto $\mathfrak{g}'_0$ that transforms $\mathfrak{k}_0$ into $\mathfrak{k}'_0$ and $\mathfrak{p}$ into $\mathfrak{p}'$. (Use Problem 15(e) and Section **20.6**, Problems 15 and 17.)

(f) Show that for each locally symmetric Riemannian manifold M (i.e., for which the Levi-Civita connection is locally symmetric) and each point $x_0 \in M$, there exists a simply connected Riemannian symmetric space N and an isometry of a neighborhood of $x_0$ onto a neighborhood of a point of N. (Use (d) and (e) above, and Problem 13(c).)

**17.** (a) Let M and M′ be two connected, simply connected, complete Riemannian manifolds (**20.18.5**) satisfying the following condition: there exists a continuous function $v: M \times M' \to \mathbf{R}$ with values $> 0$ such that for each $(x, x') \in M \times M'$ the balls $B(x; v(x, x'))$ and $B(x'; v(x, x'))$ are strictly geodesically convex (**20.17.2**) and such that each isometry of a neighborhood $V \subset B(x; v(x, x'))$ of $x$ onto a neighborhood

$$V' \subset B(x'; v(x, x'))$$

of $x'$, which maps $x$ to $x'$, extends to an isometry of $B(x; v(x, x'))$ onto

$$B(x'; v(x, x')).$$

Show that each isometry of an open subset of M onto an open subset of M extends to an isometry of M onto M′. (Let $x_0 \in M$, and suppose that there exists an isometry $f_0$ of a neighborhood of $x_0$ onto a neighborhood of a point $x'_0 \in M'$ such that $f_0(x_0) = x'_0$. Given any point $x \in M$ and a piecewise-$C^1$ path $\gamma$ from $x_0$ to $x$, define an isometry of a neighborhood of $x$ onto an open set in M′ as follows: if $r$ is the length of $\gamma$ and $c$ the infimum of $v(y, y')$ in the relatively compact set $B(x_0; 2r) \times B(x'_0; 2r)$, consider a sequence $(x_j)_{0 \leq j \leq p}$ of points of $\gamma$ such that $x = x_p$ and the arc of $\gamma$ with endpoints $x_j$ and $x_{j+1}$ has length $< c$ for $0 \leqq j \leqq p-1$. Show that for each $j$ we can define an isometry $f_j$ of $B(x_j; c)$ onto an open ball in M′, such that $f_j$ coincides with $f_{j-1}$ on the geodesically convex set $B(x_{j-1}; c) \cap B(x_j; c)$; for this purpose, use Problem 15(f) above and Section **20.6**, Problem 9(a). Then show that the isometry $f_p$, defined on $B(x, c)$, does not depend on the choice of sequence $(x_j)$ satisfying the conditions above, and consequently that $f_p(x)$ may be written as $f_\gamma(x)$, depending only on $\gamma$. Finally prove that if $\gamma'$ is another piecewise-$C^1$ path from $x_0$ to $x$, then we have $f_{\gamma'}(x) = f_\gamma(x)$, by reasoning as in (9.6.3) and using Section 20.6, Problem 9(a). We have thus defined a local isometry $f$ (20.8.1) of M into M′; proceeding in the same way but starting with $f_0^{-1}$, use Section 20.6, Problem 9(a) once again to complete the proof.)

(b) Deduce from (a) that a locally symmetric, simply connected, complete Riemannian manifold is isometric to a simply connected Riemannian symmetric space. (Use Section 20.18, Problem 9, together with Problem 13(c) above.)

18. Let G be the universal covering group of $\mathbf{SL}(2, \mathbf{R})$, and identify with $\mathbf{Z}$ the kernel of the canonical homomorphism $G \to \mathbf{SL}(2, \mathbf{R})$ (21.18.11). Let $a \in \mathbf{T}^n$ be an element whose powers form a dense set in $\mathbf{T}^n$ (Section 19.7, Problem 6). Let D be the discrete subgroup of $G \times \mathbf{T}^n$ generated by $(1, a)$, and let $H = (G \times \mathbf{T}^n)/D$. We have $\mathrm{Lie}(H) = \mathfrak{sl}(2, \mathbf{R}) \times \mathbf{R}^n$. Show that the connected Lie group H′ immersed in H, with Lie algebra $\mathfrak{sl}(2, \mathbf{R}) \times \{0\}$, is dense in H. (Cf. Section 21.6, Problem 5.)

## 19. ROOTS OF A COMPLEX SEMISIMPLE LIE ALGEBRA

**(21.19.1)** Our aim now is to show that a *complex semisimple* Lie algebra $\mathfrak{g}$ of dimension $n$ is always isomorphic to the complexification of the Lie algebra of some *compact semisimple* Lie group. The method we shall follow consists, as a first step, in constructing a *commutative* Lie subalgebra $\mathfrak{h}$ of $\mathfrak{g}$ and a direct sum decomposition of the type (21.10.1.1) possessing the properties (A), (B), and (C) of Section 21.10; from this it will follow that all the results of Sections 21.10 and 21.11 that rest only on these properties are applicable, and the second step is to show that by use of these results it is possible to construct a Lie algebra of a compact Lie group, having $\mathfrak{g}$ as complexification.

**(21.19.2)** Let $\mathfrak{g}$ be an *arbitrary* complex Lie algebra of finite dimension $n$. For each element $\mathbf{u} \in \mathfrak{g}$, the eigenvalues of the endomorphism $\mathrm{ad}(\mathbf{u})$ of the complex vector space $\mathfrak{g}$ are given by the characteristic equation

(21.19.2.1) $$\det(\mathrm{ad}(\mathbf{u}) - \xi \cdot 1_{\mathfrak{g}}) = 0,$$

the left-hand side of which is a polynomial in $\xi$ of degree $n$, with $(-1)^n$ as coefficient of $\xi^n$. Let $\mathbf{u}_0$ be an element of $\mathfrak{g}$ for which the number of *distinct* roots of (21.19.2.1) is as large as possible. Since $[\mathbf{u}_0, \mathbf{u}_0] = 0$, $\mathrm{ad}(\mathbf{u}_0)$ will always have 0 as an eigenvalue; let then $\lambda_0 = 0, \lambda_1, \ldots, \lambda_m$ denote the distinct eigenvalues of $\mathrm{ad}(\mathbf{u}_0)$, and let $\mathfrak{g}_k$ $(0 \leq k \leq m)$ denote the vector subspace $\mathrm{N}(\lambda_k)$ of $\mathfrak{g}$ on which $\mathrm{ad}(\mathbf{u}_0) - \lambda_k \cdot 1_{\mathfrak{g}}$ is nilpotent. From (11.4.1), $\mathfrak{g}$ is the *direct sum* of the $\mathfrak{g}_k$.

**(21.19.3)** *For all indices h, k in* $[0, m]$, *we have* $[\mathfrak{g}_h, \mathfrak{g}_k] = 0$ *if* $\lambda_h + \lambda_k$ *is not an eigenvalue of* $\operatorname{ad}(\mathbf{u}_0)$, *and* $[\mathfrak{g}_h, \mathfrak{g}_k] \subset \mathfrak{g}_l$ *if* $\lambda_h + \lambda_k = \lambda_l$ *for some index l. In particular,* $\mathfrak{g}_0$ *is a Lie subalgebra of* $\mathfrak{g}$, *and we have* $[\mathfrak{g}_0, \mathfrak{g}_k] \subset \mathfrak{g}_k$ *for* $1 \leqq k \leqq m$.

For all $\mathbf{x}, \mathbf{y} \in \mathfrak{g}$ we have

$$(\operatorname{ad}(\mathbf{u}_0) - (\lambda_h + \lambda_k) \cdot 1)[\mathbf{x}, \mathbf{y}] = [(\operatorname{ad}(\mathbf{u}_0) - \lambda_h \cdot 1) \cdot \mathbf{x}, \mathbf{y}] + [\mathbf{x}, (\operatorname{ad}(\mathbf{u}_0) - \lambda_k \cdot 1) \cdot \mathbf{y}]$$

from which it follows immediately by induction on $p$ that the element $(\operatorname{ad}(\mathbf{u}_0) - (\lambda_h + \lambda_k) \cdot 1)^p[\mathbf{x}, \mathbf{y}]$ is a linear combination of brackets of the form

$$[(\operatorname{ad}(\mathbf{u}_0) - \lambda_h \cdot 1)^r \cdot \mathbf{x}, (\operatorname{ad}(\mathbf{u}_0) - \lambda_k \cdot 1)^{p-r} \cdot \mathbf{y}]$$

for $0 \leqq r \leqq p$. The restriction of $\operatorname{ad}(\mathbf{u}_0) - (\lambda_h + \lambda_k) \cdot 1$ to $[\mathfrak{g}_h, \mathfrak{g}_k]$ is therefore nilpotent, and the proposition is proved.

**(21.19.4)** Since $[\mathfrak{g}_0, \mathfrak{g}_k] \subset \mathfrak{g}_k$ for $0 \leqq k \leqq m$, it follows that for each element $\mathbf{u} \in \mathfrak{g}_0$ the endomorphism $\operatorname{ad}(\mathbf{u})$ leaves *stable* each of the subspaces $\mathfrak{g}_k$. Let $\alpha_k$ denote the linear form

$$\mathbf{u} \mapsto \frac{1}{\dim \mathfrak{g}_k} \operatorname{Tr}(\operatorname{ad}(\mathbf{u}) \,|\, \mathfrak{g}_k)$$

on the vector space $\mathfrak{g}_0$. Also let $\mathrm{P}_k(\mathbf{u})$ denote the characteristic polynomial of the restriction of $\operatorname{ad}(\mathbf{u})$ to $\mathfrak{g}_k$: the coefficients of this polynomial (in $\xi$) are therefore polynomials in the coordinates of $\mathbf{u}$ with respect to any given basis of $\mathfrak{g}$. Hence the *resultant*† $\mathrm{R}_{hk}(\mathbf{u})$ of the polynomials $\mathrm{P}_h(\mathbf{u})$ and $\mathrm{P}_k(\mathbf{u})$, where $h \neq k$, is a polynomial in the coordinates of $\mathbf{u}$ that is *not identically zero*, because $\mathrm{P}_h(\mathbf{u}_0)(\xi) = (\lambda_h - \xi)^{\dim \mathfrak{g}_h}$ and $\mathrm{P}_k(\mathbf{u}_0)(\xi) = (\lambda_k - \xi)^{\dim \mathfrak{g}_k}$ have no root in common. It follows that, for each pair of distinct indices $h$, $k$, the set of elements $\mathbf{u} \in \mathfrak{g}_0$ such that $\mathrm{R}_{hk}(\mathbf{u}) = 0$ is *nowhere dense* in $\mathfrak{g}_0$ (for otherwise $\mathrm{R}_{hk}$ would be identically zero, by virtue of the principle of analytic continuation **(9.4.1)**). Let E be the dense open subset of $\mathfrak{g}_0$ in which $\mathrm{R}_{hk}(\mathbf{u}) \neq 0$ for *all* pairs $(h, k)$ of distinct indices.

**(21.19.5)** *For each* $\mathbf{u} \in \mathfrak{g}_0$, *the restriction of* $\operatorname{ad}(\mathbf{u}) - \alpha_k(\mathbf{u}) \cdot 1_{\mathfrak{g}}$ *to* $\mathfrak{g}_k$ *is a nilpotent endomorphism. Furthermore,* $\alpha_0(\mathbf{u}) = 0$ *for all* $\mathbf{u} \in \mathfrak{g}_0$.

Suppose first that $\mathbf{u} \in \mathrm{E}$. Since $\operatorname{ad}(\mathbf{u}) \cdot \mathbf{u} = 0$, it is enough to show that the restriction of $\operatorname{ad}(\mathbf{u})$ to each $\mathfrak{g}_k$ cannot have two or more district eigenvalues. Since, by the definition of E, these eigenvalues would be distinct

† See, for example, my book *Infinitesimal Calculus*, Paris (Hermann), 1968, p. 61.

from all the eigenvalues of the restrictions of $\mathrm{ad}(\mathbf{u})$ to the other $\mathfrak{g}_h$, and since there are at least $m$ of these that are all distinct, it follows that the endomorphism $\mathrm{ad}(\mathbf{u})$ of $\mathfrak{g}$ would have at least $m+2$ distinct eigenvalues, contrary to the choice of $\mathbf{u}_0$. Hence for $\mathbf{u} \in \mathrm{E}$ we have $\mathrm{P}_k(\mathbf{u})(\xi) = (\alpha_k(\mathbf{u}) - \xi)^{\dim \mathfrak{g}_k}$ and $\mathrm{P}_0(\mathbf{u})(\xi) = (-\xi)^{\dim \mathfrak{g}_0}$; by continuity, these relations hold for all $\mathbf{u} \in \mathfrak{g}_0$, because E is dense in $\mathfrak{g}_0$.

**(21.19.6)** We shall now change our notation and denote by $\mathfrak{h}$ the Lie subalgebra $\mathfrak{g}_0$ of $\mathfrak{g}$, and by $\mathbf{S}$ the set of linear forms $\alpha_1, \alpha_2, \ldots, \alpha_m$ on $\mathfrak{h}$, which are all $\neq 0$ and pairwise distinct, because they take distinct nonzero values at the point $\mathbf{u}_0$; also we shall write $\mathfrak{g}(\alpha_k)$ in place of $\mathfrak{g}_k$, and put $\mathfrak{g}(\beta) = \{0\}$ for every linear form $\beta$ on $\mathfrak{h}$ distinct from 0 and the $\alpha \in \mathbf{S}$. Then $\mathfrak{g}(\alpha)$ may also be defined as the largest vector subspace of $\mathfrak{g}$ such that for *each* $\mathbf{u} \in \mathfrak{h}$ the restriction of $\mathrm{ad}(\mathbf{u}) - \alpha(\mathbf{u}) \cdot 1_{\mathfrak{g}}$ to this subspace is nilpotent. The proof of **(21.19.3)** shows that

**(21.19.6.1)** $$[\mathfrak{g}(\alpha), \mathfrak{g}(\beta)] \subset \mathfrak{g}(\alpha + \beta)$$

for any two linear forms $\alpha$, $\beta$ on $\mathfrak{h}$.

**(21.19.7)** *For all* $\mathbf{u}, \mathbf{v} \in \mathfrak{h}$ *and all* $\alpha \in \mathbf{S}$ *we have* $\alpha([\mathbf{u}, \mathbf{v}]) = 0$.

For the trace of the restriction of $\mathrm{ad}([\mathbf{u}, \mathbf{v}])$ to $\mathfrak{g}(\alpha)$ is $\dim(\mathfrak{g}(\alpha)) \cdot \alpha([\mathbf{u}, \mathbf{v}])$. On the other hand, we have

$$\mathrm{ad}([\mathbf{u}, \mathbf{v}]) = \mathrm{ad}(\mathbf{u})\,\mathrm{ad}(\mathbf{v}) - \mathrm{ad}(\mathbf{v})\,\mathrm{ad}(\mathbf{u}),$$

and therefore the trace of the restriction of $\mathrm{ad}([\mathbf{u}, \mathbf{v}])$ to $\mathfrak{g}(\alpha)$ is zero.

**((21.19.8)** *For all elements* $\mathbf{u}, \mathbf{v} \in \mathfrak{h}$, *we have*

**(21.19.8.1)** $$\mathrm{B}_{\mathfrak{g}}(\mathbf{u}, \mathbf{v}) = \sum_{\alpha \in \mathbf{S}} (\dim \mathfrak{g}(\alpha)) \cdot \alpha(\mathbf{u})\alpha(\mathbf{v}).$$

Since $\mathrm{B}_{\mathfrak{g}}$ is bilinear and symmetric, it is sufficient to calculate $\mathrm{B}_{\mathfrak{g}}(\mathbf{u}+\mathbf{v}, \mathbf{u}+\mathbf{v})$: in other words, we need only prove **(21.19.8.1)** when $\mathbf{u} = \mathbf{v}$. But then the restriction of $\mathrm{ad}(\mathbf{u})^2$ to $\mathfrak{g}(\alpha)$ has the single eigenvalue $\alpha(\mathbf{u})^2$; since the restriction of $\mathrm{ad}(\mathbf{u})$ to $\mathfrak{h}$ is nilpotent, the result follows.

**(21.19.9)** *If* $\alpha + \beta \neq 0$, *the subspaces* $\mathfrak{g}(\alpha)$ *and* $\mathfrak{g}(\beta)$ *are orthogonal relative to the Killing form* $\mathrm{B}_{\mathfrak{g}}$.

Let $\mathbf{u} \in \mathfrak{g}(\alpha)$ and $\mathbf{v} \in \mathfrak{g}(\beta)$. Then it follows from **(21.19.6.1)** that the image of $\mathfrak{g}(\gamma)$ under $\mathrm{ad}(\mathbf{u})\,\mathrm{ad}(\mathbf{v})$ is contained in $\mathfrak{g}(\alpha+\beta+\gamma)$. If we take a basis of $\mathfrak{g}$

consisting of a basis of $\mathfrak{h} = \mathfrak{g}(0)$ and bases of each of the subspaces $\mathfrak{g}(\alpha)$, $\alpha \in \mathbf{S}$, it is clear that the diagonal elements of the matrix of $\mathrm{ad}(\mathbf{u})\,\mathrm{ad}(\mathbf{v})$ relative to this basis are all zero, and the result follows.

**(21.19.10)** *Let $\alpha$, $\beta$ be two linear forms belonging to* $\mathbf{S}$. *Let $p$* (resp. $q$) *be the smallest* (resp. *largest*) *rational integer such that $\beta + p\alpha$* (resp. $\beta + q\alpha$) *belongs to* $\mathbf{S}$. *Then, for all $\mathbf{u} \in [\mathfrak{g}(\alpha), \mathfrak{g}(-\alpha)] \subset \mathfrak{h}$, we have*

$$\sum_{k=p}^{q} (\dim \mathfrak{g}(\beta + k\alpha))(\beta(\mathbf{u}) + k\alpha(\mathbf{u})) = 0 \tag{21.19.10.1}$$

*and consequently $\beta(\mathbf{u}) = r_{\alpha\beta}\,\alpha(\mathbf{u})$, where $r_{\alpha\beta}$ is a rational number.*

Consider the subspace V of $\mathfrak{g}$ that is the direct sum of the $\mathfrak{g}(\beta + k\alpha)$ for $p \leqq k \leqq q$. It will suffice to prove the formula (21.19.10.1) for $\mathbf{u} = [\mathbf{x}, \mathbf{y}]$, where $\mathbf{x} \in \mathfrak{g}(\alpha)$ and $\mathbf{y} \in \mathfrak{g}(-\alpha)$. Since the image of $\mathfrak{g}(\gamma)$ under $\mathrm{ad}(\mathbf{x})$ is contained in $\mathfrak{g}(\gamma + \alpha)$, by (21.19.6.1), and since $\mathfrak{g}(\beta + q\alpha + \alpha) = \{0\}$, it follows that V is *stable* under $\mathrm{ad}(\mathbf{x})$. Likewise, the image of $\mathfrak{g}(\gamma)$ under $\mathrm{ad}(\mathbf{y})$ is contained in $\mathfrak{g}(\gamma - \alpha)$, and we have $\mathfrak{g}(\beta + p\alpha - \alpha) = \{0\}$, so that V is stable under $\mathrm{ad}(\mathbf{y})$ and hence also under $\mathrm{ad}([\mathbf{x}, \mathbf{y}]) = \mathrm{ad}(\mathbf{x})\,\mathrm{ad}(\mathbf{y}) - \mathrm{ad}(\mathbf{y})\,\mathrm{ad}(\mathbf{x})$. This being so, the trace of the restriction to V of $\mathrm{ad}(\mathbf{x})\,\mathrm{ad}(\mathbf{y}) - \mathrm{ad}(\mathbf{y})\,\mathrm{ad}(\mathbf{x})$ is zero. If we now observe that the restriction of $\mathrm{ad}([\mathbf{x}, \mathbf{y}])$ to $\mathfrak{g}(\beta + k\alpha)$ has only one eigenvalue, namely $\beta([\mathbf{x}, \mathbf{y}]) + k\alpha([\mathbf{x}, \mathbf{y}])$, the formula (21.19.10.1) follows immediately.

**(21.19.11)** *Suppose that the Lie algebra $\mathfrak{g}$ is semisimple. Then* (with the same notation as above):

(i) *The restriction to $\mathfrak{h}$ of the Killing form $\mathrm{B}_{\mathfrak{g}}$ is nondegenerate.*

(ii) *If $\alpha \in \mathbf{S}$, then also $-\alpha \in \mathbf{S}$.*

(iii) *If $l$ is the dimension of $\mathfrak{h}$, there exist $l$ linearly independent forms belonging to* $\mathbf{S}$.

(iv) *$\mathfrak{h}$ is a maximal commutative subalgebra of $\mathfrak{g}$.*

(v) *For each $\mathbf{u} \in \mathfrak{h}$, the restriction of* $\mathrm{ad}(\mathbf{u})$ *to $\mathfrak{g}(\alpha)$, for each $\alpha \in \mathbf{S}$, is a homothety of ratio $\alpha(\mathbf{u})$: in other words*

$$[\mathbf{u}, \mathbf{x}] = \alpha(\mathbf{u})\mathbf{x} \tag{21.19.11.1}$$

*for all $\mathbf{x} \in \mathfrak{g}(\alpha)$.*

(i) If $\mathbf{u} \in \mathfrak{h}$ is orthogonal to $\mathfrak{h}$ (relative to $\mathrm{B}_{\mathfrak{g}}$), then $\mathbf{u}$ is orthogonal to all of $\mathfrak{g}$, because by virtue of (21.19.9) it is orthogonal to each $\mathfrak{g}(\alpha)$, $\alpha \in \mathbf{S}$. Since $\mathfrak{g}$ is semisimple, it follows that $\mathbf{u} = 0$.

(ii) If we had $-\alpha \notin \mathbf{S}$, then we should have $\alpha + \beta \neq 0$ for each $\beta \in \mathbf{S}$, and therefore by virtue of **(21.19.9)** all the elements of $\mathfrak{g}(\alpha)$ would be orthogonal to each $\mathfrak{g}(\beta)$, $\beta \in \mathbf{S}$; since they are also orthogonal to $\mathfrak{h}$, they would be orthogonal to the whole of $\mathfrak{g}$, and this is impossible since $\mathfrak{g}$ is semisimple.

(iii) If the rank of $\mathbf{S}$, in the dual space of $\mathfrak{h}$, were strictly less than $l$, then there would exist an element $\mathbf{u} \neq 0$ in $\mathfrak{h}$ such that $\alpha(\mathbf{u}) = 0$ for all $\alpha \in \mathbf{S}$; hence $B_{\mathfrak{g}}(\mathbf{u}, \mathbf{v}) = 0$ for all $\mathbf{v} \in \mathfrak{h}$ by virtue of the formula **(21.19.8.1)**, and this would contradict (i).

(iv) Since $\alpha([\mathbf{u}, \mathbf{v}]) = 0$ for all $\alpha \in \mathbf{S}$ and all $\mathbf{u}, \mathbf{v} \in \mathfrak{h}$ **(21.19.7)**, it follows from (iii) that $[\mathbf{u}, \mathbf{v}] = 0$, in other words, that $\mathfrak{h}$ is commutative. Hence $\mathfrak{h}$ is the kernel of $\operatorname{ad}(\mathbf{u}_0)$ and is therefore a *maximal* commutative subalgebra.

(v) The endomorphism $\operatorname{ad}(\mathbf{u})$ decomposes uniquely as a sum $S + N$, where $S$ and $N$ are endomorphisms of the vector space $\mathfrak{g}$, which are polynomials in $\operatorname{ad}(\mathbf{u})$ with complex coefficients, such that $N$ is nilpotent, $S$ diagonalizable and $SN = NS$ **(A.25.3)**. Because $S$ is a polynomial in $\operatorname{ad}(\mathbf{u})$, it stabilizes $\mathfrak{h}$ and the $\mathfrak{g}(\alpha)$, and the triangular form **(A.6.10)** of the restriction of $\operatorname{ad}(\mathbf{u})$ to each $\mathfrak{g}(\alpha)$ shows that we have $S \cdot \mathbf{x} = \alpha(\mathbf{u})\mathbf{x}$ for all $\mathbf{x} \in \mathfrak{g}(\alpha)$. Bearing in mind **(21.19.6.1)**, we deduce that $S \cdot [\mathbf{x}, \mathbf{y}] = [S \cdot \mathbf{x}, \mathbf{y}] + [\mathbf{x}, S \cdot \mathbf{y}]$ for all $\mathbf{x} \in \mathfrak{g}(\alpha)$ and $\mathbf{y} \in \mathfrak{g}(\beta)$; and it then follows by linearity that $S$ is a *derivation* of the Lie algebra $\mathfrak{g}$. But $\mathfrak{g}$ is semisimple, hence every derivation of $\mathfrak{g}$ is inner **(21.6.7)**: that is to say, there exists $\mathbf{v} \in \mathfrak{g}$ such that $S = \operatorname{ad}(\mathbf{v}) \in \operatorname{ad}(\mathfrak{g})$. Now $\mathbf{x} \mapsto \operatorname{ad}(\mathbf{x})$ is an isomorphism of $\mathfrak{g}$ onto $\operatorname{ad}(\mathfrak{g})$ **(21.6.3)**, and therefore $\operatorname{ad}(\mathbf{u})$ commutes with $\operatorname{ad}(\mathbf{w})$ for all $\mathbf{w} \in \mathfrak{h}$; hence $S$, being a polynomial in $\operatorname{ad}(\mathbf{u})$, also commutes with $\operatorname{ad}(\mathbf{w})$ for all $\mathbf{w} \in \mathfrak{h}$. Since $\operatorname{ad}(\mathfrak{h})$ is a *maximal* commutative subalgebra of $\operatorname{ad}(\mathfrak{g})$, by (iv) above, it follows that $\mathbf{v} \in \mathfrak{h}$. Since $\operatorname{Tr}(\operatorname{ad}(\mathbf{v}) \mid \mathfrak{g}(\alpha)) = \dim(\mathfrak{g}(\alpha)) \cdot \alpha(\mathbf{v})$ and $S \cdot \mathbf{x} = \alpha(\mathbf{u})\mathbf{x}$ for all $\mathbf{x} \in \mathfrak{g}(\alpha)$, we see that $\alpha(\mathbf{u}) = \alpha(\mathbf{v})$ for all $\alpha \in \mathbf{S}$. By virtue of (iii), this implies that $\mathbf{u} = \mathbf{v}$ and shows that $\operatorname{ad}(\mathbf{u}) = S$ is diagonalizable.

Because of **(21.19.11.1)**, for a *semisimple* complex Lie algebra $\mathfrak{g}$ the linear forms $\alpha \in \mathbf{S}$ will henceforth be called the *roots* of $\mathfrak{g}$ relative to $\mathfrak{h}$.

**(21.19.12)** Since the restriction of $B_{\mathfrak{g}}$ to $\mathfrak{h}$ is nondegenerate, for each root $\alpha \in \mathbf{S}$ there exists a unique element $\mathbf{h}_\alpha^0 \in \mathfrak{h}$ such that

$$\alpha(\mathbf{u}) = B_{\mathfrak{g}}(\mathbf{u}, \mathbf{h}_\alpha^0) \qquad \textbf{(21.19.12.1)}$$

for all $\mathbf{u} \in \mathfrak{h}$.

**(21.19.13)** *Suppose that the Lie algebra $\mathfrak{g}$ is semisimple. Then, for each root $\alpha \in \mathbf{S}$:*

(i) *For each* $\mathbf{x} \in \mathfrak{g}(\alpha)$ *and* $\mathbf{y} \in \mathfrak{g}(-\alpha)$, *we have*

(21.19.13.1) $$[\mathbf{x}, \mathbf{y}] = \mathrm{B}_{\mathfrak{g}}(\mathbf{x}, \mathbf{y}) \cdot \mathbf{h}_\alpha^0.$$

(ii) $\alpha(\mathbf{h}_\alpha^0) \neq 0$.

(i) We have $[\mathbf{x}, \mathbf{y}] \in \mathfrak{h}$, and from the invariance of the Killing form (21.5.6.1)

$$\mathrm{B}_{\mathfrak{g}}([\mathbf{x}, \mathbf{y}], \mathbf{u}) = \mathrm{B}_{\mathfrak{g}}([\mathbf{u}, \mathbf{x}], \mathbf{y}) = \alpha(\mathbf{u})\mathrm{B}_{\mathfrak{g}}(\mathbf{x}, \mathbf{y}).$$

The formula (21.19.13.1) now follows from (21.19.12.1), since $\mathrm{B}_{\mathfrak{g}}$ is nondegenerate.

(ii) Let $\mathbf{x}$ be an element $\neq 0$ in $\mathfrak{g}(\alpha)$. Then $\mathbf{x}$ cannot be orthogonal to $\mathfrak{g}(-\alpha)$ relative to $\mathrm{B}_{\mathfrak{g}}$, for otherwise it would follow from (21.19.9) that $\mathbf{x}$ was orthogonal to all of $\mathfrak{g}$, contrary to the fact that $\mathfrak{g}$ is semisimple. Hence there exists an element $\mathbf{y}$ in $\mathfrak{g}(-\alpha)$ such that $[\mathbf{x}, \mathbf{y}] = \mathbf{h}_\alpha^0$, by virtue of (21.19.13.1). This being so, it follows from (21.19.10) that $\beta([\mathbf{x}, \mathbf{y}]) = r_{\alpha\beta}\,\alpha([\mathbf{x}, \mathbf{y}]) = r_{\alpha\beta}\,\alpha(\mathbf{h}_\alpha^0)$ for *each* root $\beta \in \mathbf{S}$. If we had $\alpha(\mathbf{h}_\alpha^0) = 0$, then we should have $\beta(\mathbf{h}_\alpha^0) = 0$ for *all* roots $\beta \in \mathbf{S}$; since $\mathbf{h}_\alpha^0 \neq 0$ by virtue of (21.19.12.1), this would contradict the existence of $l$ linearly independent roots (21.19.11(iii)).

## PROBLEMS

1. (a) Let $\mathfrak{g}$ be a finite-dimensional complex Lie algebra and $\mathbf{u}$ *any* element of $\mathfrak{g}$; let $\lambda_0 = 0$, $\lambda_1, \dots, \lambda_m$ be the distinct eigenvalues of $\mathrm{ad}(\mathbf{u})$, and $\mathfrak{g}_k$ $(0 \leq k \leq m)$ the subspace $\mathrm{N}(\lambda_k)$ defined in (11.4.1), so that $\mathfrak{g}$ is the direct sum of the $\mathfrak{g}_k$. If $S$ and $N$ are the diagonalizable and nilpotent endomorphisms of the vector space $\mathfrak{g}$, such that $S + N = \mathrm{ad}(\mathbf{u})$ and $SN = NS$ (A.25.2), show that $S$ and $N$ are derivations of the Lie algebra $\mathfrak{g}$. (Argue as in (21.19.11(iv)).)

   (b) Suppose that $\mathfrak{g}$ is semisimple. Then there exist uniquely determined elements $\mathbf{v}, \mathbf{w} \in \mathfrak{g}$ such that $\mathrm{ad}(\mathbf{v}) = S$ and $\mathrm{ad}(\mathbf{w}) = N$. By abuse of language, $\mathbf{v}$ and $\mathbf{w}$ are called respectively the *semisimple* and *nilpotent components* of $\mathbf{u}$; they satisfy $[\mathbf{v}, \mathbf{w}] = 0$.

   (c) Suppose that $\mathfrak{g}$ is semisimple, and let $\mathfrak{b}$ be a Lie subalgebra of $\mathfrak{g}$ that is equal to its normalizer in $\mathfrak{g}$. Show that for each $\mathbf{u} \in \mathfrak{b}$, the semisimple and nilpotent components of $\mathbf{u}$ belong to $\mathfrak{b}$.

2. Let $\mathfrak{g}$ be a complex semisimple Lie algebra. Suppose that there exist $m + 1$ distinct complex numbers $\lambda_0 = 0, \lambda_1, \dots, \lambda_m$ and a decomposition $\mathfrak{g} = \mathfrak{g}_0 \oplus \mathfrak{g}_1 \oplus \cdots \oplus \mathfrak{g}_m$ of $\mathfrak{g}$ as a direct sum of vector subspaces such that $[\mathfrak{g}_h, \mathfrak{g}_j] = 0$ if $\lambda_h + \lambda_j$ is not one of the $\lambda_l$, and $[\mathfrak{g}_h, \mathfrak{g}_j] \subset \mathfrak{g}_l$ if $\lambda_h + \lambda_j = \lambda_l$. Show that there exists an element $\mathbf{u} \in \mathfrak{g}_0$ such that $\mathrm{ad}(\mathbf{u})$ leaves stable each of the $\mathfrak{g}_j$, and such that the restriction of $\mathrm{ad}(\mathbf{u})$ to $\mathfrak{g}_j$ is multiplication by $\lambda_j$ $(0 \leq j \leq m)$. Extend this result to real semisimple Lie algebras when the $\lambda_j$ $(1 \leq j \leq m)$ are real numbers.

## 20. WEYL BASES

(21.20.1) It follows from (21.19.11) and (21.19.13) that for a complex semisimple Lie algebra $\mathfrak{g}$ there exists a commutative subalgebra $\mathfrak{h}$ of $\mathfrak{g}$ and a finite set $\mathbf{S} \subset \mathfrak{h}^* - \{0\}$ of linear forms, such that the direct sum decomposition

(21.20.1.1) $$\mathfrak{g} = \mathfrak{h} \oplus \bigoplus_{\alpha \in \mathbf{S}} \mathfrak{g}(\alpha)$$

and the Killing form $B_{\mathfrak{g}}$ satisfy conditions (A), (B), and (C) of (21.10.1). For brevity we shall call (21.20.1.1) a *root decomposition* of $\mathfrak{g}$, $\mathfrak{h}$ being the maximal commutative subalgebra and $\mathbf{S}$ the root system corresponding to this decomposition. We may then apply all the results of (21.10) and (21.11), with the (provisional) exception of those of (21.11.9), (21.11.10), and (21.11.11). The linear forms $v_\alpha$ that feature in (21.11.1) are here given by $v_\alpha(\lambda) = 2\lambda(\mathbf{h}_\alpha^0)/\alpha(\mathbf{h}_\alpha^0)$.

In particular, each subspace $\mathfrak{g}(\alpha)$ ($\alpha \in \mathbf{S}$) is one-dimensional over $\mathbf{C}$ (21.10.3), $\mathfrak{g}(\alpha) \oplus \mathfrak{g}(-\alpha)$ is nonisotropic relative to $B_{\mathfrak{g}}$ (21.10.2), and we may therefore choose in each $\mathfrak{g}(\alpha)$ a vector $\mathbf{e}_\alpha$ such that for all $\alpha \in \mathbf{S}$ we have

(21.20.1.2) $$B_{\mathfrak{g}}(\mathbf{e}_\alpha, \mathbf{e}_{-\alpha}) = 1$$

and hence, by (21.19.13.1),

(21.20.1.3) $$[\mathbf{e}_\alpha, \mathbf{e}_{-\alpha}] = \mathbf{h}_\alpha^0.$$

By virtue of (21.19.6.1), we may therefore write, for any two roots $\alpha$, $\beta$ in $\mathbf{S}$,

(21.20.1.4) $$[\mathbf{e}_\alpha, \mathbf{e}_\beta] = N(\alpha, \beta)\mathbf{e}_{\alpha+\beta}$$

if $\alpha + \beta \in \mathbf{S}$, with $N(\alpha, \beta) \in \mathbf{C}$, and

(21.20.1.5) $$[\mathbf{e}_\alpha, \mathbf{e}_\beta] = 0$$

if $\alpha + \beta \notin \mathbf{S}$ and $\alpha + \beta \neq 0$. We therefore define $N(\alpha, \beta)$ to be 0 when $\alpha + \beta \notin \mathbf{S}$ and $\alpha + \beta \neq 0$.

These formulas, together with (21.19.11.1), show that the assignment of the numbers $N(\alpha, \beta)$ determines completely (once the roots are known) the mapping $(\mathbf{x}, \mathbf{y}) \mapsto [\mathbf{x}, \mathbf{y}]$ of $\mathfrak{g} \times \mathfrak{g}$ into $\mathfrak{g}$. By expressing that this mapping defines a Lie algebra structure, we shall obtain necessary conditions relating the $N(\alpha, \beta)$.

In the first place, since the mapping $(\mathbf{x}, \mathbf{y}) \mapsto [\mathbf{x}, \mathbf{y}]$ is skew-symmetric, we must have

(21.20.1.6) $$N(\beta, \alpha) = -N(\alpha, \beta).$$

The following three lemmas follow from the Jacobi identity and properties of root systems.

(21.20.2) *Let $\alpha, \beta, \gamma \in \mathbf{S}$ be such that $\alpha + \beta + \gamma = 0$. Then*

(21.20.2.1) $$N(\alpha, \beta) = N(\beta, \gamma) = N(\gamma, \alpha).$$

We have $[\mathbf{e}_\alpha, [\mathbf{e}_\beta, \mathbf{e}_\gamma]] = N(\beta, \gamma)[\mathbf{e}_\alpha, \mathbf{e}_{-\alpha}] = N(\beta, \gamma)\mathbf{h}_\alpha^0$, so that the Jacobi identity

(21.20.2.2) $$[\mathbf{e}_\alpha, [\mathbf{e}_\beta, \mathbf{e}_\gamma]] + [\mathbf{e}_\beta, [\mathbf{e}_\gamma, \mathbf{e}_\alpha]] + [\mathbf{e}_\gamma, [\mathbf{e}_\alpha, \mathbf{e}_\beta]] = 0$$

gives the relation $N(\beta, \gamma)\mathbf{h}_\alpha^0 + N(\gamma, \alpha)\mathbf{h}_\beta^0 + N(\alpha, \beta)\mathbf{h}_\gamma^0 = 0$; consequently, by virtue of (21.19.12.1), we have

(21.20.2.3) $$N(\beta, \gamma)\alpha + N(\gamma, \alpha)\beta + N(\alpha, \beta)\gamma = 0.$$

But the subspace of $\mathfrak{h}^*$ spanned by $\alpha$, $\beta$, and $\gamma$, which is at most two-dimensional, cannot have dimension 1. For if this were so, the three roots $\alpha$, $\beta$, $\gamma$ would all be scalar multiples of one of them, say $\alpha$; but then $\beta$ and $\gamma$ would have to be equal to $\pm\alpha$ (21.10.3), and since $3\alpha \neq 0$ we should have either $\beta = -\alpha$ or $\gamma = -\alpha$, whence either $\gamma = 0$ or $\beta = 0$, both of which are impossible. Hence, replacing $\gamma$ by $-\alpha - \beta$ in (21.20.2.3), we obtain (21.20.2.1).

(21.20.3) *Let $\alpha, \beta, \gamma, \delta$ be four roots (distinct or not) such that the sum of each pair is nonzero and such that $\alpha + \beta + \gamma + \delta = 0$. Then*

(21.20.3.1) $$N(\alpha, \beta)N(\gamma, \delta) + N(\beta, \gamma)N(\alpha, \delta) + N(\gamma, \alpha)N(\beta, \delta) = 0.$$

(Observe that each term in this sum is defined, by virtue of the conditions of the lemma.)

If $\beta + \gamma \in \mathbf{S}$, then we have $[\mathbf{e}_\alpha, [\mathbf{e}_\beta, \mathbf{e}_\gamma]] = N(\beta, \gamma)[\mathbf{e}_\alpha, \mathbf{e}_{\beta+\gamma}] = N(\beta, \gamma)N(\alpha, \beta + \gamma)\mathbf{e}_{-\delta}$, because $\alpha + (\beta + \gamma) = -\delta \in \mathbf{S}$. By virtue of (21.20.2.1) applied to $\alpha$, $\beta + \gamma$, and $\delta$, we have $N(\alpha, \beta + \gamma) = N(\delta, \alpha) = -N(\alpha, \delta)$, so that $[\mathbf{e}_\alpha, [\mathbf{e}_\beta, \mathbf{e}_\gamma]] = -N(\beta, \gamma)N(\alpha, \delta)\mathbf{e}_{-\delta}$. If $\beta + \gamma \notin \mathbf{S}$, this relation still holds, because both sides are zero. By applying the Jacobi identity (21.20.2.2), we obtain (21.20.3.1).

(21.20.4) *Let* $\alpha, \beta \in \mathbf{S}$ *be two nonproportional roots, and let* $a$ (resp. $b$) *be the smallest* (resp. *largest*) *rational integer such that* $\beta + a\alpha$ (resp. $\beta + b\alpha$) *is a root. Then we have*

(21.20.4.1) $$\mathrm{N}(\alpha, \beta)\mathrm{N}(-\alpha, -\beta) = -\frac{b(1-a)}{2}\alpha(\mathbf{h}_\alpha^0).$$

Suppose first that $\alpha + \beta \in \mathbf{S}$, so that $b > 0$. In the notation of the proof of (21.10.4), we have $m = b - a$, and for $0 \leq j \leq m$ the element $\mathbf{z}_j = (j!)^{-1}\mathrm{ad}(\mathbf{x}_{-\alpha})^j \cdot \mathbf{x}_{\beta+b\alpha}$ spans $\mathfrak{g}(\beta + (b-j)\alpha)$, by virtue of (21.9.3). In particular, $\mathbf{z}_b$ spans $\mathfrak{g}(\beta)$, and by (21.9.3.1) we have

$$\mathrm{ad}(\mathbf{x}_\alpha) \cdot \mathbf{z}_b = (1-a)\mathbf{z}_{b-1}, \qquad \mathrm{ad}(\mathbf{x}_{-\alpha}) \cdot \mathbf{z}_{b-1} = b\mathbf{z}_b$$

so that $[\mathbf{x}_{-\alpha}, [\mathbf{x}_\alpha, \mathbf{e}_\beta]] = b(1-a)\mathbf{e}_\beta$. Since we may write $\mathbf{e}_\alpha = \lambda\mathbf{x}_\alpha$, $\mathbf{e}_{-\alpha} = \mu\mathbf{x}_{-\alpha}$, it follows from (21.20.1.3) and (21.10.3.2) that $\lambda\mu\mathbf{h}_\alpha = \mathbf{h}_\alpha^0$, so that $2\lambda\mu = \alpha(\mathbf{h}_\alpha^0)$ and hence

(21.20.4.2) $$[\mathbf{e}_{-\alpha}, [\mathbf{e}_\alpha, \mathbf{e}_\beta]] = \tfrac{1}{2}b(1-a)\alpha(\mathbf{h}_\alpha^0)\mathbf{e}_\beta.$$

But since $-\alpha + (\alpha + \beta) \neq 0$, it follows that $[\mathbf{e}_{-\alpha}, [\mathbf{e}_\alpha, \mathbf{e}_\beta]] = \mathrm{N}(\alpha, \beta)\mathrm{N}(-\alpha, \alpha + \beta)\mathbf{e}_\beta$; and by virtue of (21.20.2) applied to the three roots $-\alpha, \alpha + \beta, -\beta$, we have $\mathrm{N}(-\alpha, \alpha + \beta) = \mathrm{N}(-\beta, -\alpha) = -\mathrm{N}(-\alpha, -\beta)$. The relation (21.20.4.1) therefore results from (21.20.4.2). Finally, if $b = 0$, both sides of (21.20.4.1) are zero, so the relation is still true.

(21.20.5) (Weyl's theorem) *Let* $\mathfrak{g}, \mathfrak{g}'$ *be two complex semisimple Lie algebras,* $\mathfrak{g} = \mathfrak{h} \oplus \bigoplus_{\alpha \in \mathbf{S}} \mathfrak{g}(\alpha)$, $\mathfrak{g}' = \mathfrak{h}' \oplus \bigoplus_{\alpha' \in \mathbf{S}'} \mathfrak{g}'(\alpha')$ *root decompositions of* $\mathfrak{g}$ *and* $\mathfrak{g}'$ (21.20.1), $\mathbf{S} \subset \mathfrak{h}^*$ *and* $\mathbf{S}' \subset \mathfrak{h}'^*$ *the corresponding root systems,* E (resp. E$'$) *the real vector space generated by* $\mathbf{S}$ (resp. $\mathbf{S}'$) (21.11.2); *we may consider* E (resp. E$'$) *as the dual of the real vector space* $\mathfrak{h}_0$ (resp. $\mathfrak{h}'_0$) *spanned by the elements* $\mathbf{h}_\alpha^0$, $\alpha \in \mathbf{S}$ (resp. $\mathbf{h}_{\alpha'}^0$, $\alpha' \in \mathbf{S}'$) (21.11.2). *Let* $\varphi$ *be an* **R**-*linear bijection of* $\mathfrak{h}_0$ *onto* $\mathfrak{h}'_0$ *such that* ${}^t\varphi(\mathbf{S}') = \mathbf{S}$. *Then* $\varphi$ *can be extended to a* **C**-*isomorphism of the Lie algebra* $\mathfrak{g}$ *onto the Lie algebra* $\mathfrak{g}'$.

(In this statement, the elements $\mathbf{h}_{\alpha'}^0$, $\alpha' \in \mathbf{S}'$, are defined by the relations $\alpha'(\mathbf{u}') = \mathrm{B}_{\mathfrak{g}'}(\mathbf{u}', \mathbf{h}_{\alpha'}^0)$ for $\mathbf{u}' \in \mathfrak{h}'$, $\alpha' \in \mathbf{S}'$, analogous to (21.19.12.1).)

Let us first show that for $\mathbf{u}, \mathbf{v} \in \mathfrak{h}_0$ we have

(21.20.5.1) $$\mathrm{B}_\mathfrak{g}(\mathbf{u}, \mathbf{v}) = \mathrm{B}_{\mathfrak{g}'}(\varphi(\mathbf{u}), \varphi(\mathbf{v})).$$

This will follow from the relations

(21.20.5.2) $$\mathrm{B}_{\mathfrak{g}}(\mathbf{h}_\alpha^0, \mathbf{h}_\beta^0) = \mathrm{B}_{\mathfrak{g}'}(\mathbf{h}_{\alpha'}^0, \mathbf{h}_{\beta'}^0)$$

for $\alpha'$, $\beta'$ in $\mathbf{S}'$, and $\alpha = {}^t\varphi(\alpha')$, $\beta = {}^t\varphi(\beta')$.

For by definition we have

$$\beta'(\varphi(\mathbf{u})) = \beta(\mathbf{u}) = \mathrm{B}_{\mathfrak{g}}(\mathbf{u}, \mathbf{h}_\beta^0) = \mathrm{B}_{\mathfrak{g}'}(\varphi(\mathbf{u}), \mathbf{h}_{\beta'}^0)$$

so that the relation (21.20.5.2) will imply that $\mathrm{B}_{\mathfrak{g}'}(\varphi(\mathbf{h}_\alpha^0), \mathbf{h}_{\beta'}^0) = \mathrm{B}_{\mathfrak{g}'}(\mathbf{h}_{\alpha'}^0, \mathbf{h}_{\beta'}^0)$ for all roots $\alpha' \in \mathbf{S}'$. But since the $\mathbf{h}_{\beta'}^0$ span the complex vector space $\mathfrak{h}'$, and since the restriction of $\mathrm{B}_{\mathfrak{g}'}$ to $\mathfrak{h}'$ is nondegenerate, it follows that the relation $\alpha = {}^t\varphi(\alpha')$ implies that $\mathbf{h}_{\alpha'}^0 = \varphi(\mathbf{h}_\alpha^0)$ for all $\alpha' \in \mathbf{S}'$. Since the set of vectors $\mathbf{h}_{\alpha'}^0$ contains a basis of $\mathfrak{h}'_0$, the relation (21.20.5.1) indeed follows from (21.20.5.2).

To establish (21.20.5.2), we observe that the hypothesis on $\varphi$ implies that if $\alpha'$, $\beta'$ are two nonproportional roots in $\mathbf{S}'$, then $\alpha = {}^t\varphi(\alpha')$ and $\beta = {}^t\varphi(\beta')$ are nonproportional, and the rational integers $k$ for which $\beta' + k\alpha' \in \mathbf{S}'$ are exactly those for which $\beta + k\alpha \in \mathbf{S}$. By virtue of (21.10.4), we have therefore

$$\frac{\beta'(\mathbf{h}_{\alpha'}^0)}{\alpha'(\mathbf{h}_{\alpha'}^0)} = \frac{\beta(\mathbf{h}_\alpha^0)}{\alpha(\mathbf{h}_\alpha^0)}$$

or, by virtue of the definition of the $\mathbf{h}_\alpha^0$ and the $\mathbf{h}_{\alpha'}^0$,

$$\frac{\mathrm{B}_{\mathfrak{g}'}(\mathbf{h}_{\alpha'}^0, \mathbf{h}_{\beta'}^0)}{\mathrm{B}_{\mathfrak{g}'}(\mathbf{h}_{\alpha'}^0, \mathbf{h}_{\alpha'}^0)} = \frac{\mathrm{B}_{\mathfrak{g}}(\mathbf{h}_\alpha^0, \mathbf{h}_\beta^0)}{\mathrm{B}_{\mathfrak{g}}(\mathbf{h}_\alpha^0, \mathbf{h}_\alpha^0)}.$$

Since the Killing form is symmetric, this proves already that the ratio $c_{\alpha'} = \alpha'(\mathbf{h}_{\alpha'}^0)/\alpha(\mathbf{h}_\alpha^0)$ is the same for all the roots $\alpha' \in \mathbf{S}'$, and that if we denote this ratio by $c$, then we have $\beta'(\mathbf{h}_{\alpha'}^0) = c \cdot \beta(\mathbf{h}_\alpha^0)$ for all roots $\alpha', \beta' \in \mathbf{S}'$ (with $\alpha = {}^t\varphi(\alpha')$, $\beta = {}^t\varphi(\beta')$). This relation may also be written as $\mathrm{B}_{\mathfrak{g}'}(\mathbf{h}_{\alpha'}^0, \mathbf{h}_{\beta'}^0) = c \cdot \mathrm{B}_{\mathfrak{g}}(\mathbf{h}_\alpha^0, \mathbf{h}_\beta^0)$. On the other hand, the formula (21.19.8.1) applied to $\mathfrak{g}$ and to $\mathfrak{g}'$ gives, because of the hypothesis ${}^t\varphi(\mathbf{S}') = \mathbf{S}$,

$$\mathrm{B}_{\mathfrak{g}'}(\mathbf{h}_{\alpha'}^0, \mathbf{h}_{\beta'}^0) = \sum_{\gamma' \in \mathbf{S}'} \gamma'(\mathbf{h}_{\alpha'}^0)\gamma'(\mathbf{h}_{\beta'}^0) = c^2 \sum_{\gamma \in \mathbf{S}} \gamma(\mathbf{h}_\alpha^0)\gamma(\mathbf{h}_\beta^0)$$
$$= c^2 \mathrm{B}_{\mathfrak{g}}(\mathbf{h}_\alpha^0, \mathbf{h}_\beta^0).$$

By comparison with the previous result, we obtain $c = c^2$, so that $c = 1$ (because $c \neq 0$); this proves (21.20.5.1) and also establishes the relations

(21.20.5.3) $$\varphi(\mathbf{h}_\alpha^0) = \mathbf{h}_{\alpha'}^0$$

where $\alpha = {}^t\varphi(\alpha')$.

(21.20.5.4) Suppose that we have chosen in each $\mathfrak{g}(\alpha)$ a vector $\mathbf{e}_\alpha$ such that the relations (21.20.1.2)–(21.20.1.5) are satisfied. We shall show that it is possible to find in each $\mathfrak{g}'(\alpha')$ an element $\mathbf{e}'_{\alpha'}$ such that:

(1) for each pair of roots $\alpha', \beta' \in \mathbf{S}'$ such that $\alpha' + \beta' \neq 0$,

(21.20.5.5) $$[\mathbf{e}'_{\alpha'}, \mathbf{e}'_{\beta'}] = \mathrm{N}(\alpha, \beta)\mathbf{e}'_{\alpha'+\beta'}$$

where $\alpha = {}^t\varphi(\alpha')$, $\beta = {}^t\varphi(\beta')$ (which implies $\alpha + \beta = {}^t\varphi(\alpha' + \beta')$, and therefore $\alpha + \beta \neq 0$, and $\alpha + \beta \in \mathbf{S}$ if and only if $\alpha' + \beta' \in \mathbf{S}'$);

(2) for each root $\alpha' \in \mathbf{S}'$,

(21.20.5.6) $$\mathrm{B}_{\mathfrak{g}'}(\mathbf{e}'_{\alpha'}, \mathbf{e}'_{-\alpha'}) = 1$$

which, by (21.19.13.1), implies

(21.20.5.7) $$[\mathbf{e}'_{\alpha'}, \mathbf{e}'_{-\alpha'}] = \mathbf{h}^0_{\alpha'}.$$

Once the existence of these vectors $\mathbf{e}'_{\alpha'}$ has been established, the theorem will be proved by taking the extension of $\varphi$ to be the **C**-linear mapping $\tilde{\varphi}$ such that

(21.20.5.8) $$\tilde{\varphi}(\mathbf{e}_\alpha) = \mathbf{e}'_{\alpha'}$$

for $\alpha' \in \mathbf{S}'$ and $\alpha = {}^t\varphi(\alpha')$. For it will follow from (21.20.5.3), (21.20.5.5), and (21.20.5.7) that $\tilde{\varphi}([\mathbf{e}_\alpha, \mathbf{e}_\beta]) = [\tilde{\varphi}(\mathbf{e}_\alpha), \tilde{\varphi}(\mathbf{e}_\beta)]$ for all $\alpha, \beta \in \mathbf{S}$; also, by reason of (21.20.5.3) and (21.20.5.1), we shall have $\tilde{\varphi}([\mathbf{h}^0_\alpha, \mathbf{e}_\beta]) = [\tilde{\varphi}(\mathbf{h}^0_\alpha), \tilde{\varphi}(\mathbf{e}_\beta)]$ for all $\alpha, \beta \in \mathbf{S}$; and since the $\mathbf{h}^0_\alpha$ and the $\mathbf{e}_\alpha$ span $\mathfrak{g}$, it will follow that $\tilde{\varphi}$ is a Lie algebra isomorphism of $\mathfrak{g}$ onto $\mathfrak{g}'$.

(21.20.5.9) In order to define the vectors $\mathbf{e}'_{\alpha'}$, we shall begin by defining a lexicographic ordering on the real vector space E spanned by $\mathbf{S}$: we consider a basis $(\varepsilon_j)_{1 \leq j \leq l}$ of this space, and for any two elements $\xi = \sum_{j=1}^{l} x_j \varepsilon_j$, $\eta = \sum_{j=1}^{l} y_j \varepsilon_j$, we define the relation $\xi \prec \eta$ to mean:

"$\xi \neq \eta$, and if $k$ is the smallest index such that $x_k \neq y_k$, then $x_k < y_k$ in **R**."

It is immediately verified that the relation "$\xi \prec \eta$ or $\xi = \eta$" on E is a *total ordering* (called the *lexicographic* ordering), that the relation $\xi \prec \eta$ implies $\xi + \zeta \prec \eta + \zeta$ for all $\zeta \in$ E (which implies that $\xi \succ 0$ is equivalent to

$-\xi \prec 0$) and that the relations $\xi \succ 0$, $\eta \succ 0$ imply $\xi + \eta \succ 0$. (This order relation should not be confused with that defined in (21.14.5).)

We may therefore write the elements of **S** in the form of a strictly increasing sequence, relative to this lexicographic ordering:

$$-\rho_m \prec -\rho_{m-1} \prec \cdots \prec -\rho_1 \prec 0 \prec \rho_1 \prec \cdots \prec \rho_{m-1} \prec \rho_m.$$

We shall define the $\mathbf{e}'_{\alpha'}$ by the following (finite) inductive procedure: for each integer $k$ such that $1 \leq k \leq m$, assume that the $\mathbf{e}'_{\alpha'}$ have been defined for the $\alpha' \in \mathbf{S}'$ such that $\alpha = {}^t\varphi(\alpha')$ satisfies the relations $-\rho_k \prec \alpha \prec \rho_k$, that the relations (21.20.5.6) are satisfied by these roots $\alpha'$, and that the relations (21.20.5.5) are satisfied by all pairs of these roots that *also* satisfy the conditions $\alpha' + \beta' \neq 0$ and $-\rho_k \prec {}^t\varphi(\alpha' + \beta') \prec \rho_k$. The inductive step then consists in defining $\mathbf{e}'_{\rho'_k}$ and $\mathbf{e}'_{-\rho'_k}$ (where $\rho_k = {}^t\varphi(\rho'_k)$) in such a way that these conditions continue to be satisfied when we replace $k$ by $k + 1$.

(21.20.5.10) With a change of notation, our problem is reduced to the following: given a root $\rho \succ 0$ in **S**, let $\mathbf{S}_\rho$ denote the set of roots $\alpha \in \mathbf{S}$ such that $-\rho \prec \alpha \prec \rho$. Suppose that the $\mathbf{e}'_{\alpha'}$ have been determined for those $\alpha'$ such that ${}^t\varphi(\alpha') \in \mathbf{S}_\rho$, and that they satisfy (21.20.5.6) and (21.20.5.5) whenever $\alpha' + \beta' \neq 0$ and ${}^t\varphi(\alpha' + \beta') \in \mathbf{S}_\rho$. Then the problem is to define $\mathbf{e}'_{\rho'}$ and $\mathbf{e}'_{-\rho'}$, where ${}^t\varphi(\rho') = \rho$, in such a way that the same conditions are still satisfied when we replace $\mathbf{S}_\rho$ by $\mathbf{S}_\rho \cup \{-\rho, \rho\}$.

If there exists no decomposition $\rho = \alpha + \beta$ with $\alpha, \beta \in \mathbf{S}_\rho$, then we may take $\mathbf{e}'_{\rho'}$ to be an arbitrary nonzero element of $\mathfrak{g}'(\rho')$, and $\mathbf{e}'_{-\rho'}$ the unique element of $\mathfrak{g}'(-\rho')$ such that $\mathrm{B}_{\mathfrak{g}'}(\mathbf{e}'_{\rho'}, \mathbf{e}'_{-\rho'}) = 1$. If on the other hand there exist $\alpha, \beta \in \mathbf{S}_\rho$ such that $\rho = \alpha + \beta$, and if $\alpha = {}^t\varphi(\alpha')$, $\beta = {}^t\varphi(\beta')$, then we have $\mathrm{N}(\alpha, \beta) \neq 0$ (21.10.5), and we shall define $\mathbf{e}'_{\rho'}$ by the equation

$$\text{(21.20.5.11)} \qquad \mathrm{N}(\alpha, \beta)\mathbf{e}'_{\rho'} = [\mathbf{e}'_{\alpha'}, \mathbf{e}'_{\beta'}].$$

We have then $\mathbf{e}'_{\rho'} \neq 0$ (21.10.5), and we define $\mathbf{e}'_{-\rho'}$ to be the unique element of $\mathfrak{g}'(-\rho')$ such that $\mathrm{B}_{\mathfrak{g}'}(\mathbf{e}'_{\rho'}, \mathbf{e}'_{-\rho'}) = 1$. We then define $\mathbf{e}''_{\gamma'}$ for *all* $\gamma' \in \mathbf{S}'$ by taking $\mathbf{e}''_{\gamma'} = \mathbf{e}'_{\gamma'}$ if ${}^t\varphi(\gamma') \in \mathbf{S}_\rho \cup \{-\rho, \rho\}$; if not, then we take $\mathbf{e}''_{\gamma'}$ to be a nonzero element of $\mathfrak{g}'(\gamma')$, and $\mathbf{e}''_{-\gamma'}$ to be the unique element of $\mathfrak{g}(-\gamma')$ such that $\mathrm{B}_{\mathfrak{g}'}(\mathbf{e}''_{\gamma'}, \mathbf{e}''_{-\gamma'}) = 1$. We may then write $[\mathbf{e}''_{\gamma'}, \mathbf{e}''_{\delta'}] = \mathrm{N}'(\gamma, \delta)\mathbf{e}''_{\gamma'+\delta'}$ if $\gamma' + \delta' \neq 0$ (where $\gamma = {}^t\varphi(\gamma')$, $\delta = {}^t\varphi(\delta')$), and we know already that $\mathrm{N}'(\gamma, \delta) = \mathrm{N}(\gamma, \delta)$ whenever $\gamma$, $\delta$ and $\gamma + \delta$ are in $\mathbf{S}_\rho$. We have to prove that this relation remains true when $\gamma$, $\delta$, and $\gamma + \delta$ are in $\mathbf{S}_\rho \cup \{-\rho, \rho\}$. There are various cases to consider:

(a) $\gamma + \delta = \rho$, and we may assume that $\gamma$ and $\delta$ are both distinct from $\alpha$ and $\beta$. Then we have $\alpha + \beta + (-\gamma) + (-\delta) = 0$, and no two of the four roots

$\alpha, \beta, -\gamma, -\delta$ sum to 0. We may therefore apply **(21.20.3)** to $\mathfrak{g}$ and to $\mathfrak{g}'$, thus obtaining

(21.20.5.12)

$$\mathrm{N}(\alpha, \beta)\mathrm{N}(-\gamma, -\delta) = -\mathrm{N}(\beta, -\gamma)\mathrm{N}(\alpha, -\delta) - \mathrm{N}(-\gamma, \alpha)\mathrm{N}(\beta, -\delta),$$
$$\mathrm{N}'(\alpha, \beta)\mathrm{N}'(-\gamma, -\delta) = -\mathrm{N}'(\beta, -\gamma)\mathrm{N}'(\alpha, -\delta) - \mathrm{N}'(-\gamma, \alpha)\mathrm{N}'(\beta, -\delta).$$

We remark now that we must have $\alpha \succ 0$, $\beta \succ 0$, $\gamma \succ 0$, $\delta \succ 0$; for if, for example, $\alpha \prec 0$, it would follow that $\beta = \rho + (-\alpha) \succ \rho$, which is absurd. Hence $\gamma, \delta, \beta - \gamma, \alpha - \delta, \alpha - \gamma$, and $\beta - \delta$ all belong to $\mathbf{S}_\rho$, and therefore the inductive hypothesis implies that the right-hand sides of the two relations **(21.20.5.12)** are equal; since also $\mathrm{N}'(\alpha, \beta) = \mathrm{N}(\alpha, \beta)$ by **(21.20.5.11)** and $\mathrm{N}(\alpha, \beta) \neq 0$ **(21.10.5)**, it follows that $\mathrm{N}(-\gamma, -\delta) = \mathrm{N}'(-\gamma, -\delta)$. Now, by using the fact that the integers $k$ such that $\gamma' + k\delta' \in \mathbf{S}'$ are exactly those for which $\gamma + k\delta \in \mathbf{S}$, together with the relations **(21.20.5.2)**, we deduce from **(21.20.4.1)** applied to $\mathfrak{g}$ and to $\mathfrak{g}'$ that

$$\mathrm{N}(\gamma, \delta)\mathrm{N}(-\gamma, -\delta) = \mathrm{N}'(\gamma, \delta)\mathrm{N}'(-\gamma, -\delta),$$

whence finally $\mathrm{N}(\gamma, \delta) = \mathrm{N}'(\gamma, \delta)$.

(b) $\gamma + \delta = -\rho$; then $-\gamma$ and $-\delta$ belong to $\mathbf{S}_\rho$, and we have

$$(-\gamma) + (-\delta) = \rho.$$

The reasoning in (a) above proves that $\mathrm{N}(\gamma, \delta) = \mathrm{N}'(\gamma, \delta)$.

(c) One of the roots $\gamma, \delta$ is equal to $\pm\rho$, for example, $\gamma = -\rho$. Then $\delta \neq \pm\rho$, otherwise we should have either $\gamma + \delta = 0$ or else

$$\gamma + \delta \notin \mathbf{S}_\rho \cup \{-\rho, \rho\}.$$

We have $\rho = \delta + (-\gamma - \delta)$, and by hypothesis

$$-\gamma - \delta \in \mathbf{S}_\rho \cup \{-\rho, \rho\};$$

but we cannot have $-\gamma - \delta = \pm\rho$, for this would imply that $\delta = 0$ or $\delta = 2\rho$, both of which are absurd; hence $-\gamma - \delta \in \mathbf{S}_\rho$ and $\delta \in \mathbf{S}_\rho$. Consequently, by (a) above, we have $\mathrm{N}'(\delta, -\gamma - \delta) = \mathrm{N}(\delta, -\gamma - \delta)$. But since the sum of the roots $\gamma, \delta$, and $-\gamma - \delta$ is zero, we can apply **(21.20.2)** to $\mathfrak{g}$ and to $\mathfrak{g}'$, and obtain $\mathrm{N}(\gamma, \delta) = \mathrm{N}(\delta, -\gamma - \delta)$ and $\mathrm{N}'(\gamma, \delta) = \mathrm{N}'(\delta, -\gamma - \delta)$. Hence again we have $\mathrm{N}(\gamma, \delta) = \mathrm{N}'(\gamma, \delta)$, and the proof of the theorem **(21.20.5)** is now complete.

**(21.20.6)** *Let $\mathfrak{g}$ be a complex semisimple Lie algebra and let $\mathfrak{g} = \mathfrak{h} \oplus \bigoplus_{\alpha \in \mathbf{S}} \mathfrak{g}(\alpha)$ be a root decomposition of $\mathfrak{g}$. Then there exists for each $\alpha \in \mathbf{S}$ an element*

$\mathbf{e}_\alpha \in \mathfrak{g}(\alpha)$ *such that the conditions of* (21.20.1) *are satisfied, and moreover such that*

(21.20.6.1) $$N(\alpha, \beta) = -N(-\alpha, -\beta)$$

*whenever* $\alpha + \beta \neq 0$.

*For each system of elements* $(\mathbf{e}_\alpha)_{\alpha \in \mathbf{S}}$ *satisfying these conditions we have*

(21.20.6.2) $$N(\alpha, \beta)^2 = \frac{b(1-a)}{2}\alpha(\mathbf{h}_\alpha^0)$$

*whenever* $\alpha + \beta \neq 0$, *where a and b are the integers defined in* (21.20.4), *and* $N(\alpha, \beta)$ *is real.*

Let $\varphi$ be the mapping $\mathbf{u} \mapsto -\mathbf{u}$ of the real vector space $\mathfrak{h}_0$ onto itself. Clearly we have ${}^t\varphi(\lambda) = -\lambda$ for each linear form $\lambda \in E$, so that ${}^t\varphi(\mathbf{S}) = \mathbf{S}$, and we may apply (21.20.5) with $\mathfrak{g}' = \mathfrak{g}$. Let us denote by $\mathbf{z}_\alpha \in \mathfrak{g}(\alpha)$ the elements constructed in the proof of (21.20.5) (and denoted there by $\mathbf{e}'_{\alpha'}$); for the automorphism $\tilde{\varphi}$ of $\mathfrak{g}$ that extends $\varphi$, they satisfy by virtue of (21.20.5.8) the condition $\tilde{\varphi}(\mathbf{z}_\alpha) \in \mathfrak{g}(-\alpha)$, and also the relation $B_\mathfrak{g}(\mathbf{z}_\alpha, \mathbf{z}_{-\alpha}) = 1$. We may therefore write $\tilde{\varphi}(\mathbf{z}_\alpha) = c_{-\alpha}\mathbf{z}_{-\alpha}$, with $c_{-\alpha} \in \mathbf{C}$, and since $B_\mathfrak{g}$ is invariant under the automorphism $\tilde{\varphi}$, we have $c_\alpha c_{-\alpha} = 1$. Hence there exists for each $\alpha \in \mathbf{S}$ a complex number $a_\alpha$ such that $a_\alpha^2 = -c_\alpha$ and $a_\alpha a_{-\alpha} = 1$, whence $a_\alpha c_{-\alpha} = -a_{-\alpha}$. Now put $\mathbf{e}_\alpha = a_\alpha \mathbf{z}_\alpha$ for each $\alpha \in \mathbf{S}$. First of all, we have $B_\mathfrak{g}(\mathbf{e}_\alpha, \mathbf{e}_{-\alpha}) = a_\alpha a_{-\alpha} B_\mathfrak{g}(\mathbf{z}_\alpha, \mathbf{z}_{-\alpha}) = 1$. Also $\tilde{\varphi}(\mathbf{e}_\alpha) = a_\alpha \tilde{\varphi}(\mathbf{z}_\alpha) = a_\alpha c_{-\alpha}\mathbf{z}_{-\alpha} = -a_{-\alpha}\mathbf{z}_{-\alpha} = -\mathbf{e}_{-\alpha}$. If $\alpha$, $\beta$ are two roots such that $\alpha + \beta \in \mathbf{S}$, then we have $\tilde{\varphi}([\mathbf{e}_\alpha, \mathbf{e}_\beta]) = [-\mathbf{e}_{-\alpha}, -\mathbf{e}_{-\beta}] = N(-\alpha, -\beta)\mathbf{e}_{-\alpha-\beta}$, and on the other hand $\tilde{\varphi}([\mathbf{e}_\alpha, \mathbf{e}_\beta]) = N(\alpha, \beta)\tilde{\varphi}(\mathbf{e}_{\alpha+\beta}) = -N(\alpha, \beta)\mathbf{e}_{-\alpha-\beta}$, which proves the formula (21.20.6.1). The relation (21.20.6.2) then follows from (21.20.4.1). Finally, since $a \leq 0$ and $b \geq 0$, in order to show that $N(\alpha, \beta)$ is real it is enough to prove that $\alpha(\mathbf{h}_\alpha^0) > 0$. Now by (21.19.8) we have

$$\alpha(\mathbf{h}_\alpha^0) = B_\mathfrak{g}(\mathbf{h}_\alpha^0, \mathbf{h}_\alpha^0) = \sum_{\beta \in \mathbf{S}} \beta(\mathbf{h}_\alpha^0)^2 = (\alpha(\mathbf{h}_\alpha^0))^2 \sum_{\beta \in \mathbf{S}} r_{\alpha\beta}^2$$

by virtue of (21.19.10), since $\mathbf{h}_\alpha^0 \in [\mathfrak{g}(\alpha), \mathfrak{g}(-\alpha)]$. Since $\alpha(\mathbf{h}_\alpha^0) \neq 0$, it follows that $\alpha(\mathbf{h}_\alpha^0) > 0$, and the proof is complete.

A **C**-basis of $\mathfrak{g}$ which consists of an **R**-basis of $\mathfrak{h}_0$ and elements $\mathbf{e}_\alpha \in \mathfrak{g}(\alpha)$ satisfying the conditions (21.20.1.2) and (21.20.6.1), appears therefore as a generalization of the notion of a *Weyl basis* of the complexification of the Lie algebra of a compact semisimple Lie group (21.10.6). In fact, the two notions are *identical*; and it is precisely the existence of such a basis in any complex semisimple Lie algebra that will enable us to prove the result announced in (21.19.1):

(21.20.7) *Every complex semisimple Lie algebra* $\mathfrak{g}$ *is isomorphic to the complexification of the Lie algebra* $\mathfrak{k}$ *of some compact semisimple Lie group; and the Lie algebra* $\mathfrak{k}$ *having this property is unique up to isomorphism.*

Consider elements $\mathbf{e}_\alpha$ ($\alpha \in \mathbf{S}$) having the properties of (21.20.6), and put $\mathbf{y}_\alpha = \mathbf{e}_\alpha - \mathbf{e}_{-\alpha}$, $\mathbf{z}_\alpha = i(\mathbf{e}_\alpha + \mathbf{e}_{-\alpha})$; it is clear that $i\mathfrak{h}_0$ and the element $\mathbf{y}_\alpha$, $\mathbf{z}_\alpha$ ($\alpha \in \mathbf{S}$) span a *real* vector subspace $\mathfrak{k}$ of $\mathfrak{g}$, of dimension equal to $\dim_{\mathbf{C}} \mathfrak{g}$ (21.11.2), and that $\mathfrak{g} = \mathfrak{k} \oplus i\mathfrak{k}$. Moreover, by use of (21.20.6.1), the following formulas are easily verified:

(21.20.7.1) $$[i\mathbf{h}_\alpha^0, \mathbf{y}_\beta] = \beta(\mathbf{h}_\alpha^0)\mathbf{z}_\beta, \qquad [i\mathbf{h}_\alpha^0, \mathbf{z}_\beta] = -\beta(\mathbf{h}_\alpha^0)\mathbf{y}_\beta$$

(21.20.7.2) $$[\mathbf{y}_\alpha, \mathbf{z}_\alpha] = 2i\mathbf{h}_\alpha^0,$$

(21.20.7.3) $$\begin{aligned}
[\mathbf{y}_\alpha, \mathbf{y}_\beta] &= \mathrm{N}(\alpha, \beta)\mathbf{y}_{\alpha+\beta} - \mathrm{N}(\alpha, -\beta)\mathbf{y}_{\alpha-\beta},\\
[\mathbf{z}_\alpha, \mathbf{z}_\beta] &= -\mathrm{N}(\alpha, \beta)\mathbf{y}_{\alpha+\beta} - \mathrm{N}(\alpha, -\beta)\mathbf{y}_{\alpha-\beta},\\
[\mathbf{y}_\alpha, \mathbf{z}_\beta] &= \mathrm{N}(\alpha, \beta)\mathbf{z}_{\alpha+\beta} + \mathrm{N}(\alpha, -\beta)\mathbf{z}_{\alpha-\beta},
\end{aligned}$$

if $\alpha + \beta \neq 0$. These formulas show that $\mathfrak{k}$ is a *real Lie algebra*, since the $\mathrm{N}(\alpha, \beta)$ are real, and $\mathfrak{g}$ is isomorphic to the complexification of $\mathfrak{k}$. To see that $\mathfrak{k}$ is the Lie algebra of a *compact* semisimple Lie group, it is enough to show that the restriction to $\mathfrak{k}$ of the Killing form $\mathrm{B}_\mathfrak{g}$ is *negative definite* ((21.6.1) and (21.6.9)). Now we know already that the restriction of $\mathrm{B}_\mathfrak{g}$ to $\mathfrak{h}$ is nondegenerate; on the other hand, $\beta(\mathbf{h}_\alpha^0)$ is real for all $\alpha, \beta \in \mathbf{S}$ (21.19.10); and therefore, since the $\mathbf{h}_\alpha^0$ span the real vector space $\mathfrak{h}_0$, $\beta(\mathbf{u})$ is real for all $\mathbf{u} \in \mathfrak{h}_0$, and the formula $\mathrm{B}_\mathfrak{g}(\mathbf{u}, \mathbf{u}) = \sum_{\beta \in \mathbf{S}} \beta(\mathbf{u})^2$ (21.19.8.1) shows that the restriction of $\mathrm{B}_\mathfrak{g}$ to $\mathfrak{h}_0 \times \mathfrak{h}_0$ is positive definite. Consequently its restriction to $i\mathfrak{h}_0 \times i\mathfrak{h}_0$ is negative definite. Since the $\mathbf{y}_\alpha$ and $\mathbf{z}_\alpha$ are orthogonal to $i\mathfrak{h}_0$ relative to $\mathrm{B}_\mathfrak{g}$ (21.19.9), since $\mathrm{B}_\mathfrak{g}(\mathbf{e}_\alpha, \mathbf{e}_\beta) = 0$ for $\alpha + \beta \neq 0$, and since

$$\mathrm{B}_\mathfrak{g}(\mathbf{y}_\alpha, \mathbf{z}_\alpha) = i\mathrm{B}_\mathfrak{g}(\mathbf{e}_\alpha, \mathbf{e}_\alpha) - i\mathrm{B}_\mathfrak{g}(\mathbf{e}_{-\alpha}, \mathbf{e}_{-\alpha}) = 0$$

by virtue of (21.19.9), it remains to show that $\mathrm{B}_\mathfrak{g}(\mathbf{y}_\alpha, \mathbf{y}_\alpha) < 0$ and $\mathrm{B}_\mathfrak{g}(\mathbf{z}_\alpha, \mathbf{z}_\alpha) < 0$; and this follows from the formulas

$$\begin{aligned}
\mathrm{B}_\mathfrak{g}(\mathbf{y}_\alpha, \mathbf{y}_\alpha) &= -2\mathrm{B}_\mathfrak{g}(\mathbf{e}_\alpha, \mathbf{e}_{-\alpha}) = -2,\\
\mathrm{B}_\mathfrak{g}(\mathbf{z}_\alpha, \mathbf{z}_\alpha) &= -2\mathrm{B}_\mathfrak{g}(\mathbf{e}_\alpha, \mathbf{e}_{-\alpha}) = -2.
\end{aligned}$$

The uniqueness (to within isomorphism) of the Lie algebra $\mathfrak{k}$ is a consequence of the study of the real forms of $\mathfrak{g}$ undertaken in Section **21.18**. With the notation of that section, the Killing form $B_{\mathfrak{g}_0}$ is negative definite only if $\mathfrak{p}_0 = \{0\}$, that is to say, $\mathfrak{g}_0 = \mathfrak{g}_u$; hence there is no real form of $\mathfrak{g}$ that is the Lie algebra of a compact group, except for the subalgebras $\varphi(\mathfrak{g}_u)$, where $\varphi$ is an automorphism of $\mathfrak{g}$ **(21.18.3)**.

### PROBLEM

Under the conditions of **(21.20.6)**, show that there exists for each $\alpha \in \mathbf{S}$ an element $\mathbf{e}'_\alpha \in \mathfrak{g}(\alpha)$ such that if we put $[\mathbf{e}'_\alpha, \mathbf{e}'_\beta] = N'(\alpha, \beta)\mathbf{e}'_{\alpha+\beta}$ when $\alpha + \beta \in \mathbf{S}$, then the $N'(\alpha, \beta)$ are real and satisfy the condition $N'(\alpha, \beta) = -N'(-\alpha, -\beta)$, and such that for each pair $\alpha, \beta \in \mathbf{S}$ satisfying $\alpha + \beta \neq 0$ we have $|N'(\alpha, \beta)| = 1 - a$. A basis of $\mathfrak{g}$ consisting of the $\mathbf{e}'_\alpha$ and a basis of $\mathfrak{h}_0$ over $\mathbf{R}$ is called a *Chevalley basis* of $\mathfrak{g}$. (Reduce to the case where $\mathbf{S}$ is irreducible (Section **21.11**, Problem 10) and observe that, in this case if $(\lambda|\mu)$ is a scalar product on E which is invariant under the Weyl group W, then $\alpha(\mathbf{h}^0_\alpha)/\beta(\mathbf{h}^0_\beta) = (\alpha|\alpha)/(\beta|\beta)$ by using **(21.11.5.5)**; then use Section **21.11**, Problem 1(b).)

## 21. THE IWASAWA DECOMPOSITION

**(21.21.1)** Let $\mathfrak{g}$ be a complex semisimple Lie algebra, which we may, by virtue of **(21.20.7)**, consider as the complexification $(\mathfrak{g}_u)_{(\mathbf{C})} = \mathfrak{g}_u \oplus i\mathfrak{g}_u$ of the Lie algebra $\mathfrak{g}_u$ of a simply connected compact semisimple Lie group $\tilde{G}_u$. With the notation of Section **21.18**, let $c_0$ be a conjugation of $\mathfrak{g}$ that commutes with the conjugation $c_u$, and let $\mathfrak{g}_0$ be the real form of $\mathfrak{g}$ consisting of the elements of $\mathfrak{g}$ fixed by $c_0$; let $\mathfrak{g}_0 = \mathfrak{k}_0 \oplus \mathfrak{p}_0$ be the corresponding Cartan decomposition **(21.18.4.1)**, with the relations **(21.18.4.2)**, and recall that we have $\mathfrak{g}_u = \mathfrak{k}_0 \oplus i\mathfrak{p}_0$ and $\mathfrak{k}_0 = \mathfrak{g}_0 \cap \mathfrak{g}_u$.

**(21.21.2)** Let $\mathfrak{a}_0$ be a *maximal commutative (real) Lie subalgebra* contained in the real vector space $\mathfrak{p}_0$. (There exist nonzero commutative real subalgebras of $\mathfrak{p}_0$, for example, the one-dimensional subspaces; we may take $\mathfrak{a}_0$ to be such a subalgebra of largest possible dimension.) The subspace $i\mathfrak{a}_0$ of $i\mathfrak{p}_0$ is then also a maximal commutative subalgebra of $i\mathfrak{p}_0$.

**(21.21.3)** *If* $\mathfrak{t}$ *is a maximal commutative subalgebra of the real Lie algebra* $\mathfrak{g}_u$, *containing* $i\mathfrak{a}_0$, *then* $\mathfrak{t} = i\mathfrak{a}_0 \oplus (\mathfrak{t} \cap \mathfrak{k}_0)$.

Let $\mathbf{x} = \mathbf{y} + i\mathbf{z}$ be an element of $\mathfrak{t}$, with $\mathbf{y} \in \mathfrak{k}_0$ and $\mathbf{z} \in \mathfrak{p}_0$. For each $\mathbf{u} \in \mathfrak{a}_0$ we must have $[\mathbf{y}, i\mathbf{u}] + [i\mathbf{z}, i\mathbf{u}] = 0$; but $[\mathbf{y}, i\mathbf{u}] \in i\mathfrak{p}_0$ and $[i\mathbf{z}, i\mathbf{u}] \in \mathfrak{k}_0$, so that $[\mathbf{y}, i\mathbf{u}] = 0$ and $[\mathbf{z}, \mathbf{u}] = 0$. Since $\mathfrak{a}_0$ is a maximal commutative subalgebra of $\mathfrak{p}_0$, it follows that $\mathbf{z} \in \mathfrak{a}_0$ and hence $\mathbf{y} \in \mathfrak{t} \cap \mathfrak{k}_0$.

For the rest of this section, let $\mathfrak{t}$ be a maximal commutative subalgebra of $\mathfrak{g}_u$ containing $i\mathfrak{a}_0$, fixed once for all. Let

(21.21.3.1) $$\mathfrak{t} \oplus i\mathfrak{t} = \mathfrak{h},$$

then it is clear that

(21.21.3.2) $$c_0(\mathfrak{h}) = \mathfrak{h}.$$

**(21.21.4)** Let $\mathbf{S}$ be the root system of $\mathfrak{g}_u$ relative to $\mathfrak{t}$ **(21.8.1)**; we recall that the roots $\alpha \in \mathbf{S}$ are $\mathbf{R}$-linear mappings of $\mathfrak{t}$ into $i\mathbf{R}$, which may be canonically identified with linear forms on the complex vector space $\mathfrak{h} = \mathfrak{t} \oplus i\mathfrak{t}$. They take *real* values on $\mathfrak{a}_0$, and moreover **(21.8.2)** we have

(21.21.4.1) $$\alpha \circ c_u = -\alpha,$$

(21.21.4.2) $$c_u(\mathfrak{g}_\alpha) = \mathfrak{g}_{-\alpha}.$$

Let $\mathbf{S}'$ denote the set of roots *that vanish on* $i\mathfrak{a}_0$ (or on $\mathfrak{a}_0 \oplus i\mathfrak{a}_0$); it is clear that $-\mathbf{S}' = \mathbf{S}'$. Let $\varphi$ denote the *involutory automorphism* $c_0 c_u = c_u c_0$ of the complex Lie algebra $\mathfrak{g}$. Clearly $\varphi$ leaves $\mathfrak{g}_0$ and $\mathfrak{g}_u$ stable, and we have

(21.21.4.3) $$c_u | \mathfrak{g}_0 = \varphi | \mathfrak{g}_0.$$

**(21.21.5)** (i) *We have $\varphi(\mathfrak{t}) = \mathfrak{t}$ (and hence $\varphi(\mathfrak{h}) = \mathfrak{h}$), and the mapping $\alpha \mapsto \alpha \circ \varphi$ is a bijection of $\mathbf{S}$ onto itself.*

(ii) *We have $\alpha \in \mathbf{S}'$ if and only if $\alpha \circ \varphi = \alpha$.*

(iii) *For each root $\alpha \in \mathbf{S}'$ we have $\mathfrak{g}_\alpha \oplus \mathfrak{g}_{-\alpha} \subset \mathfrak{k}_0 + i\mathfrak{k}_0$.*

(i) For $\mathbf{x} \in \mathfrak{k}_0$ we have $\varphi(\mathbf{x}) = \mathbf{x}$, and for $\mathbf{x} \in i\mathfrak{p}_0$ we have $\varphi(\mathbf{x}) = -\mathbf{x}$. Hence $\varphi(\mathfrak{t}) = \mathfrak{t}$ by virtue of **(21.21.3)**, and the fact that $\alpha \mapsto \alpha \circ \varphi$ is a bijection of $\mathbf{S}$ onto itself follows from **(21.8.6)**.

(ii) If $\alpha \circ \varphi = \alpha$, then $\alpha(\mathbf{x}) = \alpha(\varphi(\mathbf{x})) = -\alpha(\mathbf{x})$ for $\mathbf{x} \in i\mathfrak{a}_0$, and therefore $\alpha(\mathbf{x}) = 0$. Conversely, if $\alpha \in \mathbf{S}'$, then $\alpha(\mathbf{x}) = \alpha(\varphi(\mathbf{x}))$ for $\mathbf{x} \in i\mathfrak{a}_0$ and for $\mathbf{x} \in \mathfrak{k}_0$, and therefore $\alpha = \alpha \circ \varphi$ by virtue of **(21.21.3)**.

(iii) If $\alpha \in \mathbf{S}'$, it is clear that $\varphi(\mathfrak{g}_\alpha) = \mathfrak{g}_\alpha$. Since the complex vector space $\mathfrak{g}_\alpha$ has dimension 1 **(20.10.3)** and since $\varphi$ is an involutory bijection of $\mathfrak{g}_\alpha$ onto itself, we must have either $\varphi(\mathbf{x}) = \mathbf{x}$ for all $\mathbf{x} \in \mathfrak{g}_\alpha$ or else $\varphi(\mathbf{x}) = -\mathbf{x}$ for all $\mathbf{x} \in \mathfrak{g}_\alpha$. In the first case, we have $\mathbf{x} \in \mathfrak{k}_0 \oplus i\mathfrak{k}_0$; in the second case,

$\mathbf{x} \in \mathfrak{p}_0 \oplus i\mathfrak{p}_0$, and by definition $[\mathbf{z}, \mathbf{x}] = \alpha(\mathbf{z})\mathbf{x} = 0$ for all $\mathbf{z} \in \mathfrak{a}_0$; if $\mathbf{x} = \mathbf{v} + i\mathbf{w}$, with $\mathbf{v}$ and $\mathbf{w}$ in $\mathfrak{p}_0$, we have therefore $[\mathbf{z}, \mathbf{v}] + i[\mathbf{z}, \mathbf{w}] = 0$, and since $[\mathbf{z}, \mathbf{v}] \in \mathfrak{k}_0$ and $i[\mathbf{z}, \mathbf{w}] \in i\mathfrak{k}_0$, we have $[\mathbf{z}, \mathbf{v}] = [\mathbf{z}, \mathbf{w}] = 0$. But since $\mathfrak{a}_0$ is a *maximal* commutative subalgebra in $\mathfrak{p}_0$, these relations imply that $\mathbf{v} \in \mathfrak{a}_0$ and $\mathbf{w} \in \mathfrak{a}_0$, whence $\mathbf{x} \in \mathfrak{h}$; and since $\mathfrak{h} \cap \mathfrak{g}_\alpha = \{0\}$ by virtue of (21.8.1), it follows that $\mathbf{x} = 0$ in this case. Hence the assumption that $\varphi(\mathbf{x}) = -\mathbf{x}$ for all $\mathbf{x} \in \mathfrak{g}_\alpha$ is untenable, and the proof is complete.

(21.21.6) Let $\mathbf{S}'' = \mathbf{S} - \mathbf{S}'$. For each root $\alpha \in \mathbf{S}''$, the set of vectors $\mathbf{z} \in \mathfrak{a}_0$ such that $\alpha(\mathbf{z}) = 0$ is a hyperplane in the real vector space $\mathfrak{a}_0$. Since $\mathbf{S}''$ is finite, there exists $\mathbf{z}_0 \in \mathfrak{a}_0$ such that $\alpha(\mathbf{z}_0) \neq 0$ for all roots $\alpha \in \mathbf{S}''$. Let $\mathbf{S}''_+$ denote the set of roots $\alpha \in \mathbf{S}''$ such that $\alpha(\mathbf{z}_0) > 0$. Since $-\mathbf{S}'' = \mathbf{S}''$, it is clear that $\mathbf{S}''$ is the union of the two disjoint sets $\mathbf{S}''_+$ and $-\mathbf{S}''_+$. Since $c_0(\mathbf{z}_0) = \mathbf{z}_0$, the set $\mathbf{S}''_+$ is *stable* under the mapping $\alpha \mapsto \alpha \circ c_0$. Since $\varphi(\mathbf{z}_0) = -\mathbf{z}_0$, the image of $\mathbf{S}''_+$ under the mapping $\alpha \mapsto \alpha \circ \varphi$ is $-\mathbf{S}''_+$.

(21.21.7) *Let* $\mathfrak{n} = \bigoplus_{\alpha \in \mathbf{S}''_+} \mathfrak{g}_\alpha$ *and* $\mathfrak{n}_0 = \mathfrak{n} \cap \mathfrak{g}_0$. *Then we have a direct sum decomposition of the real semisimple Lie algebra* $\mathfrak{g}_0$:

(21.21.7.1)
$$\mathfrak{g}_0 = \mathfrak{k}_0 \oplus \mathfrak{a}_0 \oplus \mathfrak{n}_0$$

(Iwasawa decomposition of $\mathfrak{g}_0$).

Let us first show that the sum on the right-hand side of (21.21.7.1) is direct. Suppose then that $\mathbf{x} \in \mathfrak{k}_0$, $\mathbf{y} \in \mathfrak{a}_0$, and $\mathbf{z} \in \mathfrak{n}_0$ are such that

$$\mathbf{x} + \mathbf{y} + \mathbf{z} = 0.$$

By operating with $\varphi$ we obtain $\mathbf{x} - \mathbf{y} + \varphi(\mathbf{z}) = 0$, hence

$$2\mathbf{y} + \mathbf{z} - \varphi(\mathbf{z}) = 0.$$

But $2\mathbf{y} \in \mathfrak{h}$, $\mathbf{z} \in \bigoplus_{\alpha \in \mathbf{S}''_+} \mathfrak{g}_\alpha$ and $\varphi(\mathbf{z}) \in \bigoplus_{\alpha \in \mathbf{S}''_+} \mathfrak{g}_{-\alpha}$ (21.21.6); hence, by virtue of (21.8.1), $\mathbf{y} = \mathbf{z} = 0$ and consequently also $\mathbf{x} = 0$.

It remains to prove the equality (21.21.7.1). Let $\mathbf{x} \in \mathfrak{g}_0$, then by definition $\mathbf{x} = c_0(\mathbf{x}) = \frac{1}{2}(\mathbf{x} + c_0(\mathbf{x}))$. Since also $\mathbf{x} = \mathbf{h} + \sum_{\alpha \in \mathbf{S}} \mathbf{v}_\alpha$, where $\mathbf{h} \in \mathfrak{h}$ and $\mathbf{v}_\alpha \in \mathfrak{g}_\alpha$ for each $\alpha \in \mathbf{S}$, we have $\mathbf{x} = \frac{1}{2}(\mathbf{h} + c_0(\mathbf{h})) + \frac{1}{2}\sum_{\alpha \in \mathbf{S}} (\mathbf{v}_\alpha + c_0(\mathbf{v}_\alpha))$. Since $c_0(\mathbf{h}) = \mathbf{h}$ and since the elements of $\mathfrak{h}$ fixed by $c_0$ are those which belong to $(\mathfrak{t} \cap \mathfrak{k}_0) \oplus \mathfrak{a}_0$, we have $\mathbf{h} + c_0(\mathbf{h}) \in \mathfrak{k}_0 \oplus \mathfrak{a}_0$. If $\alpha \in \mathbf{S}'$, we have $\mathbf{v}_\alpha \in \mathfrak{k}_0 \oplus i\mathfrak{k}_0$ by virtue of (21.21.5), hence $\mathbf{v}_\alpha + c_0(\mathbf{v}_\alpha) \in \mathfrak{k}_0$. If $\alpha \in \mathbf{S}''_+$, then by definition $\mathbf{v}_\alpha \in \mathfrak{n}$, hence $\mathbf{v}_\alpha + c_0(\mathbf{v}_\alpha) \in \mathfrak{n} \cap \mathfrak{g}_0 = \mathfrak{n}_0$. Finally, if $\alpha \in -\mathbf{S}''_+$, the relation $\mathbf{v}_\alpha \in \mathfrak{g}_\alpha$ implies $c_u(\mathbf{v}_\alpha) \in \mathfrak{g}_{-\alpha}$ (21.21.4.2), hence

$$c_u(\mathbf{v}_\alpha) + c_0(c_u(\mathbf{v}_\alpha)) \in \mathfrak{n}_0$$

by the preceding result. On the other hand, the sum

$$\mathbf{v}_\alpha + c_0(\mathbf{v}_\alpha) + c_u \mathbf{v}_\alpha + c_0(c_u(\mathbf{v}_\alpha))$$

is fixed by $c_0$ and by $c_u$, and therefore belongs to $\mathfrak{k}_0$. It follows that $\mathbf{v}_\alpha + c_0(\mathbf{v}_\alpha) \in \mathfrak{k}_0 \oplus \mathfrak{n}_0$, and this completes the proof of (21.21.7.1).

(21.21.8) A finite-dimensional (real or complex) Lie algebra $\mathfrak{b}$ is said to be *nilpotent* if there exists an integer $r$ such that, for all sequences $\mathbf{x}_1, \mathbf{x}_2, \ldots, \mathbf{x}_r$ of elements of $\mathfrak{b}$, we have

(21.21.8.1) $$\operatorname{ad}(\mathbf{x}_1) \circ \operatorname{ad}(\mathbf{x}_2) \circ \cdots \circ \operatorname{ad}(\mathbf{x}_r) = 0$$

in the ring $\operatorname{End}(\mathfrak{b})$ of endomorphisms of the vector space $\mathfrak{b}$. A *connected* Lie group is said to be *nilpotent* if its Lie algebra is nilpotent.

(21.21.9) (i) *In the Iwasawa decomposition* (21.21.7.1), $\mathfrak{n}$ (resp. $\mathfrak{n}_0$) *is a nilpotent complex* (resp. *real*) *Lie algebra, and* $\mathfrak{s}_0 = \mathfrak{a}_0 \oplus \mathfrak{n}_0$ *is a solvable Lie algebra in which* $\mathfrak{n}_0$ *is an ideal.*

(ii) *Relative to the hermitian scalar product* $-\mathrm{B}_\mathfrak{g}(\mathbf{x}, c_u(\mathbf{y}))$ *on* $\mathfrak{g}$ (21.17.2), *there exists an orthonormal basis for which the endomorphism* $\operatorname{ad}(\mathbf{x})$ *of the vector space* $\mathfrak{g}$ *is represented by a matrix that is*

(a) *skew-hermitian if* $\mathbf{x} \in \mathfrak{g}_u$;
(b) *lower triangular with zero diagonal if* $\mathbf{x} \in \mathfrak{n}$;
(c) *real diagonal if* $\mathbf{x} \in \mathfrak{a}_0$.

We shall prove (ii) first. Recall that for each root $\alpha \in \mathbf{S}$ we have $c_u(\mathfrak{g}_\alpha) = \mathfrak{g}_{-\alpha}$ (21.21.4.2) and therefore, in the decomposition $\mathfrak{g} = \mathfrak{h} \oplus \bigoplus_{\alpha \in \mathbf{S}} \mathfrak{g}_\alpha$, the subspaces $\mathfrak{h}$ and $\mathfrak{g}_\alpha$ ($\alpha \in \mathbf{S}$) are *pairwise orthogonal* relative to the hermitian scalar product $-\mathrm{B}_\mathfrak{g}(\mathbf{x}, c_u(\mathbf{y}))$ (21.19.9). Take in $\mathfrak{h}$ an orthonormal basis $\mathbf{h}_1, \ldots, \mathbf{h}_l$, and a unit vector $\mathbf{a}_\alpha$ in each $\mathfrak{g}_\alpha$; with the notation of (21.21.6), range the roots belonging to $\mathbf{S}' \cup \mathbf{S}''_+$ in a sequence $\alpha_1, \ldots, \alpha_r, \alpha_{r+1}, \ldots, \alpha_{r+m}$, so that $\alpha_j \in \mathbf{S}'$ for $1 \leqq j \leqq r$, $\alpha_{r+j} \in \mathbf{S}''_+$ for $1 \leqq j \leqq m$, and such that $\alpha_{r+j}(\mathbf{z}_0) \leqq \alpha_{r+j+1}(\mathbf{z}_0)$ for $1 \leqq j \leqq m-1$. Consider now the orthonormal basis of $\mathfrak{g}$ arranged in the following order:

$$\mathbf{a}_{-\alpha_{r+m}}, \ldots, \mathbf{a}_{-\alpha_{r+1}}, \mathbf{h}_1, \ldots, \mathbf{h}_l, \mathbf{a}_{\alpha_1}, \ldots, \mathbf{a}_{\alpha_r}, \mathbf{a}_{\alpha_{r+1}}, \ldots, \mathbf{a}_{\alpha_{r+m}}.$$

We shall show that this basis has the required properties. For $\mathbf{x} \in \mathfrak{g}_u$ this is clear, because every orthonormal basis will satisfy the condition (a)

(21.17.3.2). It is equally clear for $\mathbf{x} \in \mathfrak{a}_0$, because $\mathrm{ad}(\mathbf{x}) \cdot \mathbf{h}_j = 0$ for $1 \leqq j \leqq l$ and $\mathrm{ad}(\mathbf{x}) \cdot \mathbf{a}_\alpha = \alpha(\mathbf{x})\mathbf{a}_\alpha$ for all $\alpha \in \mathbf{S}$, and furthermore on $it \supset \mathfrak{a}_0$ the roots are real-valued (21.8.1). Finally, to verify that condition (b) is satisfied we may restrict attention to the case where $\mathbf{x} = \mathbf{a}_{\alpha_{r+j}}$, $1 \leqq j \leqq m$. We have $\mathrm{ad}(\mathbf{a}_{\alpha_{r+j}}) \cdot \mathbf{h}_k = \alpha_{r+j}(\mathbf{h}_k)\mathbf{a}_{\alpha_{r+j}}$, and $\mathrm{ad}(\mathbf{a}_{\alpha_{r+j}}) \cdot \mathbf{a}_{-\alpha_{r+h}}$ is either zero or belongs to $\mathfrak{g}_\beta$ if $\beta = \alpha_{r+j} - \alpha_{r+h}$ is a root; but in this latter case we have $\beta(\mathbf{z}_0) = \alpha_{r+j}(\mathbf{z}_0) - \alpha_{r+h}(\mathbf{z}_0)$, so that either $\beta \in \mathbf{S}'$ or $\beta \in \mathbf{S}''_+$ or $\beta = -\alpha_{r+k}$, but with $k < h$. Next, $\mathrm{ad}(\mathbf{a}_{\alpha_{r+j}}) \cdot \mathbf{a}_{\alpha_h}$, where $1 \leqq h \leqq r$, is either zero or belongs to $\mathfrak{g}_\beta$ if $\beta = \alpha_{r+j} + \alpha_h$ is a root; but then $\beta(\mathbf{z}_0) = \alpha_{r+j}(\mathbf{z}_0) > 0$, so that $\beta \in \mathbf{S}''_+$. Finally, $\mathrm{ad}(\mathbf{a}_{\alpha_{r+j}}) \cdot \mathbf{a}_{\alpha_{r+h}}$ is zero or belongs to $\mathfrak{g}_\beta$ if $\beta = \alpha_{r+j} + \alpha_{r+h}$ is a root; but then $\beta(\mathbf{z}_0) > \alpha_{r+h}(\mathbf{z}_0) > 0$, so that $\beta$ is of the form $\alpha_{r+k}$ with $k > h$. This completes the proof of (ii).

For the proof of (i), we observe that because $\mathbf{x} \mapsto \mathrm{ad}(\mathbf{x})$ is an isomorphism of $\mathfrak{g}$ onto $\mathrm{ad}(\mathfrak{g})$ (21.6.3), it is enough to show that $\mathrm{ad}(\mathfrak{n})$ is a complex Lie subalgebra of $\mathrm{ad}(\mathfrak{g})$ that is nilpotent, and that $\mathrm{ad}(\mathfrak{s}_0)$ is a real Lie subalgebra of $\mathrm{ad}(\mathfrak{g}_0)$ that is solvable (the fact that $\mathfrak{n}_0$ is an ideal in $\mathfrak{s}_0$ follows from the relation $[\mathfrak{a}_0, \mathfrak{g}_\alpha] \subset \mathfrak{g}_\alpha$). Since every subalgebra of a solvable (resp. nilpotent) Lie algebra is solvable (resp. nilpotent), it is enough by virtue of (ii) to consider the Lie algebra $\mathfrak{gl}(\mathfrak{g}) = \mathbf{M}_n(\mathbf{C})$ (where $n = \dim(\mathfrak{g})$) and for each integer $k$ such that $0 \leqq k \leqq n$ the vector subspace $\mathfrak{T}_k$ consisting of the matrices $(x_{kj})$ such that $x_{hj} = 0$ for $j + k > h$. It is easily verified that

(21.21.9.1) $$[\mathfrak{T}_j, \mathfrak{T}_k] \subset \mathfrak{T}_{j+k}$$

which shows that $\mathfrak{T}_0$ (the algebra of lower triangular matrices) is solvable (19.12.3) and that $\mathfrak{T}_1$ (the algebra of lower triangular matrices with zeros on the diagonal) is a nilpotent Lie algebra.

(21.21.10) *Let* $G_1$ *be a connected semisimple Lie group with Lie algebra* $\mathfrak{g}_0$, *and let* $K_1$, $A_1$, $N_1$ *be the connected Lie groups immersed in* $G_1$ *whose respective Lie algebras are* $\mathfrak{k}_0$, $\mathfrak{a}_0$ *and* $\mathfrak{n}_0$, *in the notation of* (21.21.7).

(i) *The subgroups* $K_1$, $A_1$, $N_1$ *are closed in* $G_1$, *and* $K_1$ *contains the center* $C_1$ *of* $G_1$ (21.17.11). *The mapping* $\mathbf{x} \mapsto \exp_{G_1}(\mathbf{x})$ *is an isomorphism of* $\mathfrak{a}_0$ *onto* $A_1$ *and a diffeomorphism of* $\mathfrak{n}_0$ *onto* $N_1$, *so that* $A_1$ *is a commutative group isomorphic to* $\mathbf{R}^n$ *for some n, and* $N_1$ *is a nilpotent group diffeomorphic to* $\mathbf{R}^m$ *for some m.*

(ii) *The mapping* $(x, y, z) \mapsto xyz$ *is a diffeomorphism of* $K_1 \times A_1 \times N_1$ *onto* $G_1$ (Iwasawa decomposition of $G_1$). *The image of* $\{e_1\} \times A_1 \times N_1$ *under this mapping is a closed solvable subgroup* $S_1$ *of* $G_1$.

*Furthermore, if* $m_{G_1}$ *is a Haar measure on* $G_1$, *there exists a Haar measure*

$m_{K_1}$ *on* $K_1$ *and a left Haar measure* $m_{S_1}$ *on* $S_1$ *such that, for every continuous function* $f$ *on* $G_1$ *with compact support, we have*

$$\text{(21.21.10.1)} \quad \int_{G_1} f(s)\, dm_{G_1}(s) = \iint_{K_1 \times S_1} f(xy^{-1})\, dm_{K_1}(x)\, dm_{S_1}(y).$$

(Recall that $G_1$ and $K_1$ are *unimodular* ((21.6.6) and (21.6.10)).)

(I) Consider first the case where $C_1 = \{e_1\}$, so that $G_1$ may be identified with its adjoint group $\mathrm{Ad}(G_1)$. If G is the complex semisimple subgroup that is the identity component of $\mathrm{Aut}(\mathfrak{g}) \subset \mathbf{GL}(\mathfrak{g})$ (21.17.1), whose Lie algebra $\mathrm{ad}(\mathfrak{g})$ is isomorphic to $\mathfrak{g}$, then $\mathrm{Ad}(G_1)$ is the connected Lie subgroup of $G_{|\mathbf{R}}$ whose Lie algebra is $\mathrm{ad}(\mathfrak{g}_0)$ (21.6.8). We may therefore likewise identify $K_1$, $A_1$, $N_1$ with the connected Lie groups immersed in $\mathbf{GL}(\mathfrak{g})$ corresponding to the images $\mathrm{ad}(\mathfrak{k}_0)$, $\mathrm{ad}(\mathfrak{a}_0)$, and $\mathrm{ad}(\mathfrak{n}_0)$ of the real Lie subalgebras $\mathfrak{k}_0$, $\mathfrak{a}_0$, $\mathfrak{n}_0$ under the isomorphism $\mathbf{x} \mapsto \mathrm{ad}(\mathbf{x})$ of $\mathfrak{g}$ onto $\mathrm{ad}(\mathfrak{g})$. We shall assume that an orthonormal basis of $\mathfrak{g}$ has been chosen to satisfy the conditions of (21.21.9(ii)). Since the matrices of $\mathrm{ad}(\mathfrak{a}_0)$ are real and diagonal, the group $A_1$ consists of real diagonal matrices with diagonal entries $> 0$, and it is clear that the exponential mapping of $\mathbf{GL}(\mathfrak{g})$ is an isomorphism of $\mathrm{ad}(\mathfrak{a}_0)$ onto $A_1$, and that $A_1$ is closed in $\mathbf{GL}(\mathfrak{g})$.

As to the group $N_1$, we observe that in the notation of (21.21.9) we have $\mathrm{ad}(\mathfrak{n}_0) \subset \mathfrak{T}_1$, so that $N_1$ is a connected Lie group immersed in the connected Lie group $T_1$, immersed in $\mathbf{GL}(\mathfrak{g})$ with Lie algebra $\mathfrak{T}_1$. Now we have the following proposition:

(21.21.10.2) *The group* $T_1$ *is the closed subgroup of all matrices* $I + N$, *where* $N \in \mathfrak{T}_1$, *and the mapping* $N \mapsto \exp(N)$ (*where* exp *is the exponential map of* $\mathbf{GL}(\mathfrak{g})$) *is a diffeomorphism of* $\mathfrak{T}_1$ *onto* $T_1$.

It is immediate that $N^n = 0$ for all $N \in \mathfrak{T}_1$, where $n = \dim(\mathfrak{g})$, and it is clear that the matrices $I + N$ form a closed subgroup of $\mathbf{GL}(\mathfrak{g})$. If we put

$$P_1(N) = N - \frac{1}{2}N^2 + \frac{1}{3}N^3 - \cdots + (-1)^{n-2}\frac{1}{n-1}N^{n-1},$$

$$P_2(N) = I + \frac{1}{1!}N + \frac{1}{2!}N^2 + \cdots + \frac{1}{(n-1)!}N^{n-1},$$

then we have $\exp(N) = P_2(N) \in T_1$, and $P_1(N) \in \mathfrak{T}_1$ for all $N \in \mathfrak{T}_1$. To prove the proposition it is enough to show that $P_2(P_1(N)) = I + N$ and $P_1(P_2(N) - I) = N$ for all $N \in \mathfrak{T}_1$. Now, for $x$ real and sufficiently small we have $\log(1 + x) = \sum_{n=1}^{\infty} (-1)^{n-1}x^n/n$ (9.3.7), and for all real $x$ we have

$e^x = \sum_{p=0}^{\infty} x^p/p!$; the theorem of substitution of power series in a power series **(9.2.1)** shows that

$$\sum_{\substack{n_1+n_2+\cdots+n_p=k \\ n_1\geq 1,\,\ldots,\,n_p\geq 1}} \frac{1}{p!}\frac{(-1)^{n_1+n_2+\cdots+n_p-p}}{n_1 n_2 \cdots n_p} = \begin{cases} 1 & \text{if} \quad k=1, \\ 0 & \text{if} \quad k>1, \end{cases}$$

$$\sum_{\substack{p_1+p_2+\cdots+p_n=k \\ p_1\geq 1,\,\ldots,\,p_n\geq 1}} \frac{(-1)^{n-1}}{n}\cdot\frac{1}{p_1!p_2!\cdots p_n!} = \begin{cases} 1 & \text{if} \quad k=1, \\ 0 & \text{if} \quad k>1, \end{cases}$$

and the coefficients of $N^k$ in $P_2(P_1(N))$ and $P_1(P_2(N)-I)$, for $k<n$, are precisely the left-hand sides of these two relations.

Since $\mathrm{ad}(\mathfrak{n}_0)$ is closed in $\mathfrak{T}_1$, it follows from **(21.21.10.2)** that $N_1$ is closed in $T_1$; since $T_1$ is closed in $\mathbf{GL}(\mathfrak{g})$, it follows that $N_1$ is also closed in $G_1$ (because the topology of $G_1$ is induced by that of $\mathbf{GL}(\mathfrak{g})$).

Since $\mathfrak{n}_0$ is an ideal of $\mathfrak{s}_0 = \mathfrak{a}_0 \oplus \mathfrak{n}_0$, the elements of $A_1$ normalize $N_1$ **(19.11.4)**, so that $A_1N_1 = N_1A_1$, which shows that $S_1 = A_1N_1$ is the connected group immersed in $G_1$ with Lie algebra $\mathfrak{s}_0$. Moreover, if $D \in A_1$ and $U \in N_1$, then $D$ is the diagonal of the triangular matrix $DU$. If $(D_\nu, U_\nu)$ is a sequence of matrices in $S_1$ converging to a limit in $T_1$, then the sequence $(D_\nu)$ also converges, and therefore so also does the sequence $(U_\nu)$; since $A_1$ and $N_1$ are closed, it follows that $S_1$ also is closed. Moreover, the mapping $(D, U)\mapsto DU$ of $A_1 \times N_1$ onto $S_1$ is a diffeomorphism, the inverse mapping being $X \mapsto (D(X), D(X)^{-1}X)$, where $D(X)$ is the diagonal of $X$.

In the situation under consideration, we know that $K_1$ is a subgroup of $\mathbf{GL}(\mathfrak{g})$ consisting of *unitary* matrices **(21.17.4)**. It follows that $K_1 \cap S_1 = \{I\}$: for the inverse of a lower triangular matrix is again lower triangular, and therefore cannot be unitary unless it is a diagonal matrix, with diagonal entries that are complex numbers of absolute value 1; hence $K_1 \cap S_1$ consists of diagonal matrices whose diagonal entries are simultaneously positive real numbers and complex numbers of absolute value 1, and the only such matrix is $I$.

Since $K_1$ is compact, $S_1K_1$ is closed in G **(12.10.5)**; its image under the canonical mapping $p$: $G_1 \to G_1/K_1$ is therefore closed **(12.10.5)**. But if $\{x_0\} = p(K_1)$, this image is just the orbit $S_1 \cdot x_0$ for the action of $S_1$ on the space $G_1/K_1$. Now for this action the stabilizer of $x_0$ is $S_1 \cap K_1 = \{I\}$, hence $S_1 \cdot x_0$ is a submanifold of $G_1/K_1$, of dimension equal to that of $S_1$, hence to that of $G_1/K_1$ **(16.10.7)**. It follows that $S_1 \cdot x_0$ is both open and closed in the connected space $G_1/K_1$, so that $S_1 \cdot x_0 = G_1/K_1$, or equivalently $S_1K_1 = G_1 = K_1S_1$ (since $K_1$ and $S_1$ are subgroups of $G_1$).

The $C^\infty$ mapping $(x, s)\mapsto xs$ of $K_1 \times S_1$ into $G_1$ is therefore bijective; we have to show that it is a diffeomorphism. At each point $(x_1, s_1) \in K_1 \times S_1$, every tangent vector to $K_1$ (resp. $S_1$) can be written uniquely in the form $x_1 \cdot \mathbf{v}$ (resp. $\mathbf{w} \cdot s_1$), where $\mathbf{v} \in \mathfrak{k}_0$ (resp. $\mathbf{w} \in \mathfrak{s}_0$) **(16.9.8)**. The tangent linear mapping to $(x, s) \mapsto xs$ at the point $(x_1, s_1)$ is therefore **(16.9.9)**

$$(x_1 \cdot \mathbf{v}, \mathbf{w} \cdot s_1) \mapsto x_1 \cdot (\mathbf{v} + \mathbf{w}) \cdot s_1$$

which is clearly bijective, because the sum $\mathfrak{k}_0 \oplus \mathfrak{s}_0$ is direct. The result now follows from **(16.8.8)**.

(II) We now consider the general case. Let $p_1 \colon G_1 \to G_1/C_1$ be the canonical homomorphism of $G_1$ onto its adjoint group, so that $p_1(K_1)$, $p_1(A_1)$, and $p_1(N_1)$ are connected Lie groups immersed in $p_1(G_1)$, with Lie algebras respectively $\mathfrak{k}_0$, $\mathfrak{a}_0$, and $\mathfrak{n}_0$. Using the fact that $p_1(A_1)$ and $p_1(N_1)$ are simply connected, we may repeat without any substantial change the argument of **(21.17.10)**, by showing first that $A_1$ and $N_1$ are the identity components of $p_1^{-1}(p_1(A_1))$ and $p_1^{-1}(p_1(N_1))$, respectively. Then, using as in **(21.17.10)** the fact that $C_1 \subset K_1$, we see that $(x, y, z) \mapsto xyz$ is a bijection of $K_1 \times A_1 \times N_1$ onto $G_1$, and finally, using the result of (I), that it is a diffeomorphism.

To prove **(21.21.10.1)**, denote by $z \mapsto (p(z), q(z))$ the diffeomorphism of $G_1$ onto $K_1 \times S_1$ that is the inverse of $(x, s) \mapsto xs$. Let $u \in \mathscr{K}(K_1)$ be a function with values $\geqq 0$. As $v$ runs through $\mathscr{K}(S_1)$, the mapping

$$v \mapsto \int u(p(z))v(q(z))\, dm_{G_1}(z)$$

is a positive linear form on $\mathscr{K}(S_1)$ which is *right-invariant*, because $G_1$ is unimodular and $p(zs) = p(z)$ for $s \in S_1$; it can therefore be written as $v \mapsto J(u) \int v(s^{-1})\, dm_{S_1}(s)$, where $J(u)$ is a constant $\geqq 0$ **(14.1)**. We next extend $J$ to a positive linear form on $\mathscr{K}(K_1)$ in the obvious way; since $p(xz) = p(z)$ for $x \in K_1$, this linear form is *left-invariant* and hence, by a suitable choice of $m_{K_1}$, it can be taken to be equal to $m_{K_1}$ **(14.1)**. The formula **(21.21.10.1)** is therefore established for all functions $f$ of the form $z \mapsto u(p(z))v(q(z))$. To complete the proof, we invoke **(13.21.1)** and the existence of the homeomorphism $(x, s) \mapsto xs$ of $K_1 \times S_1$ onto $G_1$.

The relations between the Iwasawa decomposition and the Cartan decomposition of $G_1$ **(21.18.8)** are described in the next proposition:

**(21.21.11)** *With the notation of* **(21.21.10)** *and Section* **21.18**, *let* $\tau_1$ *be the involutory automorphism of* $G_1$ *for which* $K_1$ *is the set of fixed points, and such that* $\tau_1(z) = z^{-1}$ *for* $z \in P_1$ **(21.18.10)**. *Then the mapping* $f_1 \colon s \mapsto \tau_1(s)s^{-1}$ *is a diffeomorphism of* $S_1$ *onto* $P_1$.

Since each $x \in G_1$ can be written uniquely as $x = zy$ with $z \in P_1$ and $y \in K_1$, we have $\tau_1(x)x^{-1} = z^{-2} \in P_1$, so that in particular $f_1(S_1) \subset P_1$. The mapping $f_1$ is a bijection of $S_1$ onto $P_1$: for the relation $\tau_1(s')s'^{-1} = \tau_1(s'')s''^{-1}$ for $s', s'' \in S_1$ implies that $\tau_1(s''^{-1}s') = s''^{-1}s'$, so that $s''^{-1}s' \in K_1$ and therefore $s'' = s'$ since $K_1 \cap S_1 = \{e_1\}$. Moreover $f_1(S_1) = P_1$, for every $z \in P_1$ is uniquely of the form $\exp_{G_1}(\mathbf{u})$ with $\mathbf{u} \in \mathfrak{p}_0$; there exist two elements $x' \in K_1$ and $s' \in S_1$ such that $z' = x's'^{-1} = \exp_{G_1}(\frac{1}{2}\mathbf{u})$ (21.21.10); since $z' \in P_1$, we have

$$z = z'^2 = \tau_1(z'^{-1})z' = \tau_1(s')\tau_1(x'^{-1})x's'^{-1} = \tau_1(s')s'^{-1} = f_1(s')$$

since $\tau_1(x') = x'$. Finally, this calculation shows that the inverse of the mapping $f_1: S_1 \to P_1$ is the mapping $z \mapsto q(\exp_{G_1}(\frac{1}{2}l(z)))$, where $q: G_1 \to S_1$ is the mapping defined in the proof of (21.21.10), and $l: P_1 \to \mathfrak{p}_0$ is the inverse of the restriction to $\mathfrak{p}_0$ of the exponential map. This shows that $f_1$ is a diffeomorphism of $S_1$ onto $P_1$.

*Remarks*

(21.21.12) (i) With the notation of (21.21.10), $K_1$ is isomorphic to the product of a *compact* group $K_1'$ and a vector group $\mathbf{R}^p$ (21.6.9), whence we recover the fact (21.18.8) that $G_1$ is *diffeomorphic to the product of a compact group* $K_1'$ *and a vector space* $\mathbf{R}^N$. Moreover, the compact subgroup $K_1'$ of $K_1$ is *maximal* in $G_1$, for the components in $A_1$ and in $N_1$ of an element of $G_1$ will generate subgroups that are noncompact if they are $\neq \{e_1\}$; hence every compact subgroup of $G_1$ containing $K_1'$ must be contained in $K_1$, and hence in $K_1'$.

(ii) It can be proved that the Lie algebras $\mathfrak{k}_0$, $\mathfrak{a}_0$, $\mathfrak{n}_0$ that figure in the Iwasawa decomposition are determined up to isomorphism. The dimension of $\mathfrak{a}_0$ is called the *rank* of the *symmetric space* $G_1/K_1$.

(iii) If $\mathfrak{g}_0$ is a *normal* real form of $\mathfrak{g}$ (21.18.9), we have $\mathfrak{a}_0 = i\mathfrak{t}$, and the rank of $\mathfrak{g}_0$ is equal to that of $\mathfrak{g}_u$. It can be shown that this condition characterizes the normal forms of $\mathfrak{g}$, which are all isomorphic. We have in this case $\mathbf{S}' = \varnothing$, and $\mathbf{S}''_+$ is the set $\mathbf{S}_+$ of all *positive* roots, relative to the ordering defined in (21.14.5).

We have then

$$\mathfrak{g}_{|\mathbf{R}} = \mathfrak{g}_u \oplus i\mathfrak{t} \oplus \mathfrak{n}_{|\mathbf{R}}. \tag{21.21.12.1}$$

For the argument of (21.21.7) shows that the sum on the right-hand side is direct. Also $\mathfrak{g}_u$ contains $\mathfrak{t}$, hence the right-hand side of (21.21.12.1) contains

$\mathfrak{h} = \mathfrak{t} \oplus i\mathfrak{t}$; also $\mathfrak{g}_u$ contains the elements $\mathbf{e}_\alpha - \mathbf{e}_{-\alpha}$ and $i(\mathbf{e}_\alpha + \mathbf{e}_{-\alpha})$ for $\alpha \in \mathbf{S}_+$, and since $\mathfrak{n}$ contains the elements $\mathbf{e}_\alpha$ and $i\mathbf{e}_\alpha$ for $\alpha \in \mathbf{S}_+$, the sum on the right-hand side of (21.21.12.1) contains $\mathbf{e}_\alpha$ and $i\mathbf{e}_\alpha$ for all $\alpha \in \mathbf{S}$, and is therefore equal to $\mathfrak{g}$. The same proof as in (21.21.10) then shows that if G is a *complex* connected Lie group with Lie algebra $\mathfrak{g}$, and if $G_u$, A, and N are the connected real Lie groups immersed in $G_{|\mathbf{R}}$ that correspond respectively to the subalgebras $\mathfrak{g}_u$, $i\mathfrak{t}$, and $\mathfrak{n}_{|\mathbf{R}}$, then $G_u$ is compact, A and N are closed subgroups of G, and the mapping $(x, y, z) \mapsto xyz$ is a diffeomorphism of $G_u \times A \times N$ onto $G_{|\mathbf{R}}$.

(21.21.13) In example (I) of (21.18.9), we may take $\mathfrak{a}_0$ to be the set of real diagonal $n \times n$ matrices with zero trace; then the $\mathfrak{g}_\alpha$ corresponding to the positive roots $\alpha$ are the spaces $\mathbf{C}E_{rs}$ for $r < s$ (21.12.1), and $\mathfrak{n}_0$ is therefore the nilpotent Lie algebra of all upper triangular real matrices with zeros on the diagonal.

(21.21.14) *With the notation of* (21.21.7) *and* (21.21.10), *for each vector* $\mathbf{x} \in \mathfrak{p}_0$ *there exists* $s \in K_1$ *such that* $\operatorname{Ad}(s) \cdot \mathbf{x} \in \mathfrak{a}_0$.

Since we may replace $K_1$ by its adjoint group, we may assume that $K_1$ is compact. We have $\operatorname{Ad}(s) \cdot \mathfrak{p}_0 = \mathfrak{p}_0$ for all $s \in K_1$ ((21.18.4.2) and (19.11.3)), and the restriction of the Killing form $B_\mathfrak{g}$ to $\mathfrak{p}_0 \times \mathfrak{p}_0$ is positive definite and invariant under the action $(s, \mathbf{z}) \mapsto \operatorname{Ad}(s) \cdot \mathbf{z}$ of $K_1$ on $\mathfrak{p}_0$. For brevity let $\|\mathbf{z}\| = (B_\mathfrak{g}(\mathbf{z}, \mathbf{z}))^{1/2}$ for $\mathbf{z} \in \mathfrak{p}_0$, and consider as in (21.7.7.1) the continuous function $s \mapsto \|\operatorname{Ad}(s) \cdot \mathbf{x} - \mathbf{z}_0\|^2$ on $K_1$, where $\mathbf{z}_0$ is the element defined in (21.21.6); this function attains its minimum at a point $s_0$, and by replacing $\mathbf{x}$ by $\operatorname{Ad}(s_0) \cdot \mathbf{x}$ we may suppose that $s_0 = e_1$, the identity element of $K_1$. By expressing that for each $\mathbf{y} \in \mathfrak{k}_0$ the derivative of the function $t \mapsto \|\operatorname{Ad}(\exp(t\mathbf{y})) \cdot \mathbf{x} - \mathbf{z}_0\|^2$ vanishes at $t = 0$ we obtain, using the invariance of $B_\mathfrak{g}$,

$$0 = 2B_\mathfrak{g}([\mathbf{y}, \mathbf{x}], \mathbf{x} - \mathbf{z}_0) = 2B_\mathfrak{g}(\mathbf{y}, [\mathbf{x}, \mathbf{x} - \mathbf{z}_0])$$

for all $\mathbf{y} \in \mathfrak{k}_0$. Since the restriction of $B_\mathfrak{g}$ to $\mathfrak{g}_0 \times \mathfrak{g}_0$ is nondegenerate, and since $\mathfrak{k}_0$ and $\mathfrak{p}_0$ are orthogonal supplements of each other relative to this form, the formula just written implies that $[\mathbf{x}, \mathbf{z}_0] \in \mathfrak{p}_0$. But now $[\mathfrak{p}_0, \mathfrak{p}_0] \subset \mathfrak{k}_0$, and hence $[\mathbf{x}, \mathbf{z}_0] \in \mathfrak{p}_0 \cap \mathfrak{k}_0 = \{0\}$; since $\alpha(\mathbf{z}_0) \neq 0$ for each root $\alpha \in \mathbf{S}''$ (21.21.6), we conclude that $\mathbf{x}$ lies in the intersection of $\mathfrak{g}_0$ and $\mathfrak{h} \oplus \bigoplus_{\alpha \in \mathbf{S}'} \mathfrak{g}_\alpha$, and because $\mathfrak{g}_\alpha \subset \mathfrak{k}_0 \oplus i\mathfrak{k}_0$ for $\alpha \in \mathbf{S}'$ by virtue of (21.21.5), we have $\mathbf{x} \in \mathfrak{k}_0 \oplus \mathfrak{a}_0$, and finally $\mathbf{x} \in \mathfrak{a}_0$ because $\mathbf{x} \in \mathfrak{p}_0$. This completes the proof.

(21.21.15) *With the notation of* (21.21.10), *we have* $G_1 = K_1 A_1 K_1$.

Since $G_1 = P_1 K_1$ (21.18.8), it is enough to remark that every element of $P_1$ is of the form $\exp(\mathbf{z})$ with $\mathbf{z} \in \mathfrak{p}_0$, and that $\mathbf{z} = \mathrm{Ad}(s) \cdot \mathbf{x}$ for some $s \in K_1$ and $\mathbf{x} \in \mathfrak{a}_0$ (21.21.14); since $\exp(\mathrm{Ad}(s) \cdot \mathbf{x}) = s(\exp(\mathbf{x}))s^{-1}$ (19.11.2.3), we have $P_1 \subset K_1 A_1 K_1$.

## PROBLEMS

1. (a) With the notation of Sections 21.18 and 21.21, let $\mathfrak{m}_0$ be the intersection of $\mathfrak{k}_0$ with the centralizer of $\mathfrak{a}_0$ in $\mathfrak{g}_0$, and let $\mathfrak{l}_0$ (resp. $\mathfrak{q}_0$) be the subspace of $\mathfrak{k}_0$ orthogonal to $\mathfrak{m}_0$ (resp. the subspace of $\mathfrak{p}_0$ orthogonal to $\mathfrak{a}_0$) relative to the form $-B_{\mathfrak{g}}(\mathbf{x}, c_u(\mathbf{y}))$. Let $\mathfrak{k} = \mathfrak{k}_0 + i\mathfrak{k}_0$, $\mathfrak{a} = \mathfrak{a}_0 + i\mathfrak{a}_0$, $\mathfrak{m} = \mathfrak{m}_0 + i\mathfrak{m}_0$, $\mathfrak{l} = \mathfrak{l}_0 + i\mathfrak{l}_0$, $\mathfrak{q} = \mathfrak{q}_0 + i\mathfrak{q}_0$. Show that $\mathfrak{m}$ is the direct sum of $\mathfrak{h} \cap \mathfrak{k}$ and the subspaces $\mathfrak{g}_\alpha + \mathfrak{g}_{-\alpha}$, $\alpha \in \mathbf{S}'$; if $\mathbf{x}_\alpha$ is a basis element of $\mathfrak{g}_\alpha$ over $\mathbf{C}$, then the elements $\mathbf{x}_\alpha + \varphi(\mathbf{x}_\alpha)$ with $\alpha \in \mathbf{S}''$ form a basis of $\mathfrak{l}$, and the elements $\mathbf{x}_\alpha - \varphi(\mathbf{x}_\alpha)$ with $\alpha \in \mathbf{S}''$ form a basis of $\mathfrak{q}$.
(b) Deduce from (a) that there exists an element $\mathbf{z}_0 \in \mathfrak{a}$ such that the centralizer of $\mathbf{R}\mathbf{z}_0$ in $\mathfrak{g}$ is $\mathfrak{a} + \mathfrak{m}$. Hence show, with the help of (21.21.14), that for each commutative subalgebra $\mathfrak{b}_0$ of $\mathfrak{p}_0$ there exists $s \in K_2$ such that $\mathrm{Ad}(s)(\mathfrak{b}_0) \subset \mathfrak{a}_0$.
(c) Show that for each $\alpha \in \mathbf{S}''$ the intersection $\mathfrak{g}''_\alpha = \mathfrak{g}_\alpha \cap \mathfrak{g}_0$ is one-dimensional over $\mathbf{R}$. Show that there exists $\mathbf{x}''_\alpha \in \mathfrak{g}''_\alpha$ such that, if we put $\mathbf{x}''_{-\alpha} = c_u(\mathbf{x}''_\alpha) \in \mathfrak{g}''_{-\alpha}$ and $\mathbf{h}''_\alpha = [\mathbf{x}''_\alpha, \mathbf{x}''_{-\alpha}]$, then we have $[\mathbf{h}''_\alpha, \mathbf{x}''_\alpha] = 2\mathbf{x}''_\alpha$ and $[\mathbf{h}''_\alpha, \mathbf{x}''_{-\alpha}] = -2\mathbf{x}''_{-\alpha}$. Also $\mathbf{y}''_\alpha = \mathbf{x}''_\alpha + \mathbf{x}''_{-\alpha} \in \mathfrak{k}_0$, and $\mathbf{z}''_\alpha = \mathbf{x}''_\alpha - \mathbf{x}''_{-\alpha} \in \mathfrak{p}_0$.
(d) Let M and M′ be, respectively, the centralizer and the normalizer of $i\mathfrak{a}_0$ in $K_2$ (i.e., the intersections of $K_2$ with the centralizer and normalizer of $i\mathfrak{a}_0$ in $G_2$, cf. (19.11.13)). Show that M and M′ have the same Lie algebra $\mathfrak{m}_0$. (Observe that if $\mathbf{u} \in \mathfrak{k}_0$ is such that $[\mathbf{u}, \mathbf{x}] \in i\mathfrak{a}_0$ for all $\mathbf{x} \in i\mathfrak{a}_0$, then $B_{\mathfrak{g}}(\mathrm{ad}(\mathbf{u}) \cdot \mathbf{x}, \mathrm{ad}(\mathbf{u}) \cdot \mathbf{x}) = 0$.) The finite group $W(G_2/K_2) = M'/M$ is called the *Weyl group* of the symmetric space $G_2/K_2$; it acts faithfully on $i\mathfrak{a}_0$. With the notation of (c), show that for each $\alpha \in \mathbf{S}''$ there exists a real number $\xi$ such that $r''_\alpha = \exp(\xi \mathbf{y}''_\alpha)$ belongs to M′ and has as image in $W(G_2/K_2)$ the orthogonal reflection $s''_\alpha$ with respect to the hyperplane with equation $\alpha(\mathbf{u}) = 0$ in $i\mathfrak{a}_0$.
(e) In order that a linear form $\lambda$ on $i\mathfrak{a}_0$ should be the restriction of a root $\alpha \in \mathbf{S}''$, it is necessary and sufficient that there exist $\mathbf{x} \neq 0$ in $i\mathfrak{a}_0$ such that $(\mathrm{ad}(\mathbf{u}))^2 \cdot \mathbf{x} = (\lambda(\mathbf{u}))^2 \mathbf{x}$ for all $\mathbf{u} \in i\mathfrak{a}_0$. Deduce that every element of the Weyl group $W(G_2/K_2)$ permutes the restrictions to $i\mathfrak{a}_0$ of the roots $\alpha \in \mathbf{S}''$.
(f) A *Weyl chamber* in $i\mathfrak{a}_0$ is any connected component of the complement of the union of the hyperplanes $\alpha(\mathbf{u}) = 0$ in $i\mathfrak{a}_0$, for all $\alpha \in \mathbf{S}''$. Show that $W(G_2/K_2)$ acts simply transitively on the set of Weyl chambers in $i\mathfrak{a}_0$, and is generated by the orthogonal reflections $s''_\alpha$, $\alpha \in \mathbf{S}''$. (If $C_1$, $C_2$ are two Weyl chambers and if W′ is the subgroup of $W(G_2/K_2)$ generated by the $s''_\alpha$, consider for $\mathbf{u}_1 \in C_1$ and $\mathbf{u}_2 \in C_2$ an element $w \in W'$ such that the distance from $\mathbf{u}_1$ to $w \cdot \mathbf{u}_2$ is as small as possible; show that this implies that $w \cdot \mathbf{u}_2 \in C_1$. To show that the action of $W(G_2/K_2)$ is simply transitive, follow the proof of (21.11.10).)
(g) The image $A_2$ of $i\mathfrak{a}_0$ under the exponential mapping $\exp_{G_2}$ is a torus contained in $P_2$, and $P_2$ is the union of the tori $sA_2s^{-1}$ for $s \in K_2$.

2. With the same notation as in Problem 1, let $\Sigma$ denote the set of restrictions to $i\mathfrak{a}_0$ of the roots $\alpha \in \mathbf{S}''$. For each $\lambda \in \Sigma$, let $\Sigma(\lambda)$ denote the set of $\alpha \in \mathbf{S}''$ of which $\lambda$ is the restriction, and $m(\lambda)$ the number of elements of $\Sigma(\lambda)$; $m(\lambda)$ is called the *multiplicity* of the linear form $\lambda$.

(a) The set $\Sigma$ satisfies the conditions $S_I$, $S_{II}$, and $S_{III}$ of (21.10.3) relative to the space $(i\mathfrak{a}_0)^*$: in other words, $\Sigma$ is a *root system*, in general *nonreduced*, i.e., not necessarily satisfying $S_{IV}$ (cf. Problem 3). The elements $\lambda \in \Sigma$ are called the *roots* of $G_2/K_2$ relative to the choice of the maximal commutative subalgebra $\mathfrak{a}_0$ of $\mathfrak{p}_0$.
(b) For each $\lambda \in \Sigma$ let $\mathfrak{k}_\lambda$ (resp. $i\mathfrak{p}_\lambda$) be the intersection of $\mathfrak{k}_0$ (resp. $i\mathfrak{p}_0$) with the sum of the spaces $\mathfrak{g}_\alpha + \mathfrak{g}_{-\alpha}$ for $\alpha \in \Sigma(\lambda)$; they are real vector spaces of dimension $m(\lambda)$. For $\lambda, \mu \in \Sigma$ we have

$$[\mathfrak{k}_\lambda, \mathfrak{k}_\mu] \subset \mathfrak{k}_{\lambda+\mu} + \mathfrak{k}_{\lambda-\mu}, \qquad [\mathfrak{k}_\lambda, i\mathfrak{p}_\mu] \subset i\mathfrak{p}_{\lambda+\mu} + i\mathfrak{p}_{\lambda-\mu},$$
$$[i\mathfrak{p}_\lambda, i\mathfrak{p}_\mu] \subset \mathfrak{k}_{\lambda+\mu} + \mathfrak{k}_{\lambda-\mu}$$

where, if $\lambda + \mu = 0$ (resp. $\lambda - \mu = 0$), $\mathfrak{k}_{\lambda+\mu}$ (resp. $\mathfrak{k}_{\lambda-\mu}$) is to be replaced by $\mathfrak{m}_0$, and $i\mathfrak{p}_{\lambda+\mu}$ (resp. $i\mathfrak{p}_{\lambda-\mu}$) by $i\mathfrak{a}_0$.
(c) For each $\lambda \in \Sigma$ and each integer $k \in \mathbf{Z}$, let $\mathfrak{u}''_{\lambda,k}$ be the affine hyperplane in $i\mathfrak{a}_0$ given by the equation $\lambda(\mathbf{u}) = 2k\pi i$; it is the intersection of $i\mathfrak{a}_0$ with the affine hyperplanes $\mathfrak{u}_{\alpha,k}$ for all roots $\alpha \in \Sigma(\lambda)$. The union of the hyperplanes $\mathfrak{u}''_{\lambda,k}$ for all $\lambda \in \Sigma$ and $k \neq 0$ is the set of $\mathbf{u} \in i\mathfrak{a}_0$ at which the restriction to $i\mathfrak{a}_0$ of the tangent linear mapping $T_{\mathbf{u}}(\exp_{G_2})$ is not bijective. If we identify $i\mathfrak{p}_0$ with the tangent space at $x_0$ to $G_2/K'_2$ (or $G_2/K_2$), the points $\mathbf{u} \in i\mathfrak{a}_0$ for which the tangent mapping $T_{\mathbf{u}}(\exp_{x_0})$ (Section **21.18**, Problem 7) is not bijective are those for which $2\mathbf{u}$ belongs to the union of the $\mathfrak{u}''_{\lambda,k}$ with $k \neq 0$.
(d) Show that, for all the points $x = \exp_{G_2}(\mathbf{u})$ such that $\mathbf{u} \in i\mathfrak{a}_0$ belongs to none of the $\mathfrak{u}''_{\lambda,k}$ but $2\mathbf{u}$ does lie in their union, the tangent mapping $T_x(\pi')$, where $\pi' \colon P_2 \to G_2/K'_2$ is the canonical mapping, is not bijective (use Section **21.18**, Problem 7(a)). Does this result remain true when $\mathbf{u}$ belongs to one of the $\mathfrak{u}''_{\lambda,k}$?
(e) A point $x \in P_2$ is said to be *singular* if the dimension of the orbit of $x$ under the action $(s, y) \mapsto sys^{-1}$ of $K_2$ on $P_2$ is strictly less than $\dim(K_2/M)$. The singular points of $P_2$ that belong to A are the points lying in the union of the $U''_\lambda$, where $U''_\lambda$ is the torus of codimension 1 in A that is the image under $\exp_{G_2}$ of any of the hyperplanes $\mathfrak{u}''_{\lambda,k}$, $\lambda \in \Sigma$; we have $U''_{-\lambda} = U''_\lambda$, but $U''_{2\lambda} \neq U''_\lambda$ if both $\lambda$ and $2\lambda$ are in $\Sigma$. A point $x$ is said to be *regular* if it is not singular. Let $A_{\text{reg}}$ denote the open set in A consisting of the regular points of A.
(f) Show that the mapping $(s, t) \mapsto sts^{-1}$ of $K_2 \times A$ into $P_2$ is a submersion at all points $(s, t)$ such that $t \in A_{\text{reg}}$ (argue as in (**21.15.1**)). Deduce that $(K_2/M) \times A_{\text{reg}}$ is a covering, with $\operatorname{card}(W(G_2/K_2))$ sheets, of an open subset V of $P_2$ consisting of regular points and such that the complement of V is negligible. This complement is the union of the sets $P''_\lambda$, where $P''_\lambda$ is the image under $(s, t) \mapsto sts^{-1}$ of the set $M''_\lambda \times U''_\lambda$, and $M''_\lambda$ is the centralizer of $U''_\lambda$ in $K_2$. Show that $P''_\lambda$ is the image under a $C^\infty$ mapping of a manifold of dimension $\leqq \dim(P_2) - (1 + r)$, where $r$ is the smallest value of the multiplicity $m(\lambda)$ for roots $\lambda \in \Sigma$ such that $2\lambda \notin \Sigma$.

3. Take $c_0$ as in example (III) of (**21.18.11**), so that $G_2$ is the quotient of $\mathbf{SU}(p + q)$ by its center, and $K_2$ is the quotient of $K_0$ by the center of $\mathbf{SU}(p + q)$; then $i\mathfrak{p}_0$ is the space of matrices of the form $\begin{pmatrix} 0_p & Z \\ -{}^t\bar{Z} & 0_q \end{pmatrix}$, where $Z$ is a $p \times q$ complex matrix.

Let $i\mathfrak{a}_0$ denote the subspace of $i\mathfrak{p}_0$ spanned (over $\mathbf{R}$) by the matrices $E_{j,p+j} - E_{p+j,j} = H_j$ for $1 \leqq j \leqq q$. Let $(\varepsilon_j)_{1 \leq j \leq q}$ be the basis dual to $(H_j)_{1 \leq j \leq q}$, so that $\varepsilon_j(H_k) = \delta_{jk}$. Show that $i\mathfrak{a}_0$ is a maximal commutative subalgebra of $i\mathfrak{p}_0$. The roots of the corresponding system $\Sigma$ are

$$\pm\varepsilon_j \; (1 \leqq j \leqq q), \qquad \pm 2\varepsilon_j \; (1 \leqq j \leqq q), \qquad \pm\varepsilon_j \pm \varepsilon_h \; (1 \leqq j < h \leqq q),$$

with multiplicities

$$m(\varepsilon_j) = 2(p - q), \qquad m(2\varepsilon_j) = 1, \qquad m(\varepsilon_j \pm \varepsilon_h) = 2.$$

Determine the subspaces $\mathfrak{k}_\lambda$ and $i\mathfrak{p}_\lambda$ corresponding to these roots.

**4.** With the notation of Problem 1, show that for the compact symmetric space $G_2/K_2$ the following properties are equivalent:

($\alpha$) For each system of four points $p, q, r, s$ in $G_2/K_2$ such that $d(p, q) = d(r, s)$, where $d$ is the Riemannian distance, there exists $x \in G_2$ such that $x \cdot p = r$ and $x \cdot q = s$.

($\beta$) The group $K_2$ acts transitively on the set of lines passing through the origin in the tangent space $i\mathfrak{a}_0$ at the point $\pi(e)$ of $G_2/K_2$.

($\gamma$) For all $\mathbf{u} \in i\mathfrak{p}_0$ we have $i\mathfrak{p}_0 = \mathbf{Ru} + [\mathfrak{k}_0, \mathbf{u}]$.

($\delta$) $G_2/K_2$ has rank 1, in other words, $i\mathfrak{a}_0$ has dimension 1.

($\varepsilon$) The sectional curvature $A(P_x)$ along a tangent plane $P_x$ to $G_2/K_2$ (**20.21.1**) is never zero.

(To prove that ($\alpha$) and ($\beta$) are equivalent, use the Hopf–Rinow theorem (**20.18.5**). To prove the equivalence of ($\beta$) and ($\gamma$), use Problem 1(e); to prove the equivalence of ($\gamma$) and ($\varepsilon$), consider an orthonormal basis of $i\mathfrak{p}_0$ consisting of eigenvectors of the endomorphism $\mathbf{v} \mapsto [[\mathbf{u}, \mathbf{v}], \mathbf{u}]$ (cf. Section **20.20**, Problem 2, and (**21.21.2**)).)

**5.** With the notation of Sections **21.18** and **21.21**, show that the following conditions are equivalent:

($\alpha$) The rank of $\mathfrak{g}_0$ (or of $G_2/K_2$) is equal to the rank of $\mathfrak{g}$ (or of $G_u$).

($\beta$) All the roots $\lambda \in \Sigma$ have multiplicity 1 (Problem 2).

($\gamma$) The rank of $\mathfrak{g}$ is equal to $2\dim(i\mathfrak{p}_0) - \dim(\mathfrak{g})$.

($\delta$) The rank of $\mathfrak{g}$ is equal to $\dim(i\mathfrak{p}_0) - \dim(\mathfrak{k}_0)$.

**6.** (a) With the notation of Section **21.21**, show that for a root $\alpha \in \mathbf{S}''$ the following conditions are equivalent:

(1) $\alpha \circ \varphi = -\alpha$, (2) $c_0(\mathfrak{g}_\alpha) = \mathfrak{g}_\alpha$, (3) the restriction of $\alpha$ to $\mathfrak{k}_0 \cap \mathfrak{t}$ is zero. When these conditions are satisfied, the subalgebra $\mathfrak{k}_0 \cap \mathfrak{t}$ of $\mathfrak{k}_0$ is not a maximal commutative subalgebra. (Observe that we also have $c_0(\mathfrak{g}_{-\alpha}) = \mathfrak{g}_{-\alpha}$ and deduce, using Problem 1(c), that $\mathbf{y}''_\alpha$ centralizes $\mathfrak{k}_0 \cap \mathfrak{t}$.)

(b) Deduce from (a) that for each $\lambda \in \Sigma$ there can exist only one root $\alpha \in \Sigma(\lambda)$ such that $\alpha \circ \varphi = -\alpha$. For such a root to exist it is necessary and sufficient that the multiplicity $m(\lambda)$ should be odd. (Observe that if $\alpha \in \Sigma(\lambda)$, then also $\alpha \circ \varphi \in \Sigma(\lambda)$.)

(c) Suppose that $m(\lambda)$ is even for all $\lambda \in \Sigma$. Show that $\mathfrak{k}_0$ is the direct sum of $\mathfrak{k}_0 \cap \mathfrak{t}$ and the distinct subspaces $\mathfrak{k}_0 \cap (\mathfrak{g}_\alpha + \mathfrak{g}_{-\alpha} + \mathfrak{g}_{\alpha \circ \varphi} + \mathfrak{g}_{-\alpha \circ \varphi})$, and that $\mathfrak{k}_0 \cap \mathfrak{t}$ is a maximal commutative subalgebra of $\mathfrak{t}_0$.

(d) If $\alpha \in \mathbf{S}''_+$ is such that $\alpha \circ \varphi = -\alpha$, show that if $\mathfrak{u}'_\alpha$ denotes the hyperplane in $\mathfrak{a}_0$ with equation $\alpha(\mathbf{x}) = 0$, then the sum of $\mathfrak{u}'_\alpha$, $\mathfrak{k}_0 \cap \mathfrak{t}$, and $\mathbf{Ry}''_\alpha$ is a commutative subalgebra of $\mathfrak{g}_0$ whose complexification is also the complexification of a maximal commutative subalgebra of $\mathfrak{g}_u$ (a "Cartan subalgebra" of $\mathfrak{g}_0$, cf. Section **21.22**, Problem 4). Deduce that if there exist $r$ roots $\alpha \in \mathbf{S}''_+$ such that $\alpha \circ \varphi = -\alpha$ and $[\mathfrak{g}_\alpha, \mathfrak{g}_\beta] = 0$ for any two of these roots, then there exist $r$ Cartan subalgebras in $\mathfrak{g}_0$ with the property that no two of them are transforms of one another by automorphisms of $\mathfrak{g}_0$ of the form $\mathrm{Ad}(s)$, where $s \in G_1$. (Consider the commutative subgroups of $\mathrm{Ad}(G_1)$ having these $r$ subalgebras as their Lie algebras.)

## 22. CARTAN'S CRITERION FOR SOLVABLE LIE ALGEBRAS

In this section and the next, the Lie algebras under consideration are Lie algebras over either **R** or **C**, unless the contrary is expressly stated.

(21.22.1) *Every finite-dimensional nilpotent* (21.21.8) *Lie algebra* $\mathfrak{g}$ *is solvable.*

It is enough to prove that $\mathfrak{D}(\mathfrak{g}) \neq \mathfrak{g}$; for every Lie subalgebra of a nilpotent Lie algebra is evidently nilpotent, so that by induction we shall have $\mathfrak{D}^{k+1}(\mathfrak{g}) \neq \mathfrak{D}^k(\mathfrak{g})$, and if $\dim(\mathfrak{g}) = n$ this will imply that $\mathfrak{D}^n(\mathfrak{g}) = 0$. By hypothesis, there exists an integer $r$ satisfying (21.21.8.1). If we had $\mathfrak{D}(\mathfrak{g}) = \mathfrak{g}$, then for a basis $(\mathbf{u}_j)_{1 \leq j \leq n}$ of $\mathfrak{g}$, the sum of the subspaces $\operatorname{ad}(\mathbf{u}_j) \cdot \mathfrak{g}$ would be equal to $\mathfrak{g}$. By induction, it would follow that the sum of the subspaces $(\operatorname{ad}(\mathbf{u}_{j_1}) \circ \operatorname{ad}(\mathbf{u}_{j_2}) \circ \cdots \circ \operatorname{ad}(\mathbf{u}_{j_r})) \cdot \mathfrak{g}$ (where $(j_k)_{1 \leq k \leq r}$ runs through the set of all sequences of $r$ elements of $[1, n]$) would be equal to $\mathfrak{g}$, contrary to (21.21.8.1).

(21.22.2) (Engel's theorem) *Let* $\mathfrak{g}$ *be a finite-dimensional Lie algebra such that for each* $\mathbf{x} \in \mathfrak{g}$ *the endomorphism* $\operatorname{ad}(\mathbf{x})$ *of the vector space* $\mathfrak{g}$ *is nilpotent. Then* $\mathfrak{g}$ *is a nilpotent Lie algebra.*

This statement is equivalent to the following:

(21.22.2.1) *If* E *is a finite-dimensional vector space and* $\mathfrak{g}$ *is a Lie subalgebra of* $\mathfrak{gl}(\mathrm{E})$ *consisting of nilpotent endomorphisms of* E, *then there exists an integer* $r$ *such that* $\mathfrak{g}^r = \{0\}$.

(If A and B are two vector subspaces of $\mathfrak{gl}(\mathrm{E}) = \operatorname{End}(\mathrm{E})$, we denote by AB the vector subspace of all linear combinations of products of elements of A with elements of B in the algebra End(E), and we define $\mathrm{A}^r$ inductively to be $\mathrm{A}^{r-1}\mathrm{A}$ for all integers $r \geqq 2$.)

The proof is by induction on $n = \dim(\mathfrak{g})$; for $n = 1$, the result is trivial. Let $\mathfrak{h} \neq \mathfrak{g}$ be a Lie subalgebra of $\mathfrak{g}$ whose dimension $m$ is maximal among proper subalgebras of $\mathfrak{g}$. The endomorphisms $\operatorname{ad}(X)$ of the vector space $\mathfrak{g}$ (where $\operatorname{ad}(X) \cdot Y = XY - YX$) as $X$ runs through $\mathfrak{h}$ form a Lie subalgebra $\mathfrak{h}_1$ of $\mathfrak{gl}(\mathfrak{g})$, of dimension $\leqq m < n$. Furthermore, each of the endomorphisms $\operatorname{ad}(X)$ is nilpotent, for it is immediate by induction on $r$ that $(\operatorname{ad}(X))^r \cdot Y$ is a linear combination of the products $X^pYX^q$ such that $p + q = r$, hence if $X^s = 0$ we have $(\operatorname{ad}(X))^{2s} = 0$. The inductive hypothesis implies that

$\mathfrak{h}_1^t = \{0\}$ for some integer $t > 0$. Let $Z$ be an element of $\mathfrak{g}$ that does not belong to $\mathfrak{h}$, and let $s \leqq t$ be the largest integer such that there exist $s$ elements $X_1, \ldots, X_s \in \mathfrak{h}$ for which $Y = (\operatorname{ad}(X_1) \operatorname{ad}(X_2) \cdots \operatorname{ad}(X_s)) \cdot Z \notin \mathfrak{h}$. Then, by definition, we have $\operatorname{ad}(X) \cdot Y = [X, Y] \in \mathfrak{h}$ for all $X \in \mathfrak{h}$, and therefore the vector subspace of $\mathfrak{g}$ spanned by $\mathfrak{h}$ and $Y$, of dimension $m + 1$, is a Lie subalgebra of $\mathfrak{g}$, and hence by the definition of $\mathfrak{h}$ is equal to $\mathfrak{g}$. Also, we have by hypothesis $Y^p = 0$ for some integer $p > 0$, and $\mathfrak{h}^q = \{0\}$ for some integer $q > 0$ by virtue of the inductive assumption. We shall deduce from this that, for each sequence $(T_j)_{1 \leq j \leq pq}$ of elements of $\mathfrak{g}$, we have $T_1 T_2 \cdots T_{pq} = 0$, in other words, that $\mathfrak{g}^{pq} = \{0\}$. Clearly we may assume that each $T_j$ is either equal to $Y$ or else belongs to $\mathfrak{h}$. Suppose first that the number $r$ of indices $j \in [1, pq]$ such that $T_j \in \mathfrak{h}$ is less than $p$. Then the set of $pq - r$ indices $j \in [1, pq]$ for which $T_j = Y$ is the union of at most $r + 1$ intervals in $\mathbf{N}$, and since $(r + 1)(q - 1) \leqq p(q - 1) < pq - r$, at least one of these intervals must contain at least $q$ numbers, so that the product $T_1 T_2 \cdots T_{pq}$ certainly vanishes in this case. Next, for each value of $r$ and each $k \geqq r$, we have

$$T_1 T_2 \cdots T_k \in \mathfrak{h}^r + Y\mathfrak{h}^r + \cdots + Y^{k-r}\mathfrak{h}^r \tag{21.22.2.2}$$

if the number of indices $j$ such that $T_j \in \mathfrak{h}$ is equal to $r$. For this is obviously true for arbitrary $r$ and $k = r$; and by induction on $k - r$ (the number of factors $T_j$ equal to $Y$) one sees immediately that **(21.22.2.2)** is true provided that $T_k \in \mathfrak{h}$. Now for each $r$ we have

$$\mathfrak{h}^r Y \subset Y\mathfrak{h}^r + \mathfrak{h}^r. \tag{21.22.2.3}$$

Indeed, this is true when $r = 1$, because $XY = YX + [X, Y]$ and $[X, Y] \in \mathfrak{h}$ for all $X \in \mathfrak{h}$; and from **(21.22.2.3)** we have

$$\begin{aligned} \mathfrak{h}^{r+1} Y \subset \mathfrak{h} Y \mathfrak{h}^r + \mathfrak{h}^{r+1} &\subset (Y\mathfrak{h} + \mathfrak{h})\mathfrak{h}^r + \mathfrak{h}^{r+1} \\ &= Y\mathfrak{h}^{r+1} + \mathfrak{h}^{r+1}, \end{aligned}$$

which proves **(21.22.2.3)** by induction on $r$. This being so, if we have $T_k = Y$ in **(21.22.2.2)**, then by hypothesis

$$T_1 T_2 \cdots T_{k-1} \in \mathfrak{h}^r + Y\mathfrak{h}^r + \cdots + Y^{k-r-1}\mathfrak{h}^r$$

from which it follows that

$$T_1 T_2 \cdots T_k \in \mathfrak{h}^r Y + Y\mathfrak{h}^r Y + \cdots + Y^{k-r-1}\mathfrak{h}^r Y,$$

which, together with **(21.22.2.3)**, implies **(21.22.2.2)**. If now $r \geqq p$, the right-hand side of **(21.22.2.2)** reduces to zero, and the proof is complete.

**(21.22.3)** (Cartan's criterion) *Let $\mathfrak{g}$ be a finite-dimensional Lie algebra such that the Killing form $B_{\mathfrak{g}}$ is identically zero on $\mathfrak{g} \times \mathfrak{g}$; then $\mathfrak{g}$ is solvable.*

By considering the complexification of $\mathfrak{g}$, if $\mathfrak{g}$ is a real Lie algebra, we may assume that $\mathfrak{g}$ is a *complex* Lie algebra **(21.6.1)**. The proof is by induction on $\dim(\mathfrak{g})$, the result being trivial when $\dim(\mathfrak{g}) = 1$. It will be enough to show that $\mathfrak{D}(\mathfrak{g}) \neq \mathfrak{g}$; for the Killing form $B_{\mathfrak{D}(\mathfrak{g})}$, being the restriction of $B_{\mathfrak{g}}$ to the ideal $\mathfrak{D}(\mathfrak{g})$ **(21.5.7)**, is identically zero and therefore by the inductive hypothesis $\mathfrak{D}(\mathfrak{g})$ will be solvable, and hence $\mathfrak{g}$ also will be solvable **(19.12.3)**.

Suppose then that $\mathfrak{D}(\mathfrak{g}) = \mathfrak{g}$. We define an element $\mathbf{u}_0 \in \mathfrak{g}$ as in **(21.19.2)** and consider the corresponding direct sum decomposition $\mathfrak{g} = \mathfrak{h} \oplus \bigoplus_{\alpha \in \mathbf{S}} \mathfrak{g}(\alpha)$. It follows that $\mathfrak{D}(\mathfrak{g}) = [\mathfrak{g}, \mathfrak{g}]$ is the sum (not in general direct) of the subspaces $[\mathfrak{h}, \mathfrak{h}]$, $[\mathfrak{h}, \mathfrak{g}(\alpha)]$, and $[\mathfrak{g}(\alpha), \mathfrak{g}(\beta)]$ for $\alpha, \beta$ in $\mathbf{S}$. But

$$[\mathfrak{g}(\alpha), \mathfrak{g}(\beta)] \subset \mathfrak{g}(\alpha + \beta)$$

and $[\mathfrak{h}, \mathfrak{g}(\alpha)] \subset \mathfrak{g}(\alpha)$ **(21.19.6.1)**, hence our assumption on $\mathfrak{g}$ implies that

**(21.22.3.1)**
$$\mathfrak{h} = [\mathfrak{h}, \mathfrak{h}] + \sum_{\alpha \in \mathbf{S}} [\mathfrak{g}(\alpha), \mathfrak{g}(-\alpha)].$$

We shall deduce from this that $\mathbf{S}$ must be empty. For every linear form $\beta \in \mathbf{S}$, the restriction of $\beta$ to $[\mathfrak{h}, \mathfrak{h}]$ is zero **(21.19.7)**; also, for each $\alpha \in \mathbf{S}$ and each $\mathbf{u} \in [\mathfrak{g}(\alpha), \mathfrak{g}(-\alpha)]$ we have $\beta(\mathbf{u}) = r_{\alpha\beta}\,\alpha(\mathbf{u})$, where $r_{\alpha\beta}$ is a rational number **(21.19.10)**. But since by hypothesis $B_{\mathfrak{g}}(\mathbf{u}, \mathbf{u}) = 0$, the relation **(21.19.8.1)** implies that

$$\dim(\mathfrak{g}(\alpha)) \cdot \alpha(\mathbf{u})^2 + \sum_{\beta \neq \alpha} r_{\alpha\beta}^2 \dim(\mathfrak{g}(\beta))\alpha(\mathbf{u})^2 = 0$$

and hence that $\alpha(\mathbf{u}) = 0$. But then $\beta(\mathbf{u}) = r_{\alpha\beta}\,\alpha(\mathbf{u}) = 0$ for all $\beta \in \mathbf{S}$. If we had $\mathbf{S} \neq \varnothing$, every linear form $\beta \in \mathbf{S}$ would be zero on all of $\mathfrak{h}$, by what has just been proved and **(21.22.3.1)**; and this contradicts the definition of $\mathbf{S}$.

Hence we have $\mathbf{S} = \varnothing$ and consequently $\mathfrak{g} = \mathfrak{h}$; but then $\operatorname{ad}(\mathbf{u})$ would be nilpotent for all $\mathbf{u} \in \mathfrak{g}$ **(21.19.5)**; by virtue of Engel's theorem **(21.22.2)**, $\mathfrak{g}$ would be nilpotent and *a fortiori* solvable **(21.22.1)**, contrary to our assumption. Q.E.D.

**(21.22.4)** *If a finite-dimensional Lie algebra $\mathfrak{g}$ contains no solvable ideal $\neq \{0\}$, then $\mathfrak{g}$ is semisimple.*

Suppose the contrary, so that the Killing form $B_{\mathfrak{g}}$ is degenerate: in other words, if $\mathfrak{n}$ is the subspace of all $\mathbf{x} \in \mathfrak{g}$ such that $B_{\mathfrak{g}}(\mathbf{x}, \mathbf{y}) = 0$ for all $\mathbf{y} \in \mathfrak{g}$, then $\mathfrak{n} \neq \{0\}$. But $\mathfrak{n}$ is an *ideal* of $\mathfrak{g}$, because by virtue of **(21.5.6.1)**, if $\mathbf{x} \in \mathfrak{n}$

and $\mathbf{z} \in \mathfrak{g}$ we have $B_{\mathfrak{g}}([\mathbf{x}, \mathbf{z}], \mathbf{y}) = B_{\mathfrak{g}}(\mathbf{x}, [\mathbf{z}, \mathbf{y}]) = 0$ for all $\mathbf{y} \in \mathfrak{g}$. This being so, the restriction of $B_{\mathfrak{g}}$ to $\mathfrak{n} \times \mathfrak{n}$ is identically zero; since this restriction is equal to $B_{\mathfrak{n}}$ (21.5.7), it follows that $\mathfrak{n}$ is solvable by Cartan's criterion (21.22.3), contrary to hypothesis.

## PROBLEMS

1. Let E be a real or complex vector space of finite dimension and G a connected Lie group immersed in **GL**(E). Let N be a normal subgroup of G; suppose that the linear representation of G on E defined by the canonical injection $G \to \mathbf{GL}(E)$ is *irreducible*, and that there exists a vector $x_0 \neq 0$ in E such that $t \cdot x_0 = \lambda(t)x_0$ for all $t \in N$. Then N is contained in the center of **GL**(E), consisting of the nonzero scalar multiples of the identity. (Observe that the mapping $t \mapsto \lambda(t)$ is continuous on N and that $\lambda(sts^{-1}) = \lambda(t)$ for all $s \in G$, and deduce that the set of $x \in E$ such that $t \cdot x = \lambda(t)x$ for all $t \in N$ is stable under G.)

2. Let E be a finite-dimensional *complex* vector space. Show that if G is a *solvable* connected Lie group immersed in **GL**(E), then there exists a basis of E such that G, identified by this choice of basis with a group of matrices, is contained in the group of lower triangular matrices (*Lie's theorem*). (It is enough to prove that there exists $x_0 \neq 0$ in E such that $s \cdot x_0 = \lambda(s)x_0$ for all $s \in G$. Do this first for G commutative, and then use Problem 1 and the definition of solvable groups.)

3. (a) Show that if $\mathfrak{g}$ is a finite-dimensional solvable (real or complex) Lie algebra, then its derived algebra $\mathfrak{D}(\mathfrak{g})$ is nilpotent. (Use Engel's theorem and Lie's theorem.)
(b) In order that $\mathfrak{g}$ should be solvable, it is necessary and sufficient that the restriction to $\mathfrak{D}(\mathfrak{g})$ of the Killing form of $\mathfrak{g}$ should be identically zero (use (a) and Cartan's criterion). Give an example of a solvable Lie algebra whose Killing form is not identically zero.
(c) If $\mathfrak{g}$ is nilpotent, then the Killing form of $\mathfrak{g}$ is identically zero. Give an example of a solvable but not nilpotent Lie algebra whose Killing form is identically zero. (Consider the solvable Lie group defined in Section **19.14**, Problem 4, and note that for the Killing form of its Lie algebra to vanish identically, it is necessary and sufficient that $\mathrm{Tr}(U^2) = 0$.)

4. (a) Let $\mathfrak{g}$ be a (real or complex) Lie algebra of finite dimension. An element $\mathbf{u} \in \mathfrak{g}$ is said to be *regular* if the multiplicity of 0 as an eigenvalue of the endomorphism $\mathrm{ad}(\mathbf{u})$ of $\mathfrak{g}$ is as small as possible. For the Lie algebra of a compact group, this notion coincides with that defined in (**21.7.13**). (Use (**21.8.4**).)
(b) Let $\mathfrak{g}$ be a real Lie algebra of finite dimension. Then an element $\mathbf{u} \in \mathfrak{g}$ is regular if and only if it is regular in the complexification $\mathfrak{g}_{(\mathbf{C})}$.
(c) Show that the set R of regular elements of $\mathfrak{g}$ is a dense open subset of $\mathfrak{g}$. If moreover $\mathfrak{g}$ is a *complex* Lie algebra, then R is *connected*. Determine R for $\mathfrak{g} = \mathfrak{sl}(2, \mathbf{R})$.
(d) Let $\mathbf{u}_0$ be a regular element of $\mathfrak{g}$, and let $\mathfrak{g}_0$ be the set of $\mathbf{x} \in \mathfrak{g}$ such that $(\mathrm{ad}(\mathbf{u}_0))^p \cdot \mathbf{x} = 0$ for a sufficiently large integer $p$. Show that $\mathfrak{g}_0$ is a nilpotent subalgebra of $\mathfrak{g}$ and is equal to its normalizer. (Use (**21.19.3**) and the fact that the endomorphism of the vector space $\mathfrak{g}/\mathfrak{g}_0$, induced by $\mathrm{ad}(\mathbf{u}_0)$ on passing to the quotient, is bijective; show that this implies that for each $\mathbf{u} \in \mathfrak{g}_0$ near enough to $\mathbf{u}_0$ we have $(\mathrm{ad}(\mathbf{u}))^p \cdot \mathbf{x} = 0$ for all $\mathbf{x} \in \mathfrak{g}_0$ and all sufficiently large integers $p$, and then use Engel's theorem.)

(e) Let $\mathfrak{a}$ be a subalgebra of $\mathfrak{g}$ and let $\mathbf{x}_0$ be an element of $\mathfrak{a}$ such that the endomorphism of the vector space $\mathfrak{g}/\mathfrak{a}$ induced by $\mathrm{ad}(\mathbf{x}_0)$ is bijective. Show that the mapping $(\mathbf{x}, \mathbf{u}) \mapsto \exp(\mathrm{ad}\,(\mathbf{u})) \cdot \mathbf{x}$ of $\mathfrak{a} \times \mathfrak{g}$ into $\mathfrak{g}$ is a submersion at the point $(\mathbf{x}_0, 0)$, and deduce with the help of (c) that $\mathfrak{a}$ contains a regular element of $\mathfrak{g}$.

(f) A nilpotent subalgebra of $\mathfrak{g}$ that is equal to its own normalizer is called a *Cartan subalgebra*. For each regular element $\mathbf{u}_0 \in \mathfrak{g}$, the subalgebra $\mathfrak{g}_0$ defined in (d) is a Cartan subalgebra; and conversely, every Cartan subalgebra of $\mathfrak{g}$ may be obtained in this way. (If $\mathfrak{n}$ is a Cartan subalgebra, show by using (e) that it contains a regular element $\mathbf{u}_0$; then we have $\mathfrak{n} \subset \mathfrak{g}_0$, and if the inclusion were strict, $\mathfrak{n}$ would not be equal to its normalizer (Section **19.14**, Problem 7).) The subalgebra $\mathfrak{g}_0$ defined in (**21.19.2**) is a Cartan subalgebra.

(g) Suppose that $\mathfrak{g}$ is a *complex* Lie algebra. Let $\Gamma$ be the connected Lie group immersed in $\mathrm{Aut}(\mathfrak{g})$ whose Lie algebra is $\mathrm{ad}(\mathfrak{g})$. Show that any Cartan subalgebra of $\mathfrak{g}$ can be transformed into any other by an automorphism $\sigma \in \Gamma$. (For each Cartan subalgebra $\mathfrak{h}$ of $\mathfrak{g}$, let $\mathfrak{h}_{\mathrm{reg}}$ denote the (open) subset of regular elements belonging to $\mathfrak{h}$, and let $\mathrm{R}_{\mathfrak{h}}$ be the image of $\Gamma \times \mathfrak{h}_{\mathrm{reg}}$ under the mapping $(\sigma, \mathbf{x}) \mapsto \sigma \cdot \mathbf{x}$ of $\Gamma \times \mathfrak{h}_{\mathrm{reg}}$ into $\mathfrak{g}$. Use (e) to show that $\mathrm{R}_{\mathfrak{h}}$ is open in $\mathfrak{g}$, and then use (c).) Compare with Section **21.21**, Problem 6.

(h) A Cartan subalgebra is a maximal nilpotent subalgebra (Section **19.14**, Problem 7). Give an example of a maximal nilpotent subalgebra that is not a Cartan subalgebra.

**5.** (a) Let $\mathfrak{g}$ be a (real or complex) Lie algebra of finite dimension. If $\mathfrak{h}$ is a Cartan subalgebra of $\mathfrak{g}$, then $\mathfrak{h}$ is also a Cartan subalgebra of every Lie subalgebra $\mathfrak{g}_1$ of $\mathfrak{g}$ that contains $\mathfrak{h}$.

(b) If $f$ is a *surjective* homomorphism of $\mathfrak{g}$ onto a Lie algebra $\mathfrak{g}_1$, then the image under $f$ of any regular element of $\mathfrak{g}$ is a regular element of $\mathfrak{g}_1$ (use the fact that the set of regular elements of $\mathfrak{g}_1$ is dense). Deduce that the image under $f$ of any Cartan subalgebra of $\mathfrak{g}$ is a Cartan subalgebra of $\mathfrak{g}_1$. If $\mathfrak{g}$ and $\mathfrak{g}_1$ are complex Lie algebras, then every Cartan subalgebra of $\mathfrak{g}_1$ is the image under $f$ of a Cartan subalgebra of $\mathfrak{g}$. (Use Problem 4(g).)

(c) If $\mathfrak{h}$ is a Cartan subalgebra of $\mathfrak{g}$ and if $\mathfrak{a}$ is a Lie subalgebra of $\mathfrak{g}$ which contains $\mathfrak{h}$, then $\mathfrak{a}$ is its own normalizer in $\mathfrak{g}$. (Apply (b) to the quotient algebra $\mathfrak{N}(\mathfrak{a})/\mathfrak{a}$.)

**6.** Let G be a connected Lie group (real or complex). A *Cartan subgroup* of G is any connected Lie group immersed in G whose Lie algebra is a Cartan subalgebra of the Lie algebra $\mathfrak{g}$ of G.

(a) Show that a Cartan subgroup is closed in G.

(b) Suppose that G is a complex semisimple group, and is the complexification of a compact connected semisimple group K (**21.17.1**). Then all the Cartan subgroups of G are conjugate in G. We obtain a Cartan subalgebra of $\mathfrak{g}$ by taking the complexification $\mathfrak{t} + i\mathfrak{t}$ of a maximal commutative subalgebra $\mathfrak{t}$ of the Lie algebra $\mathfrak{k}$ of K; the corresponding Cartan subgroup is isomorphic to $(\mathbf{C}^*)^l$, where $l$ is the rank of G.

(c) With the hypotheses of (b), show that every Cartan subgroup A of G is its own centralizer in G, and that the normalizer $\mathcal{N}(\mathrm{A})$ of A in G is such that $\mathcal{N}(\mathrm{A})/\mathrm{A}$ may be identified with the Weyl group of K. (Consider the centralizer and the normalizer of $\mathfrak{t} + i\mathfrak{t}$ in G, and argue as in Section **21.11**, Problem **12**(a).)

(d) With the same hypotheses, all elements of a Cartan subalgebra of $\mathfrak{g}$ are semisimple (Section **21.19**, Problem 1).

**7.** Let G be a complex connected Lie group and $\mathfrak{g}$ its Lie algebra. A *Borel subalgebra* of $\mathfrak{g}$ is any maximal solvable subalgebra of $\mathfrak{g}$, and a *Borel subgroup* of G is a (complex) connected Lie group immersed in G whose Lie algebra is a Borel subalgebra of $\mathfrak{g}$.

(a) Show that a Borel subgroup is closed in G.

(b) For the rest of this problem, suppose that G is a complex connected semisimple group, and keep to the notation of (**21.21.12**(iii)). Show that $\mathfrak{b} = \mathfrak{t} + i\mathfrak{t} + \mathfrak{n}$ is a Borel subalgebra of $\mathfrak{g}$, and that $\mathfrak{b}$ is the semidirect product of the Cartan subalgebra $\mathfrak{h} = \mathfrak{t} + i\mathfrak{t}$ and the nilpotent algebra $\mathfrak{n}$, which is also the derived algebra $\mathfrak{D}(\mathfrak{b})$. (Observe that a Lie subalgebra of $\mathfrak{g}_u$ that contains $\mathfrak{t}$ strictly cannot be solvable.)
(c) For each element $\mathbf{u} \in \mathfrak{b}$, the semisimple and nilpotent components of $\mathbf{u}$ (Section **21.19**, Problem 1) both belong to $\mathfrak{b}$ (cf. Problem 5(c)). Show that the nilpotent elements of $\mathfrak{b}$ are the elements of $\mathfrak{n}$.
(d) Let B be the Borel subgroup with Lie algebra $\mathfrak{b}$. If T is the maximal torus of $G_u$ with Lie algebra $\mathfrak{t}$, then $G_u \cap B = T$ (cf. (**21.21.12**)). The homogeneous space G/B is a *compact* complex manifold, and the canonical mapping of G/B onto $G_u/T$ is a diffeomorphism (Section **16.10**, Problem 3). Show that the center of B is equal to the center of G (and of $G_u$).
(e) Show that B is its own normalizer in G. (Let H be the Cartan subgroup of G with Lie algebra $\mathfrak{h}$. If $s \in \mathscr{N}(B)$, show that there exists $x \in B$ such that $\operatorname{Int}(sx)$ leaves H globally invariant, and hence also the corresponding root system **S**; furthermore, the fact that $\operatorname{Int}(sx)$ leaves B globally invariant implies that $\operatorname{Ad}(sx)$ permutes the $\mathfrak{g}_\alpha$ for $\alpha \in \mathbf{S}_+$. Deduce that $\operatorname{Ad}(sx)$ leaves each $\mathfrak{g}_\alpha$ globally invariant, and by arguing as in Section **21.11**, Problem 12(a), show that there exists $y \in H$ such that $\operatorname{Int}(sxy)$ is the identity. Complete the proof by observing that the center of G is contained in every Cartan subgroup.)

**8.** Let G be a complex connected semisimple group and $\mathfrak{g}$ its Lie algebra. A Lie subalgebra $\mathfrak{h}$ of $\mathfrak{g}$ is said to be *splittable* if for each $\mathbf{u} \in \mathfrak{h}$ the semisimple and nilpotent components of $\mathbf{u}$ belong to $\mathfrak{h}$. A subalgebra $\mathfrak{a}$ of $\mathfrak{g}$ is said to be *diagonalizable* if there exists a basis of $\mathfrak{g}$ relative to which all the endomorphisms $\operatorname{ad}(\mathbf{u})$, for $\mathbf{u} \in \mathfrak{a}$, are represented by diagonal matrices (which implies that all the elements of $\mathfrak{a}$ are semisimple).
(a) Let $\mathfrak{h}$ be a splittable subalgebra of $\mathfrak{g}$, and let $\mathfrak{a}$ be a diagonalizable subalgebra of $\mathfrak{h}$. Then there exists a direct sum decomposition of the vector space $\mathfrak{h}$: $\mathfrak{h} = \mathfrak{h}_0 \oplus \bigoplus_{\lambda \in F} \mathfrak{h}_\lambda$, where $\mathfrak{h}_0$ is the centralizer of $\mathfrak{a}$ in h, F is a set of nonzero linear forms on $\mathfrak{a}$, and for each $\lambda \in F$ we have $[\mathbf{u}, \mathbf{x}] = \lambda(\mathbf{u})\mathbf{x}$ for $\mathbf{u} \in \mathfrak{a}$ and $\mathbf{x} \in \mathfrak{h}$.
(b) With the same hypotheses, suppose in addition that $\mathfrak{a}$ is a *maximal* diagonalizable subalgebra of $\mathfrak{h}$. Show that $\mathfrak{h}_0$ is a Cartan subalgebra of $\mathfrak{h}$ and is splittable. (Observe that if two elements of $\mathfrak{g}$ commute, then so do their semisimple and nilpotent components.) Show also that $\mathfrak{a}$ is the set of semisimple elements of $\mathfrak{h}_0$, that the set $\mathfrak{n}$ of nilpotent elements of $\mathfrak{h}_0$ is an ideal in $\mathfrak{h}_0$, and that $\mathfrak{h}_0$ is the direct sum of $\mathfrak{a}$ and $\mathfrak{n}$. Consider in particular the cases where $\mathfrak{h}$ is nilpotent, and where $\mathfrak{h} = \mathfrak{g}$.
(c) Suppose now that $\mathfrak{h}$ is a solvable and splittable subalgebra of $\mathfrak{g}$. Show that the set $\mathfrak{n}$ of nilpotent elements of $\mathfrak{h}$ is an ideal in $\mathfrak{h}$. (Use Lie's theorem (Problem 2).) Show also that if $\mathfrak{a}$ is a maximal diagonalizable subalgebra of $\mathfrak{h}$, then $\mathfrak{h}$ is the semidirect product of $\mathfrak{n}$ and $\mathfrak{a}$ (**19.14.7**).

**9.** Let $\rho$ be a continuous linear representation of a Lie group G on a finite-dimensional complex vector space E. By passing to the quotient, $\rho$ induces a differentiable action of G on the associated projective space **P**(E), which is compact. Hence there exist in **P**(E) nonempty closed G-stable subsets that are *minimal* among the subsets of **P**(E) having these properties (Section **12.10**, Problem 6).

Show that if G is connected and solvable, every minimal nonempty closed G-stable subset of **P**(E) consists of a single point. (Argue by induction on $n = \dim(E)$. For $n = 2$, use Lie's theorem (Problem 2). For $n > 2$, Lie's theorem proves the existence of a point

$z_0 \in \mathbf{P}(E)$ fixed by G. If $M \subset \mathbf{P}(E)$ is closed, G-stable, and minimal, and if $z_0 \notin M$, project M from $z_0$ on a projective hyperplane not passing through $z_0$.)

**10.** With the notation of Problem 7, the group G being assumed to be semisimple, let $r = \dim(\mathfrak{b})$. The Grassmannian $\mathbf{G}_r(\mathfrak{g})$ of $r$-dimensional complex vector subspaces of $\mathfrak{g}$ (**16.11.9**) may be identified with a closed submanifold of the projective space $\mathbf{P}(\bigwedge^r \mathfrak{g})$. Show that the subset $\mathfrak{B}$ of $\mathbf{G}_r(\mathfrak{g})$ consisting of the transforms $\mathrm{Ad}(s) \cdot \mathfrak{b}$ of $\mathfrak{b}$ by the elements of G is *closed* in $\mathbf{G}_r(\mathfrak{g})$. (Observe that G acts differentiably on $\mathbf{G}_r(\mathfrak{g})$ by the action $(s, \mathfrak{m}) \mapsto \mathrm{Ad}(s) \cdot \mathfrak{m}$, and use Problem 7(d) and (**16.10.12**))

Let $\mathfrak{s}$ be a solvable subalgebra of $\mathfrak{g}$. By applying the result of Problem 9 to $E = \bigwedge^r \mathfrak{g}$, show that there exists $t \in G$ such that $\mathrm{Ad}(t) \cdot \mathfrak{s} \subset \mathfrak{b}$. In particular, any two Borel subalgebras of $\mathfrak{g}$ can be obtained one from the other by an automorphism of the form $\mathrm{Ad}(t)$, $t \in G$. Any two Borel subgroups of G are conjugate (*Borel's theorem*).

**11.** With the notation of Problem 7, show that as $w$ runs through the Weyl group W of $\mathfrak{g}_u$ with respect to $\mathfrak{t}$, the mapping $w \mapsto w(\mathfrak{b})$ is a bijection of W onto the set of Borel subalgebras of $\mathfrak{g}$ that contain $\mathfrak{h} = \mathfrak{t} + i\mathfrak{t}$. (Use Problem 10 and the fact that two Cartan subalgebras both contained in a Lie subalgebra $\mathfrak{a}$ of $\mathfrak{g}$ can be transformed one into the other by an automorphism $\mathrm{Ad}(t)$, where $t$ belongs to the connected Lie group immersed in G with Lie algebra $\mathfrak{a}$; finally use Problem 6(c).)

**12.** In a complex semisimple Lie algebra $\mathfrak{g}$, a Lie subalgebra distinct from $\mathfrak{g}$ that contains a Borel subalgebra is called *parabolic*. If G is a complex connected semisimple group, a connected Lie group immersed in G is a *parabolic subgroup* of G if its Lie algebra is parabolic.

(a) Show that a parabolic subgroup is closed in G (cf. Problem 5(c)).

(b) With the notation of Problem 7, let $\mathfrak{p}$ be a parabolic subalgebra containing $\mathfrak{b}$. Then the vector space $\mathfrak{p}$ is the direct sum of $\mathfrak{h}$ and a certain number of the $\mathfrak{g}_\alpha$, for $\alpha \in \mathbf{P}$ say, where $\mathbf{S}_+ \subset \mathbf{P} \subset \mathbf{S}$. For any two roots $\alpha, \beta \in \mathbf{P}$, if $\alpha + \beta$ is a root then $\alpha + \beta \in \mathbf{P}$. If $\mathbf{B}$ is the basis of $\mathbf{S}$ that determines the given ordering on $\mathbf{S}$, show that $\mathbf{P}$ is the union of $\mathbf{S}_+$ and $\mathbf{Q}$, where $\mathbf{Q}$ is the set of roots that are linear combinations with integral coefficients $\leqq 0$ of the roots belonging to $\mathbf{B} \cap (-\mathbf{P})$. (To show that $\mathbf{Q} \subset \mathbf{P} \cap (-\mathbf{S}_+)$, show that if $-\alpha \in \mathbf{Q}$, then $-\alpha \in \mathbf{P}$, by noting that $\alpha$ is the sum of say $n$ elements of $\mathbf{B} \cap (-\mathbf{P})$, and arguing by induction on $n$, with the help of Section **21.11**, Problem 3(c). To show that $\mathbf{P} \cap (-\mathbf{S}_+) \subset \mathbf{Q}$, show that if $-\alpha \in \mathbf{P} \cap (-\mathbf{S}_+)$, then $-\alpha \in \mathbf{Q}$, by noting that $\alpha$ is the sum of say $m$ elements of $\mathbf{B}$, and again arguing by induction.)

(c) Conversely, for each subset $\mathbf{B}_1$ of $\mathbf{B}$, if $\mathbf{Q}$ is the set of roots $\alpha \in \mathbf{S}$ that are linear combinations with integral coefficients $\leqq 0$ of the roots in $\mathbf{B}_1$, and if $\mathbf{P} = \mathbf{S}_+ \cup \mathbf{Q}$, then the direct sum of $\mathfrak{h}$ and the $\mathfrak{g}_\alpha$ for $\alpha \in \mathbf{P}$ is a parabolic subalgebra of $\mathfrak{g}$.

**13.** (a) Let G be a complex connected semisimple group, and $B_1, B_2$ two Borel subgroups of G. Show that $B_1 \cap B_2$ contains a Cartan subgroup. (Let $\mathfrak{b}_1, \mathfrak{b}_2$ be the Lie algebras of $B_1$, $B_2$, and apply Problem 8(c) to $\mathfrak{b}_1 \cap \mathfrak{b}_2$, so that we obtain $\mathfrak{b}_1 \cap \mathfrak{b}_2 = \mathfrak{a} \oplus \mathfrak{n}$, where $\mathfrak{a}$ is a maximal diagonalizable subalgebra of $\mathfrak{b}_1 \cap \mathfrak{b}_2$, and $\mathfrak{n}$ is the set of nilpotent elements of $\mathfrak{b}_1 \cap \mathfrak{b}_2$. Show that $\dim(\mathfrak{b}_1 + \mathfrak{b}_2) \leqq \dim(\mathfrak{g}) - \dim(\mathfrak{n})$ by noting that $\mathfrak{n}$ is orthogonal to $\mathfrak{b}_1$ and $\mathfrak{b}_2$, relative to the Killing form of $\mathfrak{g}$; by observing that $\dim(\mathfrak{b}_1) + \dim(\mathfrak{b}_2) = \dim(\mathfrak{g}) + \dim(\mathfrak{h})$, where $\mathfrak{h}$ is a Cartan subalgebra of $\mathfrak{g}$, conclude that $\dim(\mathfrak{a}) \geqq \dim(\mathfrak{h})$ and hence, by virtue of Problem 8(b), that $\mathfrak{a}$ is a Cartan subalgebra of $\mathfrak{g}$.)

(b) With the notation of Problem 7, for each element $w$ of the Weyl group W of $\mathfrak{g}_u$ relative to $\mathfrak{t}$, let BwB denote the double coset of any element of the normalizer $\mathcal{N}(T)$ of T in $G_u$ belonging to the class of $w$ in $\mathcal{N}(T)/T$. Show that, as $w$ runs through W, the double cosets BwB form a *partition* of G (*Bruhat decomposition*). (If $s \in G$, deduce from (a) and Problem 4(g) that there exists $x \in B$ such that $xsBs^{-1}x^{-1} \supset H$, and then use Problem 11; finally, observe that $\mathcal{N}(T) \cap B = T$.)

## 23. E. E. LEVI'S THEOREM

**(21.23.1)** Let $\mathfrak{g}$ be a (real or complex) Lie algebra of finite dimension and let $\mathfrak{a}$, $\mathfrak{b}$ be two solvable ideals in $\mathfrak{g}$. Since $(\mathfrak{a} + \mathfrak{b})/\mathfrak{a}$ is isomorphic to $\mathfrak{b}/(\mathfrak{a} \cap \mathfrak{b})$, it follows that $(\mathfrak{a} + \mathfrak{b})/\mathfrak{a}$ is a solvable Lie algebra; since the canonical image of $\mathfrak{D}^k(\mathfrak{a} + \mathfrak{b})$ in $(\mathfrak{a} + \mathfrak{b})/\mathfrak{a}$ is contained in $\mathfrak{D}^k((\mathfrak{a} + \mathfrak{b})/\mathfrak{a})$, we have $\mathfrak{D}^k(\mathfrak{a} + \mathfrak{b}) \subset \mathfrak{a}$ for sufficiently large $k$, and therefore $\mathfrak{D}^{h+k}(\mathfrak{a} + \mathfrak{b}) = \{0\}$ for sufficiently large $h$, because $\mathfrak{a}$ is solvable; hence the algebra $\mathfrak{a} + \mathfrak{b}$ is solvable. It follows that if $\mathfrak{r}$ is a solvable ideal of $\mathfrak{g}$ of maximum dimension, then *every solvable ideal of* $\mathfrak{g}$ *is contained in* $\mathfrak{r}$; for if a solvable ideal $\mathfrak{a}$ of $\mathfrak{g}$ were not contained in $\mathfrak{r}$, we should have $\dim(\mathfrak{a} + \mathfrak{r}) > \dim \mathfrak{r}$, and $\mathfrak{a} + \mathfrak{r}$ is a solvable ideal, contrary to the definition of $\mathfrak{r}$. This unique ideal $\mathfrak{r}$ of $\mathfrak{g}$, the union of all the solvable ideals in $\mathfrak{g}$, is called the *radical* of the Lie algebra $\mathfrak{g}$.

**(21.23.2)** *If* $\mathfrak{g}$ *is a finite-dimensional Lie algebra, the quotient* $\mathfrak{g}/\mathfrak{r}$ *of* $\mathfrak{g}$ *by its radical* $\mathfrak{r}$ *is a semisimple Lie algebra.*

By virtue of **(21.22.4)**, it is enough to show that the only solvable ideal $\mathfrak{a}$ of $\mathfrak{g}/\mathfrak{r}$ is $\{0\}$. Now such an ideal is of the form $\mathfrak{b}/\mathfrak{r}$, where $\mathfrak{b}$ is an ideal in $\mathfrak{g}$; since $\mathfrak{r}$ and $\mathfrak{b}/\mathfrak{r}$ are solvable, one shows as in **(21.23.1)** that $\mathfrak{b}$ is solvable. But then by definition $\mathfrak{b} \subset \mathfrak{r}$, so that $\mathfrak{a} = \{0\}$.

**(21.23.3)** (E. E. Levi's theorem) *Let* $\mathfrak{g}$ *be a finite-dimensional complex Lie algebra and* $\mathfrak{r}$ *its radical. Then there exists a semisimple subalgebra* $\mathfrak{s}$ *of* $\mathfrak{g}$ *such that* $\mathfrak{g}$ *is isomorphic to a semidirect product* $\mathfrak{r} \times_{\varphi} \mathfrak{s}$ **(19.14.7)**.

From the definition of the semidirect product of Lie algebras **(19.14.7)**, it is enough to show that there exists a semisimple Lie subalgebra $\mathfrak{s}$ of $\mathfrak{g}$ such that $\mathfrak{s} \cap \mathfrak{r} = \{0\}$ and $\mathfrak{s} + \mathfrak{r} = \mathfrak{g}$. It comes to the same thing to say that, if $p: \mathfrak{g} \to \mathfrak{g}/\mathfrak{r}$ is the canonical homomorphism, the restriction of $p$ to $\mathfrak{s}$ is an isomorphism of $\mathfrak{s}$ on $\mathfrak{g}/\mathfrak{r}$; or, again, that there exists a homomorphism $q$ of $\mathfrak{g}/\mathfrak{r}$ into $\mathfrak{g}$ such that $p \circ q = 1_{\mathfrak{g}/\mathfrak{r}}$. The proposition is therefore a particular case of the following:

**(21.23.3.1)** *Let* $\mathfrak{g}$ *be a complex semisimple Lie algebra,* $\mathfrak{e}$ *a finite-dimensional complex Lie algebra.* $p: \mathfrak{e} \to \mathfrak{g}$ *a surjective homomorphism. Then there exists a homomorphism* $q: \mathfrak{g} \to \mathfrak{e}$ *such that* $p \circ q = 1_{\mathfrak{g}}$.

Let $\mathfrak{n} = \text{Ker}(p)$, which is an ideal of $\mathfrak{e}$, and argue by induction on the dimension of $\mathfrak{n}$. If $\mathfrak{n} = \{0\}$, there is nothing to prove. If $\mathfrak{e}$ is semisimple, then $\mathfrak{e}$ is the direct sum of $\mathfrak{n}$ and an ideal $\mathfrak{n}'$ **(21.6.4)**, and the restriction of $p$ to $\mathfrak{n}'$ is an isomorphism of this Lie algebra onto $\mathfrak{g}$; we may therefore take $q$ to be the inverse of this isomorphism. Suppose therefore that $\mathfrak{e}$ is not semisimple, and let $\mathfrak{r}$ be the radical of $\mathfrak{e}$. For each ideal $\mathfrak{a}$ of $\mathfrak{e}$, it is easily seen that $\mathfrak{D}(\mathfrak{a})$ is also an ideal of $\mathfrak{e}$, by virtue of the Jacobi identity; hence $\mathfrak{D}^k(\mathfrak{r})$ is an ideal of $\mathfrak{e}$ for all $k$, and if $m$ is the smallest integer such that $\mathfrak{D}^m(\mathfrak{r}) = \{0\}$, then $\mathfrak{D}^{m-1}(\mathfrak{r})$ is a *commutative nonzero* ideal of $\mathfrak{e}$. We may therefore assume that such ideals exist in $\mathfrak{e}$; choose one, say $\mathfrak{a}$, of *smallest possible dimension*. Then $(\mathfrak{a} + \mathfrak{n})/\mathfrak{n}$ is a commutative ideal of $\mathfrak{e}/\mathfrak{n}$, isomorphic to $\mathfrak{g}$; but since $\mathfrak{g}$ is semisimple, this implies that $\mathfrak{a} + \mathfrak{n} \subset \mathfrak{n}$ **(21.6.2)**, so that $\mathfrak{a} \subset \mathfrak{n}$.

Suppose first that $\mathfrak{a} \neq \mathfrak{n}$. On passing to the quotient, $p$ gives rise to a surjective Lie algebra homomorphism $p_1: \mathfrak{e}/\mathfrak{a} \to \mathfrak{g}$, with kernel $\mathfrak{n}/\mathfrak{a}$, and the inductive hypothesis guarantees the existence of a homomorphism $q_1 : \mathfrak{g} \to \mathfrak{e}/\mathfrak{a}$ such that $p_1 \circ q_1 = 1\mathfrak{g}$. We may write $q_1(\mathfrak{g}) = \mathfrak{f}/\mathfrak{a}$, where $\mathfrak{f}$ is a Lie subalgebra of $\mathfrak{e}$ containing $\mathfrak{a}$. Since $\dim(\mathfrak{a}) < \dim(\mathfrak{n})$, we may apply the inductive hypothesis to the canonical homomorphism $p_2: \mathfrak{f} \to \mathfrak{f}/\mathfrak{a}$ with kernel $\mathfrak{a}$, and deduce that there exists a homomorphism $q_2: \mathfrak{f}/\mathfrak{a} \to \mathfrak{f}$ such that $p_2 \circ q_2 = 1_{\mathfrak{f}/\mathfrak{a}}$. It is now clear that the homomorphism $q = q_2 \circ q_1$ has the required property.

We have still to consider the case where $\mathfrak{a} = \mathfrak{n}$. We shall define canonically a Lie algebra homomorphism $\rho: \mathfrak{g} \to \mathfrak{gl}(\mathfrak{a})$ as follows: each $\mathbf{x} \in \mathfrak{g}$ is of the form $p(\mathbf{z})$ for some $\mathbf{z} \in \mathfrak{e}$; the restriction to $\mathfrak{a}$ of the endomorphism $\text{ad}(\mathbf{z})$ of $\mathfrak{e}$ is an endomorphism of the vector space $\mathfrak{a}$, because $\mathfrak{a}$ is an ideal in $\mathfrak{e}$; but since $\mathfrak{a}$ is commutative and equal to $\mathfrak{n}$, if $p(\mathbf{z}) = p(\mathbf{z}')$ then $\mathbf{z}' - \mathbf{z} \in \mathfrak{a}$, and consequently the restrictions of $\text{ad}(\mathbf{z})$ and $\text{ad}(\mathbf{z}')$ to $\mathfrak{a}$ are equal. The restriction of $\text{ad}(\mathbf{z})$ to $\mathfrak{a}$ therefore depends only on $\mathbf{x}$; if we denote it by $\rho(\mathbf{x})$, then it is clear that $\rho$ is a homomorphism of $\mathfrak{g}$ into $\mathfrak{gl}(\mathfrak{a})$, because $[p(\mathbf{z}_1), p(\mathbf{z}_2)] = p([\mathbf{z}_1, \mathbf{z}_2])$.

The vector space $\mathfrak{a}$ may therefore be regarded as a $\text{U}(\mathfrak{g})$-module by means of $\text{U}(\rho)$, and it follows from **(21.9.1)** that $\mathfrak{a}$ is a direct sum of *simple* $\text{U}(\mathfrak{g})$-submodules. But by definition a $\text{U}(\mathfrak{g})$-submodule of $\mathfrak{a}$ is an *ideal* of $\mathfrak{e}$; by virtue of the choice of $\mathfrak{a}$, we see that $\mathfrak{a}$ is necessarily a *simple* $\text{U}(\mathfrak{g})$-module.

It may happen that $\rho(\mathbf{x}) = 0$ for all $\mathbf{x} \in \mathfrak{g}$; this is the case when $\mathfrak{a} = \mathfrak{n}$ is contained in the *center* of $\mathfrak{e}$, and in fact is equal to this center, because $\mathfrak{g} = \mathfrak{e}/\mathfrak{a}$ contains no nonzero commutative ideals **(21.6.2)** (by reason of the choice of $\mathfrak{a}$, it then follows that $\dim(\mathfrak{a}) = 1$). In this case, if $\mathbf{x} = p(\mathbf{z})$, the endomorphism $\operatorname{ad}(\mathbf{z})$ itself (and not merely its restriction to $\mathfrak{a}$) depends only on $\mathbf{x}$, and if we denote it by $\rho'(\mathbf{x})$, we see as above that $\rho'$ is a homomorphism of $\mathfrak{g}$ into $\mathfrak{gl}(\mathfrak{e}) = \operatorname{End}(\mathfrak{e})$. We may therefore in this case consider the space $\mathfrak{e}$ itself as a $U(\mathfrak{g})$-module, and $\mathfrak{a}$ as a $U(\mathfrak{g})$-submodule of $\mathfrak{e}$. But then ((21.9.1) and (A.23.3)) there exists in $\mathfrak{e}$ a $U(\mathfrak{g})$-submodule *supplementary* to $\mathfrak{a}$; by definition, $\mathfrak{b}$ is an *ideal* of $\mathfrak{e}$, and the restriction of $p$ to $\mathfrak{b}$ is an isomorphism of $\mathfrak{b}$ onto $\mathfrak{g}$; we then take $q$ to be the inverse of this isomorphism.

It remains to consder the case where $\mathfrak{a} = \mathfrak{n}$ is a *simple* $U(\mathfrak{g})$-module and $\rho(\mathbf{x})$ is not zero for all $\mathbf{x} \in \mathfrak{g}$. We shall show that there exists a finite-dimensional complex vector space M, a Lie algebra homomorphism $\sigma$: $\mathfrak{e} \to \mathfrak{gl}(M) = \operatorname{End}(M)$, and an element $w \in M$ having the following properties:

(21.23.3.2) *The mapping* $\mathbf{t} \mapsto \sigma(\mathbf{t}) \cdot w$ *of* $\mathfrak{a}$ *into* M *is bijective.*

(21.23.3.3) *For each* $\mathbf{z} \in \mathfrak{e}$, *there exists* $\mathbf{t} \in \mathfrak{a}$ *such that* $\sigma(\mathbf{z}) \cdot w = \sigma(\mathbf{t}) \cdot w$.

We then define $\mathfrak{s}$ to be the set of all $\mathbf{z} \in \mathfrak{e}$ such that $\sigma(\mathbf{z}) \cdot w = 0$. For since $\sigma([\mathbf{z}_1, \mathbf{z}_2]) = \sigma(\mathbf{z}_1)\sigma(\mathbf{z}_2) - \sigma(\mathbf{z}_2)\sigma(\mathbf{z}_1)$, it is clear that $\mathfrak{s}$ is a Lie subalgebra of $\mathfrak{e}$; it follows from (21.23.3.2) that $\mathfrak{s} \cap \mathfrak{a} = \{0\}$, and from (21.23.3.3) that $\mathfrak{e} = \mathfrak{s} + \mathfrak{a}$, so that $\mathfrak{s}$ has the required properties.

We shall take M to be the vector space $\operatorname{End}(\mathfrak{e})$ and $\sigma$ to be the Lie algebra homomorphism such that, for each $f \in \operatorname{End}(\mathfrak{e})$ and each $\mathbf{z} \in \mathfrak{e}$,

$$\sigma(\mathbf{z}) \cdot f = [\operatorname{ad}(\mathbf{z}), f] = \operatorname{ad}(\mathbf{z}) \circ f - f \circ \operatorname{ad}(\mathbf{z}), \tag{21.23.3.4}$$

or, in other terms

$$(\sigma(\mathbf{z}) \cdot f)(\mathbf{y}) = [\mathbf{z}, f(\mathbf{y})] - f([\mathbf{z}, \mathbf{y}]) \tag{21.23.3.5}$$

for all $\mathbf{y} \in \mathfrak{e}$.

We shall first show that the condition (21.23.3.2) is satisfied when we take $w$ to be a *projection* of the vector space $\mathfrak{e}$ onto the subspace $\mathfrak{a}$ (so that $w(\mathbf{y}) \in \mathfrak{a}$ for all $\mathbf{y} \in \mathfrak{e}$, and $w(\mathbf{t}) = \mathbf{t}$ for all $\mathbf{t} \in \mathfrak{a}$). Indeed, it follows from (21.23.3.5) that for $\mathbf{t} \in \mathfrak{a}$ we have

$$(\sigma(\mathbf{t}) \cdot w)(\mathbf{y}) = -w([\mathbf{t}, \mathbf{y}]) = -[\mathbf{t}, \mathbf{y}] \tag{21.23.3.6}$$

because $w(\mathbf{y}) \in \mathfrak{a}$, $[\mathfrak{a}, \mathfrak{a}] = 0$, and $[\mathfrak{e}, \mathfrak{a}] \subset \mathfrak{a}$. Hence $\sigma(\mathbf{t}) \cdot w = 0$ means that $[\mathbf{t}, \mathbf{y}] = 0$ for all $\mathbf{y} \in \mathfrak{e}$, or equivalently $\rho(\mathbf{x}) \cdot \mathbf{t} = 0$ for all $\mathbf{x} \in \mathfrak{g}$. But since $\mathfrak{a}$ is a *simple* $U(\mathfrak{g})$-module and the set of $\mathbf{t} \in \mathfrak{a}$ such that $\rho(\mathbf{x}) \cdot \mathbf{t} = 0$ for all $\mathbf{x} \in \mathfrak{g}$ is a $U(\mathfrak{g})$-submodule of $\mathfrak{a}$, this submodule can only be $\mathfrak{a}$ or $\{0\}$, and the first alternative has been ruled out.

The relation (21.23.3.6) likewise shows that for (21.23.3.3) to be satisfied, it is necessary and sufficient that for each $\mathbf{z} \in \mathfrak{e}$ there should exist $\mathbf{t} \in \mathfrak{a}$ such that $\sigma(\mathbf{z}) \cdot w = -\mathrm{ad}(\mathbf{t})$ in $\mathrm{M} = \mathrm{End}(\mathfrak{e})$.

The conditions imposed on $w$ may be reformulated as follows. Let P denote the vector subspace of M that is the image of $\mathfrak{a}$ under the mapping $\mathbf{t} \mapsto \mathrm{ad}(\mathbf{t})$, and let R denote the vector subspace of M consisting of the $f \in \mathrm{End}(\mathfrak{e})$ such that (1) $f(\mathfrak{e}) \subset \mathfrak{a}$, and (2) the restriction of $f$ to $\mathfrak{a}$ is a scalar multiplication by $\lambda_f$. It is clear that R is a vector space, that $\mathrm{P} \subset \mathrm{R}$, and that $f \mapsto \lambda_f$ is a **C**-*linear form* $\lambda$ on R; the set of projections of $\mathfrak{e}$ onto $\mathfrak{a}$ is the affine hyperplane $\lambda^{-1}(1)$ in R. Observe now that if $f \in \mathrm{R}$ and $\mathbf{z} \in \mathfrak{e}$, then also $\sigma(\mathbf{z}) \cdot f \in \mathrm{R}$; for the fact that $\sigma(\mathbf{z}) \cdot f$ maps $\mathfrak{e}$ into $\mathfrak{a}$ follows from (21.23.3.5) and the fact that $\mathfrak{a}$ is an ideal of $\mathfrak{e}$, and it is immediately seen that $\lambda_{\sigma(\mathbf{z}) \cdot f} = 0$. It follows that R is a $\mathrm{U}(\mathfrak{e})$-module and that $\lambda: f \mapsto \lambda_f$ is a $\mathrm{U}(\mathfrak{e})$-module homomorphism of R into **C**, if **C** is regarded as a trivial $\mathrm{U}(\mathfrak{e})$-module. On the other hand, the Jacobi identity shows that if $f = \mathrm{ad}(\mathbf{t})$ with $\mathbf{t} \in \mathfrak{a}$, then $\sigma(\mathbf{z}) \cdot f = \mathrm{ad}([\mathbf{t}, \mathbf{z}])$, and hence P is a $\mathrm{U}(\mathfrak{e})$-submodule of R.

Next we remark that for $\mathbf{t} \in \mathfrak{a}$ and $f \in \mathrm{R}$, we have $\sigma(\mathbf{t}) \cdot f = -\lambda_f \, \mathrm{ad}(\mathbf{t})$, in other words $\sigma(\mathbf{t}) \cdot \mathrm{R} \subset \mathrm{P}$. This shows that for each $\mathbf{x} = p(\mathbf{z}) \in \mathfrak{g}$ and each $f \in \mathrm{R}$, the coset of $\sigma(\mathbf{z}) \cdot f$ modulo P depends only on $\mathbf{x}$ and the coset $\bar{f}$ of $f$ modulo P. If we denote this coset by $\bar{\sigma}(\mathbf{x}) \cdot \bar{f}$, it is immediately verified that $\bar{\sigma}$: $\mathfrak{g} \to \mathfrak{gl}(\mathrm{R}/\mathrm{P})$ is a Lie algebra homomorphism. We have thus defined a $\mathrm{U}(\mathfrak{g})$-module structure on R/P, and the mapping $\bar{\lambda}$: $\mathrm{R}/\mathrm{P} \to \mathbf{C}$ induced by $\lambda$ on passing to the quotient is a surjective $\mathrm{U}(\mathfrak{g})$-module homomorphism, if **C** is regarded as a trivial $\mathrm{U}(\mathfrak{g})$-module.

This being so, the conditions imposed on $w$ are (1) $w \in \mathrm{R}$, (2) $\lambda_w = 1$, (3) $\sigma(\mathbf{z}) \cdot w \in \mathrm{P}$ for all $\mathbf{z} \in \mathfrak{e}$. If $\bar{w}$ is the image of $w$ in R/P, these conditions are equivalent to: (1) $\bar{w} \in \mathrm{R}/\mathrm{P}$, (2) $\bar{\lambda}_{\bar{w}} = 1$, (3) $\bar{\sigma}(\mathbf{x}) \cdot \bar{w} = 0$ for all $\mathbf{x} \in \mathfrak{g}$. This implies that the one-dimensional subspace $\mathbf{C}\bar{w}$ in R/P is a $\mathrm{U}(\mathfrak{g})$-module supplementary to the $\mathrm{U}(\mathfrak{g})$-submodule $\mathrm{Ker}(\bar{\lambda})$. *Conversely*, if D is a one-dimensional subspace of R/P supplementary to $\mathrm{Ker}(\bar{\lambda})$ and is a $\mathrm{U}(\mathfrak{g})$-module, then the intersection $\{\bar{w}\}$ of D with the affine hyperplane given by the equation $\bar{\lambda}_{\bar{f}} = 1$ satisfies the conditions above, because D is then isomorphic to the $\mathrm{U}(\mathfrak{g})$-module **C**, which is trivial. Now the existence of such a $\mathrm{U}(\mathfrak{g})$-submodule supplementary to $\mathrm{Ker}(\bar{\lambda})$ is a consequence of the fact that every finite-dimensional $\mathrm{U}(\mathfrak{g})$-module is the direct sum of simple $\mathrm{U}(\mathfrak{g})$-submodules ((21.9.1) and (A.23.3)). Q.E.D.

(21.23.4) *Every finite-dimensional real* (resp. *complex*) *Lie algebra is isomorphic to the Lie algebra of a real* (resp. *complex*) *Lie group.*

As we have already remarked (19.17.4), it is enough to prove the result for a complex Lie algebra. By virtue of (21.23.3), such an algebra is the semidirect product of a solvable Lie algebra $\mathfrak{r}$ and a semisimple Lie algebra $\mathfrak{s}$. Since $\mathfrak{r}$ (resp. $\mathfrak{s}$) is the Lie algebra of a complex solvable (resp. semisimple) Lie group, by virtue of (19.14.10) and (21.6.3), the result is a consequence of (19.14.9).

(21.23.5) *Let* G *be a simply connected Lie group. For each ideal* $\mathfrak{n}_e$ *of the Lie algebra* $\mathfrak{g}_e$ *of* G, *the connected Lie group* N *immersed in* G *with Lie algebra* $\mathfrak{n}_e$ (19.7.4) *is a closed (normal) subgroup of* G.

There exists a Lie group H whose Lie algebra $\mathfrak{h}_e$ is isomorphic to $\mathfrak{g}_e/\mathfrak{n}_e$ (21.23.4), and we have a homomorphism $u: \mathfrak{g}_e \to \mathfrak{h}_e$ of Lie algebras, with kernel $\mathfrak{n}_e$. Since G is simply connected, there exists a homomorphism of Lie groups $f: G \to H$ such that $f_* = u$ (19.7.6), and N is the identity component of the kernel of $f$ (19.7.1), hence is closed in G.

## PROBLEMS

1. (a) Let E be a real or complex vector space of finite dimension; A, B two vector subspaces of End(E); T the set of $t \in \text{End}(E)$ such that $[t, A] \subset B$. Show that if $s \in T$ is such that $\text{Tr}(su) = 0$ for all $u \in T$, then $s$ is a nilpotent endomorphism of E. (Note that $\text{Tr}(s^n) = 0$ for all integers $n \geqq 2$, and deduce that the eigenvalues of $s$ are all zero, by using Newton's formulas.)
(b) Let $\mathfrak{g}$ be a finite-dimensional Lie algebra, $\rho: \mathfrak{g} \to \mathfrak{gl}(E)$ a Lie algebra homomorphism, and $B_\rho(\mathbf{u}, \mathbf{v}) = \text{Tr}(\rho(\mathbf{u})\rho(\mathbf{v}))$ the symmetric bilinear form associated with $\rho$ (21.5.5). In order that $\rho(\mathfrak{g})$ should be solvable, it is necessary and sufficient that $\mathfrak{D}(\mathfrak{g})$ should be orthogonal to $\mathfrak{g}$, relative to the form $B_\rho$. (To show that the condition is necessary, reduce to the case where $\mathfrak{g}$ and $\mathfrak{gl}(E)$ are complex Lie algebras, and use Lie's theorem. To show that the condition is sufficient, reduce to the case where $\mathfrak{g} \subset \mathfrak{gl}(E)$ and use (a), with $A = \mathfrak{g}$ and $B = \mathfrak{D}(\mathfrak{g})$.)

2. Let $\mathfrak{g}$ be a finite-dimensional complex Lie algebra.
(a) Show that if $\mathfrak{r}$ is the radical of $\mathfrak{g}$, we have $[\mathfrak{g}, \mathfrak{r}] \subset \mathfrak{D}(\mathfrak{g}) \cap \mathfrak{r}$. (Use Levi's theorem.)
(b) For each finite-dimensional complex vector space E and each Lie algebra homomorphism $\rho: \mathfrak{g} \to \mathfrak{gl}(E)$, show that $\rho([\mathfrak{g}, \mathfrak{r}])$ consists of nilpotent endomorphisms. (Observe that the elements of $\rho([\mathfrak{r}, \mathfrak{r}])$ are nilpotent by virtue of Lie's theorem, and then argue as in (21.22.2).) In particular, $[\mathfrak{g}, \mathfrak{r}]$ is a nilpotent ideal of $\mathfrak{g}$.
(c) Show that $\mathfrak{r}$ is the orthogonal supplement of $\mathfrak{D}(\mathfrak{g})$ relative to the Killing form of $\mathfrak{g}$. (To show that $\mathfrak{r}$ is contained in the orthogonal supplement $\mathfrak{r}'$ of $\mathfrak{D}(\mathfrak{g})$, use (a). Then observe that $\mathfrak{r}'$ is an ideal containing $\mathfrak{r}$, and that $\text{ad}(\mathfrak{r}')$ is solvable by virtue of Problem 1(b).)
(d) Show that for each automorphism $v$ of $\mathfrak{g}$ we have $v(\mathfrak{r}) = \mathfrak{r}$.
(e) For each ideal $\mathfrak{a}$ of $\mathfrak{g}$, show that $\mathfrak{a} \cap \mathfrak{r}$ is the radical of $\mathfrak{a}$. (Observe that, by (d) above, the radical of $\mathfrak{a}$ is an ideal of $\mathfrak{g}$.)

3. (a) Let E be a finite-dimensional complex vector space, and let $F = \mathbf{C} \times E$. Let $\sigma : \mathfrak{g} \to \mathfrak{gl}(E)$ be a homomorphism of a complex Lie algebra $\mathfrak{g}$ into $\mathfrak{gl}(E) = \text{End}(E)$. If

$f: \mathfrak{g} \to \mathrm{E}$ is a **C**-linear mapping, define a **C**-linear mapping $\rho : \mathfrak{g} \to \mathrm{End(F)}$ by the formula $\rho(\mathbf{u}) \cdot (\xi, x) = (0, \xi \cdot f(\mathbf{u}) + \sigma(\mathbf{u}) \cdot x)$ for all $\mathbf{u} \in \mathfrak{g}$. Show that for $\rho$ to be a homomorphism of $\mathfrak{g}$ into $\mathfrak{gl}(\mathrm{F})$ it is necessary and sufficient that $f$ should satisfy the condition

$$\sigma(\mathbf{u}_1) \cdot f(\mathbf{u}_2) - \sigma(\mathbf{u}_2) \cdot f(\mathbf{u}_1) = f([\mathbf{u}_1, \mathbf{u}_2])$$

for all $\mathbf{u}_1, \mathbf{u}_2 \in \mathfrak{g}$.
(b) Suppose that $\mathfrak{g}$ is semisimple. Show that if $f$ satisfies the condition in (a), there exists an element $x_0 \in \mathrm{E}$ such that $f(\mathbf{u}) = -\sigma(\mathbf{u}) \cdot x_0$ for all $\mathbf{u} \in \mathfrak{g}$. (Observe that the subspace $\{0\} \times \mathrm{E}$ of F is stable under $\rho$.)

**4.** Let $\mathfrak{g}$ be a finite-dimensional complex Lie algebra, $\mathfrak{r}$ its radical, and $\mathfrak{s}$, $\mathfrak{s}'$ two semisimple Lie subalgebras of $\mathfrak{g}$ such that $\mathfrak{g} = \mathfrak{s} + \mathfrak{r} = \mathfrak{s}' + \mathfrak{r}$ (**21.23.3**). We propose to prove that there exists an element $\mathbf{a} \in [\mathfrak{g}, \mathfrak{r}]$ such that, putting $v = \exp(\mathrm{ad}(\mathbf{a}))$ in $\mathfrak{gl}(\mathfrak{g})$, we have $v(\mathfrak{s}') = \mathfrak{s}$ (*Malcev's theorem*). Distinguish three cases:

(1) $[\mathfrak{g}, \mathfrak{r}] = \{0\}$; then $\mathfrak{r}$ is the center of $\mathfrak{g}$, and $\mathfrak{g}$ is the direct product of the ideals $\mathfrak{r}$ and $\mathfrak{D}(\mathfrak{g})$, and $\mathfrak{D}(\mathfrak{g}) = \mathfrak{s} = \mathfrak{s}'$.

(2) $\mathfrak{r}$ is commutative and $\mathfrak{r} = [\mathfrak{g}, \mathfrak{r}]$. Then for each $\mathbf{x} \in \mathfrak{s}'$ there exists a unique element $h(\mathbf{x}) \in \mathfrak{r}$ such that $\mathbf{x} + h(\mathbf{x}) \in \mathfrak{s}$. By using Problem 3(b), show that there exists an element $\mathbf{a} \in \mathfrak{r}$ such that $h(\mathbf{x}) = -[\mathbf{x}, \mathbf{a}]$ for all $\mathbf{x} \in \mathfrak{s}'$, and observe that $(\mathrm{ad}(\mathbf{a}))^2 = 0$.

(3) The general case. Observe that $[\mathfrak{g}, \mathfrak{r}]$ is a nilpotent algebra, hence has center $\mathfrak{c} \neq \{0\}$ (Section **19.12**, Problem 3). Choose in $\mathfrak{c}$ an ideal $\mathfrak{m} \neq \{0\}$ of smallest dimension. We may limit our attention to the case where $\mathfrak{m} \neq \mathfrak{r}$; consider the algebra $\mathfrak{g}/\mathfrak{m}$ and proceed by induction on the dimension of $\mathfrak{r}$ using case (2) above.

**5.** Let $\mathfrak{g}$ be a finite-dimensional complex Lie algebra and $\mathfrak{a}$ an ideal of $\mathfrak{g}$ such that $\mathfrak{g}/\mathfrak{a}$ is semisimple. Show that there exists in $\mathfrak{g}$ a semisimple Lie subalgebra $\mathfrak{s}$ supplementary to $\mathfrak{a}$. (Use (**21.23.3**) and (**21.6.4**).)

**6.** (a) Let G be a simply connected complex Lie group, H a connected *normal* Lie subgroup of G, and $p$: G $\to$ G/H the canonical mapping. Show that the principal bundle G, with base G/H and fiber H, is trivializable. (To prove the existence of a holomorphic section over G/H (**16.14.5**), proceed by induction on dim(G/H), by using (**16.14.9**) to reduce to the case where the group G/H is either 1-dimensional or almost simple.) Deduce that H is simply connected.
(b) Let G be a connected complex Lie group and H a connected normal Lie subgroup of G. Show that the canonical mapping $\pi_1(\mathrm{H}) \to \pi_1(\mathrm{G})$ (**16.27.6**) is injective. (Use (a).)

**7.** Extend the proof of Levi's theorem and the results of Problems 1 to 6 to real Lie groups.

**8.** Let G be a connected real Lie group. We propose to prove that there exists in G a maximal compact subgroup K and a finite number of closed subgroups $\mathrm{L}_1, \ldots, \mathrm{L}_m$ isomorphic to **R**, such that the mapping $(t, z_1, \ldots, z_m) \mapsto t z_1 z_2 \cdots z_m$ of the manifold $\mathrm{K} \times \mathrm{L}_1 \times \cdots \times \mathrm{L}_m$ into G is a diffeomorphism of this manifold onto G (*Iwasawa's theorem*). We proceed as follows:
(a) The theorem is true when G is semisimple (**21.21.10**) or commutative (**19.7.9**).
(b) In the general case, there exists a closed normal subgroup J of G, isomorphic to $\mathbf{R}^n$ or to $\mathbf{T}^n$, with $n > 0$ if G is not semisimple. (Use Section **19.12**, Problem 2, applied to the radical of the Lie algebra of G.)

(c) Now proceed by induction on the dimension of G. Show first that if L′ is a closed subgroup of G/J isomorphic to **R**, then there exists a closed subgroup L of G isomorphic to **R**, such that $L \cap J = \{e\}$ and such that the projection $p: G \to G/J$, restricted to L, is an isomorphism of L onto L′. (Use Section **12.9**, Problem 10.) Then observe that if J is isomorphic to $\mathbf{R}^n$ and K′ is a compact subgroup of G/J, the inverse image $p^{-1}(K')$ is the semidirect product of J and a compact subgroup (Section **19.14**, Problem 3).

**9.** (a) Let G be a connected real Lie group. The *radical* of G is defined to be the connected Lie group $R_0$ immersed in G whose Lie algebra is the radical of the Lie algebra of $\mathfrak{g}$. Show that $R_0$ is closed in G and that the quotient group $G/R_0$ is semisimple.
(b) Let Z′ be the (discrete) center of $G/R_0$. Show that every solvable normal subgroup of G is contained in the inverse image R of Z′ in G. (Use Section **12.8**, Problem 5.) The group $R_0$ is the identity component of R.
(c) If G is simply connected, then so is $R_0$ (Problem 6), and G is the semidirect product $R_0 \times_\sigma S$ of $R_0$ and a simply connected semisimple group S. Show that when $R_0$ is *commutative*, the structure of G can be completely described in terms of S and its finite-dimensional continuous linear representations.
(d) If $u: G \to G'$ is a surjective homomorphism of G onto a Lie group G′, show that $u(R_0)$ is the radical of G′ (cf. Problem 5).
(e) Show that the radical of a product $G_1 \times G_2$ of connected Lie groups is the product of the radicals of $G_1$ and $G_2$.

**10.** A connected Lie group G is said to be *reductive* if its adjoint representation $\mathrm{Ad}: G \to \mathbf{GL}(\mathfrak{g})$ is completely reducible.
(a) Show that the following conditions are equivalent:
($\alpha$) G is reductive.
($\beta$) $\mathfrak{D}(\mathfrak{g})$ is semisimple.
($\gamma$) The radical $R_0$ of G is contained in the center of G.
(b) If G is a simply connected reductive group, it is the product of a simply connected semisimple group and a group isomorphic to $\mathbf{R}^n$. Deduce from this the description of an arbitrary reductive Lie group. Give an example of a reductive group G whose commutator subgroup is not closed in G (cf. Section **21.18**, **Problem** 18).
(c) Let G be a connected Lie group. Show that for every continuous linear representation of G on a finite-dimensional complex vector space to be completely reducible, it is necessary and sufficient that G should be reductive and $G/\overline{\mathscr{D}(G)}$ compact. (To show that the condition is sufficient, consider a linear representation $\rho$ of G on a vector space V and a subspace W of V stable under $\rho$, and observe that G acts on the vector subspace E of End(V) consisting of endomorphisms $v$ such that $\rho(s) \circ v = v \circ \rho(s)$ for all $s \in \mathscr{D}(G)$ and $v(V) \subset W$, and such that the restriction of $v$ to W is a homothety.)

**11.** (a) Let G be a connected Lie group, Z its center, $Z_0$ the identity component of Z, and $\pi$: $G/Z_0 \to G/Z$ the canonical homomorphism, whose kernel $Z/Z_0$ is discrete. Identify the differential manifold G/Z with the product $K \times E$, where K is a maximal compact subgroup of G and E is diffeomorphic to a vector space $\mathbf{R}^m$ (Problem 9). Then, if $\pi^{-1}(K) = M$ and if F is the identity component of $\pi^{-1}(E)$, the manifold $G/Z_0$ may be identified with $M \times F$; M is a connected covering group of K, and F is diffeomorphic to $\mathbf{R}^m$.
(b) If G is solvable, show that M is commutative and is of the form $N/Z_0$, where N is a connected nilpotent Lie subgroup of G, containing Z. Deduce that Z is contained in a connected commutative Lie subgroup of G (cf. Section **19.14**, Problem 7(b)).
(c) If G is semisimple, so that $Z_0 = \{e\}$, then M is of the form $K_1 \times \mathbf{R}^n$, where $K_1$ is a

compact subgroup of G (21.6.9), and Z is contained in $Z_1 \times \mathbf{R}^n$, where $Z_1$ is the center of $K_1$. Deduce that Z is contained in a connected commutative Lie subgroup of G.
(d) Deduce from (b) and (c) that for *every* connected Lie group G, the center Z of G is contained in a connected commutative Lie subgroup of G. (Consider first the case where G is simply connected, and apply Problem 9(c), by observing that if A is a connected commutative Lie subgroup of S, then $R_0 A$ is a connected solvable Lie subgroup of G. Then pass to the general case by using (**20.22.5.1**).)
(e) Deduce from (d) that Z is an *elementary* commutative group (Section **19.7**, Problem 5). In particular, if Z is not compact, there exists an element $c \in Z$ such that the group generated by $c$ (consisting of the powers $c^n$ for $n \in \mathbf{Z}$) is infinite discrete.

**12.** Let G be a connected Lie group and H a connected Lie group immersed in G and *dense* in G. Then we know (Section **19.11**, Problem 3) that, if $\mathfrak{g}$ and $\mathfrak{h}$ are the Lie algebras of G and H, respectively, $\mathfrak{h}$ is an ideal in $\mathfrak{g}$, and $\mathfrak{g}/\mathfrak{h}$ is commutative.
(a) Let $\mathfrak{h} = \mathfrak{r} \oplus \mathfrak{s}$, where $\mathfrak{r}$ is the radical of $\mathfrak{h}$, and $\mathfrak{s}$ is a semisimple Lie subalgebra of $\mathfrak{h}$. Show that if $\mathfrak{r}'$ is the radical of $\mathfrak{g}$ (which contains $\mathfrak{r}$) then $\mathfrak{g} = \mathfrak{r}' \oplus \mathfrak{s}$.
(b) Let $\tilde{G}$ be the universal covering group of G and let $\pi: \tilde{G} \to G$ be the canonical homomorphism, with kernel D. We may write $\tilde{G} = R_0' \times_\sigma S$, where $\text{Lie}(R_0') = \mathfrak{r}'$ and $\text{Lie}(S) = \mathfrak{s}$ (Problem 9(c)). Let $D_1, D_2$ be the projections of D on $R_0'$ and S, respectively; then $D_1$ is contained in the centralizer of S, and $D_2$ in the center of S. Let H′ and $R_0$ be the connected Lie groups immersed in $\tilde{G}$ whose Lie algebras are $\mathfrak{h}$ and $\mathfrak{r}$, respectively. Then H′ and $R_0$ are closed in $\tilde{G}$, and we have $H' = R_0 \times_\sigma S$; $R_0$ is the radical of H′, and $\pi(R_0)$ the radical of $H = \pi(H')$.
(c) Show that $R_0' = \overline{D_1 R_0}$ (closure in $\tilde{G}$). If U is a connected Lie subgroup of S that contains the center of S, deduce that $R_0' \subset \overline{DUR_0}$, and hence that $G = \overline{\pi(U)\pi(R_0)H}$.
(d) With the same hypotheses, show that $\pi(U)\pi(R_0')$ is closed in G and that $\overline{\pi(U)\pi(R_0)} \cap H = \pi(U)\pi(R_0)$.

**13.** Let G be a connected Lie group and H a connected Lie group immersed in G. For H to be closed in G, it is necessary and sufficient that the closure in G of every one-parameter subgroup of H should be contained in H (*Malcev's theorem*). (Use Problem 12 to reduce the question to proving that $\pi(U)\pi(R_0)$ is closed in G; by Problem 11, we may take U to be commutative, and then $\pi(U)\pi(R_0)$ is solvable, and we can apply Section **19.14**, Problem 15.)

Deduce that for H to be closed in G it is necessary and sufficient that the intersection of H with every compact subgroup K of G should be closed in K. (Use Section **12.9**, Problem 10.)

**14.** Let G be a complex connected semisimple Lie group, $\mathfrak{g}$ its Lie algebra, and consider a root decomposition (**21.10.1.1**) of $\mathfrak{g}$. We shall use the notation of (**21.10.3**). Let $\mathbf{B} = \{\beta_1, \dots, \beta_l\}$ be a basis of the root system $\mathbf{S}$, and put $\mathbf{h}_j = \mathbf{h}_{\beta_j}$, $1 \leq j \leq l$. The $\mathbf{h}_j$ and the $\mathbf{x}_\alpha$, $\alpha \in \mathbf{S}$, form a basis of the vector space $\mathfrak{g}$.
(a) Let $\mathbf{B}_1$ be a subset of $\mathbf{B}$, let $\mathbf{S}_1 \subset \mathbf{S}$ be the set of roots that are linear combinations of roots belonging to $\mathbf{B}_1$, and let $\mathfrak{h}_1 \subset \mathfrak{h}$ be the subspace spanned by the $\mathbf{h}_j$ such that $\beta_j \in \mathbf{B}_1$. Show that the (direct) sum of $\mathfrak{h}_1$ and the $\mathfrak{g}_\alpha$ such that $\alpha \in \mathbf{S}_1$ is a semisimple Lie algebra. (To calculate the Killing form, use (**21.19.8.1**) and (**21.20.4.2**).)
(b) Let $\mathfrak{h}_2$ be the subspace of $\mathfrak{h}$ defined by the equations $\beta_j(\mathbf{u}) = 0$ for $\beta_j \in \mathbf{B}_1$. Show that if $\mathfrak{p}$ is the parabolic subalgebra defined in Section **21.22**, Problem 12(c) with $\mathbf{Q} = -(\mathbf{S}_1)_+$, then the radical of $\mathfrak{p}$ is the direct sum $\mathfrak{r}_1$ of $\mathfrak{h}_2$ and the $\mathfrak{g}_\alpha$ such that $\alpha \in \mathbf{S}_+ \cap \complement(\mathbf{S}_1)_+$, and $\mathfrak{p} = \mathfrak{g}_1 \oplus \mathfrak{r}_1$ is a Levi decomposition of $\mathfrak{p}$.

Show that $\mathfrak{D}(\mathfrak{p}) = [\mathfrak{p}, \mathfrak{p}]$ is the direct sum of $\mathfrak{g}_1$ and the $\mathfrak{g}_\alpha$ such that $\alpha \in \mathbf{S}_+ \cap \complement(\mathbf{S}_1)_+$, and that the sum $\mathfrak{r}_2$ of these latter subspaces is the radical of $[\mathfrak{p}, \mathfrak{p}]$. Show that $\mathfrak{p}$ is the normalizer of $\mathfrak{r}_2$ in $\mathfrak{g}$.

**15.** (a) With the hypotheses and notation of Problem 14, let $\lambda_j$ be the linear form on $\mathfrak{g}$ such that $\lambda_j(\mathbf{h}_k) = \delta_{jk}$ and $\lambda_j(\mathbf{x}_\alpha) = 0$ for all $\alpha \in \mathbf{S}$. Show that for each element $\mathbf{x} \in \mathfrak{g}$ there exists an automorphism $v = \exp(\operatorname{ad}(\mathbf{u}))$ of $\mathfrak{g}$ and an index $j$ such that $\lambda_j(v(\mathbf{x})) = 0$. (Reduce to the case where $\mathbf{x} \notin \mathfrak{h}$.)
(b) Deduce from (a) that there exists an automorphism $v_1 = \exp(\operatorname{ad}(\mathbf{u}_1))$ of $\mathfrak{g}$ such that $v_1(\mathbf{x})$ belongs to the vector subspace that is the sum of the $\mathfrak{g}_\alpha$, $\alpha \in \mathbf{S}$. (Argue by induction on the dimension of $\mathfrak{g}$, using Problem 14.)
(c) Deduce from (b) that there exist two elements $\mathbf{y}, \mathbf{z} \in \mathfrak{g}$ such that $\mathbf{x} = [\mathbf{y}, \mathbf{z}]$. (Cf. (21.7.6.3).)

**16.** In the group $\mathbf{GL}(n, \mathbf{C})$, let $\mathrm{I}(n)$ denote the subgroup of all lower triangular matrices with all diagonal elements equal to 1 (in other words, matrices $(a_{ij})$ such that $a_{ij} = 0$ for $i < j$ and $a_{ii} = 1$ for all $i$). Also let $\mathrm{S}(n)$ denote the subgroup of all upper triangular matrices (i.e., matrices $(a_{ij})$ such that $a_{ij} = 0$ for $i > j$). For each matrix $X = (x_{ij}) \in \mathbf{GL}(n, \mathbf{C})$ and each integer $k \in [1, n]$, let $X_k$ denote the matrix $(x_{ij})_{1 \le i, j \le k}$, and put $\Delta_k(X) = \det(X_k)$ (the "principal minors" of $X$). The set $\Omega$ of matrices $X \in \mathbf{GL}(n, \mathbf{C})$ such that $\Delta_k(X) \neq 0$ for $1 \leqq k \leqq n - 1$ is a dense *connected* open set in $\mathbf{GL}(n, \mathbf{C})$ (Section 16.3, Problem 3). Show that the mapping $(Y, Z) \mapsto YZ$ of $\mathrm{I}(n) \times \mathrm{S}(n)$ into $\mathbf{GL}(n, \mathbf{C})$ is an isomorphism of complex manifolds of $\mathrm{I}(n) \times \mathrm{S}(n)$ onto $\Omega$; the inverse mapping $X \mapsto (i(X), s(X))$ is such that the entries of the matrices $i(X)$ and $s(X)$ are rational functions of the $x_{ij}$. (Observe that $\Delta_k(s(X)) = \Delta_k(X)$ for $1 \leqq k \leqq n$.)

**17.** (a) With the hypotheses and notation of Problems 14 and 16, put $n = \dim(\mathfrak{g})$ and suppose that G has trivial center, so that G may be identified with Ad(G). Then identify G with a subgroup of $\mathbf{GL}(n, \mathbf{C})$ by identifying the canonical basis of $\mathbf{C}^n$ with the basis of $\mathfrak{g}$ ranged in the following order:

$$\mathbf{x}_{-\alpha_m}, \ldots, \mathbf{x}_{-\alpha_1}, \quad \mathbf{h}_1, \ldots, \mathbf{h}_l, \quad \mathbf{x}_{\alpha_1}, \ldots, \mathbf{x}_{\alpha_m},$$

where the positive roots $\alpha_1, \ldots, \alpha_m$ are ordered so that if $\alpha_i + \alpha_j = \alpha_k$ is a root, then $i < k$ and $j < k$. Let $\mathfrak{b} = \mathfrak{h} \oplus \mathfrak{n}_+$ be the Borel subalgebra spanned by the $\mathbf{h}_j$ and the $\mathbf{x}_\alpha$ with $\alpha \in \mathbf{S}_+$, and let $\mathfrak{n}_-$ be the nilpotent subalgebra spanned by the $\mathbf{x}_{-\alpha}$ for $\alpha \in \mathbf{S}_+$, so that $\mathfrak{g} = \mathfrak{b} \oplus \mathfrak{n}_-$. Let B and $\mathrm{N}_-$ be the connected Lie subgroups of G having $\mathfrak{b}$ and $\mathfrak{n}_-$ as their respective Lie algebras. With the notation of Problem 16, show that $\mathrm{B} = \mathrm{G} \cap \mathrm{S}(n)$ and $\mathrm{N}_- = \mathrm{G} \cap \mathrm{I}(n)$; furthermore, if $\Omega_0 = \Omega \cap \mathrm{G}$, then $\Omega_0$ is a dense connected open set in G, and B (resp. $\mathrm{N}_-$) is the image of $\Omega_0$ under the mapping $X \mapsto s(X)$ (resp. $X \mapsto i(X)$); also the mapping $(Y, Z) \mapsto YZ$ of $\mathrm{N}_- \times \mathrm{B}$ into G is an isomorphism of complex manifolds of $\mathrm{N}_- \times \mathrm{B}$ onto $\Omega_0$.
(b) Let $\mathfrak{n}'_-$ (resp. $\mathfrak{n}''_-$) be the sum of the $\mathfrak{g}_{-\alpha}$ for $\alpha \in (\mathbf{S}_1)_+$ (resp. $\alpha \in \mathbf{S}_+ \cap \complement(\mathbf{S}_1)_+$). Then we have $\mathfrak{n}_- = \mathfrak{n}'_- \oplus \mathfrak{n}''_-$, and $\mathfrak{n}'_-$ is a Lie subalgebra of $\mathfrak{n}_-$, and $\mathfrak{n}''_-$ is an ideal of $\mathfrak{n}_-$. If $\mathrm{N}'_-$ and $\mathrm{N}''_-$ are the connected Lie subgroups having $\mathfrak{n}'_-$, $\mathfrak{n}''_-$ as their Lie algebras, then the mapping $(Y', Y'') \mapsto Y'Y''$ of $\mathrm{N}'_- \times \mathrm{N}''_-$ into $\mathrm{N}_-$ is an isomorphism of complex manifolds. If P is the parabolic subgroup of G with Lie algebra $\mathfrak{p}$, we have $\mathrm{N}'_-\mathrm{B} \subset \mathrm{P}$, and $\Omega_0 \subset \mathrm{N}''_-\mathrm{P}$.
(c) Let $\mathrm{P}(n)$ be the normalizer in $\mathbf{GL}(n, \mathbf{C})$ of $\mathfrak{r}_2$, considered as a Lie subalgebra of $\mathfrak{gl}(n, \mathbf{C})$. Show that the normalizer $\mathscr{N}(\mathrm{P})$ of P in G is equal to $\mathrm{G} \cap \mathrm{P}(n)$.
(d) Show that $\mathscr{N}(\mathrm{P}) \cap \mathrm{N}''_- = \{e\}$. (Use the fact that the exponential mapping of $\mathfrak{n}''_-$ into

$N''_-$ is surjective (Section **19.14**, Problem 6), and that if $\mathbf{u} \in \mathfrak{n}$, $\mathrm{ad}(\mathbf{u})$ is a polynomial with respect to $\exp(\mathrm{ad}(\mathbf{u}))$ (21.21.10.2). Deduce that $\mathcal{N}(P) = P$. (Observe that otherwise $N''_- \mathcal{N}(P)$ would contain two dense open subsets of G, disjoint from each other.)

(e) Show that there exists a complex-analytic isomorphism of G/P onto a submanifold of the Grassmannian $\mathbf{G}_{n,p}(\mathbf{C})$, where $p$ is the number of roots of $\mathbf{S}_+$ that do not belong to $(\mathbf{S}_1)_+$. (Observe that G/P is isomorphic to $\mathbf{GL}(n, \mathbf{C})/P(n)$.)

**18.** Let $\mathfrak{L}$ be a finite-dimensional real Lie algebra, and let G be a connected Lie subgroup of $\mathrm{Aut}(\mathfrak{L})$. Suppose that there exists a decreasing sequence $\mathfrak{L} = \mathfrak{L}_{-1} \supset \mathfrak{L}_0 \supset \mathfrak{L}_1 \supset \cdots \supset \mathfrak{L}_r$ of G-stable vector subspaces of $\mathfrak{L}$, satisfying the following conditions:

(1) $[\mathfrak{L}_p, \mathfrak{L}_q] \subset \mathfrak{L}_{p+q}$, with the conventions that $\mathfrak{L}_{-2} = \mathfrak{L}$ and $\mathfrak{L}_k = 0$ for $k \geqq r + 1$.

(2) If there exists $\mathbf{y} \in \mathfrak{L}_p$, $p \geqq 0$, such that $\mathbf{y} \notin \mathfrak{L}_{p+1}$, then there exists $\mathbf{x} \in \mathfrak{L}$ such that $[\mathbf{y}, \mathbf{x}] \notin \mathfrak{L}_p$.

(3) $\mathfrak{L}_{-1} \neq \{0\}$.

(4) If V is a G-stable vector subspace of $\mathfrak{L}$ containing $\mathfrak{L}_0$ and such that $[\mathfrak{L}_0, V] \subset V$, then either $V = \mathfrak{L}$ or $V = \mathfrak{L}_0$.

(a) Show that $\mathfrak{L}_j \neq \mathfrak{L}_{j+1}$ for $-1 \leqq j \leqq r$.

(b) For each nonzero G-stable ideal $\mathfrak{J}$ of $\mathfrak{L}$, show that $\mathfrak{L} = \mathfrak{J} + \mathfrak{L}_0$. (Show that the assumption that $\mathfrak{J} \subset \mathfrak{L}_0$ would contradict property (2).) Deduce that if there exists $\mathbf{y} \in \mathfrak{L}_p$, with $p \geqq 0$, such that $\mathbf{y} \notin \mathfrak{L}_{p+1}$, then there exists $\mathbf{u} \in \mathfrak{J}$ such that $[\mathbf{y}, \mathbf{u}] \notin \mathfrak{L}_p$; consequently we have $\mathfrak{J} \cap \mathfrak{L}_p \neq 0$ if $\mathfrak{L}_{p+1} \neq 0$.

(c) If $\mathfrak{J}_1$, $\mathfrak{J}_2$ are two G-stable ideals of $\mathfrak{L}$ such that $\mathfrak{J}_1 \cap \mathfrak{L}_0 \neq \{0\}$ and $\mathfrak{J}_2 \neq \{0\}$, show that $[\mathfrak{J}_1, \mathfrak{J}_2] \neq \{0\}$ (use (b)). Hence show that the only commutative G-stable ideal of $\mathfrak{L}$ is $\{0\}$.

(d) Show that $\mathfrak{L}$ is a *simple* Lie algebra. (Use (c) by considering the derived algebras $\mathfrak{D}^k(\mathfrak{R})$, where $\mathfrak{R}$ is the radical of $\mathfrak{L}$; then observe that if $\mathfrak{L}$ is semisimple, every connected Lie subgroup of $\mathrm{Aut}(\mathfrak{L})$ leaves stable the simple components of $\mathfrak{L}$.)

(e) Show that $\mathfrak{L}_2 = \{0\}$. (Prove that $\mathfrak{L}_2$ is orthogonal to $\mathfrak{L}$ relative to the Killing form.)

(f) Put $\mathfrak{H}_{-1} = \mathfrak{L}_{-1}/\mathfrak{L}_0$, $\mathfrak{H}_0 = \mathfrak{L}_0/\mathfrak{L}_1$, $\mathfrak{H}_1 = \mathfrak{L}_1$, $\mathfrak{H} = \mathfrak{H}_{-1} \oplus \mathfrak{H}_0 \oplus \mathfrak{H}_1$. Show that there exists a unique structure of Lie algebra on $\mathfrak{H}$ such that $[\mathfrak{H}_p, \mathfrak{H}_q] \subset \mathfrak{H}_{p+q}$ (with $\mathfrak{H}_{-2} = \mathfrak{H}_2 = \{0\}$) and such that if $\mathbf{x} \in \mathfrak{L}_p$, $\mathbf{y} \in \mathfrak{L}_q$ and if $\bar{\mathbf{x}} \in \mathfrak{H}_p$, $\bar{\mathbf{y}} \in \mathfrak{H}_q$ are the classes of $\mathbf{x}$ and $\mathbf{y}$, then $[\bar{\mathbf{x}}, \bar{\mathbf{y}}]$ is the class of $[\mathbf{x}, \mathbf{y}]$. Show that the Lie algebra $\mathfrak{H}$ and its vector subspaces $\mathfrak{H}'_p = \mathfrak{H}_p + \mathfrak{H}_{p+1} + \cdots$ for $p \geqq -1$ satisfy the same conditions as $\mathfrak{L}$ and the $\mathfrak{L}_p$, and hence that $\mathfrak{H}$ is a simple Lie algebra.

(g) Show that there exists a unique element $\bar{\mathbf{e}}$ in H such that $\mathrm{ad}(\bar{\mathbf{e}})$ leaves stable $\mathfrak{H}_{-1}$, $\mathfrak{H}_0$, and $\mathfrak{H}_1$, and such that its restriction to $\mathfrak{H}_p$ ($p = -1, 0, 1$) is the homothety with ratio $p$ (cf. Section **21.19**, Problem 2). Deduce that $\mathfrak{L}$ is isomorphic to $\mathfrak{H}$.

(h) Show that, for the Killing form B of $\mathfrak{H}$, the restriction of B to $\mathfrak{H}_0$ is nondegenerate, $\mathfrak{H}_{-1}$ and $\mathfrak{H}_1$ are totally isotropic, and $\mathfrak{H}_{-1} \oplus \mathfrak{H}_1$ is orthogonal to $\mathfrak{H}_0$.

(i) Show that there exists a Cartan decomposition $\mathfrak{H} = \mathfrak{k} \oplus \mathfrak{p}$ such that $\bar{\mathbf{e}} \in \mathfrak{p}$. (Consider the involutory automorphism $\sigma$ of $\mathfrak{H}$ such that $\sigma(\mathbf{x}) = \mathbf{x}$ for $\mathbf{x} \in \mathfrak{H}_0$, and $\sigma(\mathbf{x}) = -\mathbf{x}$ for $\mathbf{x} \in \mathfrak{H}_{-1} \oplus \mathfrak{H}_1$.) Conversely, give the description of $\mathfrak{L}$ and the $\mathfrak{L}_p$ starting from a Cartan decomposition $\mathfrak{H} = \mathfrak{k} \oplus \mathfrak{p}$ and an element $\bar{\mathbf{e}} \in \mathfrak{p}$ such that the eigenvalues of $\mathrm{ad}(\bar{\mathbf{e}})$ are $-1$, 0, and 1.

**19.** With the notation and hypotheses of Section **19.3**, Problem 5(d), suppose that the kernel of the linear representation $\rho$ of H on $T_{x_0}(M)$ is not discrete, and that $\rho$ is *irreducible*; suppose moreover that H contains no nontrivial normal subgroups of G, and that the center of G is finite. Then G is a noncompact simple group, and there exists a maximal compact connected subgroup K of G that acts transitively on M, so that M may be identified with $(K \cap H)\backslash K$ and is a Riemannian symmetric space. (Use Problem 18.)

**APPENDIX**

# MODULES

(The numbering of the sections in this Appendix continues that of the Appendix to Volume IV.)

## 22. SIMPLE MODULES

(A.22.1) The notion of a module over a commutative ring (A.8.1) may be generalized. If M is a *commutative* group, written *additively*, an *action* of a set $\Omega$ on M is any mapping $(\alpha, x) \mapsto \alpha \cdot x$ of $\Omega \times M$ into M such that $\alpha \cdot (x + y) = \alpha \cdot x + \alpha \cdot y$; in other words, for each $\alpha \in \Omega$ the mapping $x \mapsto \alpha \cdot x$ is an endomorphism of the group M. By abuse of language, the group M together with an action of $\Omega$ on M is called an *$\Omega$-module.* A *homomorphism* of an $\Omega$-module M into an $\Omega$-module N is any mapping $f: M \to N$ such that $f(x + y) = f(x) + f(y)$ and $f(\alpha \cdot x) = \alpha \cdot f(x)$ for all $x, y \in M$ and $\alpha \in \Omega$. An *isomorphism* of $\Omega$-modules is a bijective homomorphism; the inverse mapping is then also an isomorphism.

(A.22.2) If M is an $\Omega$-module, a subgroup N of M is said to be *stable* for the action of $\Omega$ (or *$\Omega$-stable*) if for all $x \in N$ and $\alpha \in \Omega$ we have $\alpha \cdot x \in N$; the subgroup N is also said to be an *$\Omega$-submodule* of M. Intersections and sums of $\Omega$-submodules of M are again $\Omega$-submodules. If $f$ is any homomorphism of M into an $\Omega$-module M′, and if N (resp. N′) is any $\Omega$-submodule of M (resp. M′), then $f(N)$ is an $\Omega$-submodule of M′, and $f^{-1}(N')$ is an $\Omega$-submodule of M. In particular, the *image* $f(M)$ is an $\Omega$-submodule of M′, and the *kernel* $f^{-1}(0)$ is an $\Omega$-submodule of M.

(A.22.3) In any $\Omega$-module M, $\{0\}$ and M are always $\Omega$-submodules, called the *trivial* submodules. An $\Omega$-module M is called *simple* if $M \neq \{0\}$ and there exist no $\Omega$-submodules of M except for the trivial ones.

(A.22.4) (Schur's lemma) *Let f be a homomorphism of an $\Omega$-module M into an $\Omega$-module N. If M is simple, then f is either injective or identically zero. If N is simple, then f is either surjective or identically zero. If both M and N are simple, then f is either bijective or identically zero.*

For if M is simple, $f^{-1}(0)$ can only be M or $\{0\}$; and if N is simple, $f(M)$ can only be N or $\{0\}$.

## 23. SEMISIMPLE MODULES

(A.23.1) The notion of *direct sum* of $\Omega$-modules is defined as in (A.1.5). If $M = \bigoplus_{\lambda \in L} M_\lambda$ is the direct sum of a family $(M_\lambda)$ of $\Omega$-modules, we define as in (A.2.3) the canonical injection $j_\lambda: M_\lambda \to M$ and the canonical projection $p_\lambda: M \to M_\lambda$ for each index $\lambda$; they are $\Omega$-module homomorphisms. All the results of (A.3.1)–(A.3.5) remain valid without modification if we replace "subspace" by "$\Omega$-submodule" and "linear mapping" by "homomorphism."

(A.23.2) An $\Omega$-module M is said to be *semisimple* if it is a *direct sum of a family of simple $\Omega$-modules.* We shall limit our attention to semisimple $\Omega$-modules that are direct sums of *at most denumerable* families of simple $\Omega$-modules.

(A.23.3) *Let* M *be an $\Omega$-module that is the sum (not necessarily direct) of a finite or infinite sequence $(N_k)_{0 \leq k < \omega}$ (where $\omega$ is an integer or $+\infty$) of simple $\Omega$-submodules. Let* E *be an $\Omega$-submodule of* M. *Then:*

(a) *There exists a subset* J *of the set* $[0, \omega[$ *such that* M *is the direct sum of* E *and the* $N_k$ *with* $k \in J$ *(so that* E *has as a supplement in* M *the semisimple $\Omega$-submodule* F *which is the direct sum of the* $N_k$ *with* $k \notin J$*).*

(b) *There exists a subset* H *of the set* $[0, \omega[$ *such that* $J \cap H = \varnothing$ *and such that* M *is the direct sum of the* $N_k$ *with* $k \in J \cup H$ *(and therefore* M *is semisimple);* E *is isomorphic to the direct sum of the* $N_k$ *with* $k \in H$.

(a) We shall define J to be the set of elements of a (finite or infinite) sequence $(k_m)$ that is constructed inductively as follows: $k_m$ is the smallest integer (if it exists) such that $N_{k_m}$ is not contained in the sum

$$E + N_{k_1} + \cdots + N_{k_{m-1}}$$

(when $m = 1$, this sum is replaced by E). The construction stops if $E + N_{k_1} + \cdots + N_{k_{m-1}}$ contains all the $N_k$ and hence is equal to M. If $k_m$ is defined, the intersection $N_{k_m} \cap (E + N_{k_1} + \cdots + N_{k_{m-1}})$, being a submodule of $N_{k_m}$ distinct from $N_{k_m}$, must be zero; hence (A.3.3) the sum of E and the $N_k$ such that $k \in J$ is direct. It remains to be shown that, when J is infinite, this sum M′ is equal to M. If not, there would exist at least one index $h \notin J$ such that $N_h \not\subset M'$; but if $m$ is the smallest integer such that $k_m > h$, then $N_h$ is not contained in $E + N_{k_1} + \cdots + N_{k_{m-1}}$, contrary to the definition of $k_m$.

(b) The set H is defined by applying (a) to the Ω-submodule F of M. The isomorphism of E with the direct sum of the $N_k$ such that $k \in H$ then follows from (A.3.5).

(A.23.4) A semisimple Ω-module M is said to be *isotypic* if it is a direct sum of *isomorphic* simple Ω-submodules. It follows from (A.23.3) that any two simple Ω-submodules of M are *isomorphic* (since a simple Ω-module cannot be isomorphic to a direct sum of two nonzero submodules). Two isotypic semisimple Ω-modules are said to be *of the same type* if every simple submodule of one is isomorphic to every simple submodule of the other. It follows from (A.23.3) that every submodule of an isotypic semisimple Ω-module is isotypic semisimple and of the same type.

(A.23.5) Let M be a semisimple Ω-module, the direct sum of a (finite or infinite) sequence $(N_k)_{0 \le k < \omega}$ of simple Ω-submodules of M. We define by induction a sequence of submodules $N'_k$ of M as follows. We take $N'_0 = N_0$; $N'_{k+1}$ is equal to $N_m$ for the smallest index $m$ such that $N_m$ is not isomorphic to any of the Ω-modules $N'_0, \ldots, N'_k$. (If all the $N_h$ are isomorphic to one or other of these Ω-modules, the induction stops at $N'_k$.) Let J be the set of indices $k \in [0, \omega[$ so obtained, and for each $k \in J$ let $I_k$ be the set of integers $m$ such that $N_m$ is isomorphic to $N'_k$. If $P_k$ is the direct sum of the Ω-submodules $N_m$ for $m \in I_k$, then it is clear that $P_k$ is an *isotypic* semisimple Ω-module, that $P_h$ and $P_k$ are not of the same type if $h \neq k$, and that M is the direct sum of the $P_k$ for $k \in J$.

(A.23.6) *Every simple Ω-submodule* N *of* M *is contained in one of the* $P_k$ *(and hence is of the same type as* $P_k$*).*

It follows from (A.23.3) that N must be isomorphic to one of the $N_m$, hence to one of the $N'_k$; if N were not contained in $P_k$, we should have $N \cap P_k = \{0\}$ because N is simple, and the projection N′ of N on the direct sum $M_k$ of the $P_h$ such that $h \neq k$ would be isomorphic to N (A.22.4); but by reason of (A.23.3) applied to $M_k$, N′ would be isomorphic to one of the modules $N'_h$ with $h \neq k$, which is absurd.

We may therefore define the $P_k$ independently of any decomposition of M as a direct sum of simple $\Omega$-submodules: $P_k$ is the sum of *all* the simple $\Omega$-submodules of M that are isomorphic to $N'_k$. The $P_k$ are called the *isotypic components* of M. The result of (A.23.6) then generalizes as follows:

(A.23.7) *Every* $\Omega$*-submodule* N *of* M *is the direct sum of the* $N_k = M \cap P_k$ *that are not equal to* 0, *and these are the isotypic components of* N.

This is an immediate consequence of (A.23.3) and (A.23.6), because N is a direct sum of simple $\Omega$-submodules of M.

## 24. EXAMPLES

(A.24.1) Let K be a (commutative) field and E a vector space over K. It is clear that E is a simple K-module if and only if E is *one-dimensional* (i.e., a "line"). Every K-vector space that is spanned by an at most denumerable family of vectors is therefore an isotypic semisimple K-module (there is only one "type"), and the results of (A.23) therefore include as particular cases the propositions (A.4.6) and (A.4.5) for vector spaces that admit an at most denumerable basis.

(A.24.2) Now let $(u_\lambda)_{\lambda \in L}$ be a set of *endomorphisms* of the K-vector space E (A.2.1) and let $\Omega$ be the sum (1.8) of the sets L and K. If we define $\lambda \cdot x = u_\lambda(x)$, then $\Omega$ acts on E. It is clear that the notions of *linear representation* introduced in (15.5) and (21.1) are particular cases of this general notion, and correspond to taking L to be an algebra or a group. The notion of *equivalent* linear representations corresponds to that of *isomorphism* of $\Omega$-modules (subjected to supplementary conditions when E is infinite-dimensional and endowed with a topology); the notion of a finite Hilbert sum of representations corresponds to a particular case of the notion of direct sum of $\Omega$-modules. Finally, if E is finite-dimensional, it comes to the same thing to say that a representation is *irreducible* or that the corresponding $\Omega$-module is *simple*.

(A.24.3) Consider in particular the case of a *single* endomorphism $u$ of the vector space E; it is said to be *semisimple* when the corresponding $\Omega$-module E is semisimple. The nature of the simple $\Omega$-modules will depend on the field

K. For example, if $K = \mathbf{R}$ and E is a plane, a rotation $u$ (relative to the usual scalar product) other than $\pm 1$ will make E a simple $\Omega$-module, and it is easy to give examples of vector spaces E of any given finite dimension $n$ over a suitable field K that are simple $\Omega$-modules for certain endomorphisms.

The most important case in analysis is that in which K is *algebraically closed* and E is finite-dimensional. The simple $\Omega$-modules corresponding to an endomorphism $u$ are then the "lines" $Kx$, where $x$ is an *eigenvector* of $u$ **(A.6.9)**, because the restriction of $u$ to a subspace F of E that is stable under $u$ always admits at least one eigenvector in F. Hence, in this case, to say that $u$ is *semisimple* means that there exists a basis $(e_j)_{1 \leq j \leq n}$ of E consisting of eigenvectors for $u$, or equivalently that the matrix of $u$ with respect to this basis is diagonal; for this reason the endomorphism $u$ is also said to be *diagonalizable.* If $\lambda_1, \ldots, \lambda_r$ are the distinct eigenvalues of $u$, the *isotypic components* of E (for $u$) are the *eigenspaces* $E(\lambda_j; u)$ for $1 \leqq j \leqq r$ **(11.1)**. The description of the vector subspaces of E that are *stable* under $u$ can be read off immediately from **(A.23.7)**: they are the direct sums $F_1 \oplus F_2 \oplus \cdots \oplus F_r$, where $F_j$ is any vector subspace of $E(\lambda_j; u)$ for $1 \leqq j \leqq r$.

**(A.24.4)** The example of a semisimple $\Omega$-module that comes up in the theory of Lie groups is that in which K is an algebraically closed field, E a K-vector space with an at most denumerable basis, and $\Omega$ the sum (or disjoint union) of K and a K-*vector space* L. Suppose that E is the *sum* of a finite or infinite sequence $(E_n)_{0 \leq n < \omega}$ (where $\omega$ is an integer or $+\infty$) of subspaces having the following property: for each $n$ such that $0 \leqq n < \omega$ there exists a linear form $p_n\colon L \to K$ on L such that $u \cdot x = p_n(u)x$ for all $u \in L$ and all $x \in E_n$, the forms $p_n$ being all distinct. It is then clear that every "line" $Kx$ contained in some $E_n$ is a *simple* $\Omega$-module, that two lines $Kx$, $Ky$ contained in the same $E_n$ are isomorphic $\Omega$-modules, and that if $Kx \subset E_n$, $Ky \subset E_m$ with $m \neq n$, then $Kx$ and $Ky$ are not isomorphic as $\Omega$-modules. It follows immediately from **(A.23.3)** and **(A.23.6)** that in fact E is the *direct sum* of all the $E_n$, which are the *isotypic components* of the semisimple $\Omega$-module E. Moreover, every simple $\Omega$-submodule of E is a line $Kx$ contained in one of the $E_n$, and more generally every $\Omega$-submodule F of E is the direct sum of the $F \cap E_n$ **(A.23.7)**.

## 25. THE CANONICAL DECOMPOSITION OF AN ENDOMORPHISM

**(A.25.1)** Part of the proof of **(11.3.3)** for a compact operator on a normed space can be adapted to the case of an arbitrary endomorphism $u$ of a *finite*-dimensional vector space E over an *algebraically closed* field K, and gives a more precise result than **(A.6.10)**. Namely,

**(A.25.2)** *If $\lambda_1, \ldots, \lambda_r$ are the distinct eigenvalues of $u$, the space $E$ is the direct sum of $r$ subspaces $E_1, \ldots, E_r$ each stable under $u$, such that the restriction of $u$ to $E_j$ has only the eigenvalue $\lambda_j$. Moreover, if $v$ is the diagonalizable endomorphism of $E$ such that $v(x) = \lambda_j x$ for $x \in E_j$ $(1 \leqq j \leqq r)$, then $w = u - v$ is nilpotent, and both $u$ and $w$ may be written as polynomials in $u$ with zero constant terms, say $v = b_1 u + \cdots + b_p u^p$, $w = c_1 u + \cdots + c_q u^q$, with coefficients in $K$ (and hence $vw = wv$).*

We proceed by induction on the dimension of E and, by replacing $u$ by $u - \lambda_1 \cdot 1_E$, we may assume that $\lambda_1 = 0$. We then form as in **(11.3.3)** the sequences of subspaces $N_1 = u^{-1}(0)$, $N_k = u^{-1}(N_{k-1})$ for $k > 1$, $F_1 = u(E)$, $F_k = u(F_{k-1})$ for $k > 1$. The dimensions of the $N_k$ (resp. $F_k$) form an increasing (resp. decreasing) sequence, and it is clear that there is a smallest integer $n$ such that $N_{k+1} = N_k$ for $k \geqq n$, and a smallest integer $m$ such that $F_{k+1} = F_k$ for $k \geqq m$. We have $N_n \cap F_n = \{0\}$, because if $y \in F_n \cap N_n$ there exists $x \in E$ such that $y = u^n(x)$, and on the other hand $u^n(y) = 0$, so that $u^{2n}(x) = 0$; which implies that $x \in N_{2n} = N_n$ and hence that $y = u^n(x) = 0$. Next, we have $F_m \subset F_n$, and indeed $F_m = F_n$. For otherwise we should have $m > n$; let $z \in F_{m-1} \subset F_n$ be such that $z \notin F_m$; since $u(z) \in F_m = u(F_m)$, there exists $t \in F_m$ such that $u(z) = u(t)$, hence $z - t \in N_1 \subset N_n$; but since $z - t \in F_n$, it follows that $z = t$, which contradicts the choice of $z$. Finally, for each $x \in E$, we have $u^n(x) \in F_n = F_m$, and since $u^n(F_n) = F_n$ by the definition of $m$, there exists $y \in F_n$ such that $u^n(x) = u^n(y)$, hence $x - y \in N_n$, and consequently $E = F_n + N_n$, the sum being direct because $F_n \cap N_n = \{0\}$. The restriction of $u$ to $F_n$ is surjective, with kernel

$$F_n \cap N_1 \subset F_n \cap N_n = \{0\},$$

hence is bijective; and the restriction of $u$ to $N_n$ is nilpotent by definition. We now apply the inductive hypothesis to $u|F_n$ in order to obtain the decomposition $u = v + w$, where $v$ is diagonalizable and $w$ nilpotent; the restriction of $v$ to each $E_j$ is scalar multiplication by $\lambda_j$, hence commutes with every endomorphism of $E_j$, and in particular with $w|E_j$, from which it follows that $vw = wv$.

It remains to prove the last assertion. We distinguish three cases:

(1) $r = 1$, so that $u$ has only one eigenvalue $\lambda$. If $\lambda = 0$, then $v = 0$ and $w = u$. If $\lambda \neq 0$, then $w^n = 0$, where $n = \dim(E)$, and $v = \lambda \cdot 1_E$, so that $(u - \lambda \cdot 1_E)^n = 0$ and therefore

$$v = \binom{n}{1} u - \binom{n}{2} \lambda^{-1} u^2 + \cdots + (-1)^{n-1} \lambda^{-n+1} u^n$$

and $w = u - v$, which proves the result in this case.

(2) Let $u_j = u|E_j$ and suppose that $u_h = 0$ for all $h \neq j$. We can then

apply the result of (1) to $E_j$ and $u_j$. If $u_j = P(u_j) + Q(u_j)$, where P and Q are polynomials with zero constant terms, $P(u_j)$ being diagonalizable and $Q(u_j)$ nilpotent, it is clear that $v = P(u)$ and $w = Q(u)$.

(3) General case: suppose first that $\lambda_j \neq 0$. Put

$$f_j = u \prod_{h \neq j} (u - \lambda_h \cdot 1_E)^{n_h}$$

where $n_h = \dim(E_h)$; it is clear that the restriction $f_j | E_h$ is 0 for $h \neq j$, and $f_j | E_j$ has only one eigenvalue, namely $\mu_j = \lambda_j \prod_{h \neq j} (\lambda_j - \lambda_h)^{n_h} \neq 0$. By case (2), there exists a polynomial $R_j$ with zero constant term such that $R_j(f_j) | E_j = 1_{E_j}$ and $R_j(f_j) | E_h = 0$ if $h \neq j$; but by definition $f_j = S_j(u)$, where $S_j$ is a polynomial with zero constant term; hence $P_j(u) = R_j(S_j(u))$ is a polynomial in $u$ with zero constant term, such that $P_j(u) | E_j = 1_{E_j}$ and $P_j(u) | E_h = 0$ for $h \neq j$. If $\lambda_j = 0$, then we take $P_j(u) = 0$. We have then $v = \sum_{j=1}^{r} \lambda_j P_j(u)$, and $w = u - v$, and the proof is complete.

**(A.25.3)** *There exists only one decomposition $u = f + g$ of an endomorphism $u$ of* E *such that $f$ is diagonalizable, $g$ nilpotent, and such that $fg = gf$.*

Let $\mu_k$ $(1 \leqq k \leqq s)$ be the distinct eigenvalues of $f$. Then E is the direct sum of the eigenspaces $L_k = E(\mu_k; f)$ **(A.24.3)**. Let us first show that $g(L_k) \subset L_k$ for all $k$. Indeed, let $x \in L_k$; we may write $g(x) = \sum_{h=1}^{s} z_h$, with $z_h \in L_h$ for $1 \leqq h \leqq s$, whence $f(g(x)) = \sum_{h=1}^{s} \mu_h z_h$; on the other hand, $f(x) = \mu_k x$, so that $g(f(x)) = \sum_{h=1}^{s} \mu_k z_h$, and the relation $f(g(x)) = g(f(x))$ therefore takes the form $\sum_{h=1}^{s} (\mu_k - \mu_h) z_h = 0$, so that we have $z_h = 0$ for all $h \neq k$, because the $\mu_h$ are all distinct. We may therefore **(A.6.12)** take a basis of E that is the union of bases of the $L_k$, such that the restriction of $u$ to each $L_k$ is represented by a lower triangular matrix with diagonal elements all equal to $\mu_k$. Consequently the $\mu_k$ are the eigenvalues of $u$, and $L_k$ is the subspace $N(\mu_k)$ consisting of vectors $x \in E$ such that $(u - \mu_k \cdot 1_E)^m(x) = 0$ for large enough $m$. This proves that $f = v$ and $g = w$.

## 26. FINITELY GENERATED Z-MODULES

**(A.26.1)** A **Z**-module M **(A.8.1)** that admits a system of generators $(a_j)_{1 \leq j \leq n}$ consisting of $n$ elements is isomorphic to a quotient $\mathbf{Z}^n/N$ of a finitely generated free **Z**-module. For if $(u_1, \ldots, u_n)$ is the canonical basis of

$\mathbf{Z}^n$, we may define a surjective homomorphism $f: \mathbf{Z}^n \to \mathrm{M}$ by putting $f(u_j) = a_j$ for $1 \leqq j \leqq n$. The study of finitely generated $\mathbf{Z}$-modules is therefore reduced to that of the submodules and quotient modules of a finitely generated *free* $\mathbf{Z}$-module.

(A.26.2) *Every submodule of a finitely generated free* $\mathbf{Z}$*-module is finitely generated and free.*

Let L be a $\mathbf{Z}$-module having a basis $(a_1, \dots, a_n)$ and let M be a submodule of L. Denote by $\mathrm{L}_j$ $(1 \leqq j \leqq n)$ the submodule of L with basis $a_1, \dots, a_j$, and put $\mathrm{M}_j = \mathrm{M} \cap \mathrm{L}_j$. If $(a_1^*, \dots, a_n^*)$ is the basis dual to $(a_1, \dots, a_n)$ in the dual $\mathbf{Z}$-module $\mathrm{L}^*$ (A.9.2), then the set of integers $\langle x, a_j^* \rangle$, where $x \in \mathrm{M}_j$, is evidently an ideal in $\mathbf{Z}$, hence is of the form $m_j \mathbf{Z}$, where $m_j$ is an integer $\geqq 0$. Hence there exists an element $b_j \in \mathrm{M}_j$ such that $\langle b_j, a_j^* \rangle = m_j$; if $m_j = 0$, we take $b_j = 0$. Let $\mathrm{M}'_j$ be the submodule of M generated by $b_1, \dots, b_j$; we shall show that $\mathrm{M}'_j = \mathrm{M}_j$. This is obvious when $j = 1$, since $\mathrm{M}_1$ is the set of multiples $p \cdot a_1$ of $a_1$ that belong to M, and $\langle pa_1, a_1^* \rangle = p$. We proceed by induction on $j$. If $x \in \mathrm{M}_j$, we have $\langle x, a_j^* \rangle = pm_j$ for some $p \in \mathbf{Z}$, hence $\langle x - pb_j, a_j^* \rangle = 0$, and since by hypothesis $x$ and $b_j$ are linear combinations of $a_1, \dots, a_j$, it follows that $x - pb_j \in \mathrm{M}_{j-1}$, which by the inductive hypothesis is equal to $\mathrm{M}'_{j-1}$; hence $x \in \mathrm{M}'_j$ as required. Taking $j = n$, we see that M is generated by $b_1, \dots, b_n$, and it remains to show that the nonzero $b_j$ form a free system. Suppose then that we have a relation $\sum_{j=1}^{n} k_j b_j = 0$, where the $k_j b_j$ are not all zero. If $h$ is the largest of the indices $j$ such that $k_j b_j \neq 0$, we have $\langle k_h b_h, a_h^* \rangle = \langle \sum_j k_j b_j, a_h^* \rangle = 0$, because $k_j b_j = 0$ for $j > h$, and $k_j b_j$ is a linear combination of $a_1, \dots, a_j$ for $j < h$. Hence we have $k_h m_h = 0$, contradicting $k_h b_h \neq 0$.

(A.26.3) *Let* L *be a finitely generated free* $\mathbf{Z}$*-module and* M *a submodule of* L. *Then there exists a basis* $(e_1, \dots, e_n)$ *of* L *and* $r \leqq n$ *integers* $\alpha_1, \dots, \alpha_r$ *which are* $> 0$ *such that* $\alpha_j$ *divides* $\alpha_{j+1}$ *for* $1 \leqq j \leqq r - 1$, *and such that* $\alpha_1 e_1, \dots, \alpha_r e_r$ *form a basis of* M. *Furthermore, the numbers* $r$, $n$ *and* $\alpha_j$ $(1 \leqq j \leqq r)$ *are uniquely determined by these properties.*

Let $(a_1, \dots, a_n)$ be a basis of L and $(a_1^*, \dots, a_n^*)$ the dual basis of $\mathrm{L}^*$. We may assume that $\mathrm{M} \neq \{0\}$. Consider the integers $\langle x, y^* \rangle$ for $x \in \mathrm{M}$ and $y^* \in \mathrm{L}^*$; by hypothesis, they are not all 0, and since $\langle -x, y^* \rangle = -\langle x, y^* \rangle$, there exists $x_1 \in \mathrm{M}$ and $y_1^* \in \mathrm{L}^*$ such that $\langle x_1, y_1^* \rangle = \alpha_1$ is the *smallest* of the nonzero integers $|\langle x, y^* \rangle|$ for $x \in \mathrm{M}$ and $y \in \mathrm{L}^*$. We deduce first that for each $x \in M$, the integer $\beta = \langle x, y_1^* \rangle$ is a *multiple of* $\alpha_1$; for otherwise the highest common factor $\delta$ of $\beta$ and $\alpha_1$ would be such that $0 < \delta < \alpha_1$, and we should have $\delta = \lambda\beta + \mu\alpha$, by Bézout's identity, where $\lambda$ and $\mu$ are suitable integers; but this would imply $\langle \lambda x + \mu x_1, y_1^* \rangle = \delta$, contradicting the definition of $\alpha_1$. One proves in the same way that $\langle x_1, y^* \rangle$ is a *multiple of* $\alpha_1$ for each $y^* \in L^*$. In particular, all the integers $\langle x_1, a_j^* \rangle$ are multiples of $\alpha_1$, and hence there exists $e_1 \in L$ such that $x_1 = \alpha_1 e_1$. Let $L_1 = \text{Ker}(y_1^*)$; we shall show that L is the *direct sum* of $\mathbf{Z}e_1$ and $L_1$. We have $\langle e_1, y_1^* \rangle = 1$ by definition; hence, for any $y \in L$, if $\langle y, y_1^* \rangle = \gamma$, we have $\langle y - \gamma e_1, y_1^* \rangle = 0$, that is to say, $y - \gamma e_1 \in L_1$. Also, we cannot have $\lambda e_1 \in L_1$ unless $\lambda = 0$, by the definition of $L_1$, and this establishes our assertion. Likewise, M is the *direct sum* of $\mathbf{Z}\alpha_1 e_1$ and $M_1 = M \cap L_1$. Namely, for each $x \in M$ we have $\langle x, y_1^* \rangle = \mu\alpha_1$ for some $\mu \in \mathbf{Z}$, hence $\langle x - \mu\alpha_1 e_1, y_1^* \rangle = 0$, that is to say, $x - \mu\alpha_1 e_1 \in L_1 \cap M = M_1$.

By virtue of **(A.26.2)**, $L_1$ admits a basis, and from the previous paragraph and the invariance of the number of elements in a basis of L **(A.8.3)**, any basis of $L_1$ must have $n - 1$ elements. By induction on $n$, we may assume that there is a basis $(e_2, \dots, e_n)$ of $L_1$ and $r - 1 \leqq n - 1$ integers $\alpha_2, \dots, \alpha_r$ such that $\alpha_j$ divides $\alpha_{j+1}$ for $2 \leqq j \leqq r - 1$ and such that $\alpha_2 e_2, \dots, \alpha_r e_r$ form a basis of $M_1$. It remains therefore to prove that $\alpha_1$ divides $\alpha_2$. If $(e_1^*, \dots, e_n^*)$ is the basis of $L^*$ dual to $(e_1, \dots, e_n)$, we have $\langle \alpha_1 e_1, e_1^* \rangle = \alpha_1$ and $\langle \alpha_2 e_2, e_2^* \rangle = \alpha_2$. If $\alpha_2$ were not a multiple of $\alpha_1$, there would exist $\lambda, \mu \in \mathbf{Z}$ such that $\delta = \lambda\alpha_1 + \mu\alpha_2$ satisfied $0 < \delta < \alpha_1$, and since

$$\langle \alpha_1 e_1 + \alpha_2 e_2, \lambda e_1^* + \mu e_2^* \rangle = \delta,$$

this would contradict the definition of $\alpha_1$.

It is clear that the quotient $\mathbf{Z}$-module L/M is isomorphic to the direct sum of $\mathbf{Z}^{n-r}$ and $r$ cyclic groups $\mathbf{Z}/\alpha_j\mathbf{Z}$ $(1 \leqq j \leqq r)$; the integers $\lambda$ such that $\lambda(L/M)$ is free are therefore exactly the multiples of $\alpha_r$, and for these integers $\lambda(L/M)$ is isomorphic to $\mathbf{Z}^{n-r}$. This already shows **(A.8.3)** that the integers $n$ and $r$ are well determined, as is the submodule T of L/M consisting of the elements of finite order in this group. Observe next that if $\mathbf{Z}/m\mathbf{Z}$ is a cyclic group and $p$ a prime number, we have $p^k(\mathbf{Z}/m\mathbf{Z}) = \mathbf{Z}/m\mathbf{Z}$ if $p$ does not divide $m$; whereas if $m = p^h m'$, where $p$ does not divide $m'$, we have $p^k(\mathbf{Z}/m\mathbf{Z}) = \mathbf{Z}/(p^{h-k}m')\mathbf{Z}$ if $k < h$, and $p^k(\mathbf{Z}/m\mathbf{Z}) = \mathbf{Z}/m'\mathbf{Z}$ if $k \geqq h$. If $p_1, \dots, p_h$ are the prime numbers dividing $\alpha_r$, it follows that the orders of the groups $p_j^k T$ determine the exponents of the $p_j$ in the $\alpha_i$: the order of $p_j^k T$ is the product of the order of $p_j^{k+1}T$ by $p_j^v$, where $v$ is the number of $\alpha_i$ divisible by $p_j^{k+1}$. This shows that the $\alpha_i$ are well determined.

The numbers $\alpha_i$ are called the *invariant factors* of M with respect to L.

We have also established:

(A.26.4) *Every finitely generated* **Z**-*module* N *is isomorphic to the product of a free* **Z**-*module* $\mathbf{Z}^s$ *and r cyclic groups* $\mathbf{Z}/\alpha_j\mathbf{Z}$ $(1 \leqq j \leqq r)$ *such that* $\alpha_j$ *divides* $\alpha_{j+1}$ *for* $1 \leqq j \leqq r-1$; *and the numbers s, r, and* $\alpha_j$ $(1 \leqq j \leqq r)$ *are uniquely determined by these properties.*

(A.26.5) Keeping the notation of (A.26.3), suppose in addition that $r = n$, so that L/M is a *finite* group. If we put $f_j = \alpha_j e_j$, the $f_j$ $(1 \leqq j \leqq n)$ form a basis of M, and the matrix of the canonical injection $u$: M → L relative to the bases $(f_j)$ and $(e_j)$ (A.5.2) is the diagonal matrix $\operatorname{diag}(\alpha_1, \ldots, \alpha_n)$. This is therefore also the matrix of the transpose ${}^t u$: L* → M* relative to the dual bases $(e_j^*)$ and $(f_j^*)$ (A.9.4). Hence L* may be canonically identified with the submodule of M* having as basis the $\alpha_j f_j^*$, and M*/L* is *isomorphic to* L/M.

(A.26.6) *In order that a submodule* M *of a finitely generated free* **Z**-*module* L *should admit a supplement in* L, *it is necessary and sufficient that the invariant factors of* M *with respect to* L *should all be equal to* 1.

The condition is clearly sufficient by virtue of (A.26.3): the $e_j$ such that $r+1 \leqq j \leqq n$ form a basis of a supplement of M. Conversely, if M admits a supplement N, then L/M is isomorphic to N and hence is a free **Z**-module (A.26.2); this implies that all the cyclic modules $\mathbf{Z}/\alpha_j\mathbf{Z}$ must be trivial, hence $\alpha_j = 1$ for $1 \leqq j \leqq r$.

# REFERENCES

## VOLUME I


[1] Ahlfors, L., "Complex Analysis," McGraw-Hill, New York, 1953.

[2] Bachmann, H., "Transfinite Zahlen" (Ergebnisse der Math., Neue Folge, Heft 1). Springer, Berlin, 1955.

[3] Bourbaki, N., "Eléments de Mathématique," Livre I, "Théorie des ensembles" (Actual. Scient. Ind., Chaps. I, II, No. 1212; Chap. III, No. 1243). Hermann, Paris, 1954–1956.

[4] Bourbaki, N., "Eléments de Mathématique," Livre II, "Algèbre" (Actual. Scient. Ind., Chap. II, Nos. 1032, 1236, 3rd ed.). Hermann, Paris, 1962.

[5] Bourbaki, N., "Eléments de Mathématique," Livre III, "Topologie générale" (Actual. Scient. Ind., Chaps. I, II, Nos. 858, 1142, 4th ed.; Chap. IX, No. 1045, 2nd ed.; Chap. X, No. 1084, 2nd ed.). Hermann, Paris, 1958–1961.

[6] Bourbaki, N., "Eléments de Mathématique," Livre, V, "Espaces vectoriels topologiques" (Actual. Scient. Ind., Chap. I, II, No. 1189, 2nd ed.; Chaps. III–V, No. 1229). Hermann, Paris, 1953–1955.

[7] Cartan, H., Séminaire de l'Ecole Normale Supérieure, 1951–1952: "Fonctions analytiques et faisceaux analytiques."

[8] Cartan, H., "Théorie Élémentaire des Fonctions Analytiques." Hermann, Paris, 1961.

[9] Coddington, E., and Levinson, N., "Theory of Ordinary Differential Equations." McGraw-Hill, New York, 1955.

[10] Courant, R., and Hilbert, D., "Methoden der mathematischen Physik," Vol. I, 2nd ed. Springer, Berlin, 1931.

[11] Halmos, P., "Finite Dimensional Vector Spaces," 2nd ed. Van Nostrand-Reinhold, Princeton, New Jersey, 1958.

[12] Ince, E., "Ordinary Differential Equations," Dover, New York, 1949.

[13] Jacobson, N., "Lectures in Abstract Algebra," Vol. II, "Linear algebra." Van Nostrand-Reinhold, Princeton, New Jersey, 1953.

[14] Kamke, E., "Differentialgleichungen reeller Funktionen." Akad. Verlag, Leipzig, 1930.

[15] Kelley, J., "General Topology." Van Nostrand-Reinhold, Princeton, New Jersey, 1955.

[16] Landau, E., "Foundations of Analysis." Chelsea, New York, 1951.

[17] Springer, G., "Introduction to Riemann Surfaces." Addison-Wesley, Reading, Massachusetts, 1957.

[18] Weil, A., "Introduction à l'Étude des Variétés Kählériennes" (Actual. Scient. Ind., No. 1267). Hermann, Paris, 1958.

[19] Weyl, H., "Die Idee der Riemannschen Fläche," 3rd ed. Teubner, Stuttgart, 1955.

## VOLUME II

[20] Akhiezer, N., "The Classical Moment Problem." Oliver and Boyd, Edinburgh–London, 1965.
[21] Arnold, V. and Avez, A., "Théorie Ergodique des Systèmes Dynamiques." Gauthier-Villars, Paris, 1967.
[22] Bourbaki, N., "Eléments de Mathématique," Livre VI, "Intégration" (Actual. Scient. Ind., Chap. I–IV, No. 1175, 2nd ed., Chap. V, No. 1244, 2nd ed., Chap. VII–VIII, No. 1306). Hermann, Paris, 1963–67.
[23] Bourbaki, N., "Eléments de Mathématique: Théories Spectrales" (Actual. Scient. Ind., Chap. I, II, No. 1332). Hermann, Paris, 1967.
[24] Dixmier, J., "Les Algèbres d'Opérateurs dans l'Espace Hilbertien." Gauthier-Villars, Paris, 1957.
[25] Dixmier, J., "Les C*-Algèbres et leurs Représentations." Gauthier-Villars, Paris, 1964.
[26] Dunford, N. and Schwartz, J., "Linear Operators. Part II: Spectral Theory." Wiley (Interscience), New York, 1963.
[27] Hadwiger, H., "Vorlesungen über Inhalt, Oberfläche und Isoperimetrie." Springer, Berlin, 1957.
[28] Halmos, P., "Lectures on Ergodic Theory." Math. Soc. of Japan, 1956.
[29] Hoffman, K., "Banach Spaces of Analytic Functions." New York, 1962.
[30] Jacobs, K., "Neuere Methoden und Ergebnisse der Ergodentheorie" (Ergebnisse der Math., Neue Folge, Heft 29). Springer, Berlin, 1960.
[31] Kaczmarz, S. and Steinhaus, H., "Theorie der Orthogonalreihen." New York, 1951.
[32] Kato, T., "Perturbation Theory for Linear Operators." Springer, Berlin, 1966.
[33] Montgomery, D. and Zippin, L., "Topological Transformation Groups." Wiley (Interscience), New York, 1955.
[34] Naimark, M., "Normal Rings." P. Nordhoff, Groningen, 1959.
[35] Rickart, C., "General Theory of Banach Algebras." Van Nostrand-Reinhold, New York, 1960.
[36] Weil, A., "Adeles and Algebraic Groups." The Institute for Advanced Study, Princeton, New Jersey, 1961.

## VOLUME III

[37] Abraham, R., "Foundations of Mechanics." Benjamin, New York, 1967.
[38] Cartan, H., Séminaire de l'École Normale Supérieure, 1949–50: "Homotopie; espaces fibrés."
[39] Chern, S. S., "Complex Manifolds" (Textos de matematica, No. 5). Univ. do Recife, Brazil, 1959.
[40] Gelfand, I. M. and Shilov, G. E., "Les Distributions," Vols. 1 and 2. Dunod, Paris, 1962.
[41] Gunning, R., "Lectures on Riemann Surfaces." Princeton Univ. Press, Princeton, New Jersey, 1966.
[42] Gunning, R., "Lectures on Vector Bundles over Riemann Surfaces." Princeton Univ. Press, Princeton, New Jersey, 1967.
[43] Hu, S. T., "Homotopy Theory." Academic Press, New York, 1969.
[44] Husemoller, D., "Fiber Bundles." McGraw-Hill, New York, 1966.

[45] Kobayashi, S., and Nomizu, K., "Foundations of Differential Geometry," Vols. 1 and 2. Wiley (Interscience), New York, 1963 and 1969.

[46] Lang, S., "Introduction to Differentiable Manifolds." Wiley (Interscience), New York, 1962.

[47] Porteous, I. R., "Topological Geometry." Van Nostrand-Reinhold, Princeton, New Jersey, 1969.

[48] Schwartz, L., "Théorie des Distributions," New ed. Hermann, Paris, 1966.

[49] Steenrod, N., "The Topology of Fiber Bundles." Princeton Univ. Press, Princeton, New Jersey, 1951.

[50] Sternberg, S., "Lectures on Differential Geometry." Prentice-Hall, Englewood Cliffs, New Jersey, 1964.


## VOLUME IV


[51] Abraham, R. and Robbin, J., "Transversal Mappings and Flows." Benjamin, New York, 1967.

[52] Berger, M., "Lectures on Geodesics in Riemannian Geometry." Tata Institute of Fundamental Research, Bombay, 1965.

[53] Carathéodory, C., "Calculus of Variations and Partial Differential Equations of the First Order," Vols. 1 and 2. Holden-Day, San Francisco, 1965.

[54] Cartan, E., "Oeuvres Complètes," Vols. $1_{I}$ to $3_{II}$. Gauthier-Villars, Paris, 1952–1955.

[55] Cartan, E., "Leçons sur la Théorie des Espaces à Connexion Projective." Gauthier-Villars, Paris, 1937.

[56] Cartan, E., "La Théorie des Groupes Finis et Continus et la Géométrie Différentielle traitées par la Méthode du Repère Mobile." Gauthier-Villars, Paris, 1937.

[57] Cartan, E., "Les Systèmes Différentiels Extérieurs et leurs Applications Géométriques." Hermann, Paris, 1945.

[58] Gelfand, I. and Fomin, S., "Calculus of Variations." Prentice Hall, Englewood Cliffs, New Jersey, 1963.

[59] Godbillon, C., "Géométrie Différentielle et Mécanique Analytique." Hermann, Paris, 1969.

[60] Gromoll, D., Klingenberg, W. and Meyer, W., "Riemannsche Geometrie im Grossen," Lecture Notes in Mathematics No. 55. Springer, Berlin, 1968.

[61] Guggenheimer, H., "Differential Geometry." McGraw-Hill, New York, 1963.

[62] Helgason, S., "Differential Geometry and Symmetric Spaces." Academic Press, New York, 1962.

[63] Hermann, R., "Differential Geometry and the Calculus of Variations." Academic Press, New York, 1968.

[64] Hochschild, G., "The Structure of Lie Groups." Holden-Day, San Francisco, 1965.

[65] Klötzler, R., "Mehrdimensionale Variationsrechnung." Birkhäuser, Basel, 1970.

[66] Loos, O., "Symmetric Spaces," Vols. 1 and 2. Benjamin, New York, 1969.

[67] Milnor, J., "Morse Theory," Princeton University Press, Princeton, New Jersey, 1963.

[68] Morrey, C., "Multiple Integrals in the Calculus of Variations." Springer, Berlin, 1966.

[69] Reeb, G., "Sur les Variétés Feuilletées." Hermann, Paris, 1952.

[70] Rund, H., "The Differential Geometry of Finsler Spaces." Springer, Berlin, 1959.

[71] Schirokow, P. and Schirokow, A., "Affine Differentialgeometrie." Teubner, Leipzig, 1962.

[72] Serre, J. P., "Lie Algebras and Lie Groups." Benjamin, New York, 1965.

[73] Wolf, J., "Spaces of Constant Curvature." McGraw-Hill, New York, 1967.

## VOLUMES V AND VI

[74] "Algebraic Groups and Discontinuous Subgroups" (Proceedings of Symposia in Pure Mathematics, Vol. IX), American Math. Soc., Providence, 1966.
[75] Bellman, R., "A Brief Introduction to Theta Functions." Holt, Rinehart and Winston, New York, 1961.
[76] Bernat, P. *et al.*, "Représentations des Groupes de Lie Résolubles" (Monographies de la Soc. math. de France, n° 4), Dunod, Paris, 1972.
[77] Borel, A., "Linear Algebraic Groups." Benjamin, New York-Amsterdam, 1969.
[78] Borel, A., "Introduction aux Groupes Arithmétiques." Hermann, Paris, 1969.
[79] Bourbaki, N., "Éléments de Mathématique, Groupes et Algèbres de Lie" (Actual. Scient. Ind., Chap. I, n° 1285, Chap. II–III, n° 1349, Chap. IV–V–VI, n° 1337). Hermann, Paris, 1960–1972.
[80] Carter, R., "Simple Groups of Lie Type." Wiley, New York, 1972.
[81] Chevalley, C., "Classification des Groupes de Lie algébriques," 2 vol.. Séminaire de l'École Normale Supérieure 1956–1958, Paris (Secr. math., 11, Rue P.-Curie).
[82] Conference on Harmonic Analysis (College Park, 1971), "Lecture Notes in Math.," n° 266, Springer, Berlin-Heidelberg-New York, 1972.
[83] Edwards, R., "Fourier Series," 2 vol.. Holt, Rinehart and Winston, New York, 1967.
[84] Gunning, R., "Lectures on Modular Forms." Princeton Univ. Press, 1962.
[85] Hausner, M., Schwartz, J., "Lie Groups, Lie algebras." Gordon Breach, New York, 1968.
[86] Igusa, J., "Thêta Functions." Springer, Berlin-Heidelberg-New York, 1972.
[87] Kahane, J.-P., Salem, R., "Ensembles Parfaits et Séries Trigonométriques." Hermann, Paris, 1963.
[88] Katznelson, Y., "An Introduction to Harmonic Analysis." Wiley, New York, 1968.
[89] Kawata, T., "Fourier Analysis in Probability Theory." Academic Press, New York, 1972.
[90] Meyer, Y., "Trois Problèmes sur les Sommes Trigonométriques," Astérisque, n° 1, 1973.
[91] Miller, W., "Lie Theory and Special Functions." Academic Press, New York, 1968.
[92] Pukansky, L., "Leçons sur les Représentations des Groupes" (Monographies de la Soc. math. de France, n° 2). Dunod, Paris, 1967.
[93] Rudin, W., "Fourier Analysis on Groups." Interscience, New York, 1968.
[94] Serre, J.-P., "Cours d'Arithmétique" (Collection SUP). Presses Univ. de France, Paris, 1970.
[95] Stein, E., Weiss, G., "Introduction to Fourier Analysis on Euclidean Spaces." Princeton Univ. Press, 1971.
[96] Vilenkin, N., "Fonctions Spéciales et Théorie de la Représentation des Groupes." Dunod, Paris, 1969.
[97] Warner, G., "Harmonic Analysis on Semi-simple Lie Groups," Vols. I and II. Springer, Berlin-Heidelberg-New Hork, 1972.
[98] Weil, A., "Basic Number Theory," Springer, Berlin-Heidelberg-New York, 1967.
[99] Zygmund, A., "Trigonometric Series," 2ᵉ éd., 2 vol.. Cambridge Univ. Press, 1968.

# INDEX

In the following index the first reference number refers to the chapter and the second to the section within the chapter.

E



F



H



I



K



L



M



N



P

# Pure and Applied Mathematics

A Series of Monographs and Textbooks



RECENT TITLES

E. R. Kolchin. Differential Algebra and Algebraic Groups

Gerald J. Janusz. Algebraic Number Fields

A. S. B. Holland. Introduction to the Theory of Entire Functions

Wayne Roberts and Dale Varberg. Convex Functions

A. M. Ostrowski. Solution of Equations in Euclidean and Banach Spaces, Third Edition of Solution of Equations and Systems of Equations

H. M. Edwards. Riemann's Zeta Function

Samuel Eilenberg. Automata, Languages, and Machines: Volumes A and B

Morris Hirsch and Stephen Smale. Differential Equations, Dynamical Systems, and Linear Algebra

Wilhelm Magnus. Noneuclidean Tesselations and Their Groups

François Treves. Basic Linear Partial Differential Equations

William M. Boothby. An Introduction to Differentiable Manifolds and Riemannian Geometry

Brayton Gray. Homotopy Theory: An Introduction to Algebraic Topology

Robert A. Adams. Sobolev Spaces

John J. Benedetto. Spectral Synthesis

D. V. Widder. The Heat Equation

Irving Ezra Segal. Mathematical Cosmology and Extragalactic Astronomy

J. Dieudonné. Treatise on Analysis: Volume II, enlarged and corrected printing; Volume IV; Volume V.

Werner Greub, Stephen Halperin, and Ray Vanstone. Connections, Curvature, and Cohomology: Volume III, Cohomology of Principal Bundles and Homogeneous Spaces

I. Martin Isaacs. Character Theory of Finite Groups

James R. Brown. Ergodic Theory and Topological Dynamics

C. Truesdell. A First Course in Rational Continuum Mechanics: Volume 1, General Concepts

George Gratzer. Lattice Theory

K. D. Stroyan and W. A. J. Luxemburg. Introduction to the Theory of Infinitesimals

B. M. Puttaswamaiah and John D. Dixon. Modular Representations of Finite Groups

Melvyn Berger. Nonlinearity and Functional Analysis: Lectures on Nonlinear Problems in Mathematical Analysis

Jan Mikusinski. The Bochner Integral

*In preparation*

Michiel Hazewinkel. Formal Groups and Applications

A B C D E F G H I J